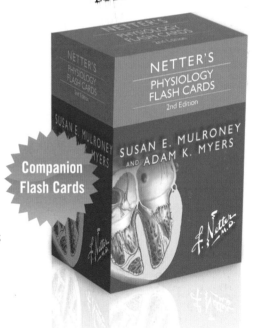

NETTER'S ESSENTIAL PHYSIOLOGY

SECOND EDITION

SUSAN E. MULRONEY, PhD

Professor of Pharmacology & Physiology
Director, Special Master's Program
Georgetown University Medical Center

ADAM K. MYERS, PhD

Professor of Pharmacology & Physiology
Associate Dean and Assistant Vice President for Special Graduate Programs
Georgetown University Medical Center

Illustrations by

FRANK H. NETTER, MD

Contributing Illustrators
Carlos A.G. Machado, MD
John A. Craig, MD
James A. Perkins, MS, MFA
Tiffany Slaybaugh DaVanzo, MA, CMI

ELSEVIER

ELSEVIER

1600 John F. Kennedy Blvd.
Ste 1800
Philadelphia, PA 19103-2899

NETTER'S ESSENTIAL PHYSIOLOGY, SECOND EDITION   ISBN: 978-0-323-35819-4

Notices

Knowledge and best practice in this field are constantly changing. As new research and experience broaden our understanding, changes in research methods, professional practices, or medical treatment may become necessary.

Practitioners and researchers must always rely on their own experience and knowledge in evaluating and using any information, methods, compounds, or experiments described herein. In using such information or methods they should be mindful of their own safety and the safety of others, including parties for whom they have a professional responsibility.

With respect to any drug or pharmaceutical products identified, readers are advised to check the most current information provided (i) on procedures featured or (ii) by the manufacturer of each product to be administered, to verify the recommended dose or formula, the method and duration of administration, and contraindications. It is the responsibility of practitioners, relying on their own experience and knowledge of their patients, to make diagnoses, to determine dosages and the best treatment for each individual patient, and to take all appropriate safety precautions.

To the fullest extent of the law, neither the Publisher nor the authors, contributors, or editors, assume any liability for any injury and/or damage to persons or property as a matter of products liability, negligence or otherwise, or from any use or operation of any methods, products, instructions, or ideas contained in the material herein.

Previous edition copyrighted 2009.

ISBN: 978-0-323-35819-4

Senior Content Strategist: Elyse O'Grady
Content Development Specialist: Stacy Eastman
Publishing Services Manager: Patricia Tannian
Senior Project Manager: Carrie Stetz
Design Direction: Julia Dummitt

Working together to grow libraries in developing countries

www.elsevier.com • www.bookaid.org

Printed in China
Last digit is the print number: 9 8 7 6 5 4 3 2 1

*We dedicate this book to our families, for their love and support.
We dedicate it also to the students of Georgetown University, who
are exceptional in their character and their love of learning.*

PREFACE

Human physiology is the study of the functions of our bodies at all levels: whole organism, systems, organs, tissues, cells, and physical and chemical processes. Physiology is a complex science, incorporating concepts and principles from biology, chemistry, biochemistry, and physics; often, a true appreciation of physiological concepts requires multiple learning modalities, beyond standard texts or lectures. This second edition of *Netter's Essential Physiology* has been prepared with this in mind. Its generous illustrations and concise, bulleted, and highlighted text are designed to draw the student in and to focus the student's efforts on understanding the essential aspects of difficult concepts. It is not intended to be a detailed reference book, but rather a guide to learning the essentials of the field, in conjunction with classroom work and other resources when necessary.

This book is organized in the classical order in which subdisciplines of physiology are taught. Beginning with fluid compartments, transport mechanisms, and cell physiology, it progresses through neurophysiology, cardiovascular physiology, the respiratory system, renal physiology, the gastrointestinal system, and endocrinology. It is ideal for the visual learner. Each section is thoroughly illustrated with the great drawings of the late Frank Netter as well as the more recent, beautiful work of Carlos Machado, John Craig, James Perkins, and Tiffany DaVanzo.

In this second edition, we have expanded some content and clarified some areas based on feedback from students and teachers over the past several years. In addition to updating information, we have added a number of clinical correlates to help reinforce the material and provide pathophysiological context. To complement the cardiovascular section, we have added a new chapter, The Blood, which includes the important topics of hemostasis and blood coagulation. The changes and additions in this edition are also incorporated into the second edition of *Netter's Physiology Flash Cards*.

Recognizing that physiology, cell biology, and anatomy go hand in hand in the modern, integrated curriculum of many institutions, we have included more than the usual number of illustrations relevant to anatomy and histology. By reading the text, studying the illustrations, and taking advantage of the review questions, the student will become familiar with the important concepts in each subdiscipline and gain the essential knowledge required in medical, dental, upper level undergraduate, or nursing courses in human physiology.

Too many textbooks, although very useful reference works, go for the most part unread by students. It is our hope that students will find this book enriching and stimulating and that it will inspire them to thoroughly learn this fascinating field.

Susan E. Mulroney, PhD
Adam K. Myers, PhD

Acknowledgments

The preparation of this textbook has benefited from the efforts of numerous colleagues and students who reviewed various sections of the work and offered valuable criticisms and suggestions. We especially wish to thank Charles Read, Henry Prange, Stefano Vicini, Jagmeet Kanwal, Peter Kot, Edward Inscho, Jennifer Rogers, Adam Mitchell, Milica Simpson, Lawrence Bellmore, and Joseph Garman for their critical reviews of the first edition. In addition, we express our appreciation to Adriane Fugh-Berman, Stefano Vicini, and Aruna Natarajan for their invaluable advice during preparation of the current edition; Amy Richards, for her constant good humor and willingness to help; and all our colleagues and coworkers for their friendship, collegiality, and encouragement during this project.

Our special thanks go to the dedicated team at Elsevier, particularly Stacy Eastman, Carrie Stetz, Marybeth Thiel, and Elyse O'Grady. We also acknowledge Jim Perkins and Tiffany DaVanzo for their talented work on the new illustrations in the first and second editions, which gracefully complement the original drawings of the master illustrator, Frank Netter.

Finally, we acknowledge the role of our students in this project. We learn from them, as they learn from us; their enthusiasm for learning is the greatest inspiration for our work.

About the Authors

Photo by Ulf Wallin Photography

SUSAN E. MULRONEY, PhD, is an award-winning teacher and researcher at Georgetown University Medical Center, where she is Professor of Pharmacology & Physiology and Director of the highly acclaimed Physiology Special Master's Program. Dr. Mulroney directed the Medical Human Physiology course for the first-year medical students at the School of Medicine and currently directs the Medical Gastrointestinal module in the systems-based medical curriculum. Dr. Mulroney lectures to medical and graduate students in multiple areas of human physiology, including renal, gastrointestinal, and endocrine physiology, and is recognized for her expertise in curricular innovation in medical education. She has won several Golden Apple Awards for teacher of the year given by Georgetown University Medical Center classes and numerous other teaching awards. Dr. Mulroney received the 2015 Arthur C. Guyton Physiology Educator of the Year Award from the American Physiological Society. Dr. Mulroney is a well-known researcher in renal and endocrine physiology, has published extensively in these areas, and was director of the Physiology PhD program for 12 years. She is also coeditor of *RNA Binding Proteins: New Concepts in Gene Regulation.*

Photo by Ulf Wallin Photography

ADAM K. MYERS, PhD, is Professor of Pharmacology & Physiology and Associate Dean and Assistant Vice President for Special Graduate Programs at Georgetown University Medical Center. Dr. Myers was director of the Special Master's Program at Georgetown for 12 years and has developed and directed other graduate programs over the years. He has won numerous teaching awards from the students and faculty of Georgetown University School of Medicine, where he teaches extensively in several medical and graduate courses in various areas of human physiology. Dr. Myers is recognized for his extensive experience in educational program development and administration as well as implementation of new educational technologies. His research in platelet and vascular biology has resulted in numerous publications. He is also author of the textbook *Crash Course: Respiratory System* and coeditor of *Alcohol and Heart Disease.*

VIDEO ANIMATION CONTENTS

Video animations are available online at ExpertConsult.com.

About the Artists

FRANK H. NETTER, MD, was born in 1906 in New York City. He studied art at the Art Student's League and the National Academy of Design before entering medical school at New York University, where he received his MD degree in 1931. During his student years, Dr. Netter's notebook sketches attracted the attention of the medical faculty and other physicians, allowing him to augment his income by illustrating articles and textbooks. He continued illustrating as a sideline after establishing a surgical practice in 1933, but he ultimately opted to give up his practice in favor of a full-time commitment to art. After service in the United States Army during World War II, Dr. Netter began his long collaboration with the CIBA Pharmaceutical Company (now Novartis Pharmaceuticals). This 45-year partnership resulted in the production of the extraordinary collection of medical art so familiar to physicians and other medical professionals worldwide.

In 2005, Elsevier Inc. purchased the Netter Collection and all publications from Icon Learning Systems. There are now more than 50 publications featuring the art of Dr. Netter available through Elsevier Inc. (in the United States: www.us.elsevierhealth.com/Netter and outside the United States: www.elsevierhealth.com).

Dr. Netter's works are among the finest examples of the use of illustration in the teaching of medical concepts. The 13-book *Netter Collection of Medical Illustrations*, which includes the greater part of the more than 20,000 paintings created by Dr. Netter, became and remains one of the most famous medical works ever published. *The Netter Atlas of Human Anatomy*, first published in 1989, presents the anatomical paintings from the Netter Collection. Now translated into 16 languages, it is the anatomy atlas of choice among medical and health professions students the world over.

The Netter illustrations are appreciated not only for their aesthetic qualities but, more importantly, for their intellectual content. As Dr. Netter wrote in 1949, "… clarification of a subject is the aim and goal of illustration. No matter how beautifully painted, how delicately and subtly rendered a subject may be, it is of little value as a *medical illustration* if it does not serve to make clear some medical point." Dr. Netter's planning, conception, point of view, and approach are what informs his paintings and what makes them so intellectually valuable.

Frank H. Netter, MD, physician and artist, died in 1991.

Learn more about the physician-artist whose work has inspired the Netter Reference collection: http://www.netterimages.com/artist/netter.htm.

CARLOS A.G. MACHADO, MD, was chosen by Novartis to be Dr. Netter's successor. He continues to be the main artist who contributes to the Netter collection of medical illustrations.

Self-taught in medical illustration, cardiologist Carlos Machado has contributed meticulous updates to some of Dr. Netter's original plates and has created many paintings of his own in the style of Netter as an extension of the Netter collection. Dr. Machado's photorealistic expertise and his keen insight into the physician/patient relationship inform his vivid and unforgettable visual style. His dedication to researching each topic and subject he paints places him among the premier medical illustrators at work today.

Learn more about his background and see more of his art at http://www.netterimages.com/artist/machado.htm.

CONTENTS

NETTER'S ESSENTIAL PHYSIOLOGY

SECOND EDITION

Section 1

CELL PHYSIOLOGY, FLUID HOMEOSTASIS, AND MEMBRANE TRANSPORT

Physiology is the study of how the systems of the body work, not only on an individual basis, but also in concert to support the entire organism. Medicine is the application of physiologic principles, and understanding these principles gives us insight into the development of disease. The terms "regulation" and "integration" will keep surfacing as you learn more about how each system functions. Because of these building interactions, the field of physiology is always expanding. As we discover more about the genes, molecules, and proteins that regulate other factors, we see that the discipline of physiology is far from static. Each new discovery gives us more insight into how our impossibly complex organism exists and how we might intercede when pathophysiology occurs. This text explores essential elements in each of the body's systems; it is not intended to be comprehensive; rather, it focuses on ensuring a solid understanding of the principles related to the regulation and integration of these systems.

Chapter 1

The Cell and Fluid Homeostasis

CELL STRUCTURE AND ORGANIZATION

Organisms evolved from single cells floating in the primordial sea (Fig. 1.1). A key to appreciating how multicellular organisms exist is through understanding how the single cells maintained their internal fluid environment when exposed directly to the outside environment, with the only barrier being a semipermeable membrane. Nutrients from the "sea" entered the cell, diffusing down their concentration gradients through channels or pores, and waste was transported out through exocytosis. In this simple system, if the external environment changed (e.g., if salinity increased due to excess heat and evaporation of sea water or if the water temperature changed), the cell adapted or perished. To evolve to multicellular organisms, cells developed additional barriers to the outside environment to allow better regulation of the intracellular environment.

In multicellular organisms, cells undergo differentiation, developing discrete intracellular proteins, metabolic systems, and products. The cells with similar properties aggregate and become tissues and organ systems (cells → tissues → organs → systems).

Various tissues serve to support and produce movement (muscle tissue), initiate and conduct electrical impulses (nervous tissue), secrete and absorb substances (epithelial tissue), and join other cells together (connective tissue). These tissues combine and support organ systems that control other cells (nervous and endocrine systems), provide nutrient input and continual excretion of waste (respiratory and gastrointestinal systems), circulate the nutrients (cardiovascular system), filter and monitor fluid and electrolyte needs and rid the body of waste (renal system), provide structural support (skeletal system), and provide a barrier to protect the entire structure (integumentary system [skin]) (Fig. 1.2).

THE CELL MEMBRANE

The human body is composed of eukaryotic cells (those that have a true nucleus) containing various organelles (e.g., mitochondria, smooth and rough endoplasmic reticulum, Golgi apparatus) that perform specific functions. The cell, with its nucleus and organelles, is surrounded by a **plasma membrane** consisting of a lipid bilayer primarily made of phospholipids, with varying amounts of glycolipids, cholesterol, and proteins. The lipid bilayer is positioned with the hydrophobic fatty acid tails of phospholipids oriented toward the middle of the membrane and the hydrophilic polar head groups oriented toward the extracellular or intracellular space. The fluidity of the membrane is maintained in large part by the amount of short-chain and unsaturated fatty acids incorporated within the phospholipids; incorporation of cholesterol into the lipid bilayer reduces fluidity (Fig. 1.3). The oily, hydrophobic interior region makes the bilayer an effective barrier to fluid (on either side), with permeability only to some small hydrophobic solutes, such as ethanol, that can diffuse through the lipids.

To accommodate multiple cellular functions, the membranes are actually **semipermeable** because of a variety of proteins inserted in the lipid bilayer. These proteins are in the form of ion channels, ligand receptors, adhesion molecules, and cell recognition markers. Transport across the membrane can involve passive or active mechanisms and is dictated by the membrane composition, concentration gradient of the solute, and availability of transport proteins (see Chapter 2). If the integrity of the membrane is disrupted by changing fluidity, protein concentration, or thickness, transport processes will be impaired.

FLUID COMPARTMENTS: SIZE AND CONSTITUTIVE ELEMENTS

Fluid Compartments and Size

The typical adult body is approximately 60% water, which, in a 70-kilogram (kg) person, equals 42 L (Fig. 1.4). The actual size of all the fluid compartments depends on a variety of factors, including the person's size and body mass index. In a normal 70-kg adult:

- **Intracellular fluid (ICF)** constitutes two thirds of the **total body water** (**TBW**; 28 L), and the **extracellular fluid (ECF)** accounts for the other third of TBW (14 L).
- The ECF compartment is composed of the *plasma* (i.e., blood without red blood cells [RBCs]) and the

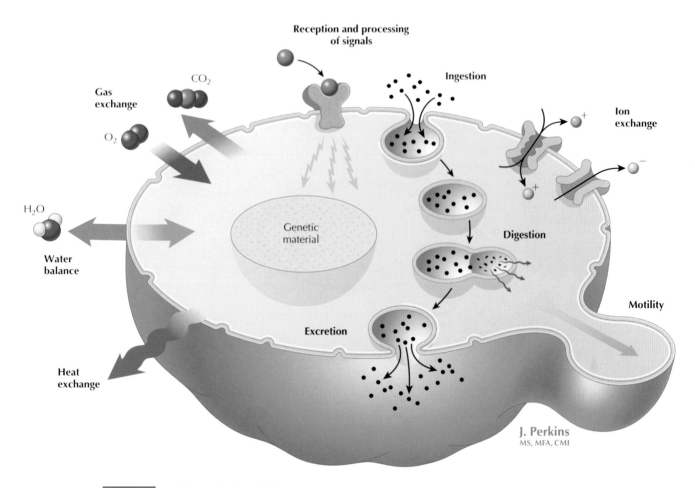

Figure 1.1 **Cell in the Primordial Sea**   The first single-celled organisms had to perform basic functions and be able to adapt to changes in their immediate external environment. The semipermeable cell membrane facilitated the processes that provided nutrients to the cell, using diffusion, endocytosis and exocytosis, and protein transporters to maintain homeostasis.

interstitial fluid (ISF), which is the fluid that bathes cells (outside the vascular system), as well as the fluid in bone and connective tissue. Plasma constitutes one fourth of ECF (3.5 L), and ISF constitutes the other three fourths of ECF (10.5 L).

The amount of **TBW** differs with age and general body type. TBW in rapidly growing infants is approximately 75% of body weight, whereas older adults have a lower percentage. In addition, body fat plays a role; obese individuals have lower TBW than age-matched individuals, and, in general, women have less TBW than age-matched men. The amount of TBW is especially relevant for drug dosages. Because fat solubility varies with the type of drug, body water composition (relative to body fat) can affect the effective concentration of the drug (Fig. 1.5).

Intracellular and Extracellular Compartments

The intracellular and extracellular compartments are separated by the cell membrane. Within the ECF, the plasma and ISF are separated by the endothelium and basement

membranes of the capillaries. The ISF surrounds the cells and is in close contact with both the cells and the plasma.

The ICF has different solute concentrations than the ECF, primarily because of the sodium-potassium–adenosine triphosphatase (Na^+/K^+ ATPase, or "sodium pump"), which maintains an ECF high in Na^+ and an ICF high in potassium (K^+) (Fig. 1.6). The maintenance of different solute concentrations is also highly dependent on the selective permeability of cell membranes separating the extracellular and intracellular spaces. The cations and anions in our body are in balance, with the number of positive charges in each compartment equaling the number of negative charges (see Fig. 1.6). Because the ion flow across the membrane is responsive to both the electrical charge and the solute gradient, the overall environment is controlled by maintenance of this **electrochemical equilibrium**.

The osmolarity (total concentration of solutes) of fluids in our bodies is ~290 milliosmoles (mOsm)/L (generally rounded to 300 mOsm/L for calculations). This is true for all of the fluid compartments (see Fig. 1.6). The basolateral sodium ATPase pumps (seen on cell membranes) are instrumental in

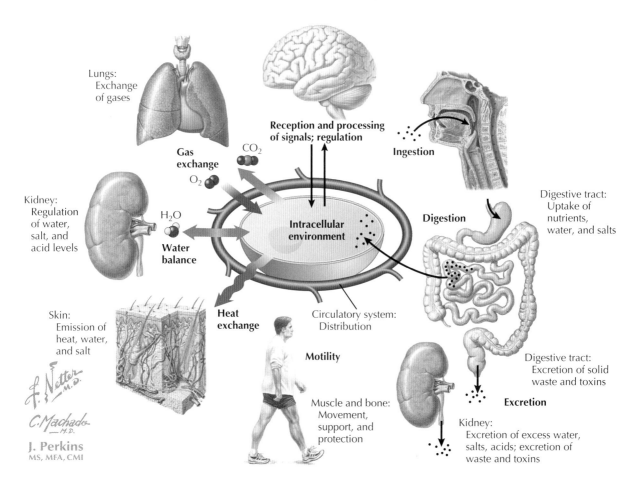

Figure 1.2 **Buffering the External Environment** In multicellular organisms, the basic homeostatic mechanisms of single-celled organisms are mirrored by integration of specialized organ systems to create a stable environment for the cells. This system allows specialization of cellular functions and a layer of protection for the systems.

establishing and maintaining the intracellular and extracellular environments. Intracellular Na^+ is maintained at a low concentration (which drives the Na^+-dependent transport into the cells) compared with the high Na^+ in ECF. The **extracellular sodium** (and the small amount of other positive ions) is balanced by chloride and bicarbonate anions and anionic proteins. For the most part, the concentration of solutes between plasma and ISF is similar, with the exception of proteins (indicated as A^-), which remain in the vascular space (under normal conditions, they cannot pass through the capillary membranes). The high ECF Na^+ concentration drives Na^+ leakage into cells, as well as many other transport processes.

The primary intracellular cation is the *potassium ion*, which is balanced by phosphates, proteins, and small amounts of other miscellaneous anions. Because of the high concentration gradients for sodium, potassium, and chloride, passive leakage of these ions occurs down their gradients. The leakage of potassium out of the cell through specific K^+ channels is the key factor contributing to the resting membrane potential. The differential sodium, potassium, and chloride concentrations across the cell membrane are crucial for the generation of electrical potentials (see Chapter 3).

Osmolarity describes the number of dissolved particles present per liter of solution. For nonelectrolytes such as the sugar sucrose, 1 millimole (mmol) of the substance is equal to a 1 mOsm solution. However, with electrolytes, each molecule needs to be accounted for, so a 1 mM solution of NaCl equals a 2 mOsm/L solution, a 1 mM solution of $CaCl_2$ equals a 3 mOsm/L solution, and so on. In addition, solutions can be described as isosmotic (~300 mOsm/L), hypo-osmotic (<300 mOsm/L) or hyperosmotic (>300 mOsm/L) compared with normal plasma osmolarity (300 mOsm/L).

OSMOSIS, STARLING FORCES, AND FLUID HOMEOSTASIS

Osmosis

Membranes are **selectively permeable (semipermeable)**, meaning they allow some (but not all) molecules to pass through. Membranes of tissues vary in their permeability to specific solutes. This tissue specificity is critical to function, as seen in the variation in cell solute permeability through a renal nephron (see Chapters 18 and 19). Factors on either side of

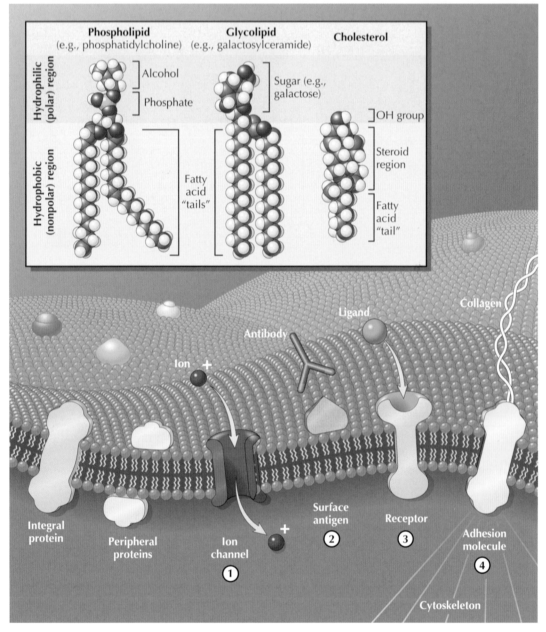

Figure 1.3 **The Eukaryotic Plasma Membrane**   The plasma membrane is a lipid bilayer, with hydrophobic ends oriented inward and hydrophilic ends oriented outward. Primary constituents of the membrane are phospholipids, glycolipids, and cholesterol. A wide variety of proteins are associated with the membrane, including *(1)* ion channels, *(2)* surface antigens, *(3)* receptors, and *(4)* adhesion molecules.

the membrane oppose and facilitate movement of water and solutes out of the compartments. These factors include:

- Concentration of specific solutes. Higher concentration of a solute on one side of the membrane will favor movement of that solute to the other side by diffusion.
- Overall concentration of solutes. Higher osmolarity on one side provides osmotic pressure that "pulls" water into that space (diffusion of water).
- Concentration of proteins. Because the membrane is impermeable to proteins, protein concentration establishes an oncotic pressure that "pulls" water into the space with higher concentration.

- Hydrostatic pressure, which is the force "pushing" water out of the space—for example, from capillaries to ISF (when capillary hydrostatic pressure exceeds ISF hydrostatic pressure).

If the membrane is permeable to a solute, **diffusion** of the solute will occur down the concentration gradient (see Chapter 2). However, if the membrane is not permeable to the solute, the solvent (in this case water) will be "pulled" across the membrane toward the compartment with higher solute concentration until the concentration reaches equilibrium across the membrane. The movement of water across the membrane by diffusion is termed **osmosis**, and the permeability of the

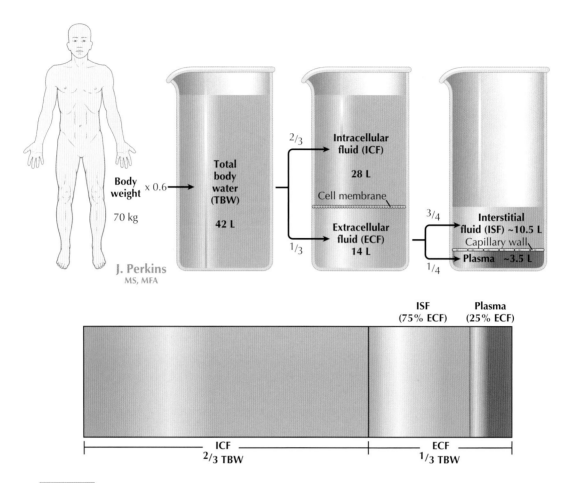

Figure 1.4 **Body Fluid Compartments**   Under normal conditions, the total volume of water in the human body (TBW) is about 60% of the body weight. Of TBW, most (two thirds) is intracellular fluid (ICF), and one third is extracellular fluid (ECF). The ECF is made up of plasma and interstitial fluid (ISF).

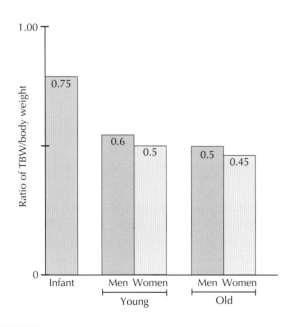

Figure 1.5 **Total Body Water as Function of Body Weight**
Under normal conditions, total body water (TBW) is most affected by the amount of body fat, and there is more body water as a percentage of body weight in infants and men. Aging decreases the ratio because of reduced muscle mass.

membrane determines whether diffusion of solute or osmosis (water movement) occurs. The concentration of the impermeable solute will determine how much water will move through the membrane to achieve **osmolar equilibration** between ECF and ICF.

Whereas osmolarity of a solution describes the concentration of dissolved particles and a solution can be described as hypo-osmotic, isosmotic, or hyperosmotic relative to another solution, whether fluid shifts will occur between two isosmotic solutions across a membrane depends on whether the solutes are permeant. The monosaccharide sucrose is impermeant to cells, and if infused into the plasma (ECF), it will stay in the ECF compartment. Thus a 300 mOsm/L solution of sucrose is isotonic relative to cells with normal osmolarity of 300 mOsm/L, and no fluid shift occurs. A sucrose solution of more than 300 mOsm/L is hypertonic, and a sucrose solution of less than 300 mOsm/L is hypotonic relative to normal cellular osmolarity. In contrast to impermeant solutes, a permeant solute, such as urea, will diffuse freely into cells until it reaches equilibrium. Thus a 300 mOsm/L solution of urea will be hypotonic even though the solution is isosmotic. When this solution is infused into the ECF, it will cause expansion of the ICF compartment.

Figure 1.6 **Electrolyte Concentration in Extracellular and Intracellular Fluid** The primary extracellular fluid cation is Na^+, and the primary intracellular fluid cation is K^+. This difference is maintained by the basolateral Na^+/K^+ ATPase, which transport three Na^+ molecules out of the cell in exchange for two K^+ molecules transported into the cell. A balance of positive and negative charges is maintained in each compartment, but by different ions. (Values are approximate.) A^-, proteins.

Osmosis occurs when **osmotic pressure** is present. Osmotic pressure is equivalent to the hydrostatic pressure necessary to prevent movement of fluid through a semipermeable membrane by osmosis. The idea can be illustrated by using a U-shaped tube with different concentrations of solute on either side of an **ideal semipermeable membrane** (i.e., the membrane is permeable to water but is impermeable to solute; Fig. 1.7, *A*).

Because of the unequal solute concentrations, fluid will move to the side with the higher solute concentration (the right side of tube), against the gravitational force (hydrostatic pressure) that opposes it, until the **hydrostatic pressure** generated is equal to the **osmotic pressure**. In the example, at equilibrium, solute concentration is nearly equal and water level is unequal, and the displacement of water is due to osmotic pressure (Fig. 1.7, *B*).

In the plasma, the presence of proteins also produces a significant **oncotic pressure** that opposes hydrostatic pressure (filtration out of the compartment) and is considered the **effective osmotic pressure of the capillary**.

Starling Forces

The oncotic and hydrostatic pressures are key components of the **Starling forces**. Starling forces are the pressures that control fluid movement across the capillary wall. Net movement of water out of the capillaries is **filtration**, and net movement into the capillaries is **absorption**. As seen in Figure 1.8, four forces control fluid movement:

- HP_c, the **capillary hydrostatic pressure**, favors movement out of the capillaries and is dependent on both arterial and venous blood pressures (generated by the heart).
- π_c, the **capillary oncotic pressure**, opposes filtration out of the capillaries and is dependent on the protein concentration in the blood. The only effective oncotic agent in capillaries is protein, which is ordinarily impermeable across the vascular wall.
- P_i, the **interstitial hydrostatic pressure**, opposes filtration out of capillaries, but normally this pressure is low.
- π_i, the **interstitial oncotic pressure**, favors movement out of the capillaries, but under normal conditions there is little loss of protein out of the capillaries, and this value is near zero.

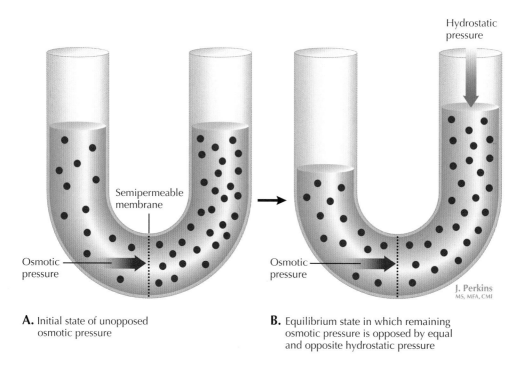

A. Initial state of unopposed osmotic pressure

B. Equilibrium state in which remaining osmotic pressure is opposed by equal and opposite hydrostatic pressure

Figure 1.7 **Osmosis and Osmotic Pressure**   When a semipermeable membrane separates two compartments in a "U tube," fluid will move through the membrane toward the higher solute concentration **(A)**, until near equilibrium of solutes is reached and the remaining osmotic pressure is opposed by the hydrostatic pressure difference between the two sides of the tube **(B)**. In blood vessels, hydrostatic pressure is generated by gravity and the pumping of the heart, and the osmotic pressure is measured as the force needed to oppose the hydrostatic pressure. Osmotic pressure works toward equalization of solute concentrations on either side of the membrane. Oncotic pressure (π) is the osmotic pressure produced by impermeable proteins. In the plasma, oncotic pressure is considered the effective osmotic pressure of the capillary.

Movement of fluid through capillary beds can differ as a result of physical factors particular to the capillary wall (e.g., pore size and fenestration) and its relative permeability to protein, but in general these factors are considered constant for most tissues.

These forces are used to describe net filtration using the *Starling equation*:

$$\text{Net filtration} = K_f[(HP_c - P_i) - \sigma(\pi_c - \pi_i)] \quad \textbf{Eq. 1.1}$$

in which the constant, $\mathbf{K_f}$, accounts for the physical factors affecting permeability of the capillary wall, and σ (the reflection coefficient) describes the permeability of the membrane to proteins (where $0 < \sigma < 1$). The liver capillaries (sinusoids) are highly permeable to proteins, and $\sigma = 0$. Thus bulk movement in the liver sinusoids is controlled by hydrostatic pressure. In contrast, capillaries in most tissues have low

permeability to proteins, and $\sigma = \sim1$, so the Starling equation can be viewed easily as the pressures governing filtration minus those favoring absorption:

$$\begin{array}{c} \text{(Filtration)} - \text{(Absorption)} \\ \text{Net Filtration} = K_f[(HP_c + \pi_i) - (P_i + \pi_c)] \end{array} \quad \textbf{Eq. 1.2}$$

Although K_f is a "constant," it differs among systemic, cerebral, and renal glomerular capillaries, with cerebral capillaries having a lower K_f (limiting filtration) and glomerular capillaries having a greater K_f (promoting filtration) compared with systemic capillaries. Thus filtration will be determined by the difference in hydrostatic pressure between the capillary and interstitium, minus the difference between capillary and interstitial oncotic pressure (corrected for the protein reflection coefficient). It should be clear that under normal conditions, the forces that are most variable are the HP_c and the π_c because those forces can reflect changes in plasma volume.

CLINICAL CORRELATE 1.1
Effect of Adding Solutes to the Extracellular
Fluid on Compartment Size

Although the endocrine and renal systems work in concert to ensure the rapid regulation of fluid and electrolyte homeostasis, it is important to understand the potential impact of the addition of water and hypertonic solutions to the ECF on ICF volume and osmolarity. To illustrate the changes that would occur (without renal compensation), an individual who weighs 60 kg would have TBW of 36 L (60 kg × 60%). Under these conditions, ICF is two thirds (24 L) and ECF is one third (12 L) of the TBW. Osmolarity will be 300 mOsm/L, with 7200 mOsm of solutes in the ICF and 3600 mOsm in ECF (total body solutes = 10,800 mOsm).

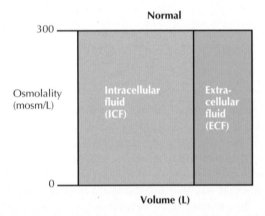

Keeping in mind that the final osmolarity will be equilibrated between ICF and ECF compartments, addition of solutions to the plasma (ECF) will have the following effects on compartment size and osmolarity:

500 mL of isotonic saline (Na⁺Cl⁻, 300 mOsm/L) to ECF

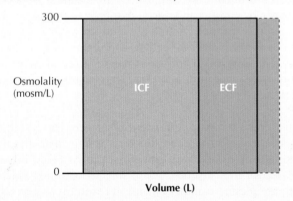

Because Na^+Cl^- will remain in the ECF and expand the compartment by 500 mL (from 12 L to 12.5 L) and 150 mOsm (300 mOsm/L × 500 mL), ICF volume remains the same, and ICF osmolarity remains normal (300 mOsm/L).

2 L of hypertonic (400 mOsm/L) saline to ECF

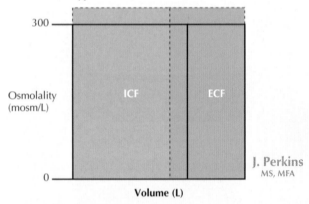

J. Perkins
MS, MFA

Again, because the solution was Na^+Cl^-, the solutes (osmoles) remain in the ECF, raising total ECF solutes to 4400 mOsm (initial 3600 mOsm + infused 800 mOsm). This elevated osmolarity draws fluid out of the ICF (i.e., the ICF "shrinks") and the ECF expands until osmolar equilibrium is achieved. Final osmolarity is determined from the total body mOsm (7200 + 3600 + 800 = 11,600 mOsm) divided by the TBW (36 + 2 = 38 L), or 11,600 mOsm ÷ 38 L = 305 mOsm/L. ECF volume is total ECF solute divided by osmolarity, or 4400 mOsm ÷ 305 mOsm/L = 14.4 L. ICF volume was reduced to 23.6 L (38 L − 14.4 L).

2 L of H₂O

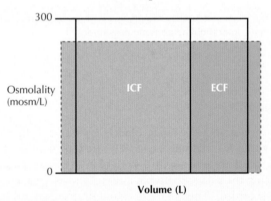

Adding solute-free water to plasma will immediately dilute the ECF osmolarity, making the ECF hypotonic relative to ICF and causing fluid to move from the ECF to the ICF. After equilibration, each compartment will have a lower osmolarity and a greater volume. Final osmolarity will be 284 mOsm/L (10,800 mosm ÷ 38 L), with compartment volumes of 25.3 L (ICF, two thirds TBW) and 12.7 L (ECF, one third TBW).

As you will learn in Chapter 20, the kidneys will rapidly respond to changes in ECF volume and osmolarity to ensure homeostasis of ICF.

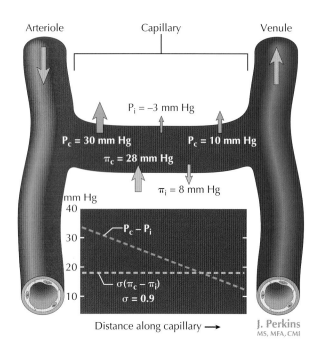

Figure 1.8 **Starling Forces Across the Capillary** The Starling forces (hydrostatic and oncotic pressures) allow for bulk flow of fluid and nutrients across the capillary wall. The permeability of the membranes of the capillary wall to proteins is usually very low in most tissues and is reflected by a protein reflection coefficient (σ) of ~1. The inset panel illustrates that, as fluid moves through the capillaries and diffusion into tissues occurs, the Starling forces change and the forces favoring net filtration (especially P_c [HP_c, capillary hydrostatic pressure]) decrease *(dotted blue line)*. π_c, **Capillary oncotic pressure;** π_i, **interstitial oncotic pressure;** P_i, **interstitial hydrostatic pressure.**

Homeostasis

French physiologist Claude Bernard first articulated the concept that maintaining a constant internal environment, or **milieu intérieur**, was essential to good health. In multicellular organisms, the balance between internal and external environments is critical. The ability to maintain a constant internal function during changes in the external environment is termed **homeostasis**. Homeostasis is accomplished through integrated regulation of the internal environment by the multiple organ systems (see Fig. 1.2).

On the cellular level, homeostasis is possible as a result of expandable semipermeable membranes that can accommodate small changes in osmolarity via osmosis. However, for proper cellular function, the ICF, and thus osmolarity, must be kept under tight control.

The plasma is the interface between the internal and the external environment; therefore, maintaining plasma osmolarity is an important key to cell homeostasis. For this reason,

Examples of the effects of Starling forces on fluid shifts can be illustrated by changes in fluid volume, as well as changes in physical factors. A severely dehydrated person will have a decreased blood volume, which would potentially lower blood pressure (e.g., HP_c) and increase π_c. According to the Starling equation, those changes will decrease filtration force and increase the absorptive force, causing an overall decrease in net filtration, which will keep fluid in the vascular space.

When physical attributes of the membrane change, Starling forces are also affected: K_f can change if the capillary membrane is damaged, such as by toxins or disease. If the clefts between endothelial cells or fenestrations expand (as seen in diseased renal glomeruli), plasma proteins may pass into the interstitial space and alter the Starling forces by increasing π_i; in peripheral capillaries, this mechanism causes **edema.** Starling forces are also affected in persons with congestive heart failure, cirrhosis, and sepsis.

many systems play a role in controlling plasma osmolarity. Both **thirst** and the **salt appetite** are behavioral responses that can be stimulated by dehydration and/or blood loss. These responses serve to stimulate specific ingestive behaviors (e.g., drinking, eating salty foods that will also stimulate drinking) that will increase the input of fluid and salt to the system. On a minute-to-minute basis, the endocrine and sympathetic nervous systems work to regulate the amount of sodium and water retained by the kidneys, thus controlling plasma osmolarity (see Chapter 20). Normally, changes in plasma osmolarity are well controlled and homeostasis is maintained as a result of hypothalamic osmoreceptors as well as sensing of fluid composition by the kidneys, sensing of pressure by carotid and aortic baroreceptors, release of hormones in response to altered pressure and osmolarity, and the actions of the kidney to regulate sodium and water reabsorption. This integrated control is the key to fluid homeostasis. Control of renal fluid and electrolyte handling is covered in Section 5.

Fluid intake and output must be in balance (Fig. 1.9). If water intake (through food and fluids) is greater than the output (through urine and insensible losses from sweat, breathing, and feces), the organism has a surplus of fluid, which will decrease plasma osmolarity, and the kidneys will excrete the excess fluid (see Chapter 20). Conversely, if the intake is less than the output, the organism has a deficit of fluid and plasma osmolarity will increase. In this situation, the thirst response will be activated and the kidneys will conserve fluid, producing less urine. This idea of balance is expanded upon in the following sections, and the integration of the endocrine, cardiovascular, and renal systems in regulation of fluid and electrolyte homeostasis is discussed more fully.

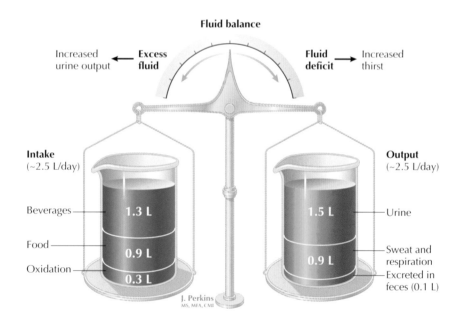

Figure 1.9 **Net Fluid Balance**   To maintain fluid balance, fluid input needs to match fluid output. If intake (from food and beverages) exceeds output (through urine, fluid in feces, and insensible losses), the organism is in positive balance and urine volume will increase to eliminate excess fluid. Negative fluid balance occurs when intake is less than output. In this case, integrated responses will increase thirst and decrease additional fluid losses until homeostasis is restored.

CLINICAL CORRELATE 1.2
Measurement of Fluid Compartment Size

The indicator-dilution method is used to determine the volume of fluid in the different fluid compartments. Indicators specific to each compartment are used. A known quantity of the substance is infused into the bloodstream of the subject and allowed to disperse. A plasma sample is then obtained, and the amount of indicator is determined. The compartment volume is then calculated with use of the following formula:

$$\text{Volume (in liters)} = \frac{\text{Amount of indicator injected (mg)}}{\text{Final concentration of indicator (mg/L)}}$$

Compartment	Indicator
TBW	Antipyrine or tritiated water because both of these substances will diffuse through *all* compartments.
ECF	Inulin, which will diffuse throughout plasma and ISF. Inulin is a large sugar (molecular weight of 5000) that cannot cross cell membranes and is not metabolized.
Plasma volume	Evans blue dye, which binds to plasma proteins. The total blood volume is composed of the plasma and RBCs, and the hematocrit is the percentage of RBCs in the whole blood. Hematocrit is ~0.42 (42% RBCs) in normal adult men and ~0.38 in women.

By extrapolation, the other compartments can be determined by the following formula:

Since TBW = ECF + ICF,

$$ICF = TBW - ECF$$

$$ISF = (ECF - \text{Plasma volume})$$

Because blood volume = plasma volume + RBC volume (see earlier), it can be calculated by the following formula:

$$\text{Blood volume} = (\text{Plasma volume} \div [1 - \text{hematocrit}])$$

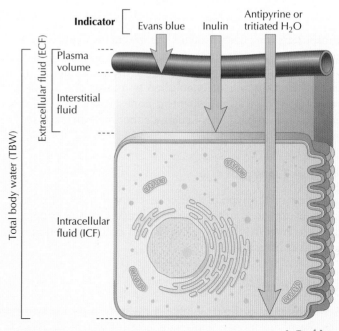

Substances Used to Determine Fluid Compartment Size

Chapter 2

Membrane Transport

CELLULAR TRANSPORT: PASSIVE AND ACTIVE MECHANISMS

Ions and solutes move through several different types of carrier proteins and channels that allow solute movement through the plasma membrane in several different ways. The carriers and channels include:

- **Ion channels** and **pores** that allow diffusion of solutes between compartments.
- **Uniporters**, which are membrane transport proteins that recognize specific molecules, such as fructose.
- **Symporters** that transport a cation (or cations) down its concentration gradient with another molecule (either another ion or a sugar, amino acid, or oligopeptide).
- **Antiporters** that transport an ion down its concentration gradient while another substance is transported in the opposite direction. This type of transport is often associated with Na^+ transport, or it may be dependent on gradients of other ions, as in the case of the HCO_3^-/Cl^- exchanger.

Transporters (or carrier proteins) can be energy dependent or independent. Movement through channels and uniporters follows the concentration gradient of the molecule or the electrochemical gradient established by movement of other ions. However, most movement in nonexcitable cells occurs though some expenditure of energy, either by primary or secondary active transport.

Passive Transport

Regardless of the type of carrier or channel involved, if no energy is expended in the transport process, it is considered **passive transport**. Passive transport can occur via **simple** or **facilitated diffusion**.

Simple Diffusion

If a substance is lipid soluble (which is a property of gases, some hormones, and cholesterol), it will move down its concentration gradient through the cell membrane by **simple diffusion** (Fig. 2.1). This movement is described by *Fick's law*.

Fick's Law
$$J_i = D_i \times A(1/X) \times (C_1 - C_2)$$

Eq. 2.1

Where:

- J_i represents net flux
- D_i is the coefficient of diffusion
- A is the area
- X is the distance through the membrane
- $(C_1 - C_2)$ is the difference in concentration across the membrane

Thus passive diffusion of a molecule across a membrane will be directly proportional to the surface area of the membrane and the difference in concentration of the molecule and inversely proportional to the thickness of the membrane.

Facilitated Diffusion

Facilitated diffusion can occur through either **gated channels** or carrier proteins in the membrane. Gated channels are pores that have "doors" that can open or close in response to external elements, regulating the flow of the solute (Fig. 2.2, *A*). Examples include Ca^{2+}, K^+, and Na^+. This type of gated transport into and out of the cell is critical to most membrane potentials, except the resting potential (see Chapter 3). When facilitated diffusion of a substance involves a carrier protein, binding of the substance to the carrier results in a conformational change in the protein and translocation of the substance to the other side of the membrane (Fig. 2.2, *B*).

Simple and facilitated diffusion do not require expenditure of energy, but they do depend on the size and composition of the membrane and the concentration gradient for the solute. The following key differences exist between these two types of diffusion:

- **Simple diffusion** occurs over all concentration ranges at a rate linearly related to the concentration gradient; if the concentration gradient increases, the rate of diffusion from the compartment with high concentration to the compartment with low concentration will increase.

■ **Facilitated diffusion** is subject to a maximal rate of uptake (V_{max}). The rate of facilitated diffusion is greater than that of passive diffusion at lower solute concentrations. However, at higher solute concentrations the rate of facilitated uptake reaches its V_{max} (i.e., the carrier is saturated), whereas the rate of passive diffusion is not limited by a carrier. Another characteristic of facilitated diffusion is that the V_{max} can be increased by adding transport proteins to the membrane, which is a key regulatory aspect of the transport process.

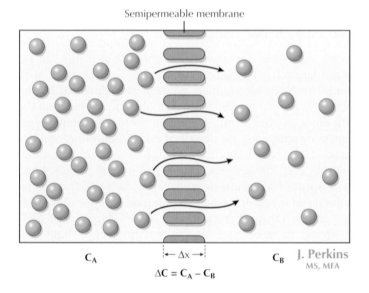

Figure 2.1 **Diffusion Through a Semipermeable Membrane** If the membrane is permeable to a solute, diffusion can occur down the solute's concentration gradient. The rate of diffusion is contingent on the solute gradient (ΔC) and the distance through the membrane (Δx).

Active Transport

Primary Active Transport

Primary active transport involves the direct expenditure of energy in the form of **adenosine triphosphate (ATP)** to transport an ion into or out of the cell (Fig. 2.3). Although the elements depicted in Figure 2.3 are important in many cells, the most ubiquitous is the *Na⁺ pump (Na⁺/K⁺ ATPase)*. The Na^+ pump uses ATP to drive Na^+ out of the cells and K^+ into the cells, which establishes the essential intracellular and extracellular ion environments (Fig. 2.4). Three molecules of Na^+ are transported out of the cell in exchange for two molecules of K^+ transported into the cell, which helps establish an electrical gradient (slightly negative inside the cell), in addition to the effects of ion diffusion due to concentration gradients (discussed in Chapter 3). The ability of the Na^+ pump to maintain the internal and external cell Na^+ and K^+ milieu is essential to cell function. If the Na^+ pump is blocked (e.g., by the drug ouabain), intracellular and extracellular Na^+ and K^+ would equilibrate, affecting membrane transport and electrical potentials (see Chapter 3).

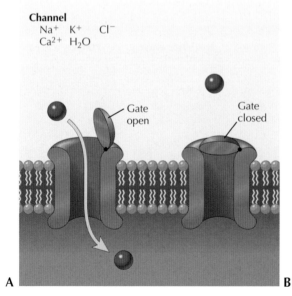
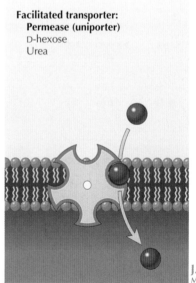

Figure 2.2 **Passive Membrane Transport** Passive transport of substances through the membrane can occur via specific channels or transport proteins. Channels **(A)** can be opened or closed depending on the position of the "gate." The conformational change to open and close gates can be stimulated by ligand binding or changes in voltage. Specific transport proteins **(B)** bind substances, undergo conformational change, and release the substance on the other side of the membrane.

ATP-powered transporters are critical for moving select ions against their concentration gradients. Whereas the ubiquitous Na$^+$/K$^+$ ATPase (sodium pump) is necessary for maintaining intracellular and extracellular sodium and potassium concentrations, other pumps aid in critical physiologic functions as well. For example, H$^+$/K$^+$ ATPases allow for formation of stomach acid, H$^+$ ATPases contribute to the urinary excretion of acid in acid-base homeostasis, and Ca^{2+} ATPases pump calcium out of cells to maintain the ultra-low intracellular calcium environment.

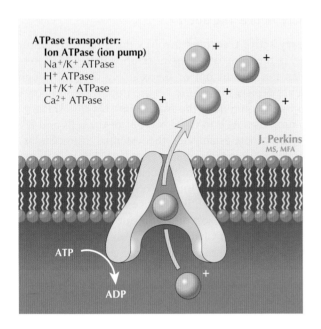

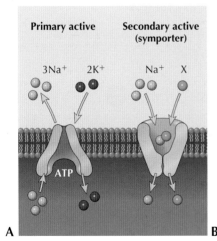

Figure 2.3 **Primary Active Transport** The proteins involved in primary active transport require energy in the form of adenosine triphosphate (ATP) to transport substances against their concentration gradients. *ADP,* adenosine diphosphate.

Secondary Active Transport

Many substances are transported into or out of the cell via **secondary active transport** (also known as **cotransport**) with Na$^+$. The Na$^+$ concentration gradient is maintained by the active Na$^+$/K$^+$ pump, which results in diffusion of Na$^+$ into the cell down its concentration gradient through a specific symporter or antiporter (as described earlier), allowing simultaneous transport of another molecule into or out of the cell (see Fig. 2.4). The *active* portion of this process is the original transport of Na$^+$ against its gradient by the Na$^+$/K$^+$ pump; the subsequent events are *secondary*. A typical example of secondary active transport by symport is Na$^+$-glucose and Na$^+$-galactose transport across the intestinal epithelium. An example of antiport is Na$^+$/H$^+$ exchange that occurs in many cells, including renal and intestinal cells, in which Na$^+$ enters the cells along its concentration gradient through the antiporter while H$^+$ leaves the cells. The Na$^+$ pump also results in *passive movement* of ions through channels: Na$^+$ (down its concentration gradient), Cl$^-$ (following Na$^+$ to preserve electroneutrality), and H$_2$O (following the osmotic pressure gradient; see Fig. 2.4, C).

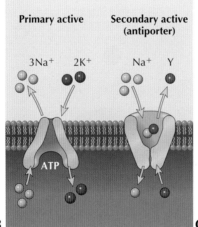

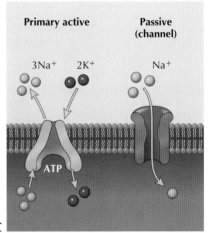

Figure 2.4 **Secondary Active Transport** Although energy is not directly expended, secondary active transporters use a concentration gradient (usually for Na$^+$) established by primary active transporters to move another substance in the same direction (symport **[A]**; e.g., Na$^+$-glucose, Na$^+$-Cl$^-$, Na$^+$-PO$_4^{2-}$, or Na$^+$-amino acid transport in the renal tubules and small intestine), the opposite direction (antiport **[B]**, e.g., Na$^+$/H$^+$ or Na$^+$/K$^+$ transport in the renal tubules and small intestine), or down the concentration gradient of the ion **(C)**.

Ouabain and digoxin are plant-derived cardiac glyco-sides that are used medicinally to treat heart failure and cardiac arrhythmias. Cardiac glycosides block Na$^+$/K$^+$ ATPase, raising intracellular Na$^+$. In the presence of high intracellular Na$^+$, the activity of a second transporter that pumps Ca^{++} out of the cell in exchange for Na$^+$ is inhibited, raising Ca^{++} levels in intracellular stores. Under these circumstances, depolarization of cardiac myocytes results in greater release of Ca^{++} and therefore stronger muscle contraction. Plants sources of cardiac glycosides include strophanthus (ouabain), foxglove (digitalis), and lily of the valley. Historically, plant extracts containing cardiac glycosides have been used as arrow tip poisons, heart tonics, and diuretics.

ION CHANNELS

Movement of ions occurs through channels as well as through membrane carrier–mediated processes. **Ion channels** show high selectivity and allow specific ions to pass down their concentration gradient (e.g., Na$^+$, Cl$^-$, K$^+$, and Ca^{2+}) (see Fig. 2.2, *A*). Selectivity depends on the size of the ion as well as its charge. Gated channels can open or close in response to different stimuli. Stimuli such as sound, light, mechanical stretch, chemicals, and voltage changes can affect control of the ion flux by controlling the gating systems.

Types of channels include the following:

- **Ligand-gated channels** are opened by the binding of a ligand specific for that channel, such as acetylcholine. Binding of the ligand to its receptor causes the channel to open, allowing ion movement. These are tetrameric or pentameric (four- or five-protein subunit) channels.
- **Voltage-gated channels** open in response to a change in membrane voltage. These channels are ion specific and are composed of several subunits, with transmembrane domains forming a pathway for ion flux across the membrane.
- **Gap junction channels** (also called *hemichannels*) are formed between two adjacent cells and open to allow passage of ions and small molecules between the cells. The hemichannels are generally hexameric (i.e., consisting of six subunits, or connexins).

Vesicular Membrane Transport

In addition to movement through channels and transporters, certain substances can enter or be expelled from the cell through **exocytosis, endocytosis,** or **transcytosis**. These types of movement through the cell membrane require ATP and involve packaging of the substances into lipid membrane vesicles for transport (Fig. 2.5).

- Exocytosis involves fusion of vesicles to the cell membrane for extrusion of substances contained in the vesicles.
- Endocytosis is the process by which a substance or particle outside the cell is engulfed by the cell membrane, forming a vesicle within the cell. Phagocytosis is endocytosis of large particles; pinocytosis ("cell drinking") is endocytosis of fluid and small particles associated with the engulfed fluid (Video 2.1).
- Transcytosis occurs in capillary endothelial cells and intestinal epithelial cells to move material across the cell via endocytosis and exocytosis.

Vesicular packaging and transport is especially important when the material needs to be isolated from the intracellular environment because of toxicity (e.g., antigens, waste, or iron) or the potential for altering signaling pathways (e.g., Ca^{2+}).

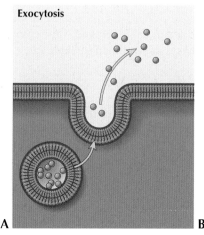
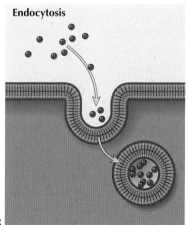
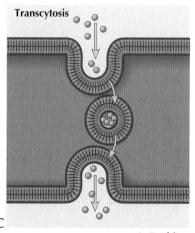

J. Perkins
MS, MFA, CMI

Figure 2.5 **Vesicular Transport Through the Membrane A,** Exocytosis. **B,** Endocytosis. **C,** Transcytosis.

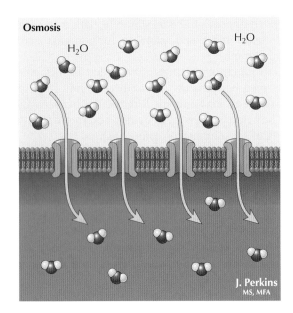

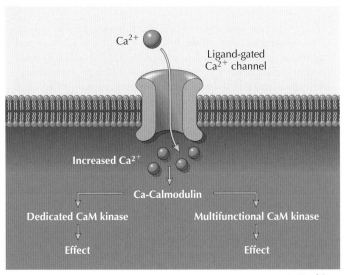

Figure 2.6 **Water Channels** Water flux follows the osmotic pressure gradient and takes place through specific water channels, or aquaporins. Water movement through aquaporins can be regulated by insertion or removal of the proteins from the cell membrane.

Figure 2.7 **Ca²⁺-Calmodulin Signal Transduction** A good example of Ca^{2+}-calmodulin signal transduction is in stimulation of smooth muscle contraction, in which a neurotransmitter (e.g., tachykinin in gut smooth muscle) will act on its receptors to open Ca^{2+} channels. The increased Ca^{2+} in the cytosol binds to calmodulin (CaM), which activates specific myosin kinases. In this case, phosphorylation of myosin results in binding to actin, causing crossbridge formation and contraction of the muscle. CaM kinase, Ca^{2+}-calmodulin-dependent protein kinase.

Apoproteins are required for assembly of vesicles that contain lipid and facilitate their export from cells via exocytosis. **Apo B** is an apoprotein that is important in chylomicron formation in intestinal epithelial cells. The lack of this protein (abetalipoproteinemia) results in malabsorption of lipids and fat-soluble vitamins as a result of the inability to export the chylomicrons from enterocytes to lymph lacteals. Abetalipoproteinemia also impairs very-low-density lipoprotein exocytosis from hepatocytes to blood (see Chapter 26).

Aquaporins

In addition to ion channels, specific water channels, or **aquaporins** (AQPs), exist that allow water to pass through the hydrophobic cell membrane, following the osmotic pressure gradient (Fig. 2.6). Many types of AQPs have been identified; the channels can be constitutively expressed in the membranes, or their insertion into the membrane can be regulated (e.g., by antidiuretic hormone; see Chapter 19). AQP-3 is always present in the *basolateral* membranes of principal cells, whereas regulation of water flux is through insertion of AQP-2 into the *apical* (luminal) membranes, as exemplified in the renal cortical collecting ducts.

SIGNAL TRANSDUCTION MECHANISMS

Much of the basic regulation of cellular processes (e.g., secretion of substances, contraction, relaxation, production of enzymes, cell growth) occurs by the binding of a regulatory substance to its receptor and the coupling of the receptor to effector proteins within the cell.

Agonists, such as neurotransmitters, steroids, and peptide hormones, stimulate different transduction pathways. The pathways frequently include activation of **second messenger systems** such as cyclic adenosine monophosphate (cAMP), cyclic guanosine monophosphate (cGMP), Ca^{2+}, and inositol trisphosphate. The second messengers can activate protein kinases, or, in the case of Ca^{2+}, calmodulin. The pathways can end in the secretion of substances, release of ions, contraction or relaxation of muscle, or regulation of transcription of specific genes, as well as other processes.

Examples of some G protein–coupled receptors using these pathways are provided in Table 2.1, and some common signal transduction pathways are outlined in Table 2.2.

Protein Kinases

Many transduction pathways work through the phosphorylation of proteins by protein kinases. Protein kinase C (PK-C) can be activated by Ca^{2+}, diacylglycerol, and certain membrane phospholipids. Protein kinases such as PK-A can be activated by the second messenger cAMP and are designated "cAMP-dependent kinases"; cGMP-dependent kinases also exist.

Another critical pathway occurs via influx of Ca^{2+} through ligand-gated channels (Fig. 2.7), which results in activation of Ca^{2+}-calmodulin–dependent kinases. These kinases are important in smooth muscle contraction, hormone secretion, and neurotransmitter release.

Table 2.1 G Proteins

G Protein	Activated by Receptors for	Effectors	Signaling Pathways
G_s	Epinephrine, norepinephrine, histamine, glucagon, ACTH, luteinizing hormone, follicle-stimulating hormone, thyroid-stimulating hormone, others	Adenylyl cyclase Ca^{2+} channels	↑ Cyclic AMP ↑ Ca^{2+} influx
G_{olf}	Odorants	Adenylyl cyclase	↑ AMP (olfaction)
G_{t1} (rods)	Photons	Cyclic GMP phosphodiesterase	↓ Cyclic GMP (vision)
G_{t2} (cones)	Photons	Cyclic GMP phosphodiesterase	↓ Cyclic GMP (color vision)
G_{i1}, G_{i2}, G_{i3}	Norepinephrine, prostaglandins, opiates, angiotensin, many peptides	Adenylyl cyclase Phospholipase C Phospholipase A_2 K^+ channels	↓ Cyclic AMP ↑ Inositol 1,4,5-trisphosphate, diacylglycerol, Ca^{2+} Membrane polarization
G_q	Acetylcholine, epinephrine	Phospholipase Cβ	↑ Inositol 1,4,5-trisphosphate, diacylglycerol, Ca^{2+}

From Hansen J: Netter's Atlas of Human Physiology, Philadelphia, Elsevier, 2002.
Note: There is more than one isoform of each class of a subunit. More than 20 distinct α subunits have been identified.
ACTH, adrenocorticotropic hormone; *AMP*, adenosine monophosphate; *GMP*, guanosine monophosphate.

Table 2.2 Signal Transduction Pathways

Adenylyl Cyclase (cAMP)	Phospholipase C ($IP_3 - Ca^{2+}$)	Cytoplasmic/Nuclear Receptor	Tyrosine Kinase	Guanylate Cyclase (cGMP)
ACTH	GnRH*	Cortisol	Insulin	ANP
LH	TRH*	Estradiol	IGFs	Nitric oxide
FSH	GHRH*	Progesterone	GH	
ADH (V_2 receptor)	CRH*	Testosterone		
PTH	Angiotensin II	Aldosterone		
Calcitonin	ADH (V_1 receptor)	Calcitriol		
Glucagon	Oxytocin	Thyroid hormones		
β-Adrenergic agonists	α-Adrenergic agonists			

From Hansen J: Netter's Atlas of Human Physiology, Philadelphia, Elsevier, 2002.
Summary of some hormones, neurotransmitters, and drugs and the signal transduction pathways involved in their actions on cells.
*Also increases intracellular cAMP.
ACTH, adrenocorticotropic hormone; *ADH*, antidiuretic hormone (vasopressin); *ANP*, atrial natriuretic peptide; *cAMP*, cyclic adenosine monophosphate; *cGMP*, cyclic guanosine monophosphate; *CRH*, corticotropin-releasing hormone; *FSH*, follicle-stimulating hormone; *GH*, growth hormone; *GHRH*, growth hormone–releasing hormone; *GnRH*, gonadotropin-releasing hormone; *IGF*, insulin-like growth factor; *IP₃*, inositol trisphosphate; *LH*, luteinizing hormone; *PTH*, parathyroid hormone; *TRH*, thyrotropin-releasing hormone.

G Proteins (Heterotrimeric Guanosine Triphosphate–Binding Proteins)

Most membrane receptors are associated with G proteins (heterotrimeric guanosine triphosphate [GTP]-binding proteins). Ligand binding to the membrane-bound receptor–G-protein complex will cause phosphorylation of guanosine diphosphate (GDP) → GTP, allowing the α subunit of the G protein to interact with and activate membrane bound enzymes—for example, adenyl cyclase or phospholipase C. These second messenger systems will then cause activation of specific effector proteins such as PK-A or PK-C (Fig. 2.8). The activated G proteins also have guanosine triphosphatase (GTPase) activity to inactivate the process. In the case of adenyl cyclase, the GTPase will produce GDP + P_i, and the presence of GDP on the α subunit will cause $G_α$ to rebind to the βγ subunits and thereby result in inactivation of the adenyl cyclase. Many hormones and peptides act through this general mechanism. Examples of several G protein–coupled receptors are provided in Table 2.1.

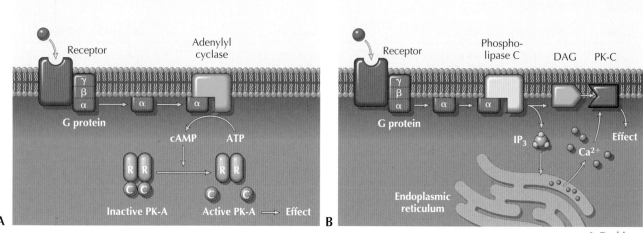

Figure 2.8 **G Protein–Coupled Receptors** Many ligands bind to receptors associated with membrane-bound G proteins to initiate transduction pathways. These G proteins interact with other membrane-bound proteins that activate second messenger systems. The second messengers represented (cyclic adenosine monophosphate [cAMP] and inositol trisphosphate [IP₃]) activate the effector proteins phosphokinase A (PK-A) **(A)** and phosphokinase C (PK-C) **(B)**, respectively. *ATP,* adenosine triphosphate; *DAG,* diacylglycerol.

Nuclear Receptors

Several ligands, such as steroid hormones and thyroid hormone, bind directly to their **nuclear receptors**, interact with deoxyribonucleic acid, and increase or decrease transcription of messenger ribonucleic acid from target genes (Fig. 2.9). When this pathway is stimulated to increase protein synthesis, a delay occurs in presentation of the end protein because the process involves gene transcription and translation. This delayed action is in contrast to that of other hormones and ligands that release proteins from storage vesicles and thus can have a rapid effect.

Simple Versus Complex Pathways

Transduction pathways can be relatively simple and fast acting, as is usually the case for the guanylate cyclase system (cGMP). This type of rapid effect is illustrated in the smooth muscle relaxation response to nitric oxide released by vascular endothelial cells (Fig. 2.10, *A*). This system is in contrast to the complex, slower transduction system observed in the multiple steps involved in growth factor signal transduction, in which ligand binding initiates a multistep process ending in nuclear transcription and protein synthesis (see Fig. 2.10, *B*).

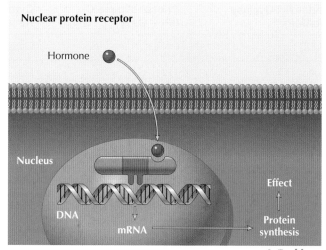

Figure 2.9 **Nuclear Protein Receptors** Several lipophilic hormones do not bind to the cell membrane but instead diffuse through the membrane and are translocated through the cytosol to the nucleus. In the nucleus, they bind to receptors associated with the deoxyribonucleic acid (DNA) and regulate ribonucleic acid (RNA) synthesis. Because this process involves transcription and translation of proteins, producing the end effect takes time.

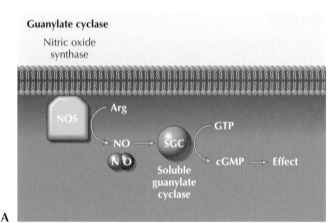

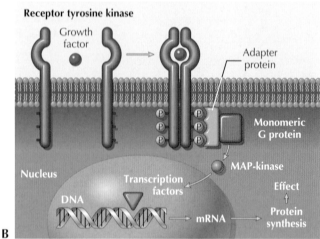

J. Perkins
MS, MFA

Figure 2.10 **Simple and Complex Transduction Systems** **A,** Some transduction pathways have immediate effects, such as those observed with guanylate cyclase activation and cyclic guanosine monophosphate (cGMP) synthesis. **B,** Other pathways are far more complex, involving multiple effectors and nuclear transcription, which prolongs the time for the final effect. *Arg,* arginine; *DNA,* deoxyribonucleic acid; *GTP,* guanosine triphosphate; *MAP,* mitogen-activated protein; *mRNA,* messenger ribonucleic acid; *NO,* nitric oxide; *NOS,* nitric oxide synthase.

CLINICAL CORRELATE 2.1
Cystic Fibrosis

The importance of **membrane transporters** is illustrated in cystic fibrosis disease, the most common lethal genetic disease affecting white persons (1 in ~2000 live births). Cystic fibrosis is caused by a defect in the *cystic fibrosis transmembrane regulator (CFTR)* gene that regulates specific apical (luminal) electrogenic chloride channels (such as in Fig. 2.2, *A*). The defect has profound effects on ion and fluid transport, primarily in the lungs and pancreas. In these tissues, it is critical for Cl⁻ to be secreted into the lumen of the conducting airways and pancreatic acini and ducts, drawing Na⁺ and water. In persons with cystic fibrosis, the CFTR proteins are significantly reduced, decreasing Cl⁻ secretion and resulting in thick secretions. In the lungs, the thick, dry mucus layer contributes to increased infections, and in the pancreas, the ducts from the acini are clogged with mucus and unable to secrete proper amounts of the buffers and enzymes necessary for proper digestion. The pancreatic insufficiency can result in gastrointestinal complications such as meconium ileus in newborns and maldigestion, malabsorption, and weight loss as the child grows older.

Cystic fibrosis is usually diagnosed by age 2 years, and recently the mean age of survival was determined to be 37 years (as cited by the Cystic Fibrosis Foundation). Although antibiotics are used to treat the frequent lung infections, currently the disease has no cure. Treatment includes physical therapy in which the patient's chest and back are pounded to loosen and expel the mucus. New methods have been developed to simulate the percussive action on the mucus, such as the intrapulmonary percussive ventilator and biphasic cuirass ventilation. As lung disease worsens, patients may need to use bilevel positive airway pressure to assist in the ventilation of clogged airways.

In addition to lung disease, the effects on the gastrointestinal tract caused by pancreatic dysfunction can necessitate surgical removal of areas of the small intestine that are amotile and usually requires supplementation of pancreatic enzymes that are reduced. In addition, reduction in endocrine pancreas function (insulin, glucagon, and somatostatin) can increase the incidence of diabetes in persons with cystic fibrosis.

Morbidity is high and life span is considerably shortened in persons with cystic fibrosis. If the patient does not succumb to lung infection, the progressive decrease in lung function and exercise intolerance usually leads to the necessity for a lung transplant.

CLINICAL CORRELATE 2.1
Cystic Fibrosis—cont'd

Clinical Features of Cystic Fibrosis

Infancy	Meconium ileus
	Meconium peritonitis
	Jaundice
Beyond infancy	Failure to thrive
Childhood	Steatorrhea
	Recurrent bronchopulmonary
	infections
	Intestinal obstructions
Adult	Hepatic steatosis
	Gallstones
	Biliary stricture
	CBD obstruction
	Pancreatitis

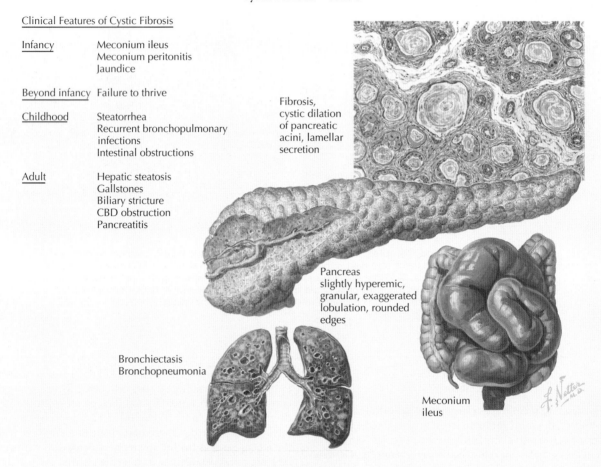

Fibrosis, cystic dilation of pancreatic acini, lamellar secretion

Pancreas slightly hyperemic, granular, exaggerated lobulation, rounded edges

Bronchiectasis
Bronchopneumonia

Meconium ileus

Congenital Cystic Fibrosis *CBD*, common bile duct.

Review Questions

CHAPTER 1: THE CELL AND FLUID HOMEOSTASIS

1. Antipyrine and inulin are injected into a 60-kg man. After equilibrium is achieved, blood is drawn, and the concentrations of the substances are determined.

	Amount of Indicator Injected	Concentration in Blood
Antipyrine	50 mg	1.39 mg/L
Inulin	20 mg	1.67 mg/L

Using these values, select the best answer:

A. The intracellular fluid (ICF) volume is 28 L.
B. The interstitial fluid (ISF) volume is 9 L.
C. The plasma volume is 5 L.
D. The total body water (TBW) is 42 L.
E. The extracellular fluid (ECF) volume is 28 L.

2. Determine the pressure and direction of fluid movement (in or out of the capillary) given the following Starling forces in a capillary bed where σ is approximately 1:

HP_c = 30 mm Hg
HP_i = 3 mm Hg
π_c = 28 mm Hg
π_i = 8 mm Hg

A. 3 mm Hg, into capillary
B. 3 mm Hg, out of capillary
C. 7 mm Hg, into capillary
D. 7 mm Hg, out of capillary
E. 19 mm Hg, out of capillary

3. Addition of pure water to the ECF will have what effect on ICF and ECF compartment volume and osmolarity after steady state is achieved? Assume no excretion of water and an original plasma osmolarity of 300 mOsm/L.

A. ICF volume decreases, ECF volume increases
B. ICF osmolarity decreases, ECF osmolarity increases
C. ICF osmolarity increases, ECF osmolarity increases
D. ICF volume increases, ECF volume increases
E. ICF volume and osmolarity decrease

4. Transferring red blood cells from an isotonic solution to which of the following solutions would cause the red blood cells to contract in volume?

A. 300 mM NaCl
B. 150 mM NaCl
C. 300 mM urea
D. 300 mM sucrose
E. 150 mM sucrose

5. An 80-kg runner drinks 2 L of a physiologic salt solution (300 mOsm/L NaCl). After equilibration (and assuming no urinary losses), what will be the most likely changes in his fluid status?

A. His TBW will be 42 L.
B. The 2 L will distribute evenly between ICF and ECF.
C. One third of the solution will be in the ECF and two thirds will be in the ICF.
D. The ECF volume will be 18 L.
E. The ICF volume will be 34 L.

CHAPTER 2: MEMBRANE TRANSPORT

6. Select the **TRUE** statement about cell transport processes:

A. Simple (passive) diffusion of a molecule is not dependent on the thickness of the cell membrane.
B. Ion channels are relatively nonselective and allow movement of multiple electrolytes through a single channel.
C. Voltage-gated channels are ion specific and open in response to a change in membrane voltage.
D. Secondary active transport directly uses adenosine triphosphate to move substances in and out of cells.
E. Facilitated diffusion requires energy to move substances in and out of cells.

7. The membrane transporter directly responsible for maintenance of low intracellular sodium concentration in intestinal or kidney epithelial cells is the:

A. basolateral Na^+/H^+ antiporter.
B. basolateral Na^+/Ca^{2+} antiporter.
C. apical Na^+/H^+ antiporter.
D. basolateral Na^+/K^+ adenosine triphosphatase (ATPase).
E. apical Na^+/K^+ ATPase.

8. Ouabain and digoxin work by:

A. stimulating cyclic guanosine monophosphate (cGMP).
B. blocking Na^+/K^+ ATPase.
C. stimulating Na^+/K^+ ATPase.
D. blocking cGMP.
E. stimulating H^+/K^+ ATPase.

9. The pathologic consequences of cystic fibrosis are initiated by a defect:

A. in a calcium channel, increasing calcium transport out of cells.
B. in a G-protein–coupled calcium channel, increasing intracellular calcium.
C. in a G-protein–coupled chloride channel, increasing chloride transport out of cells.
D. in a chloride channel, decreasing chloride transport out of cells.
E. in Na^+/K^+ ATPase, increasing the extracellular K^+ concentration.

10. Smooth muscle contraction after elevation of intracellular free Ca^{2+} is directly dependent on the presence of:

A. cyclic adenosine monophosphate.
B. calmodulin.
C. phospholipase C.
D. cGMP.
E. nuclear proteins.

Section 2

THE NERVOUS SYSTEM AND MUSCLE

In complex organisms such as humans, homeostasis requires communication between distant parts of the body, continuous monitoring of changing internal and external conditions, and coordinated, often complex responses to these conditions. These actions are accomplished in an integrated manner by highly differentiated tissues, organs, and systems. This demanding set of tasks is orchestrated in large part by the nervous system, along with various motor systems that receive its input.

Chapter 3

Nerve and Muscle Physiology

To understand the physiology of the nervous system and muscle, it is necessary to examine function in terms of relatively simple cellular processes as well as more complicated interactions between parts of the central nervous system, peripheral nerves and receptors, and muscle. This chapter discusses basic principles of neuronal and muscular function. The role of the central nervous system in the integration of neural and muscular function is considered in Chapter 4.

RESTING MEMBRANE POTENTIALS

Communication through the nervous system requires generation and transmission of electrical impulses, which in turn depend on the ability of cells to maintain resting membrane potentials. The term **resting membrane potential** is synonymous with **steady-state potential**. A resting membrane potential is created by passive diffusion of ions through a selectively permeable membrane, producing charge separation. An example using a hypothetical cell is illustrated in Figure 3.1.

In the simplest theoretical case, if a cell membrane is permeable to only one ion and that ion is present in a higher concentration inside the cell compared with outside the cell, that ion will diffuse out of the cell until sufficient membrane potential is established to oppose further net flux of the ion. For example, if the membrane is permeable only to K^+ and the intracellular K^+ concentration is higher than the extracellular concentration, then outward net flux of K^+ will occur, resulting in a negative membrane potential, in which the intracellular compartment is electrically negative relative to the outside of the cell. Only a minute fraction of the ions will diffuse out of the cell, with no appreciable change in ion concentration in the compartments, before the established electrical gradient will be sufficient to oppose further outward net flux of the ion. At this point, a resting membrane potential will be established. Because of the difference in potential between the two compartments (inside vs. outside), the excess positive charges remain close to the membrane in the extracellular compartment, whereas negative charges line up along the inside of the membrane.

Nernst Equation

The electrical potential difference between the inside and the outside of a cell (E_X) can be predicted if the membrane is permeable to only one ion, using the **Nernst equation** (see Fig. 3.1):

$$E_X = (RT/ZF)\ln([X]_o/[X]_i) \qquad \textbf{Eq. 3.1}$$

Where:

- E_x is the Nernst potential or equilibrium potential
- $\ln([X]_o/[X]_i)$ is the natural log of the ratio of the concentration of ion X outside the compartment ($[X]_o$) to the concentration of the ion inside the compartment ($[X]_i$)
- R is the ideal gas constant
- T is absolute temperature
- Z is the charge of the ion
- F is Faraday's number

In biological systems at 37°C, this equation can be simplified to:

$$E_X = (61\,mV/Z)\log([X]_o/[X]_i) \qquad \textbf{Eq. 3.2}$$

Thus, in a simple hypothetical situation in which a single, monovalent cation (e.g., K^+) is permeable and its concentration inside the cell is 10-fold higher than outside (see Fig. 3.1), this equation becomes:

$$E_{K^+} = (61\,mV/+1)\log(0.1\,mM/1.0\,mM)$$
$$E_{K^+} = (61\,mV)\log(0.1) = -61\,mV \qquad \textbf{Eq. 3.3}$$

The electrochemical equilibrium will be greatly changed by altering the concentration of the permeable ion inside or outside the cell. For a system in which only one ion is permeable, the Nernst potential for the permeable ion is equal to the resting membrane potential. In actual cells, more than one ion is permeable, and thus the resting membrane potential is a result of different permeabilities (conductances) of the ions present and the concentration differences of those ions across the cell membrane. The Nernst potential (equilibrium

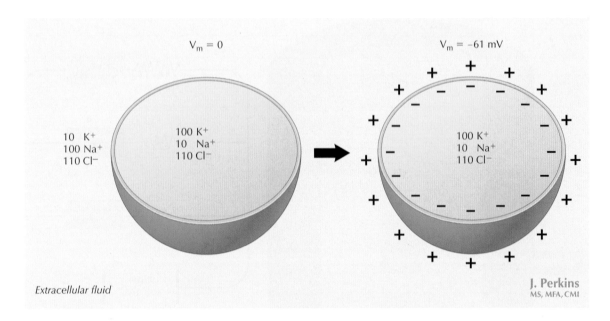

$V_m = 0$

$V_m = -61$ mV

10 K$^+$
100 Na$^+$
110 Cl$^-$

100 K$^+$
10 Na$^+$
110 Cl$^-$

100 K$^+$
10 Na$^+$
110 Cl$^-$

Extracellular fluid

J. Perkins
MS, MFA, CMI

Figure 3.1 **Membrane Potential of a Hypothetical Cell** When the cell with the indicated intracellular ion concentrations (mEq/L) is placed into a solution with different ion concentrations, there is a gradient for diffusion of ions. In this example, the membrane separating inside and outside compartments is selectively permeable; it is permeable only to K$^+$. Because the cell is selectively permeable to K$^+$, diffusion of K$^+$ from inside to outside the cell immediately begins, resulting in negative charge on the inside of the membrane and positive charge outside the membrane. An electrochemical equilibrium is established, in which the gradient for diffusion of K$^+$ is opposed by the electrical gradient, and the membrane is negatively charged. In a simple system with only one permeable ion, V$_m$ (the membrane potential) can be calculated by the Nernst equation.

Table 3.1 **Approximate Concentrations and Nernst Potentials for Major Ions in Cytosol and Interstitial Fluid**

Ion	Concentration (mM)		Nernst Potential (mV)
	Cytosolic	Interstitial	
Na$^+$	14	140	61
K$^+$	140	4	−94
Cl$^-$	8	108	−68

potential) represents the *theoretical* membrane potential that would occur if the cell membrane were to be selectively permeable to only a specific ion. Given the normal values for concentrations of various ions inside cells and in the interstitial fluid, the approximate Nernst potentials for the major permeable ions are listed in Table 3.1.

Goldman-Hodgkin-Katz Equation

The actual resting membrane potential (V$_m$) for a system involving more than one permeable ion is calculated by the **Goldman-Hodgkin-Katz equation (G-H-K equation)**, which takes into account the permeabilities and concentrations of the multiple ions:

$$V_m = \left(\frac{RT}{F}\right) \ln \left(\frac{P_{K^+}[K^+_o] + P_{Na^+}[Na^+_o] + P_{Cl^-}[Cl^-_i]}{P_{K^+}[K^+_i] + P_{Na^+}[Na^+_i] + P_{Cl^-}[Cl^-_o]} \right)$$ **Eq. 3.4**

Where:

- P$_X$ is the membrane permeability to ion X
- [X]$_i$ is the concentration of X inside the cell
- [X]$_o$ is the concentration of X outside the cell
- R is the ideal gas constant
- T is absolute temperature
- F is Faraday's number

Although cells contain many ions, this simplified G-H-K equation omits ions that are much less permeable to the cell membrane than K$^+$, Na$^+$, and Cl$^-$ because their contribution to resting membrane potential is usually negligible. The concentration of Cl$^-$ inside appears in the top of the right-most term and the concentration of Cl$^-$ outside appears in the bottom, whereas the situation for [K$^+$] and [Na$^+$] is opposite that of [Cl$^-$] because of the difference in charge of these ions (negative vs. positive).

The resting membrane potential for most cells is approximately −70 millivolts (mV); in nerve cells it is approximately −90 mV. K$^+$ contributes most to the resting membrane potential because cytosolic K$^+$ concentration is high and extracellular K$^+$ concentration is low, and permeability of the plasma membrane to K$^+$ is high relative to other ions. Therefore,

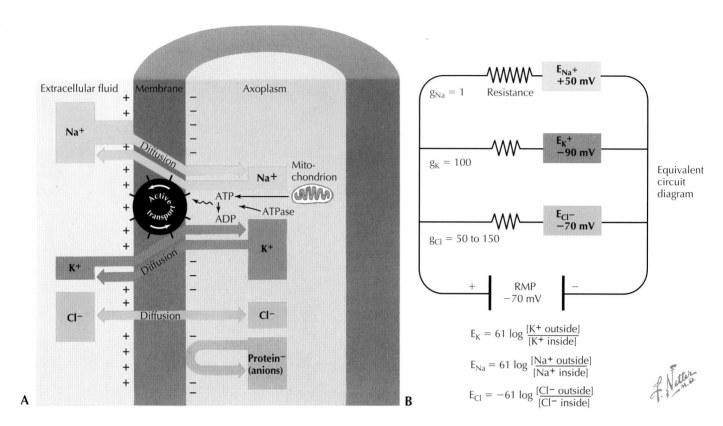

Figure 3.2 **The Resting Membrane Potential** Ion concentration differences between cytosol (in this example, axoplasm) and interstitial fluid are produced by the activity of transporters, mainly the Na^+/K^+ ATPase, which pumps Na^+ out of the cell in exchange for K^+, which is pumped into the cell **(A)**. (The size of the rectangles indicates the relative concentration of ions.) The cell membrane has differential permeabilities to the ions, resulting in leakage across the membrane at different rates. The membrane is not permeable to proteins, which are negatively charged and are mainly intracellular. Leakage of K^+ out of the cell is the major factor in setting the resting membrane potential because it is the most permeable ion. This situation can be modeled as an electrical circuit **(B)**, in which potential differences for the ions are expressed (in millivolts), based on the values obtained from the Nernst equation, and the permeability of the ions is expressed in terms of conductance (g).

although the resting membrane potential is *similar* to the Nernst potential for K^+, other ions contribute to the resting membrane potential. Specifically, leaking of Na^+ through Na^+ channels along its electrochemical gradient contributes to the fact that the resting membrane potential of cells is less negative (more positive) than the Nernst potential for K^+. The gradients of Na^+ and K^+ are maintained in living cells by active transport via Na^+/K^+ ATPase, counteracting the constant leak of Na^+ (and thus K^+; Fig. 3.2).

A cell is said to be **hyperpolarized** when its membrane potential is *more negative* than the normal resting membrane potential for the cell, whereas a cell at membrane potential *less negative* than its usual resting membrane potential is said to be **depolarized**. The membrane potential is a function of the individual concentration gradients for ions across the cell membrane and the permeabilities of the membrane to those ions.

ELECTROPHYSICAL PRINCIPLES

Some key electrophysical principles apply to movement of ions in cellular systems and also to conduction of electrical impulses in the nervous system. According to **Ohm's law:**

$$V = IR \qquad \text{Eq. 3.5}$$

Where:

- V is the potential difference across a conductor, measured in volts (V) or mV
- I is the current, measured in amperes (A), nanoamperes (10^{-9}A), or picoamperes (10^{-12}A)
- R is resistance, measured in ohms or megaohms (10^6 ohms)

In any electrical system, including movement of permeable ions across a cell membrane, current will occur (i.e., charged particles will flow) driven by the potential difference in the system and impeded by the resistance of the system.

Conductance (G) is the inverse of resistance and is measured in Siemens (S), nanoSiemens, or picoSiemens:

$$G = 1/R \qquad \textbf{Eq. 3.6}$$

In cells, conductance for various ions is a function of "leakiness" of the cell membrane to the ions (related to permeability, or the presence of open channels). Ohm's law can be restated to predict a specific ion current based on membrane potential and conductance of the ion:

$$I_X = G(V_m - E_x) \qquad \textbf{Eq. 3.7}$$

In other words, an ion current across a cell membrane will equal the product of the conductance for the ion and the difference between the membrane potential and the equilibrium (Nernst) potential for the ion. Thus the positive ion K^+ continues to leak out of the cell (along its concentration gradient) at resting membrane potential because the resting membrane potential (V_m; approximately −70 mV in most cells) is less negative than E_{K+} (−94 mV). Na^+ will continually leak into a cell because its equilibrium potential is 61 mV compared with its resting membrane potential of −70 mV. By convention, the current associated with this outward flux of K^+, I_{K+}, is a positive current, whereas the inward Na^+ current, I_{Na+}, is a negative current. Again, the concentration gradients of these ions are maintained in the face of this constant but slow leak by active transport by Na^+/K^+ ATPase.

ACTION POTENTIALS

An **action potential** is a rapid depolarization that occurs in an excitable cell (neurons and muscle cells); it is initiated by an electrical event or chemical stimulation that causes increased ion permeability in the cell membrane (Fig. 3.3 and Video 3.1). The following characteristics of action potentials are important:

- A **threshold potential**, which is the potential at which an action potential will be initiated by increased conductance of ions (Na^+ in many cases). For example, a cell with a resting membrane potential of −70 mV might have a threshold potential of −50 mV; if the cell is depolarized to that threshold, an action potential will occur.
- An **"all-or-none"** response, with a fixed amplitude (voltage change) and stereotypic shape of the action potential for a given cell type.
- **Nondecremental propagation**, in which an action potential at one point on a cell surface will be conducted to adjacent sites along the membrane.
- A **refractory period** beginning with the initiation of the action potential, when it is impossible to elicit a second action potential.
- A **relative refractory period** following the absolute refractory period, when a second potential can be evoked but requires a larger than normal stimulus.

The threshold potential results from the balance between outward K^+ leakage and inward Na^+ current through voltage-gated Na^+ channels. When a stimulus produces depolarization that reaches this threshold, voltage-gated Na^+ channels open, and the inward flux of Na^+ now exceeds the ability of K^+ leakage to maintain a steady state. The further depolarization of the membrane opens more voltage-gated Na^+ channels. This positive feedback continues until all voltage-gated Na^+ channels are open, producing the rapid, all-or-none depolarization characteristic of an action potential. These "fast channels" are rapidly inactivated as well.

Along with these changes in Na^+ conductance, a delayed, slower, and smaller increase in conductance of K^+ occurs during the action potential (see Fig. 3.3). This phenomenon is caused by opening of voltage-gated K^+ channels; along with the fall in Na^+ conductance, it is responsible for the repolarization of the membrane. These K^+ channels remain open until the membrane finally returns to the equilibrium potential and are therefore responsible for the hyperpolarization or "undershoot" phase of the action potential, during which the membrane is at a more negative potential than the resting potential. This "undershoot" contributes to the relative refractoriness of the membrane to the generation of another action potential because a greater stimulus will be required to displace this hyperpolarized membrane potential to the threshold potential. A second factor contributing to refractoriness, earlier in the relative refractory period, is that some Na^+ channels still remain inactivated.

Extracellular Ca^{2+} concentration affects threshold potential by altering the sensitivity of voltage-gated Na^+ channels. At high Ca^{2+} concentrations, greater membrane depolarization is required to activate the channels; thus the excitability of cells is reduced at high Ca^{2+} concentrations.

An **action potential** is initiated in a cell when the membrane potential is sufficiently depolarized to reach the **threshold potential**. When this phenomenon occurs, opening of specific voltage-gated channels results in increased conductance of one or more ions, resulting in a rapid depolarization of the cell. In many excitable cells, the upstroke of the action potential is a result of opening of voltage-gated Na^+ channels. When the cell reaches threshold potential, a sufficient number of **activation gates** of the Na^+ channels are opened to allow Na^+ flux through the membrane, which causes the upstroke of the action potential. Within milliseconds of opening of the activation gates, time-dependent inactivation gates close, producing an inactive state of the Na^+ channels. The inactivation gates reopen when the cell returns to the resting membrane potential.

ACTION POTENTIAL CONDUCTION

In excitable cells, an action potential is rapidly propagated over the cell membrane. In a neuron, when **threshold potential** is reached, an action potential is generated at the axon

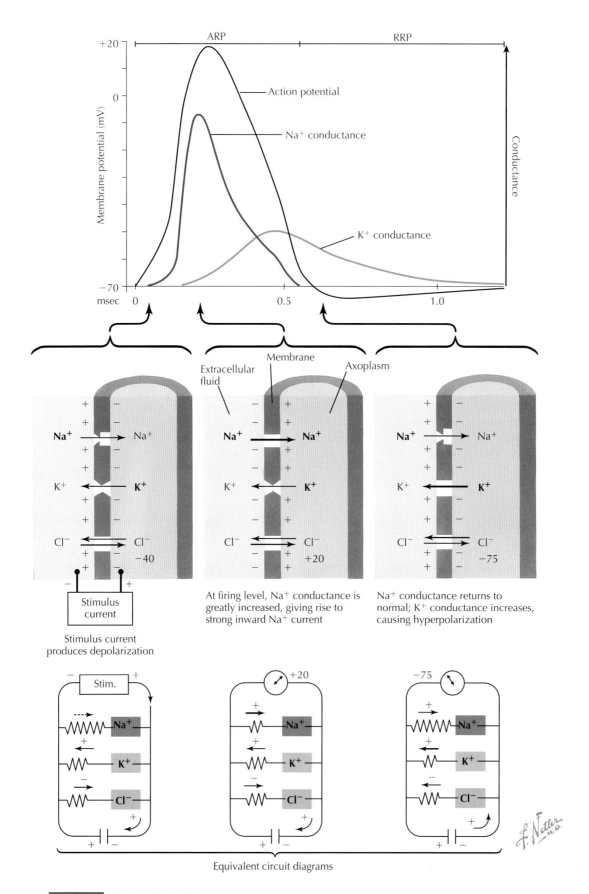

Figure 3.3 **Action Potential** The action potential in most excitable cells is initiated when a local potential on a cell reaches a threshold at which voltage-gated Na⁺ channels open, greatly increasing Na⁺ conductance. The top panel illustrates membrane potential and relative permeability of Na⁺ and K⁺. Changes in Na⁺ and K⁺ conductance during the action potential are shown in the middle panels, along with the equivalent circuit diagrams (*bottom panels*). ARP, absolute refractory period; RRP, relative refractory period.

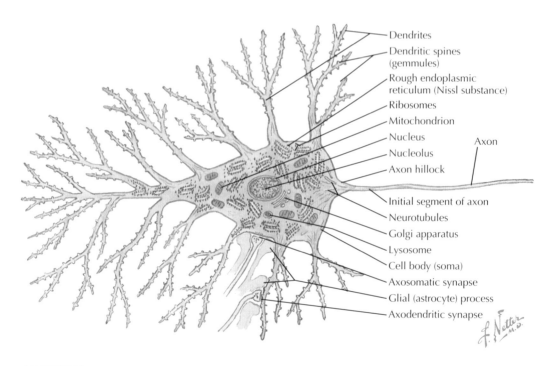

Figure 3.4 **Neurons** Neurons consist of the soma (cell body), multiple dendrites, and an axon. Neurons receive signals from other neurons at synapses, resulting in changes in local membrane potential. When the membrane potential is sufficiently depolarized, an action potential is initiated at the axon hillock and is propagated down the length of the axon to synapses with other neurons or muscle cells.

hillock, spreads to the axon, and is conducted to the axon terminal (Fig. 3.4; conduction *between* cells is discussed later). As depolarization occurs at a point on the membrane, Na^+ flows through the membrane from the extracellular fluid. Loss of positive charge at the point of depolarization causes **local currents**, in which positive charges flow from adjacent regions along the membrane to the area of depolarization (Fig. 3.5A). Local current results in depolarization of the adjacent regions; when threshold is reached, the action potential is propagated. An important characteristic of **action potential propagation** is that it occurs away from the point of initiation; it cannot travel back toward its origin. As the action potential is conducted, the area of the membrane directly behind the action potential is still in the absolute refractory state as a result of Na^+ channel inactivation, preventing retrograde conduction.

An essential property of neurons is the ability to rapidly propagate an action potential, analogous to an electrical cable. Conduction velocity is a function of:

- **Internal resistance (R_i)**, which is the impedance to current flow within the cytoplasm (e.g., resistance within the axon).
- **Membrane resistance (R_m)**, which is the impedance to current flow through the membrane.
- **Capacitance (C_m)**, which is the ability of the membrane to store charge.

The relationships between these variables result in the **cable properties** of cells—that is, the **space constant** (λ) and **the**

time constant (τ)—which are useful in describing conduction of electrical activity in an axon. The time it takes for membrane potential to change to 63% of its final value when a current is applied is $\tau(\tau = R_m \times C_m)$. It is a measure of how slowly or quickly the membrane depolarizes or repolarizes and is therefore inversely related to action potential conduction velocity. Because τ is inversely related to conduction velocity, low membrane resistance and low membrane capacitance favor faster conduction.

The space constant (λ, also known as the *length constant*) is the distance along the membrane over which a potential change will fall to 63% of its value and therefore describes how far a depolarizing current will spread. With greater λ, depolarizing currents will spread further, resulting in greater conduction velocity. λ is proportional to the ratio of R_m to R_i:

$$\lambda = \sqrt{(R_m/R_i)} \qquad \textbf{Eq. 3.8}$$

The importance of internal resistance (R_i) in action potential conduction velocity can be inferred from the anatomy of axons of large invertebrates. The giant axon of the squid has a diameter of 500 micrometer (μm); with this increased diameter, λ is high because R_i is low, resulting in the rapid conduction necessary in a large animal. The large size of these axons made them ideal subjects for the pioneering work of Nobel laureates Alan Hodgkin and Andrew Huxley on neuronal conduction.

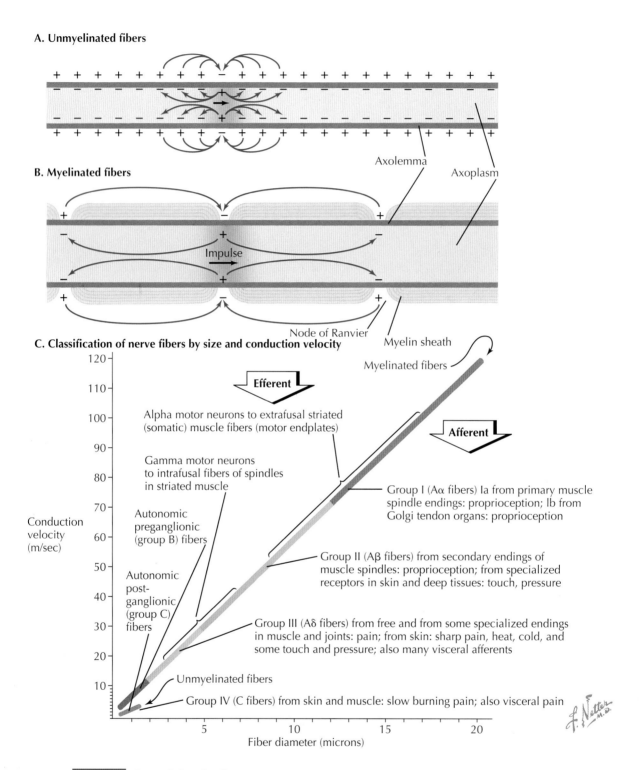

A. Unmyelinated fibers

Axolemma

Axoplasm

B. Myelinated fibers

Impulse

Node of Ranvier   Myelin sheath

C. Classification of nerve fibers by size and conduction velocity

Efferent

Myelinated fibers

Alpha motor neurons to extrafusal striated (somatic) muscle fibers (motor endplates)

Afferent

Gamma motor neurons to intrafusal fibers of spindles in striated muscle

Group I (Aα fibers) Ia from primary muscle spindle endings: proprioception; Ib from Golgi tendon organs: proprioception

Autonomic preganglionic (group B) fibers

Group II (Aβ fibers) from secondary endings of muscle spindles: proprioception; from specialized receptors in skin and deep tissues: touch, pressure

Autonomic post-ganglionic (group C) fibers

Group III (Aδ fibers) from free and from some specialized endings in muscle and joints: pain; from skin: sharp pain, heat, cold, and some touch and pressure; also many visceral afferents

Unmyelinated fibers

Group IV (C fibers) from skin and muscle: slow burning pain; also visceral pain

Conduction velocity (m/sec)

Fiber diameter (microns)

Figure 3.5   **Axonal Conduction   A,** An action potential is propagated and conducted along an axon as adjacent portions of the membrane are depolarized by local currents. **B,** In myelinated axons, the action potential "jumps" between nodes of Ranvier. This process, called *saltatory conduction,* greatly increases conduction velocity. **C,** Myelination and axonal diameter are associated with higher conduction velocity.

The nervous system of vertebrates is much more complex and is composed of a much larger number of neurons than that of invertebrates. Many nerve cells in the vertebrate nervous system are **myelinated** (Fig. 3.5B)—that is, they are covered by multiple layers of an insulating sheath of phospholipid membrane, formed by Schwann cells in the peripheral nervous system and oligodendrocytes in the central nervous system. This insulation greatly decreases capacitance and also increases membrane resistance, such that current travels through the interior of the axon but not across the membrane. To allow

CLINICAL CORRELATE 3.1
Multiple Sclerosis

Multiple sclerosis (MS) is an inflammatory neurodegenerative disease believed to be of autoimmune etiology. MS is more common in women than in men, and the usual age of onset is between 20 and 40 years. The inflammatory process of MS results in the gradual destruction of myelin sheaths around myelinated axons of the brain and spinal cord. A wide variety of symptoms may occur, including muscle weakness and paralysis; impaired coordination and poor balance; depression; impaired speech; memory problems; visual problems; altered sensory perception; pain; fatigue; and bowel, bladder, and sexual dysfunctions, resulting from the deficits in neural conduction caused by damage to myelin sheaths and underlying axons in the central nervous system. Diagnosis is based on clinical findings, magnetic resonance imaging evidence of demyelinating lesions of the brain and spinal cord, and characteristic oligoclonal bands of γ-globulins in cerebrospinal fluid.

Cerebrospinal fluid electrophoresis

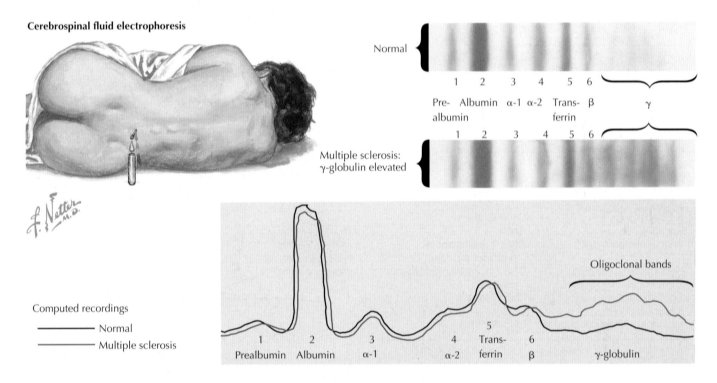

Multiple Sclerosis: Diagnostic Tests—Spinal Fluid Cerebrospinal fluid samples obtained by lumbar puncture display characteristic elevation of γ-globulins (oligoclonal bands) upon gel electrophoresis in approximately 90% of patients.

propagation of the action potential, breaks called **nodes of Ranvier** occur in the myelin sheath. They are present at 1- to 2-mm intervals along the axon, and the action potential thus "jumps" rapidly from node to node, bypassing the myelinated areas. Essentially, the nodes of Ranvier represent multiple capacitors in series. The process of conduction by which the action potential "jumps" between nodes is known as **saltatory conduction** and allows very rapid propagation of the action potential despite small axon diameter. Local currents generated at one node during depolarization result in depolarization at the next node, and the action potential skips along the axon, bypassing the highly insulated segments. Fibers throughout the nervous system vary in terms of diameter and conduction velocity and may be classified into various types on this basis (Fig. 3.5C). In contrast to saltatory conduction, in unmyelinated axons, action potentials are conducted as slower waves of depolarization as they are propagated along the axon.

SYNAPTIC TRANSMISSION

A **synapse** is a site at which an electrical response in one cell is transmitted to another cell. This transmission can occur between some cell types through **electrical synapses**, when electrical current passes directly from one cell to another through **gap junctions**. Cardiac myocytes and some types of smooth muscle communicate through this mechanism (some neurons communicate both through electrical and chemical transmission). At electrical synapses between cells, intramembrane proteins called **connexons** form channels that allow flow of ions between cells, thus allowing current transmission.

At a **chemical synapse**, release of a neurotransmitter by a neuron results in electrical stimulation of the postsynaptic cell. A wide variety of neurotransmitters are found in various regions of the central and peripheral nervous system

Table 3.2 **Summary of Some Neurotransmitters and Where They Are Found Within the Central and Peripheral Nervous System**

Transmitter	Location
Acetylcholine	Neuromuscular junction, autonomic endings and ganglia, CNS
Biogenic amines	
Norepinephrine	Sympathetic endings, CNS
Dopamine	CNS
Serotonin	CNS, GI tract
Amino acids	
γ-Aminobutyric acid (GABA)	CNS
Glutamate	CNS
Purines	
Adenosine	CNS
Adenosine triphosphate	CNS
Gas	
Nitric oxide	CNS, GI tract
Peptides	
β-Endorphins	CNS, GI tract
Enkephalins	CNS
Antidiuretic hormone	CNS (hypothalamus/posterior pituitary)
Hypothalamic releasing hormones	CNS (hypothalamus/anterior pituitary)
Somatostatin	CNS, GI tract
Neuropeptide Y	CNS
Vasoactive intestinal peptide	CNS, GI tract

From Hansen J: Netter's Atlas of Human Physiology, Philadelphia, Elsevier, 2002.
CNS, central nervous system; *GI,* gastrointestinal.

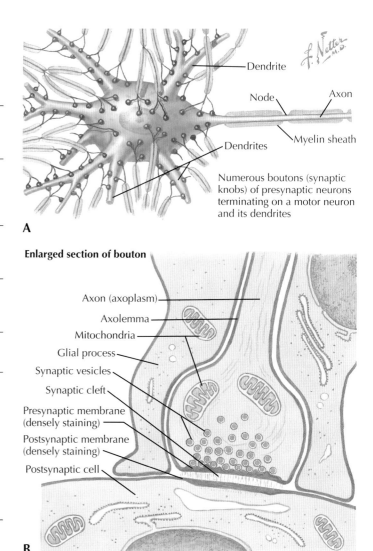

A

Enlarged section of bouton

B

Figure 3.6 **Morphology of Synapses** An action potential in one neuron is transmitted to another neuron or an effector tissue at the synapses. The terminal regions of axons branch and form synaptic boutons. The myelination is lost where the bouton is in close contact with a dendrite or cell body of another neuron **(A)**. Neurotransmitters are released into the synaptic cleft when vesicles of the synaptic bouton fuse with the presynaptic membrane **(B)**. The released neurotransmitters diffuse and bind to receptors on the postsynaptic membrane, resulting in excitatory or inhibitory effects at the postsynaptic membrane.

(Table 3.2). The structure of a chemical synapse is illustrated in Figure 3.6. The following sequence of events in transmission occurs at this type of synapse:

1. Depolarization of the neuron reaches the **synaptic bouton** (axon terminal), where it causes opening of Ca^{2+} channels.
2. The influx of Ca^{2+} results in the release of neurotransmitter, stored in presynaptic vesicles, into the **synaptic cleft** (the gap between the cells).
3. The transmitter diffuses across the synaptic cleft and binds to specific membrane receptors on the postsynaptic membrane.
4. Binding of the neurotransmitter produces a change in the membrane potential of the postsynaptic membrane.

Depending on the presynaptic neuron involved (and thus the specific neurotransmitter released), the chemical transmission may result in either an **excitatory postsynaptic potential** (depolarization) or an **inhibitory postsynaptic potential** (hyperpolarization) as a result of Na^+ or Cl^- influx at the postsynaptic membrane, respectively (Fig. 3.7 and Video 3.2). Chemical transmission of this type is **unidirectional** between a presynaptic fiber and a postsynaptic cell. A given neuron typically receives synaptic input from multiple inhibitory and excitatory neurons; therefore, generation of an action potential depends on summation of these multiple inputs. An action potential may occur as a result of **temporal summation** when a series of impulses is generated by an excitatory presynaptic fiber, or as a result of **spatial summation** of local potentials generated by multiple excitatory fibers in the postsynaptic cell (Fig. 3.8). Synaptic inhibition may be either presynaptic or postsynaptic; some inhibitory fibers act directly on an excitatory fiber (**presynaptic inhibition**), whereas others produce **postsynaptic inhibition** (Fig. 3.9). Summation of local

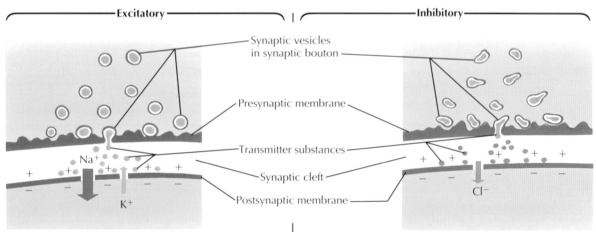

——————— Excitatory ——————— | ——————— Inhibitory ———————

Synaptic vesicles in synaptic bouton
Presynaptic membrane
Transmitter substances
Synaptic cleft
Postsynaptic membrane

Na⁺
K⁺
Cl⁻

When impulse reaches the excitatory synaptic bouton, it causes release of a transmitter substance into the synaptic cleft. This increases permeability of the postsynaptic membrane to Na⁺ and K⁺. More Na⁺ moves into the postsynaptic cell than K⁺ moves out, due to greater electrochemical gradient.

At the inhibitory synapse, transmitter substance released by an impulse increases permeability of the postsynaptic membrane to Cl⁻.

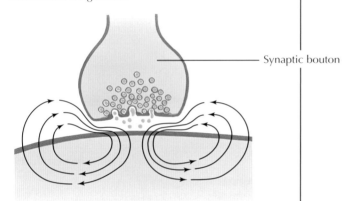

——— Synaptic bouton ———

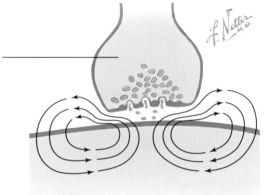

Resultant net ionic current flow is in a direction that tends to depolarize the postsynaptic cell. If depolarization reaches firing threshold, an impulse is generated in the postsynaptic cell.

Resultant ionic current flow is in the direction that tends to hyperpolarize the postsynaptic cell. This makes depolarization by excitatory synapses more difficult—more depolarization is required to reach the threshold.

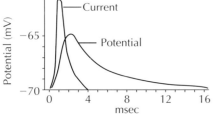

Current
Potential

Current flow and potential change

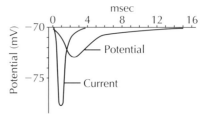

Potential
Current

Current flow and potential change

Figure 3.7 **Chemical Synaptic Transmission**   Synaptic transmission may result in either excitatory or inhibitory effects at the postsynaptic membrane. In excitatory transmission *(left)*, a depolarizing change in local potential (excitatory postsynaptic potential) is produced. In inhibitory transmission *(right)*, a hyper-polarizing change in local potential (inhibitory postsynaptic potential) is produced. An action potential occurs when the temporal and spatial summation of local potentials reaches the threshold for action potential generation (see Fig. 3.8).

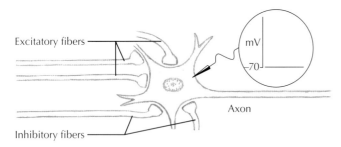

A. Resting state: motor nerve cell shown with synaptic boutons of excitatory and inhibitory nerve fibers ending close to it

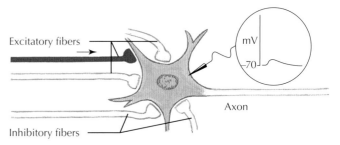

B. Partial depolarization: impulse from one excitatory fiber has caused partial (below firing threshold) depolarization of a motor neuron

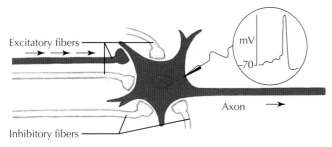

C. Temporal excitatory summation: a series of impulses in one excitatory fiber together produce a suprathreshold depolarization that triggers an action potential

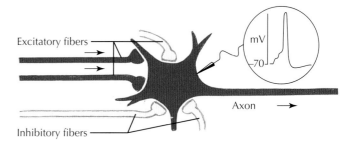

D. Spatial excitatory summation: impulses in two excitatory fibers cause two synaptic depolarizations that together reach the firing threshold, triggering an action potential

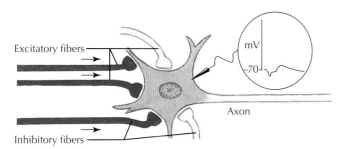

E. Spatial excitatory summation with inhibition: impulses from two excitatory fibers reach a motor neuron, but impulses from inhibitory fiber prevent depolarization from reaching the threshold

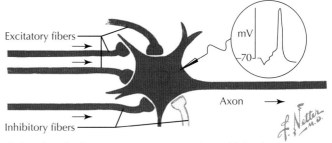

E. (continued): the motor neuron now receives additional excitatory impulses and reaches firing threshold despite a simultaneous inhibitory impulse; additional inhibitory impulses might still prevent firing

Figure 3.8 **Temporal and Spatial Summation** In the resting state **(A)**, the nerve cell is at its resting membrane potential. Neurons are subject to multiple inhibitory and stimulatory signals that produce local potential changes, which are subject to both temporal and spatial summation. The sum of these influences determines whether an action potential is produced (**B** to **E**).

potentials in the postsynaptic membrane depends on the space constant and time constant of the membrane.

NEUROMUSCULAR JUNCTION

Motor neurons (motoneurons) are efferent nerves that originate in the central nervous system and communicate with skeletal muscle fibers at specialized synapses known as **motor end plates** or **neuromuscular junctions** (Fig. 3.10). Among the several types of motor neurons, the most common is the α-motor neuron. Branches of an α-motor neuron may form multiple neuromuscular junctions in depressions in the **sarcolemmae** (muscle cell membranes) of muscle fibers. Each α-motor neuron may thus innervate multiple muscle fibers, although each fiber is innervated by only one α-motor neuron.

An α-motor neuron and the fibers it innervates is called a **motor unit**. All the motor neurons innervating a muscle are collectively known as a **motor neuron pool**.

The synaptic boutons of α-motor neurons are rich in vesicles containing the neurotransmitter acetylcholine. As is the case for other neurons, an action potential in an α-motor neuron produces Ca^{2+} influx at the synaptic bouton, which results in release of neurotransmitter from vesicles of the synaptic bouton. Vesicular release of neurotransmitter is **quantal** in nature and is conceptualized as the release of packets (quanta) of neurotransmitter, with the quanta being defined by the response to the neurotransmitter contained in a single vesicle. At the postsynaptic membrane of a neuromuscular junction, acetylcholine binds to a specific subtype of acetylcholine receptor known as the *nicotinic receptor*. The action of

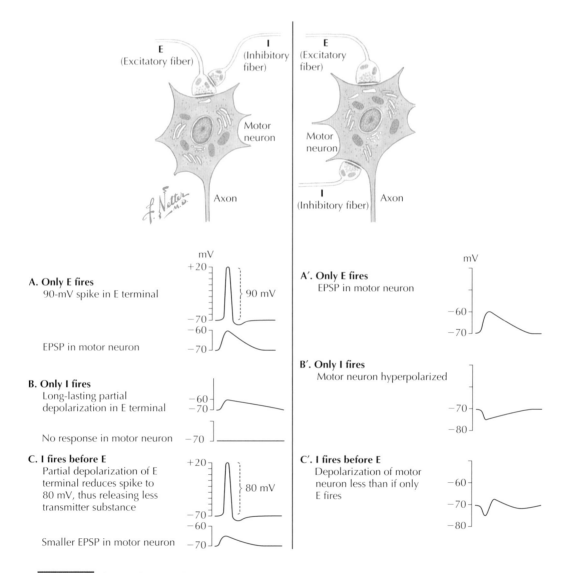

Figure 3.9 **Synaptic Inhibitory Mechanisms** Inhibition can occur through either presynaptic or postsynaptic mechanisms. In presynaptic inhibition, illustrated for a motor neuron *(top left)*, an inhibitory fiber of one nerve cell has a synapse on an excitatory axon of another neuron before the latter communicates with the motor neuron; in postsynaptic inhibition *(top right)*, the inhibitory and excitatory fibers both synapse directly with the target neuron. **A, B,** and **C** illustrate the changes in potential at the excitatory terminal and in the motor neuron when only the excitatory, only the inhibitory, or both inhibitory and excitatory fibers fire. **A′, B′,** and **C′** illustrate membrane potential changes in the motor neuron when only the excitatory, only the inhibitory, or both excitatory and inhibitory fibers fire. *EPSP,* excitatory postsynaptic potential.

acetylcholine is rapid; it is also short lived because acetylcholine diffuses out of the synaptic cleft or is rapidly broken down by acetylcholinesterase localized on the basement membrane of the muscle fiber.

Binding of postsynaptic nicotinic receptors by acetylcholine causes the opening of ligand-gated channels. Influx of Na^+ and K^+ through these channels produces an excitatory postsynaptic potential (end plate potential). When this end plate potential reaches threshold, an action potential is generated, ultimately producing contraction of the muscle fiber. Pharmacologically, the actions of acetylcholine can be blocked by **curare**, a competitive antagonist that binds reversibly to the nicotinic receptor, thereby blocking acetylcholine binding, or

by **cobra venom** (α**-bungarotoxin**), a noncompetitive antagonist that binds irreversibly to the receptor.

End plate potentials differ from neuronal action potentials in several important ways:

- End plate potentials are produced by a ligand-gated channel; neuronal action potentials are caused by voltage-gated channels. In both cases, the action potential is generated in the postsynaptic cell when a threshold is reached.
- Rapid depolarization, to a potential of 0 mV, compared with rapid depolarization to +40 mV during a neuronal action potential.

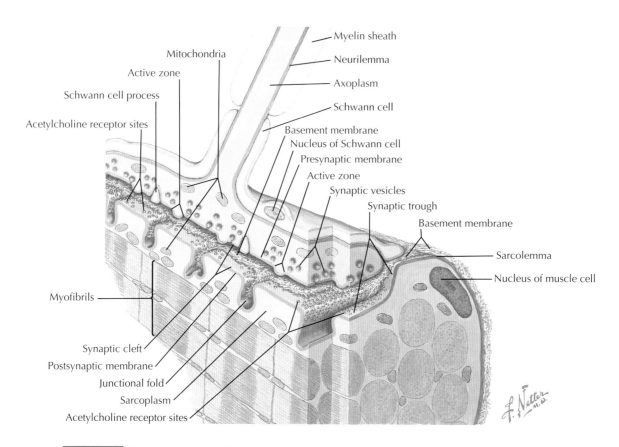

Figure 3.10 **Structure of the Neuromuscular Junction** At the neuromuscular junction, the axon of a motor nerve synapses with skeletal muscle at a site known as the *motor end plate*. Stimulation of a motor nerve results in the release of acetylcholine from vesicles at the presynaptic membrane; acetylcholine diffuses and binds to postsynaptic receptors, producing depolarization of the sarcolemma and leading to an action potential.

■ A single, large channel for Na^+ and K^+ carries the charge during an end plate potential. Multiple ion channels are involved in a neuronal action potential, which is mainly produced by Na^+ influx.

■ Repolarization of the end plate is passive, whereas increased K^+ conductance is responsible for repolarization during a neuronal action potential.

Lidocaine and similar drugs are injected locally to produce anesthesia during dental or medical procedures and can be applied topically to relieve cutaneous itching and pain. Lidocaine has this anesthetic effect because it blocks voltage-gated Na^+ channels in neuronal membranes, thereby blocking signal transmission in sensory neurons. Intravenous lidocaine is also used under specific circumstances as an antiarrhythmic agent; this effect is related to the blockage of voltage-gated Na^+ channels in cardiac cells.

SKELETAL MUSCLE ORGANIZATION

Skeletal muscle contraction is the basis for voluntary muscle movement. Multinucleated cells contain **sarcomeres**, which are specialized structures that produce contraction upon stimulation of the muscle. Sarcomeres are contained within **myofibrils**, which are further organized as **muscle fibers**; groups of muscle fibers form **muscle fascicles** (Fig. 3.11). Microscopically, the arrangement of sarcomeres and myofibrils results in the striated appearance of skeletal muscle. Contraction is based on sliding of the thin and thick filaments of sarcomeres (see "Excitation-Contraction Coupling" below).

Within the skeletal muscle cell, the **sarcoplasmic reticulum** composes a complex network surrounding the myofibrils (Fig. 3.12). This specialized form of smooth endoplasmic reticulum is the site for the storage of high concentrations of intracellular Ca^{2+} and contains Ca^{2+}-ATPase and calsequestrin (a low-affinity Ca^{2+}-binding protein) for sequestration of this ion as well as L-type (ligand-gated) Ca^{2+} channels. The **transverse tubules** (**T tubules**) are deep invaginations of the muscle cell membrane (sarcolemma). They form **triads** with two terminal cisternae of the sarcoplasmic reticulum; these triads are arranged perpendicularly to the muscle fiber. The T tubules extend into the muscle fiber from the surface, allowing close communication between the interior of the cell and the extracellular fluid. They are responsible for conducting the action potential to the cisternae of the sarcoplasmic reticulum.

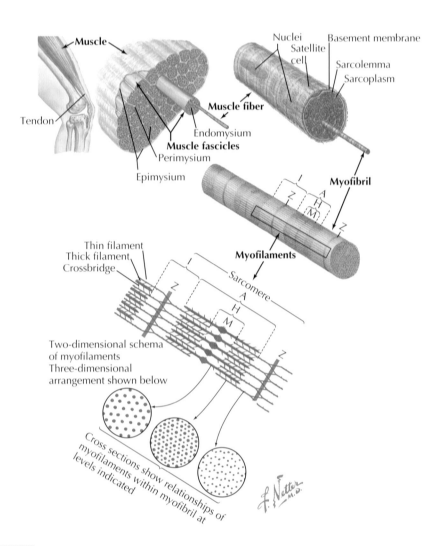

Figure 3.11 **Organization of Skeletal Muscle**   Skeletal muscle is composed of fascicles that are in turn composed of multinucleated muscle fibers. These fibers are composed of smaller myofibrils, which contain sarcomeres, the site at which sliding of actin and myosin filaments produces contraction. The organization of sarcomeres within the skeletal muscle produces its striated appearance. The Z line marks the boundary between two sarcomeres. The I band contains only the actin thin filaments, which extend from the Z line toward the center of the sarcomere. Myosin thick filaments are found in the dark A band. At the H zone, there is no overlap between actin and myosin. The M line is at the center of the sarcomere and is the site at which the thick filaments are linked with each other.

EXCITATION-CONTRACTION COUPLING

As depolarization spreads across the sarcoplasmic reticulum, it is conducted into the T tubules (Fig. 3.13). The T tubule membrane contains voltage-gated Ca^{2+} channels, also known as **dihydropyridine receptors**. Although the dihydropyridine receptor is a voltage-gated Ca^{2+} channel, ion flux through this channel is not required for contraction of skeletal muscle. Rather, a conformational change in the dihydropyridine receptor, caused by depolarization of the T tubule, is required. These receptors are in close apposition to calcium channel proteins known as **ryanodine receptors**, which are large proteins of the sarcoplasmic reticulum that extend into the gap between the cisternae of the sarcoplasmic reticulum and the T tubules. Conformational change of the dihydropyridine receptors is believed to produce a subsequent conformational change in the ryanodine receptors, allowing stored Ca^{2+} to be released from the sarcoplasmic reticulum, initiating the contraction process. The term **excitation-contraction coupling** refers to this linking of depolarization to Ca^{2+} release.

To summarize these events:

- Depolarization of the motor neuron terminal results in Ca^{2+} influx.
- Vesicles of the axon terminal release acetylcholine.
- Binding of acetylcholine by nicotinic receptors results in an end plate potential.
- An action potential is initiated and is propagated along the sarcolemma and down the T tubules.
- A conformational change in the dihydropyridine receptor of the T tubule is transduced to a conformational change in the ryanodine receptor of the sarcoplasmic reticulum.
- Ca^{2+} is released from the sarcoplasmic reticulum, initiating contraction.

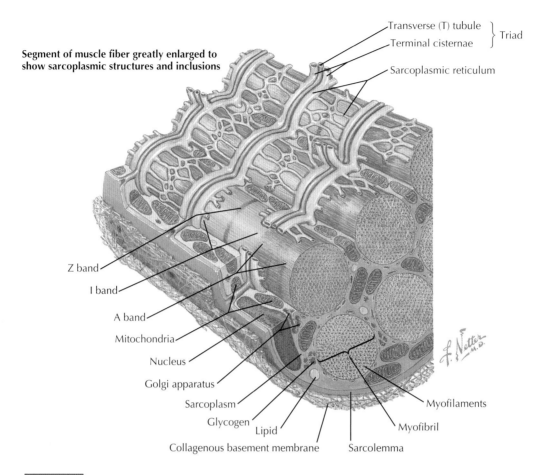

Segment of muscle fiber greatly enlarged to
show sarcoplasmic structures and inclusions

Transverse (T) tubule ⎫
Terminal cisternae ⎬ Triad
Sarcoplasmic reticulum

Z band
I band
A band
Mitochondria
Nucleus
Golgi apparatus
Sarcoplasm
Glycogen
Lipid
Collagenous basement membrane
Sarcolemma
Myofibril
Myofilaments

Figure 3.12 **Sarcoplasmic Reticulum** The sarcoplasmic reticulum is a complex network surrounding the myofibrils and storing high concentrations of Ca^{2+}, sequestered from the sarcoplasm. The membrane of the sarcoplasmic reticulum contains Ca^{2+}-ATPase, which is essential for this sequestration. The transverse tubules are deep invaginations of the sarcolemma and form triads with the terminal cisternae of the sarcoplasmic reticulum. These transverse tubules conduct the action potential from the sarcolemma to the cisternae, causing release of Ca^{2+}.

THE SLIDING FILAMENT THEORY

The sliding filament theory describes the processes that take place in the sarcomeres that account for skeletal muscle contraction. The sliding of the interdigitated, anchored thick and thin filaments is the basis of contraction (Fig. 3.14). The thick filaments of the sarcomeres are composed of the protein **myosin** and are anchored to the **M line**; the thin filaments are composed of **actin, tropomyosin,** and **troponin** and are anchored to the **Z line** (Fig. 3.15). Within the thin filament, globular G-actin is polymerized to form an α-helical filament (F-actin). Actin has binding sites for myosin along the groove of its helix; these sites are covered by the protein **tropomyosin**. Three forms of **troponin** are also incorporated at regular intervals (troponin C, troponin I, and troponin T).

In the relaxed state, in which cytosolic Ca^{2+} concentration is extremely low (10^{-7} M), binding of myosin to actin is blocked because myosin binding sites on actin are covered by tropomyosin. Partially hydrolyzed ATP (i.e., ADP) is bound to

the myosin head groups (see Fig. 3.15). When an action potential causes release of Ca^{2+} from the sarcoplasmic reticulum, binding of Ca^{2+} to troponin causes troponin to move into the groove of the actin α-helix. This action results in exposure of the myosin binding sites, thus promoting the binding of the myosin head group to actin, forming a **crossbridge** (Video 3.3). This binding is followed by a ratcheting action of the myosin head group, shortening the sarcomere as the actin and myosin slide past each other. ADP and inorganic phosphate (P_i) are released. Next, binding of ATP to the myosin head group causes detachment of actin and myosin, after which ATP is partially hydrolyzed by ATPase, which causes "recocking" of the head group. If cytosolic Ca^{2+} is still elevated, myosin rapidly binds to actin, and crossbridge cycling in this manner causes the contraction to continue. Crossbridge cycling occurs simultaneously at multiple sites, producing shortening of the sarcomere. When Ca^{2+} is resequestered into the sarcoplasmic reticulum, muscle relaxes as a result of low cytosolic Ca^{2+} concentration.

CLINICAL CORRELATE 3.2
Myasthenia Gravis

Myasthenia gravis is an autoimmune, neuromuscular disease that affects the acetylcholine receptors of the neuromuscular junction. In this disease, autoantibodies are produced that block or damage the **acetylcholine receptor** of the motor end plate, inhibiting the normal effects of acetylcholine at these receptors and thus producing muscle weakness. Muscles of the eye and face and muscles involved in swallowing, talking, and chewing are most commonly affected, although weakness of other muscles may also occur. Episodes of myasthenia gravis may occur suddenly, often after a period of high physical activity, and may abate after a period of rest. A myasthenic crisis is a medical emergency that may be caused by an infectious disease or adverse drug reaction. In a crisis, muscles of respiration are weakened, making breathing difficult. Such cases require assisted ventilation, such as positive airway pressure ventilation. In myasthenia gravis, the appearance of symptoms is usually episodic, and treatment often involves an immunosuppressive drug such as corticosteroids, as well as a **cholinesterase inhibitor** (e.g., neostigmine). Cholinesterase inhibitors inhibit acetylcholinesterase, prolonging the half-life of acetylcholine in the neuromuscular junction. In some cases, plasmapheresis may be used to remove autoantibodies, or a thymectomy may be performed.

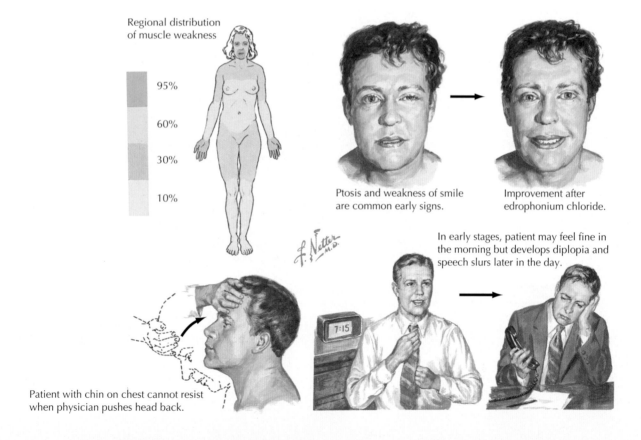

Regional distribution of muscle weakness

95%

60%

30%

10%

Ptosis and weakness of smile are common early signs.

Improvement after edrophonium chloride.

In early stages, patient may feel fine in the morning but develops diplopia and speech slurs later in the day.

7:15

Patient with chin on chest cannot resist when physician pushes head back.

Clinical Manifestations of Myasthenia Gravis In persons with myasthenia gravis, an autoimmune disease, antibodies block or reduce the number of nicotinic acetylcholine receptors at the neuromuscular junction, resulting in muscle fatigability. Diagnostic tests include the edrophonium test. Edrophonium chloride is a cholinesterase inhibitor and thus increases acetylcholine at the neuromuscular junction. Administered intravenously, it temporarily relieves symptoms of muscle weakness, including diplopia (double vision), in persons with myasthenia gravis.

Skeletal muscle fibers can be categorized as **fast-twitch** and **slow-twitch** fibers, also known as *white muscle* and *red muscle*, respectively. Most muscles in the body are composed of a mixture of the two types, but individual motor units contain only one fiber type. The oxygen-binding protein **myoglobin** is found only in slow-twitch fibers and is responsible for its red color. Fast-twitch (white muscle) fibers use anaerobic glycolysis as an energy source and therefore are high in glycogen content. This type of muscle is used for rapid, powerful movements; ATPase activity and contraction speed are high, and resistance to fatigue is low. Slow-twitch (red muscle) fibers use oxidative phosphorylation, and thus glycogen content is low. Slow-twitch fibers are used for activities requiring endurance. ATPase activity and contraction speed are low, and resistance to fatigue is high.

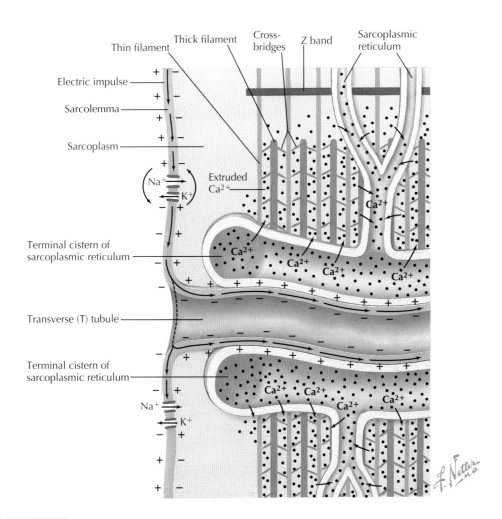

Thin filament · Thick filament · Cross-bridges · Z band · Sarcoplasmic reticulum

Electric impulse

Sarcolemma

Sarcoplasm

Na⁺ K⁺

Extruded Ca²⁺

Terminal cistern of sarcoplasmic reticulum

Transverse (T) tubule

Terminal cistern of sarcoplasmic reticulum

Na⁺ K⁺

Figure 3.13 **Excitation-Contraction Coupling**   When an action potential in a motor neuron results in the release of acetylcholine at the neuromuscular junction, binding of acetylcholine on the sarcolemma opens a cation channel, permitting influx of Na⁺. An action potential is produced and spreads into the transverse tubules, resulting in the release of Ca²⁺ stored in the sarcoplasmic reticulum, initiating crossbridge formation and muscle contraction. Ca²⁺ is resequestered into the sarcoplasmic reticulum by Ca²⁺-ATPase to terminate the contraction.

During crossbridge cycling, each cycle requires one ATP molecule to dissociate each myosin head group. Furthermore, relaxation requires ATP for sequestration of Ca²⁺ into the sarcoplasmic reticulum. In **rigor mortis**, because ATP is depleted, myosin remains bound to actin, causing stiffness of the muscles.

MECHANICAL CONSIDERATIONS IN SKELETAL MUSCLE CONTRACTION

Generation of force by skeletal muscle is controlled by a number of factors (Fig. 3.16). For example, variation in the size of motor units is consistent with the function of the motor units. Small motor units are effective in performing fine movements, as in movements of the fingers and eyes, whereas large motor units perform coarser movements (see

Fig. 3.16A). The force of contraction of skeletal muscles can be increased by **recruitment** of motor units and **summation of twitches**. As noted earlier, all the muscle fibers associated with a single motor unit will contract simultaneously. Greater contraction can be achieved by recruiting more motor units (**spatial summation**). With repeated stimulation of a muscle, **temporal summation** may occur. In this case, individual twitches of the muscle occur without full relaxation between twitches, with greater force of contraction resulting as Ca²⁺ is released from the sarcoplasmic reticulum at a higher rate than it can be resequestered between the twitches (see Fig. 3.16B). In **tetanus** (tetany), a sustained, forceful contraction occurs as a result of high-frequency stimulation.

Finally, the muscle tension generated during contraction depends on the resting tension or the degree of stretch of the muscle at rest. In other words, there is a **length-tension**

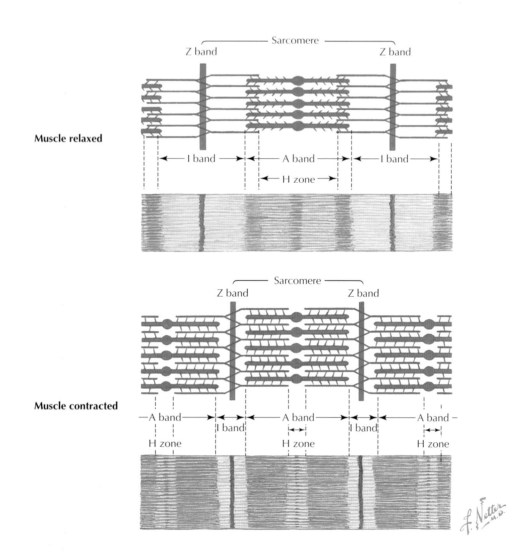

Figure 3.14 **Muscle Contraction and Relaxation** During muscle contraction, thin filaments of each myofibril slide deeply between thick filaments, bringing Z bands closer together and shortening sarcomeres. A bands remain the same width, but I bands narrow. H zones also narrow or disappear as thin filaments encroach upon them. Myofibrils and, consequently, muscle fibers (muscle cells), fascicles, and muscles as a whole grow thicker. During relaxation, the reverse occurs.

relationship for skeletal muscle (see Fig. 3.16C). The extent of overlap of the thin and thick filaments of the sarcomeres affects the number of crossbridges formed during contraction. With greater stretch before stimulation, more force is generated, up to an optimal sarcomere length.

Muscle contractions can also be characterized as isometric or isotonic. In **isometric contraction**, muscle length is constant but a change in tone is observed. An example is the contraction that occurs when we exert force against an immovable object. Force is generated, but length remains unchanged. Sarcomere shortening is accompanied by stretch of connective tissue and cytoskeletal components of the muscle (**series elastic elements**). During **isotonic contraction**, muscle tone remains constant while the muscle shortens. An example is the

contraction associated with lifting a book. In reality, all muscle movements are a combination of isometric and isotonic contractions. For example, pressing of heavy free weights involves a large isometric component but also involves an isotonic component.

In general, isometric exercises such as weight lifting are useful for bodybuilding. With repeated isometric exercise, **hypertrophy** of muscle occurs, in which the number of sarcomeres is increased (**hyperplasia**, or increased number of muscle cells, does not normally occur). Muscle **atrophy** or wasting (i.e., a decreased number of sarcomeres) occurs with disuse, such as during prolonged bed rest.

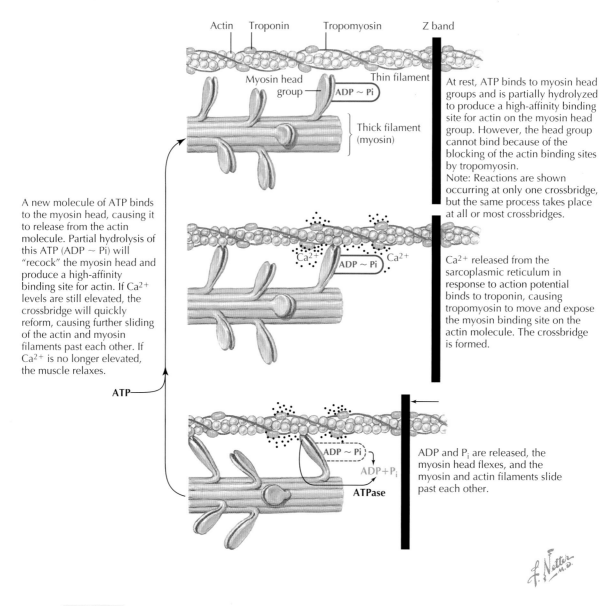

At rest, ATP binds to myosin head groups and is partially hydrolyzed to produce a high-affinity binding site for actin on the myosin head group. However, the head group cannot bind because of the blocking of the actin binding sites by tropomyosin.
Note: Reactions are shown occurring at only one crossbridge, but the same process takes place at all or most crossbridges.

A new molecule of ATP binds to the myosin head, causing it to release from the actin molecule. Partial hydrolysis of this ATP (ADP ~ Pi) will "recock" the myosin head and produce a high-affinity binding site for actin. If Ca^{2+} levels are still elevated, the crossbridge will quickly reform, causing further sliding of the actin and myosin filaments past each other. If Ca^{2+} is no longer elevated, the muscle relaxes.

Ca^{2+} released from the sarcoplasmic reticulum in response to action potential binds to troponin, causing tropomyosin to move and expose the myosin binding site on the actin molecule. The crossbridge is formed.

ADP and P_i are released, the myosin head flexes, and the myosin and actin filaments slide past each other.

Figure 3.15 **Biochemical Mechanics of Muscle Contraction** Muscle contraction is produced when actin and myosin form and recycle crossbridges. The process depends on the presence of free intracellular Ca^{2+} and the availability of ATP.

SMOOTH MUSCLE

Smooth muscle is a type of nonstriated muscle found within organs. Contractile proteins are not organized as sarcomeres in smooth muscle; rather, actin is anchored to the cell membrane and to dense bodies within the cell. As in other types of muscle, actin-myosin interactions are the basis of contraction (Fig. 3.17). A comparison of skeletal, smooth, and cardiac muscle is shown in Table 3.3.

Types of Smooth Muscle

Smooth muscle is further classified as unitary or multiunit types. Of these types, **unitary smooth muscle** is far more abundant and is found in the walls of blood vessels and in the bladder and the gut, among other organs. This type of smooth muscle is capable of sustained and often powerful contractions. The cells have gap junctions between them, allowing rapid and direct spread of action potentials because electrical potentials are directly conducted between cells through ion fluxes. As a result, many smooth muscle cells act as a single unit, producing a synchronous contraction. In contrast, **multiunit smooth muscle** is organized into motor units similar to those in skeletal muscle. Cells are electrically isolated from each other (there are no gap junctions), allowing fine motor control. This type of muscle is found in a few specific regions such as the ciliary body of the eye, the vas deferens, and the piloerector muscles in the skin.

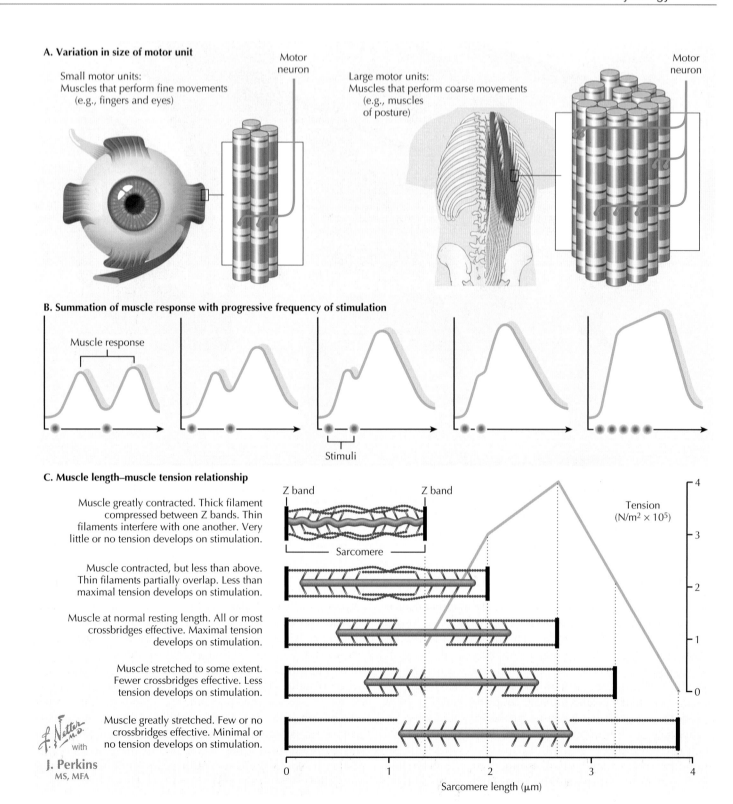

A. Variation in size of motor unit

Small motor units:
Muscles that perform fine movements
(e.g., fingers and eyes)

Motor neuron

Large motor units:
Muscles that perform coarse movements
(e.g., muscles of posture)

Motor neuron

B. Summation of muscle response with progressive frequency of stimulation

Muscle response

Stimuli

C. Muscle length–muscle tension relationship

Muscle greatly contracted. Thick filament compressed between Z bands. Thin filaments interfere with one another. Very little or no tension develops on stimulation.

Muscle contracted, but less than above. Thin filaments partially overlap. Less than maximal tension develops on stimulation.

Muscle at normal resting length. All or most crossbridges effective. Maximal tension develops on stimulation.

Muscle stretched to some extent. Fewer crossbridges effective. Less tension develops on stimulation.

Muscle greatly stretched. Few or no crossbridges effective. Minimal or no tension develops on stimulation.

Z band   Z band

Sarcomere

Tension
$(N/m^2 \times 10^5)$

Sarcomere length (μm)

J. Perkins
MS, MFA

Figure 3.16 **Grading of Muscle Tension and Length-Tension Relationship**   The force generated when skeletal muscle is stimulated is related to the size of the motor units stimulated **(A)**, the number of motor units activated and the frequency of stimulation of the muscle fibers **(B)**, and the resting length of muscle fibers **(C)**. Small motor units are involved in fine motor control, for example, in fingers and eyes, whereas larger motor units are involved in coarser movements **(A)**. Skeletal muscles also produce more force when more motor units are recruited. With repeated stimulation of a muscle fiber, summation occurs, in which individual twitches occur without complete relaxation between twitches **(B)**. A length-tension relationship also exists, whereby greater resting sarcomere length (stretch of the muscle before contraction) is associated with greater force of contraction, up to an optimal resting length **(C)**.

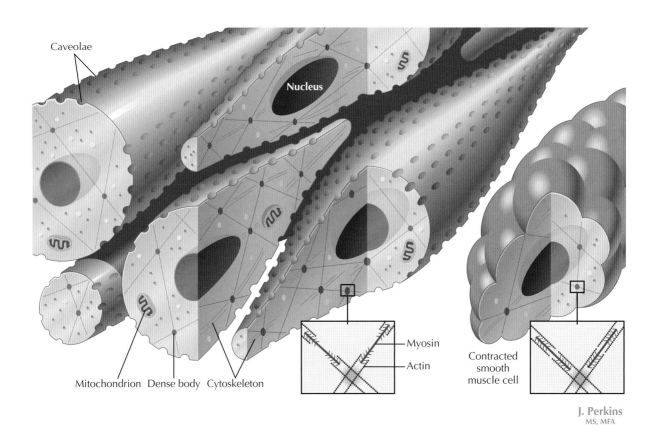

Caveolae

Nucleus

Mitochondrion   Dense body   Cytoskeleton

Myosin

Actin

Contracted
smooth
muscle cell

J. Perkins
MS, MFA

Figure 3.17 **Smooth Muscle Structure** Actin and myosin filaments of spindle-shaped smooth muscle cells are arranged quite differently than contractile proteins in skeletal muscle. Actin is anchored to dense bodies within the muscle cell and to the plasma membrane, and the sliding of myosin and actin filaments produces contraction when intracellular Ca^{2+} is elevated, either by release from intracellular stores or flux through Ca^{2+} channels. Caveolae are invaginations in the cell membrane and are a site of Ca^{2+} flux.

Contraction of Smooth Muscle

Smooth muscle contraction is controlled by multiple neurotransmitters and other chemical ligands that affect cytosolic Ca^{2+} concentration. Some of these substances produce depolarization of the cell membrane, resulting in opening of voltage-gated membrane Ca^{2+} channels and release of Ca^{2+} from intracellular stores, in a process similar to that in skeletal muscle. In **pharmacomechanical coupling**, binding of a ligand to a membrane receptor produces an increase in intracellular Ca^{2+} concentration—and thus smooth muscle contraction—without altering membrane potential. An example is contraction of vascular smooth muscle by binding of norepinephrine or epinephrine to α-adrenergic receptors.

> In **phasic smooth muscle**, contractions are short in duration, as in skeletal muscle. In **tonic smooth muscle**, tension may be maintained for long periods with little ATP utilization. In this case, myosin is dephosphorylated while attached to actin, forming the **"latch state"** (see Fig. 3.18). In this state, crossbridge recycling is very slow and ATP utilization is reduced. This latch state is important in maintaining tension in blood vessels and sphincters.

Depolarization or pharmacomechanical coupling leads to elevation of intracellular Ca^{2+}, the common signal in smooth muscle contraction (Fig. 3.18). In the latter case, ligand binding activates membrane phospholipase C, which cleaves phosphatidylinositol bisphosphate, thereby producing inositol trisphosphate (IP_3), which releases Ca^{2+} from intracellular stores. Whether elevated by membrane depolarization or activation of the inositol pathway, Ca^{2+} binds to the protein **calmodulin**, and the Ca^{2+}-calmodulin complex activates the enzyme **myosin kinase**, allowing interaction between myosin and actin and producing contraction as these proteins slide past each other. The contraction cycle continues as long as intracellular Ca^{2+} is elevated.

Relaxation of Smooth Muscle

Smooth muscle *relaxation* may also be induced by pharmacomechanical coupling, in which case various substances stimulate a rise in intracellular cAMP or cGMP. Ultimately, intracellular Ca^{2+} is reduced, causing relaxation. An example is relaxation produced by binding of epinephrine or norepinephrine to β-adrenergic receptors on vascular smooth muscle.

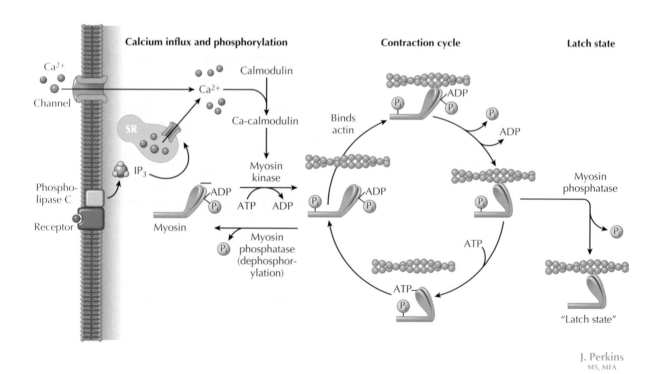

Figure 3.18 **Excitation-Contraction Coupling of Smooth Muscle** Binding of a ligand to the sarcolemma results in elevation of free intracellular Ca^{2+} through either depolarization of the cell membrane and opening of Ca^{2+} channels or activation of the enzyme phospholipase C. In the latter case, cleavage of phosphatidylinositol bisphosphate by phospholipase C produces IP_3, which binds to sites on the sarcoplasmic reticulum (SR), causing release of stored Ca^{2+}. In either case, Ca^{2+} binds to the protein calmodulin, which activates myosin kinase, initiating actin-myosin interaction. The contraction cycle continues as long as Ca^{2+} is elevated. The latch state occurs when myosin is dephosphorylated by myosin phosphatase. In this state, contraction can be maintained without further ATP hydrolysis and thus without further energy expenditure.

CARDIAC MUSCLE

Cardiac muscle is similar to skeletal muscle or smooth muscle in some respects and dissimilar in others (Fig. 3.19; see Table 3.3). Contraction of skeletal muscle is under the voluntary control of the central nervous system, whereas contraction of cardiac and smooth muscle is involuntary. Both unitary smooth muscle and cardiac muscle have the capability for spontaneous electrical activity; cardiac contraction is normally under the control of cardiac pacemaker cells in the sinoatrial node. Gap junctions in cardiac muscle, as in unitary smooth muscle, allow synchronous contraction. These gap junctions in cardiac muscle are found in the **intercalated disks** between cells. Cardiac and skeletal muscles have highly organized sarcomeres, leading to a striated appearance. The velocity of contraction of skeletal muscle depends on muscle fiber type (fast-twitch vs. slow-twitch fibers). Cardiac muscle contraction is slower than that of skeletal muscle but more rapid than contraction in smooth muscle. Cardiac and smooth muscle use both intracellular and extracellular sources of Ca^{2+}, whereas the only Ca^{2+} source for contraction of skeletal muscle is intracellular (from the sarcoplasmic reticulum). As in skel-

etal muscle, Ca^{2+} in cardiac muscle binds to troponin to initiate crossbridge formation.

Other aspects of cardiac muscle function, including mechanical function and regulation of pacemaker activity, are considered in Section 3.

Calcium-channel blockers are drugs that block voltage-dependent Ca^{2+} channels (L-type Ca^{2+} channels). These drugs are often used in antihypertensive therapy for their effects on vascular smooth muscle. However, because they also block voltage-dependent Ca^{2+} channels of the heart, they slow conduction in the heart and reduce myocardial contractility. Calcium-channel blockers may belong to several classes of chemicals, including dihydropyridines. Nifedipine and amlodipine are examples of the dihydropyridine class. Voltage-gated Ca^{2+} channels of T tubules in cardiac and skeletal muscle are also referred to as *dihydropyridine receptors* for their ability to bind these drugs. Note, however, that although cardiac muscle contractility is reduced by calcium blockers, skeletal muscle contractility is unaffected. Skeletal muscle does not depend on extracellular Ca^{2+} for contraction.

Table 3.3 Comparison of Muscle Structure and Function

	Skeletal Muscle	Cardiac Muscle	Smooth Muscle
STRUCTURE			
Morphology	Long; cylindrical	Branched	Spindle or fusiform
Nuclei	Multiple; located peripherally	One (sometimes two); located centrally	One; located centrally
Sarcomere	Yes; striated pattern	Yes; striated pattern	No
T tubules	Yes; forms triad with sarcoplasmic reticulum	Yes; forms dyad with sarcoplasmic reticulum	No; caveolae
Electrical coupling of cells	No	Yes; intercalated disks contain gap junctions	Yes; gap junctions
Regeneration	Yes; via satellite cells	No	Yes
Mitosis	No	No	Yes
PHYSIOLOGY			
Extracellular Ca^{2+} required for contraction	No	Yes	Yes
Regulation of crossbridge formation	Ca^{2+} binding to troponin	Ca^{2+} binding to troponin	Ca^{2+}-calmodulin activation of myosin kinase and phosphorylation of myosin
Control of contraction	Motor neurons	Autonomic nerves; β-adrenergic agonists	Autonomic nerves; hormones
Summation of twitches by increased stimulus frequency	Yes	No*	Yes
Tension varies with filament overlap	Yes	Yes	Yes

From Hansen J: *Netter's Atlas of Human Physiology,* Philadelphia, Elsevier, 2002.
Major differences in structure and function of skeletal, cardiac, and smooth muscle are indicated.
*Cardiac muscle cannot be tetanized, but the force of contraction will increase at high stimulus frequency because of an increase in intracellular $[Ca^{2+}]$, a phenomenon termed "Treppe."

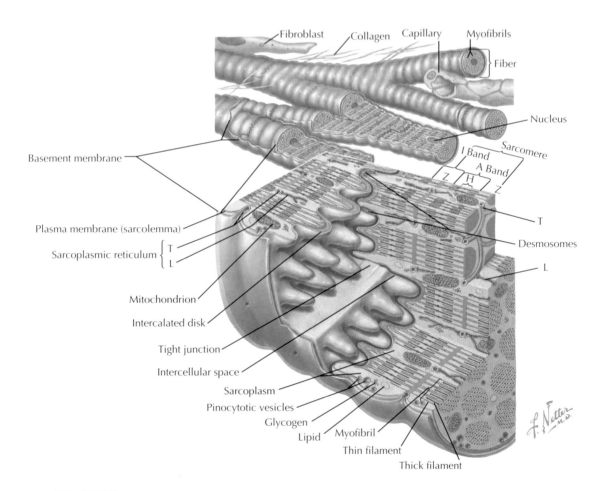

Figure 3.19 **Schema of Structure of Cardiac Muscle**   The striated appearance of cardiac muscle is associated with the arrangement of the contractile proteins actin and myosin into sarcomeres, as in skeletal muscle. A notable difference between these types of muscles is that cardiac muscle uses extracellular Ca^{2+}, as well as intracellular stores, to initiate contraction. Another difference is that transverse tubules of cardiac muscle form dyads with the terminal cisternae of the sarcoplasmic reticulum, as opposed to the triads found in skeletal muscle. In addition, unlike skeletal muscle, cardiac muscle fibers are connected by gap junctions at intercalated disks between the cells, allowing the spread of depolarization from cell to cell, producing synchronous contraction of muscle. *T,* transverse tubule; *L,* longitudinal tubule.

Chapter 4

Organization and General Functions of the Nervous System

The nervous system consists of the **central nervous system (CNS)** and the **peripheral nervous system**. The CNS includes the brain and spinal cord. The peripheral nervous system includes nerves, ganglia, and sensory receptors outside the CNS. The peripheral nervous system can also be subdivided into **sensory** and **motor** divisions. Sensory nerves transmit information from various sensory receptors to the CNS; motor nerves transmit information from the CNS to muscles and glands, thereby controlling their activity.

CENTRAL NERVOUS SYSTEM

The general structure of the CNS and brain is illustrated in Figure 4.1.

Brain

The brain can be subdivided into the **telencephalon** (also known as the *cerebrum* or *cerebral hemispheres*), **diencephalon** (thalamus and hypothalamus), **cerebellum**, and **brainstem** (midbrain, pons, and medulla).

Blood-Brain Barrier

The environment of neurons within the CNS is maintained in part by the **blood-brain barrier**. Endothelial cells of capillaries within the CNS are joined by tight junctions, preventing the movement of water-soluble substances, highly charged molecules, and cells between blood and brain. Astrocytes (nonneuronal cells of the CNS) are also involved in maintaining the integrity of the blood-brain barrier.

Cerebrospinal Fluid

The formation, circulation, and regulation of the composition of **cerebrospinal fluid (CSF)** constitute a second important factor in the homeostasis of the environment within the CNS. CSF, which is secreted by ependymal (epithelial) cells of the choroid plexus, differs from blood plasma somewhat in composition and circulates through the ventricles of the brain and the subarachnoid space around the brain and spinal cord (Table 4.1 and Figs. 4.2 and 4.3). Specifically, the CSF has a lower concentration of bicarbonate anion than does plasma,

resulting in lower pH. This difference makes the CSF sensitive to changes in P_{CO_2} of the blood, an important factor in regulation of respiration by the brain. Substances within the CSF are freely exchanged with the interstitial fluid of the brain, in contrast to substances within blood plasma. Because CSF is secreted at a rate of approximately 500 mL/day and the total volume of CSF is approximately 150 mL, it turns over at a rate of three to four times per day.

Telencephalon

In the telencephalon, the right and left cerebral hemispheres consist of the outer cerebral cortex (gray matter) and inner white matter. (Gray matter contains unmyelinated axons, and white matter contains myelinated axons.) The hemispheres are linked anatomically and functionally by bundles of nerves known as *commissures*: the large corpus callosum and smaller **anterior, posterior**, and **hippocampal commissures** (Fig. 4.4).

The cerebral cortex consists of five major areas based on gross anatomic features: the frontal, parietal, temporal, and occipital lobes and the insula (Fig. 4.5). The cerebral cortex has afferent and efferent connections to the thalamus and basal ganglia (two sets of nuclei deep in the brain with origins in the cerebrum, thalamus, and brainstem) and many other regions of the brain. Regions within the cerebral cortex receive and integrate sensory information, integrate motor function, and perform other high-level functions such as learning and reasoning (see Fig. 4.4). Much of the sensory information is received indirectly, passing through the thalamus, except in the case of olfactory signals. In general, the right and left cerebral hemispheres receive input from the contralateral side of the body.

The term **neocortex** refers to the outermost layer of the cerebral hemispheres. This area, which is the most recently evolved part of the cerebral cortex, is only a few millimeters thick. The neocortex is considered the highest center in the brain; it is where the most complex functions ascribed to the cerebral cortex are performed, including sensory perception, initiation of motor function, language skills, and conscious thought.

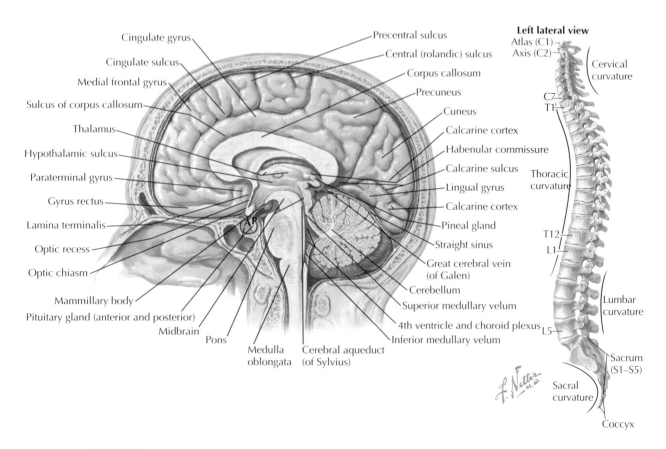

Figure 4.1 **Major Parts of the Central Nervous System and the Vertebral Column** The brain is illustrated as a midsagittal section in which the frontal, parietal, and occipital lobes of the cerebral cortex (outer layer of the cerebral hemispheres) are visible, as well as other major divisions of the brain. The temporal lobe of the cerebral cortex is not seen in this section (see Fig. 4.5). The vertebral column consists of cervical, thoracic, lumbar, and sacral vertebrae. The vertebral column houses the spinal cord, which extends from the medulla oblongata to lumbar level L1-L2.

Table 4.1 **Comparison of Cerebrospinal Fluid and Blood Composition**

Parameter	Cerebrospinal Fluid	Blood Plasma
Na^+ (mEq/L)	140-145	135-147
K^+ (mEq/L)	3	3.5-5.0
Cl^- (mEq/L)	115-120	95-105
HCO_3^- (mEq/L)	20	22-28
Glucose (mg/dL)	50-75	70-110
Protein (g/dL)	0.05-0.07	6.0-7.8
pH	7.3	7.35-7.45

From Hansen J: *Netter's Atlas of Human Physiology,* Philadelphia, Elsevier, 2002.

The basal ganglia are nuclei located deep within the cerebral hemispheres; they are involved in the regulation of movement, among other functions. Two additional deep formations, the **hippocampus** and **amygdala**, are located within the medial temporal lobes. Both are part of the broader limbic system,

are involved in emotion and long-term memory, and affect the endocrine and autonomic nervous systems (Fig. 4.6). The hippocampus has an important role in memory and spatial navigation, and the amygdala is involved in emotion. Through its connections to the hypothalamus, the hippocampus also affects sympathetic nervous system function.

Diencephalon

The **diencephalon**, consisting of the thalamus and hypothalamus, is located between the cerebral hemispheres and the brainstem and is part of the limbic system (see Fig. 4.6). The thalamus processes sensory input and relays it to the cerebral cortex; it also processes motor signals leaving the cerebral cortex. The hypothalamus has a central role in the regulation of body temperature, the reproductive system, hunger and thirst, salt and water balance, circadian rhythms, the autonomic nervous system, and endocrine function (Table 4.2).

Cerebellum

The **cerebellum** is located between the cerebral cortex and the spinal cord and is in close proximity to the brainstem (see Fig. 4.1). The cerebellum integrates sensory and motor

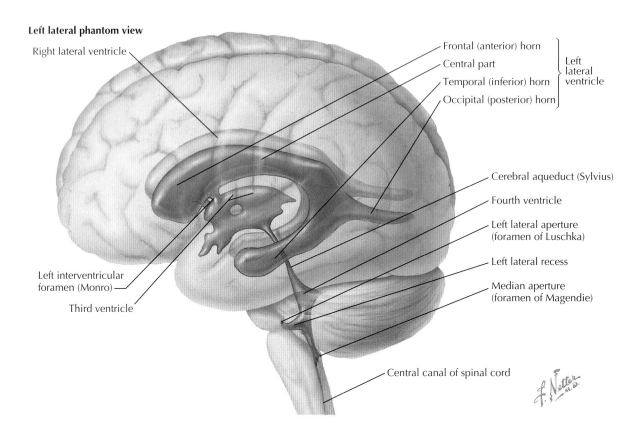

Left lateral phantom view

Right lateral ventricle

Frontal (anterior) horn

Central part

Temporal (inferior) horn

Occipital (posterior) horn

Left lateral ventricle

Cerebral aqueduct (Sylvius)

Fourth ventricle

Left lateral aperture (foramen of Luschka)

Left lateral recess

Median aperture (foramen of Magendie)

Left interventricular foramen (Monro)

Third ventricle

Central canal of spinal cord

Figure 4.2 **Brain Ventricles and Cerebrospinal Fluid Composition**   Circulation of cerebrospinal fluid through the four ventricles of the brain and in the subarachnoid space is essential for maintaining the environment of the brain and spinal cord. The composition of cerebrospinal fluid, which is secreted by the choroid plexus, differs from that of plasma (see Table 4.1).

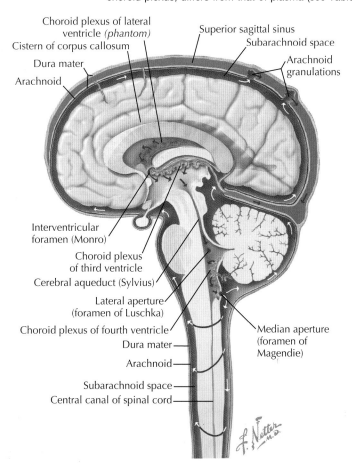

Choroid plexus of lateral ventricle *(phantom)*

Cistern of corpus callosum

Dura mater

Arachnoid

Superior sagittal sinus

Subarachnoid space

Arachnoid granulations

Interventricular foramen (Monro)

Choroid plexus of third ventricle

Cerebral aqueduct (Sylvius)

Lateral aperture (foramen of Luschka)

Choroid plexus of fourth ventricle

Dura mater

Arachnoid

Subarachnoid space

Central canal of spinal cord

Median aperture (foramen of Magendie)

Figure 4.3 **Circulation of Cerebrospinal Fluid**   Cerebrospinal fluid, which is produced by the choroid plexus, circulates through the two lateral ventricles and the third and fourth ventricles; it leaves the fourth ventricle through the lateral and medial apertures and enters the subarachnoid space. Much of the fluid is reabsorbed at the arachnoid granulations, into the venous system, and into the capillaries of the central nervous system and pia mater.

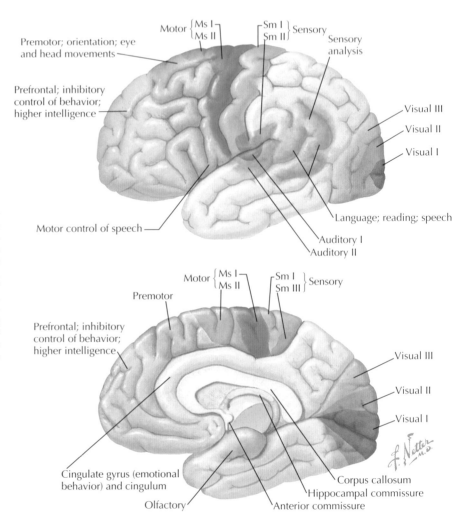

Figure 4.4 **Cerebral Cortex: Association Pathways and Localization of Function** The two cerebral hemispheres are linked functionally and anatomically by the corpus callosum and anterior, posterior, and hippocampal commissures. In the cerebral cortex, functional regions involved in sensory and motor activity, as well as integration of information from various sources, have been identified. The cortex is essential for high-level intellectual functions including learning, reasoning, memory and language skills, and conscious thought, and it is involved in the planning and execution of fine motor activity. *Ms I,* primary motor; *Ms II,* supplementary motor; *Sm I,* primary sensory; *Sm II,* secondary sensory; *Sm III,* tertiary sensory.

information, as well as information from the inner ear, and has an important role in **proprioception** (i.e., unconscious perception of posture, spatial orientation, and movement). Proprioception is based on input from proprioceptors within muscles, joints, tendons, and the inner ear.

Brainstem

The **brainstem** is the lowest portion of the brain, consisting of the midbrain, pons, and medulla oblongata (medulla). The medulla, which is anatomically continuous with the spinal cord, contains centers that regulate autonomic functions, including the centers involved in the regulation and integration of cardiovascular and respiratory functions, as well as swallowing, vomiting, and coughing reflexes. The pons, situated above (rostral to) the medulla, is also involved in regulation of breathing and relays sensory information from the cerebrum to the cerebellum. The midbrain, also known as the **mesencephalon**, is the most rostral portion of the brainstem. It is involved in eye movement and transmission of visual and auditory information. It is also the location of the **substantia nigra**, which, as part of the basal ganglia, has a role in regulation of motor activity. **Cranial nerves III to XII** originate from the brainstem.

The Spinal Cord

The spinal cord originates at the medulla at the base of the skull and extends within the vertebral column from the medulla oblongata to the lumbar region (Fig. 4.7). The spinal cord consists of neural tissues covered by three meninges (membranes) that are continuous with those of the brain:

- The inner pia mater
- The middle arachnoid membrane
- The outer dura mater

CSF is found between the arachnoid membrane and pia mater. The spinal cord gives rise to 31 pairs of spinal nerves, organized into three plexuses:

- The **cervical plexus**, which forms nerves distributed to the back of the head and the neck
- The **brachial plexus**, which forms nerves distributed to the upper limbs
- The **lumbosacral plexus**, which forms nerves distributed to the pelvis and lower limbs

The anatomy of spinal membranes and nerve roots is illustrated in Figure 4.8.

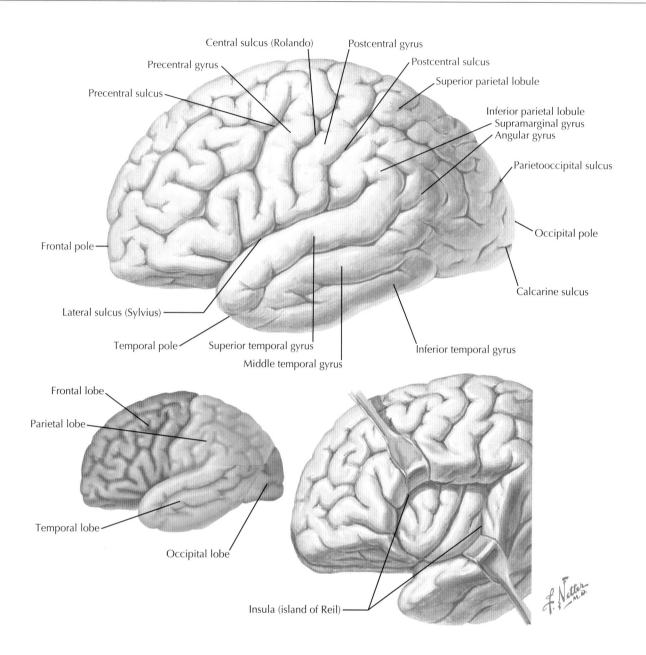

Central sulcus (Rolando)
Postcentral gyrus
Precentral gyrus
Postcentral sulcus
Precentral sulcus
Superior parietal lobule
Inferior parietal lobule
Supramarginal gyrus
Angular gyrus
Parietooccipital sulcus
Frontal pole
Occipital pole
Calcarine sulcus
Lateral sulcus (Sylvius)
Temporal pole
Superior temporal gyrus
Inferior temporal gyrus
Middle temporal gyrus

Frontal lobe
Parietal lobe
Temporal lobe
Occipital lobe
Insula (island of Reil)

Figure 4.5 **Organization of the Brain: Surface Anatomy of the Cerebral Cortex** The outer layers of the cerebrum (telencephalon) are the gray matter, which contains neurons with unmyelinated fibers. The cerebral cortex is only 2 to 4 mm thick, but it is highly convoluted in primates, and thus most of the cortex is actually buried in sulci (grooves). Four lobes are discernible from the surface; the fifth lobe, the insula, can be viewed by retracting the outer cortex at the lateral fissure. The cerebral cortex is the highest center of the brain and is involved in learning, memory, and reasoning, as well as motor and sensory control and integration.

THE PERIPHERAL NERVOUS SYSTEM

The peripheral nervous system is the portion of the nervous system that lies beyond the brain and spinal cord. The peripheral nervous system consists of the following neurons:

- Motor neurons that convey impulses from the CNS to effector tissues or organs
- Sensory neurons that convey impulses from peripheral sensory receptors to the CNS

The peripheral nervous system is subdivided into somatic and autonomic components. In the somatic portion of the peripheral nervous system, motor fibers of the spinal nerves innervate skeletal muscles, and sensory information from receptors in the skeletal muscles, joints, and skin is then transmitted through sensory fibers to the CNS (Fig. 4.9). The autonomic component of the peripheral nervous system is composed of motor and sensory fibers involved in the involuntary control of homeostatic mechanisms through actions on visceral organs. The autonomic nervous system and its sympathetic and parasympathetic divisions are discussed further in Chapter 7.

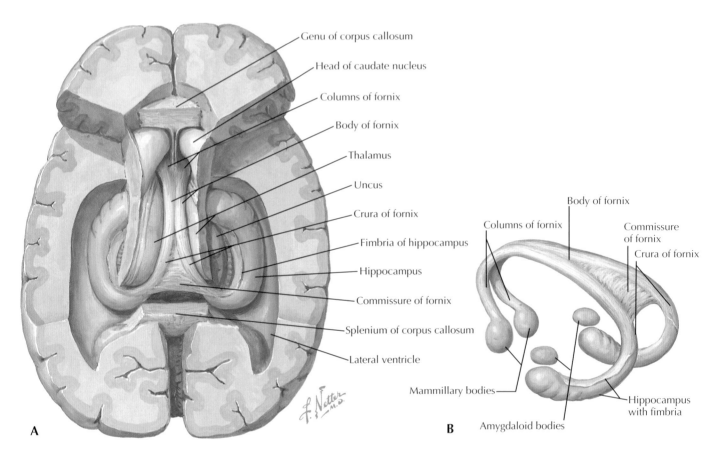

Figure 4.6 **The Limbic System** The limbic system consists of the hypothalamus and several structures that form a border or ring (limbus) around the diencephalon. The limbic system includes the cingulate, parahippocampal, and subcallosal gyri, as well as the amygdala, hypothalamus, hippocampi, and mammillary bodies. **A,** Cerebral cortical and white matter have been removed, along with the corpus callosum and part of the thalamus and caudate nucleus. **B,** The fornix consists of curved bundles of fibers connecting the mammillary bodies and hippocampi. The limbic system has an important role in emotion and the linking of emotion with memory, as well as endocrine and autonomic function.

Table 4.2 **Major Functions of the Hypothalamus**

Hypothalamic Nuclei	Major Functions
Medial preoptic	Gonadotropin-releasing hormone secretion
Anterior	Thermoregulation
Posterior	Thermoregulation
Ventromedial	Satiety center: inhibits eating behavior
Dorsomedial	Feeding
Suprachiasmatic	Circadian rhythm
Supraoptic	ADH and oxytocin secretion (sensation of thirst)
Paraventricular	Oxytocin and ADH secretion; corticotropin-releasing hormone and thyrotropin-releasing hormone secretion
Arcuate	Growth hormone–releasing hormone secretion

ADH, antidiuretic hormone.

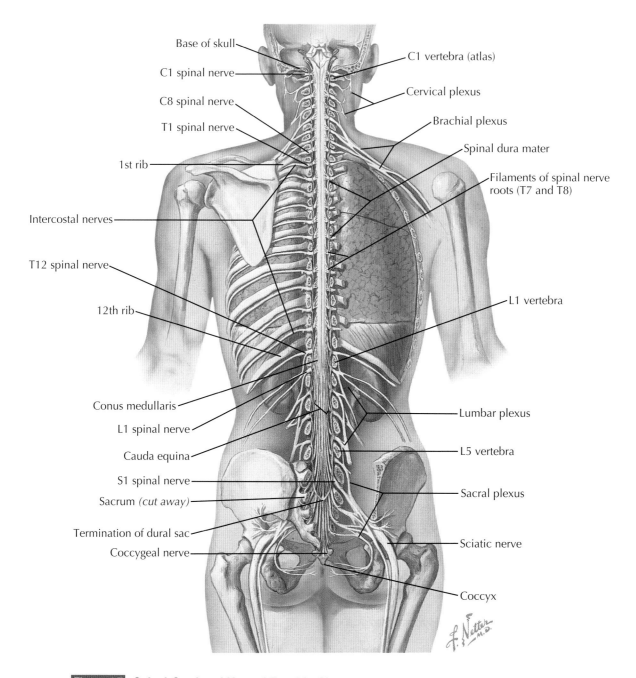

Base of skull

C1 spinal nerve

C8 spinal nerve

T1 spinal nerve

1st rib

Intercostal nerves

T12 spinal nerve

12th rib

Conus medullaris

L1 spinal nerve

Cauda equina

S1 spinal nerve

Sacrum (cut away)

Termination of dural sac

Coccygeal nerve

C1 vertebra (atlas)

Cervical plexus

Brachial plexus

Spinal dura mater

Filaments of spinal nerve roots (T7 and T8)

L1 vertebra

Lumbar plexus

L5 vertebra

Sacral plexus

Sciatic nerve

Coccyx

Figure 4.7 **Spinal Cord and Ventral Rami In Situ** The 31 pairs of spinal nerves originating from the spinal cord are organized into the cervical, brachial, and lumbosacral plexuses. Fibers from the plexuses are distributed, respectively, to the neck, upper limbs, and pelvis and lower limbs. Efferent fibers innervate skeletal muscle, whereas afferent fibers carry sensory information from the skin, muscle, and joints to the central nervous system.

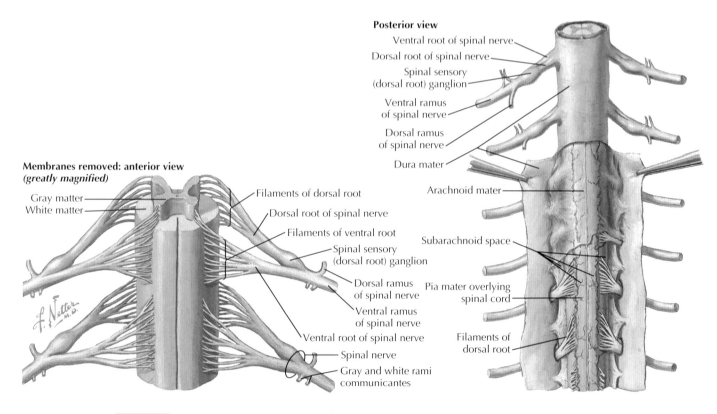

Posterior view

Ventral root of spinal nerve

Dorsal root of spinal nerve

Spinal sensory (dorsal root) ganglion

Ventral ramus of spinal nerve

Dorsal ramus of spinal nerve

Dura mater

Arachnoid mater

Subarachnoid space

Pia mater overlying spinal cord

Filaments of dorsal root

Membranes removed: anterior view *(greatly magnified)*

Gray matter

White matter

Filaments of dorsal root

Dorsal root of spinal nerve

Filaments of ventral root

Spinal sensory (dorsal root) ganglion

Dorsal ramus of spinal nerve

Ventral ramus of spinal nerve

Ventral root of spinal nerve

Spinal nerve

Gray and white rami communicantes

Figure 4.8   Spinal Membranes and Nerve Roots   Motor fibers and sensory fibers pass through 31 pairs of spinal nerves. The outer dura mater, arachnoid mater, and inner pia mater are the three coverings of the spinal cord, with cerebrospinal fluid circulating in the subarachnoid space.

Figure 4.9   Somatic Component of the Peripheral Nervous System   The peripheral nervous system can be subdivided into somatic and autonomic components. The somatic nervous system contains motor nerves and sensory nerves that innervate skin and muscle. The soma (cell bodies) of motor nerves and sensory nerves are located in the gray matter of the anterior horn of the spinal cord and in the dorsal root ganglia, respectively.

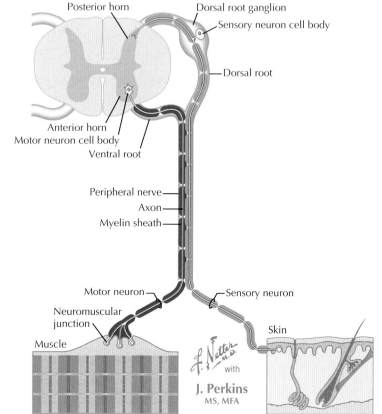

Posterior horn

Dorsal root ganglion

Sensory neuron cell body

Dorsal root

Anterior horn

Motor neuron cell body

Ventral root

Peripheral nerve

Axon

Myelin sheath

Motor neuron

Sensory neuron

Neuromuscular junction

Skin

Muscle

with

J. Perkins
MS, MFA

CLINICAL CORRELATE 4.1
Spinal Cord Injury

Injury to the spinal cord results in deficits in motor and sensory function below the point of injury. The most common cause of spinal cord injury is trauma; other causes include neurodegenerative diseases as well as tumors and vascular diseases affecting the spine. Spinal cord injury is classified as complete when no neurologic function is retained beyond the point of damage and incomplete when some sensory or motor function remains. Effects of spinal cord damage correspond to the level of injury, with increasing severity at higher levels:

■ Sacral and lumbar injuries are associated with problems in sexual, bladder, and bowel function, as well as motor and sensory deficits in the lower limbs.

■ Thoracic injuries are associated with paraplegia (paralysis of the lower body).

■ Cervical injuries result in some degree of quadriplegia (paralysis below the neck), depending on the level of injury. Patients with injuries at spinal cord level C3 or above require mechanical ventilation.

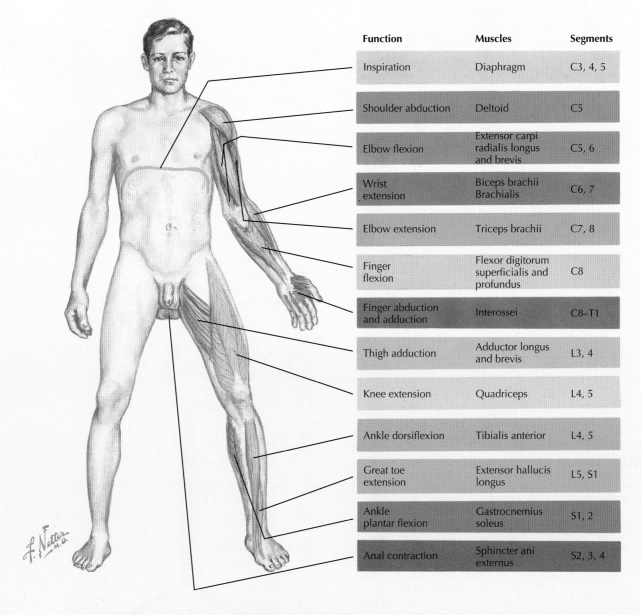

Function	Muscles	Segments
Inspiration	Diaphragm	C3, 4, 5
Shoulder abduction	Deltoid	C5
Elbow flexion	Extensor carpi radialis longus and brevis	C5, 6
Wrist extension	Biceps brachii Brachialis	C6, 7
Elbow extension	Triceps brachii	C7, 8
Finger flexion	Flexor digitorum superficialis and profundus	C8
Finger abduction and adduction	Interossei	C8–T1
Thigh adduction	Adductor longus and brevis	L3, 4
Knee extension	Quadriceps	L4, 5
Ankle dorsiflexion	Tibialis anterior	L4, 5
Great toe extension	Extensor hallucis longus	L5, S1
Ankle plantar flexion	Gastrocnemius soleus	S1, 2
Anal contraction	Sphincter ani externus	S2, 3, 4

Motor Impairment Related to Level of Spinal Injury

Chapter 5

Sensory Physiology

Sensory systems allow us to maintain homeostasis and respond to the external environment in an adaptive manner. Much of the awake brain is devoted to transducing, encoding, sensing, and eventually perceiving and reacting to or trying to make sense of both our external and internal environments. An important task of the brain is to filter out the large amounts of incoming information to prevent information overload. During sleep, the incoming sensory information is drastically reduced as the brain continues to try to make sense of the stored information and "reset" itself for the next day's influx of sensory inputs.

SENSORY RECEPTORS

Specialized receptors detect stimuli of various types, including visual, olfactory, gustatory (taste), auditory, and somatosensory (mechanical, thermal, and painful) stimuli. Stimulation of sensory receptors results in the opening or closing of ion channels and generation of a **receptor or generator potential** (a membrane potential change in the receptor). When threshold is reached, the stimulus has been transduced into an electrical signal, and information is transmitted via afferent pathways to the central nervous system, which integrates such information and transmits signals back to effector systems. The **somatosensory system** includes **mechanoreceptors, thermal receptors**, and **nociceptors** (pain receptors) that respond to stimuli in the skin or at the body surface. The term **somatovisceral sensory system** is also used because such receptors are also located in visceral organs, muscle, and joints.

The receptive field of a sensory neuron is the area over which a stimulus will trigger firing of the neuron. A rapidly adapting receptor responds quickly to a stimulus but adapts rapidly to a constant stimulus, returning to its normal firing rate.

Several types of sensory receptors are found in the skin (and other tissues). These receptors can be differentiated by the stimuli to which they respond as well as by the degree of adaptation they exhibit to intensity, duration, and change in stimulus (Fig. 5.1):

- **Meissner's corpuscles** are mechanoreceptors located in the dermal papillae, especially in the fingertips, palms, soles, lips, face, tongue, and genital skin (nonhairy skin). These corpuscles are rapidly adapting receptors with small receptive fields that allow point discrimination and detection of low-frequency stimuli such as flutter.
- **Pacinian corpuscles** are rapidly adapting mechanoreceptors that detect pressure and vibration. The lamellated capsule of these receptors allows them to respond specifically to rapid changes in pressure and vibration, as opposed to continual pressure or slow changes in pressure (Fig. 5.2).
- **Merkel's disks** are slowly adapting mechanoreceptors that respond through small receptive fields to pressure and touch, particularly to indentation of the skin.
- **Hair follicle receptors** consist of nerve endings wrapped around the base of hair follicles. These rapidly adapting mechanoreceptors detect movement across the surface of the skin.
- **Ruffini's corpuscles** are mechanoreceptors found in the dermis and in joints; they respond to stretch and are slowly adapting.
- **Thermal receptors** are sensory nerves with endings in the skin that respond to temperature; they are subdivided into warm and cold receptors. Both are slowly adapting.
- **Nociceptors** are sensory nerves with endings in skin, cornea, muscle, joints, and visceral organs. Various types of nociceptors respond to temperature and to mechanical and chemical stimuli.
- **Muscle and joint receptors** include muscle spindles and Golgi tendon organs. These receptors are important in proprioception and coordination of motor activity and are considered in more detail in Chapter 6.

Transduction of Somatosensory Signals

Somatosensory (i.e., pain, touch, pressure, temperature) signals originating below the head and in the head are ultimately conveyed to the primary somatosensory area, which is located in the postcentral gyrus of the parietal lobe of the

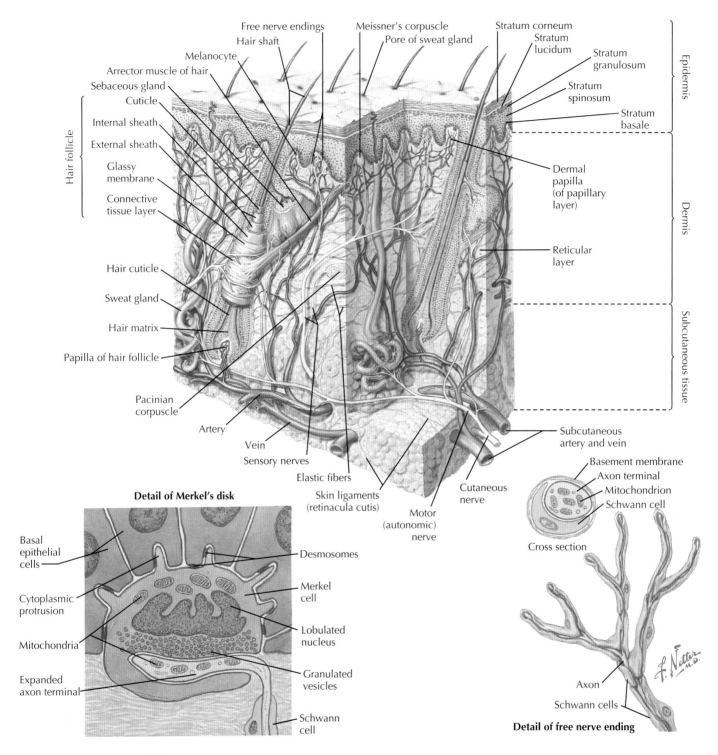

Free nerve endings
Hair shaft
Melanocyte
Arrector muscle of hair
Sebaceous gland
Cuticle
Internal sheath
External sheath
Glassy membrane
Connective tissue layer
Hair follicle
Hair cuticle
Sweat gland
Hair matrix
Papilla of hair follicle
Pacinian corpuscle
Artery
Vein
Sensory nerves
Elastic fibers
Skin ligaments (retinacula cutis)
Motor (autonomic) nerve
Cutaneous nerve
Meissner's corpuscle
Pore of sweat gland
Stratum corneum
Stratum lucidum
Stratum granulosum
Stratum spinosum
Stratum basale
Epidermis
Dermal papilla (of papillary layer)
Reticular layer
Dermis
Subcutaneous tissue
Subcutaneous artery and vein
Basement membrane
Axon terminal
Mitochondrion
Schwann cell
Cross section

Detail of Merkel's disk

Basal epithelial cells
Cytoplasmic protrusion
Mitochondria
Expanded axon terminal
Desmosomes
Merkel cell
Lobulated nucleus
Granulated vesicles
Schwann cell

Axon
Schwann cells
Detail of free nerve ending

Figure 5.1 **Skin and Cutaneous Receptors** Mechanoreceptors, thermoreceptors, and nociceptors transduce the effects of touch, temperature, and painful stimuli into neural signals. Within the skin, mechanoreceptors known as *Meissner's corpuscles, pacinian corpuscles,* and *Merkel's disks* respond to various types of mechanical stimulation. Temperature and painful stimuli are detected by free nerve endings.

cerebral cortex. The afferent somatosensory signals originating below the head are conveyed to the dorsal root ganglia and subsequently through the spinothalamic and spinoreticular tracts of the anterolateral system, ultimately reaching the primary somatosensory cortex (Fig. 5.3). Signals involved in

proprioception, as well as signals generated by vibratory and tactile stimuli, are carried through the fasciculus gracilis and fasciculus cuneatus to the ventral posterolateral nucleus of the thalamus. The lateral cervical system carries some proprioceptive, vibratory, and tactile signals as well. These

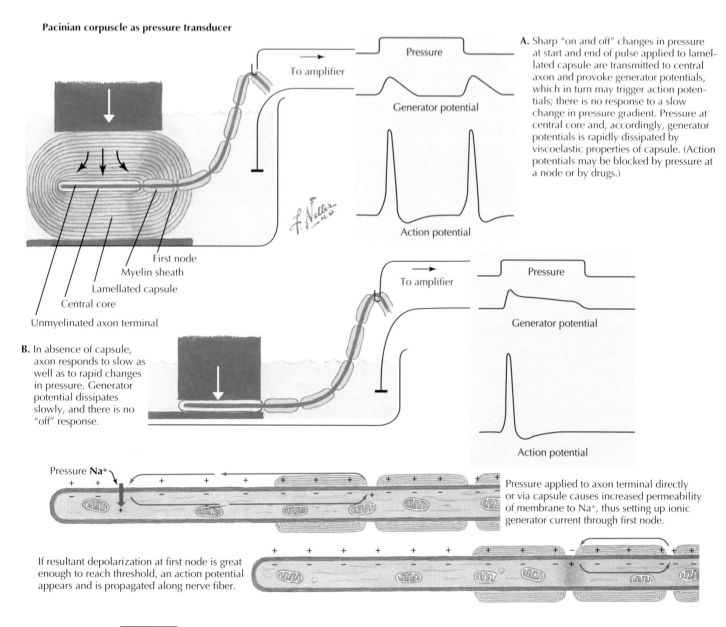

Pacinian corpuscle as pressure transducer

Pressure

To amplifier

Generator potential

Action potential

A. Sharp "on and off" changes in pressure at start and end of pulse applied to lamellated capsule are transmitted to central axon and provoke generator potentials, which in turn may trigger action potentials; there is no response to a slow change in pressure gradient. Pressure at central core and, accordingly, generator potentials is rapidly dissipated by viscoelastic properties of capsule. (Action potentials may be blocked by pressure at a node or by drugs.)

First node
Myelin sheath
Lamellated capsule
Central core
Unmyelinated axon terminal

B. In absence of capsule, axon responds to slow as well as to rapid changes in pressure. Generator potential dissipates slowly, and there is no "off" response.

To amplifier

Pressure

Generator potential

Action potential

Pressure Na⁺

Pressure applied to axon terminal directly or via capsule causes increased permeability of membrane to Na⁺, thus setting up ionic generator current through first node.

If resultant depolarization at first node is great enough to reach threshold, an action potential appears and is propagated along nerve fiber.

Figure 5.2 **Pacinian Corpuscle** Pressure and vibration are transduced into electrical potential changes by pacinian corpuscles, which are rapidly adapting mechanoreceptors. The lamellated capsule surrounding the axon terminal dissipates slow pressure changes, and thus the axon responds only to rapid changes in pressure **(A)**, in contrast to an axon lacking a capsule **(B)**. The action potentials produced are carried through afferent fibers to the dorsal root ganglia (see Fig. 4.9).

pathways reach synapses in the thalamus before extending to the cerebral cortex.

In the head, touch, pressure, pain, and temperature result in afferent nerve traffic to nerve cell bodies in the trigeminal (semilunar) ganglia of the trigeminal nerve. Signals involved in proprioception are carried to neuronal cell bodies in the mesencephalic nucleus of the trigeminal nerve (cranial nerve V). Projections in this system are mainly to the contralateral ventral posterolateral nucleus of the thalamus, subsequently reaching the primary somatosensory cortex (postcentral gyrus; Fig. 5.4).

THE VISUAL SYSTEM

The **visual system** performs the remarkable task of detecting light stimuli between 400 and 700 nanometers in wavelength, transducing the stimuli into electrical signals, and transmitting these signals to the central nervous system, where they are used to construct a three-dimensional representation of objects in the visual field that includes color and brightness and detects motion.

Light passes through the cornea of the eye and the pupil (the circular opening of the iris) and is then focused by the lens

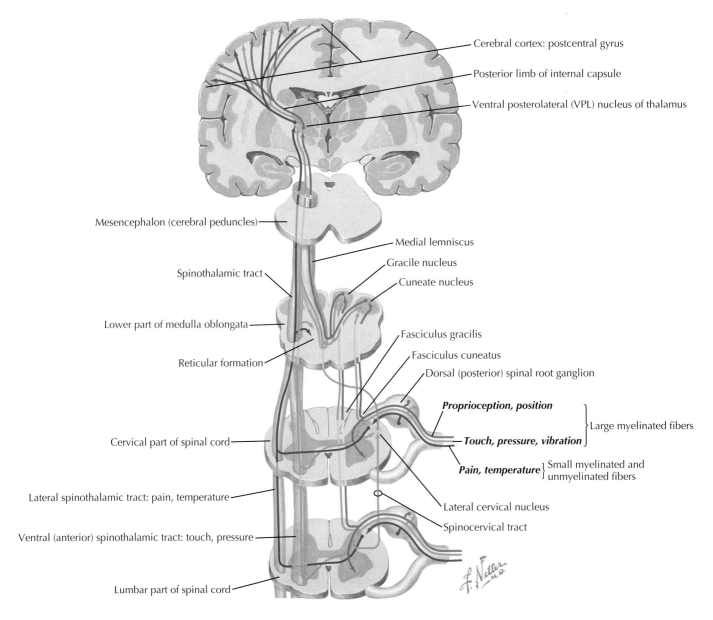

Cerebral cortex: postcentral gyrus

Posterior limb of internal capsule

Ventral posterolateral (VPL) nucleus of thalamus

Mesencephalon (cerebral peduncles)

Medial lemniscus

Gracile nucleus

Cuneate nucleus

Spinothalamic tract

Lower part of medulla oblongata

Fasciculus gracilis

Fasciculus cuneatus

Dorsal (posterior) spinal root ganglion

Reticular formation

Proprioception, position

Large myelinated fibers

Cervical part of spinal cord

Touch, pressure, vibration

Pain, temperature   Small myelinated and unmyelinated fibers

Lateral spinothalamic tract: pain, temperature

Lateral cervical nucleus

Spinocervical tract

Ventral (anterior) spinothalamic tract: touch, pressure

Lumbar part of spinal cord

Figure 5.3 **Somesthetic System of the Body** The primary somatosensory cortex (postcentral gyrus) receives information regarding proprioception, touch, pressure, vibration, pain, and temperature that originates at receptors below the head through the illustrated tracts by way of the thalamus. Pain-, temperature-, and pressure-related signals are conducted through the spinothalamic and spinoreticular tracts of the anterolateral system *(red and blue lines)*. The fasciculus gracilis and fasciculus cuneatus, as well as the lateral cervical system, convey proprioception-, touch-, and vibration-related signals *(blue and purple lines)*.

onto the retina (Fig. 5.5). Within the retina, photoreceptors known as *rods and cones* contain the pigment **rhodopsin**, which absorbs photons. Color perception is mainly mediated by cones, whereas rods are sensitive photoreceptors adapted to perceive light at low intensity. The visual fields of the two eyes overlap, with the center being projected, upside down, on the sensitive macular zone of the retina (Fig. 5.6). Within the macula, the **fovea** is the region of greatest visual acuity; it contains only cones.

Reception of light by the photoreceptors initiates a sequence of events that results in reduced Na⁺ permeability and hyperpolarization of the photoreceptor cell membrane. This situation results in inhibition of the release of either inhibitory or excitatory neurotransmitters at synapses between photoreceptors and **bipolar** and **horizontal cells**. Ultimately, the generated signals are transmitted by the **ganglion cells**, whose axons form the **optic nerves**. Nerve fibers originating in the nasal hemiretina (i.e., the nasal half of the retina) of each eye cross at the **optic chiasm** (the site of partial crossing of the two optic nerves), whereas fibers from the temporal hemiretina remain in the ipsilateral tract (on the same side). The optic tracts carry the signals to synapses at the **lateral geniculate nuclei**, from which signals are transmitted to the visual cortex in the occipital lobe of the cerebral cortex.

CLINICAL CORRELATE 5.1
Dermatomes

Dermatomes are areas of the skin associated with a specific pair of dorsal nerve roots. In other words, sensory signals originating in the skin within a dermatome are largely transmitted to the spinal cord through one of a specific pair of afferent sensory nerve roots, although significant overlap occurs between adjacent dermatomes. Clinically, dermatomes are useful in identifying the site of spinal nerve damage, based on the region of skin in which sensation is affected. For example, sensations in the feet are conveyed to the spinal cord through nerve roots at L4, L5, or S1, depending on the area of the foot, with some overlap in regions. In shingles (herpes zoster), the herpes virus usually resides dormant in nerve cell bodies of the dorsal root ganglia, but when the virus is active, it specifically affects the skin in the corresponding dermatome on one or both sides (depending on which ganglia are infected).

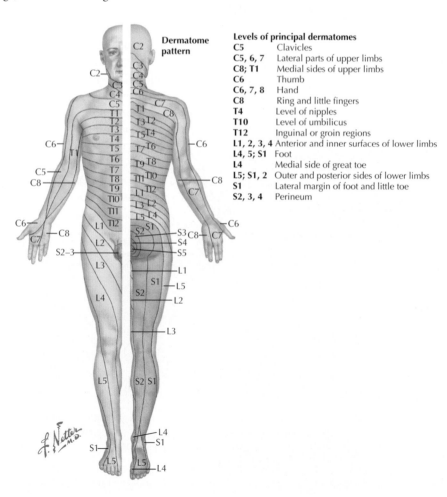

Levels of principal dermatomes

C5	Clavicles
C5, 6, 7	Lateral parts of upper limbs
C8; T1	Medial sides of upper limbs
C6	Thumb
C6, 7, 8	Hand
C8	Ring and little fingers
T4	Level of nipples
T10	Level of umbilicus
T12	Inguinal or groin regions
L1, 2, 3, 4	Anterior and inner surfaces of lower limbs
L4, 5; S1	Foot
L4	Medial side of great toe
L5; S1, 2	Outer and posterior sides of lower limbs
S1	Lateral margin of foot and little toe
S2, 3, 4	Perineum

CLINICAL CORRELATE 5.2
Cataracts

A cataract is the clouding of the lens that is most often associated with aging and denaturation of proteins in the lens. Cataract formation is accelerated by diabetes, hypertension, and smoking. The lens becomes opacified, resulting in poor vision and potentially leading to blindness, because light reaching the retina is reduced, affecting the activity of photoreceptors. Cataracts are usually treated surgically through extraction of the opacified lens and replacement with an artificial lens. In 90% of cases, vision is improved to 20/40 or better with this treatment.

THE AUDITORY SYSTEM

The **auditory system** detects sound, transduces it into electrical signals, and transmits it to the brain, where the complex task of interpretation of sound—for example, as speech—takes place. The process begins at the external ear, which directs sound to the tympanic membrane (eardrum) (Fig. 5.7). This membrane separates the external ear from the air-filled middle ear, where sound is amplified. Vibration of the tympanic membrane is conducted to the malleus, incus, and stapes, the small bones of the middle ear known collectively as the **ossicles**. The stapes is connected to the **oval window**, one of two membranous "windows" between the middle ear

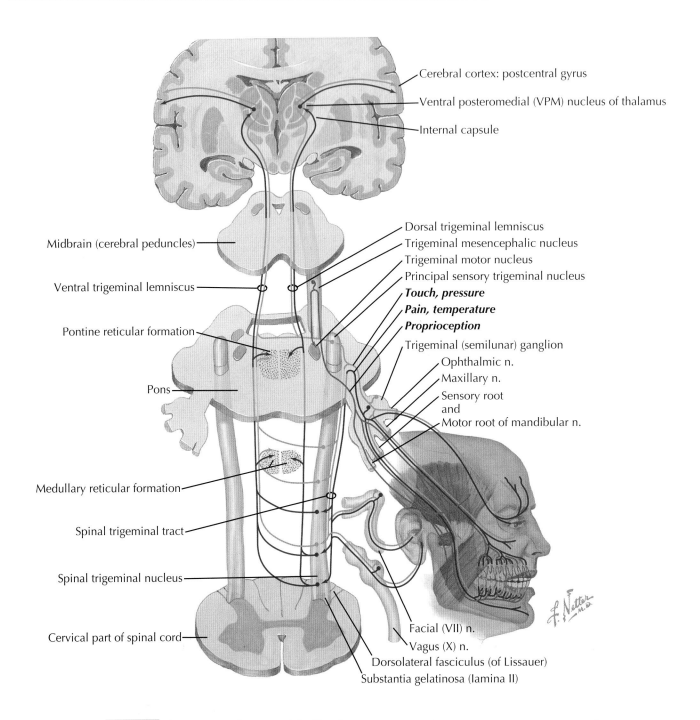

Figure 5.4 Somesthetic System of the Head   The primary somatosensory cortex (postcentral gyrus) receives information regarding proprioception, touch, pressure, pain, and temperature that originates at receptors in the head through the illustrated tracts by way of the thalamus. Proprioceptive signals are carried by nerve fibers *(purple lines)* originating in cell bodies in the mesencephalic nucleus of the trigeminal nerve (cranial nerve V), whereas touch, pressure, pain, and temperature produce signals carried by nerve fibers *(blue and red lines)* to cell bodies in the trigeminal ganglion of cranial nerve V. *n,* nerve.

and inner ear. Thus, vibration of the ossicles is transmitted to the inner ear by the stapes.

The fluid-filled inner ear is a complex structure that includes the vestibule, the cochlea, and the semicircular canals. The vestibule and cochlea consist of a **membranous labyrinth** within a **bony labyrinth** (Fig. 5.8). The membranous labyrinth of the cochlea includes three ducts known as the *scala vestibuli, scala tympani,* and *scala media* (cochlear duct). The fluid of the scala media, the **endolymph,** resembles intracellular fluid in its composition, whereas the **perilymph** of the other two ducts is similar to other extracellular fluids.

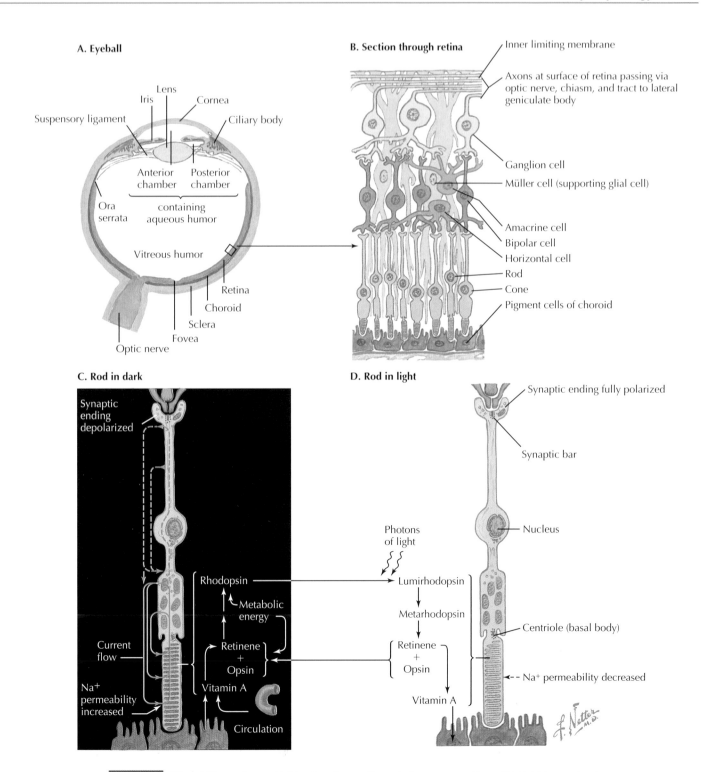

A. Eyeball

Suspensory ligament
Iris
Lens
Cornea
Ciliary body
Anterior chamber
Posterior chamber
Ora serrata
containing aqueous humor
Vitreous humor
Retina
Choroid
Sclera
Fovea
Optic nerve

B. Section through retina

Inner limiting membrane
Axons at surface of retina passing via optic nerve, chiasm, and tract to lateral geniculate body
Ganglion cell
Müller cell (supporting glial cell)
Amacrine cell
Bipolar cell
Horizontal cell
Rod
Cone
Pigment cells of choroid

C. Rod in dark

Synaptic ending depolarized
Current flow
Na+ permeability increased
Rhodopsin
Metabolic energy
Retinene + Opsin
Vitamin A
Circulation

D. Rod in light

Synaptic ending fully polarized
Synaptic bar
Nucleus
Photons of light
Lumirhodopsin
Metarhodopsin
Retinene + Opsin
Vitamin A
Centriole (basal body)
Na+ permeability decreased

Figure 5.5 **Visual Receptors** Light enters the eye through the cornea and lens **(A)**. Light focused onto the retina by the lens is absorbed by rhodopsin in photoreceptors (rods and cones, **B**). In the light, the cGMP second messenger system is activated, resulting in closure of membrane Na+ channels, thus hyperpolarizing the cell (illustrated for rod cells in **D**). Subsequent inhibition of neurotransmitter release at synapses with bipolar and horizontal cells is the next step in the transmission of light-generated signals to the optic nerve. In darkness, the rod or cone is depolarized **(C)**.

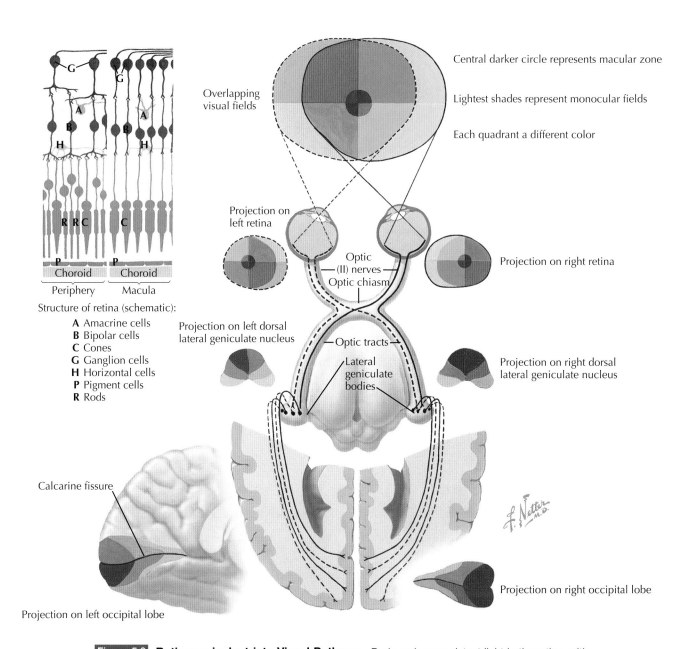

Central darker circle represents macular zone

Lightest shades represent monocular fields

Each quadrant a different color

Overlapping visual fields

Projection on left retina

Optic (II) nerves
Optic chiasm

Projection on right retina

Projection on left dorsal lateral geniculate nucleus

Optic tracts

Lateral geniculate bodies

Projection on right dorsal lateral geniculate nucleus

Calcarine fissure

Projection on left occipital lobe

Projection on right occipital lobe

Structure of retina (schematic):
- **A** Amacrine cells
- **B** Bipolar cells
- **C** Cones
- **G** Ganglion cells
- **H** Horizontal cells
- **P** Pigment cells
- **R** Rods

Choroid
Choroid
Periphery
Macula

Figure 5.6 **Retinogeniculostriate Visual Pathway** Rods and cones detect light in the retina, with cones specifically mediating color perception and rods detecting light. In the central macular zone of the retina, only cones are found; this zone is the region with greatest acuity *(top left)*. Axons of ganglion cells carry afferent signals through the optic nerve. At the optic chiasm, fibers originating in the nasal half of the retina cross to the contralateral tract, whereas fibers from the temporal half remain in the ipsilateral tract. Fibers project to synapses in the lateral geniculate nucleus (visual fields are inverted at this point). Signals are conveyed from the lateral geniculate nucleus to the primary visual cortex.

Transduction of Sound into Electrical Signals

The **spiral organ of Corti** is the site of transduction of sound into electrical signals. Mechanoreceptors known as **hair cells** have **stereocilia,** with their distal ends extending to and embedded in the **tectorial membrane** (see Fig. 5.8). When sound is conducted into the fluid of the inner ear, vibration of the basilar membrane results in bending of these stereocilia. The endolymph is unusual in terms of its electrical potential (approximately +80 millivolts [mV] relative to the perilymph), resulting in a charge difference across the apical membrane of

approximately 170 mV. Bending of the stereocilia results in changes in cation conductance of the apical cell membrane of hair cells; bending in one direction produces hyperpolarization, and bending in the other direction causes depolarization. As a result, an oscillatory release of neurotransmitter occurs at the synapses between hair cells and afferent nerve fibers, with depolarization of hair cells producing increased neurotransmitter release and hyperpolarization causing reduced release. When the threshold is reached in the afferent nerve fibers (during depolarizations of the hair cells), action potentials are produced. Within regions of the organ of Corti, the

Frontal section

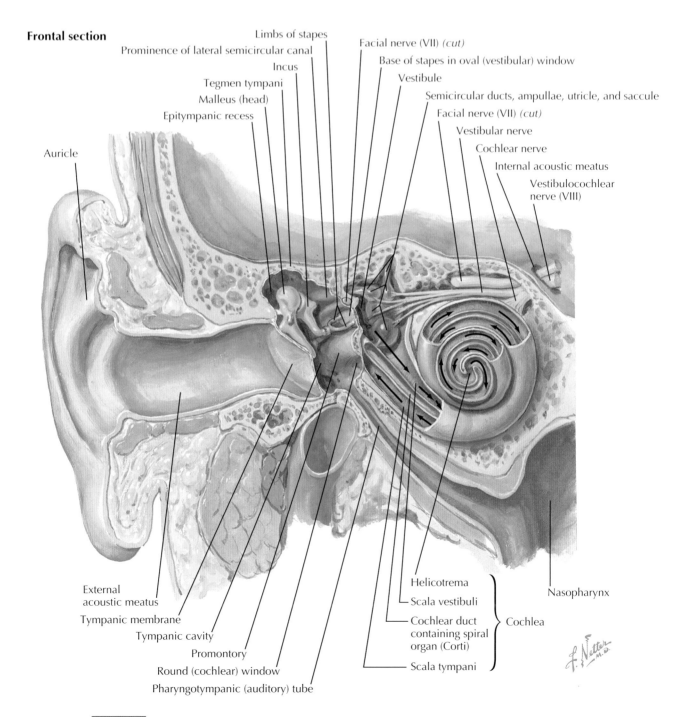

Auricle

Limbs of stapes

Prominence of lateral semicircular canal

Incus

Tegmen tympani

Malleus (head)

Epitympanic recess

Facial nerve (VII) *(cut)*

Base of stapes in oval (vestibular) window

Vestibule

Semicircular ducts, ampullae, utricle, and saccule

Facial nerve (VII) *(cut)*

Vestibular nerve

Cochlear nerve

Internal acoustic meatus

Vestibulocochlear nerve (VIII)

External acoustic meatus

Tympanic membrane

Tympanic cavity

Promontory

Round (cochlear) window

Pharyngotympanic (auditory) tube

Helicotrema

Scala vestibuli

Cochlear duct containing spiral organ (Corti)

Scala tympani

⎫
⎬ Cochlea
⎭

Nasopharynx

Figure 5.7 **Peripheral Pathways for Sound Reception**   The process of auditory perception begins when sound reaches the auricle (external ear) and is directed through the external acoustic meatus to the tympanic membrane, causing it to vibrate. The vibration is transmitted by the bones of the air-filled middle ear (malleus, incus, and stapes) to the fluid-filled inner ear via the oval window. Within the inner ear, the organ of Corti of the cochlea contains auditory receptors (see Fig. 5.8). *Arrows* indicate the course of sound waves.

amplitude by which the basilar membrane is repeatedly displaced as it vibrates is dependent on sound frequency. Thus the organ of Corti is tonotopically organized, with high-frequency sounds causing the largest displacement in the basal region of the cochlea and low-frequency sounds causing the largest displacement in the apical region (Fig. 5.9). As a result, cochlear nerve fibers respond to different frequencies of sound according to the **characteristic frequency** of the fiber (the

sound frequency of lowest intensity that produces a response in the fiber).

Impulses pass through the **auditory nerve** to the brainstem, where neurons compute various sound parameters (see Fig. 5.9). For example, the **superior olivary complex** detects differences in timing and intensity of sound signals originating from the two ears, allowing us to detect the location of a sound

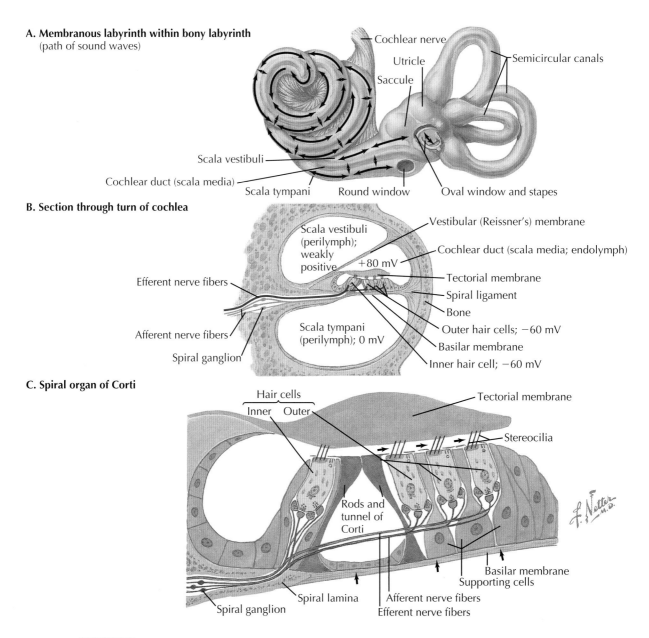

A. Membranous labyrinth within bony labyrinth (path of sound waves)

Cochlear nerve

Utricle

Saccule

Semicircular canals

Scala vestibuli

Cochlear duct (scala media)

Scala tympani

Round window

Oval window and stapes

B. Section through turn of cochlea

Scala vestibuli (perilymph); weakly positive   +80 mV

Vestibular (Reissner's) membrane

Cochlear duct (scala media; endolymph)

Tectorial membrane

Spiral ligament

Bone

Outer hair cells; −60 mV

Basilar membrane

Inner hair cell; −60 mV

Efferent nerve fibers

Afferent nerve fibers

Spiral ganglion

Scala tympani (perilymph); 0 mV

C. Spiral organ of Corti

Hair cells

Inner   Outer

Tectorial membrane

Stereocilia

Rods and tunnel of Corti

Basilar membrane

Supporting cells

Spiral lamina

Afferent nerve fibers

Spiral ganglion

Efferent nerve fibers

Figure 5.8 **Cochlear Receptors** Sound waves reaching the inner ear travel through the membranous labyrinth of the cochlea **(A)**, which consists of three ducts **(B)**. When vibration caused by sound reaches the cochlea of the inner ear, periodic displacement of the basilar membrane relative to the tectorial membrane causes bending of stereocilia and depolarizes hair cells within the spiral organ of Corti, resulting in transduction of sound into neural signals **(C)**.

source. Once the information reaches the thalamus (**medial geniculate body**), it is further processed and relayed to the amygdala and **auditory areas** of the temporal lobe. Within these areas, there is correspondence of tonotopic organization to that within the cochlea. Other parameters, such as amplitude and pitch of a sound, are also mapped there.

THE VESTIBULAR SYSTEM

In addition to its role in audition, the inner ear has an important role in balance and equilibrium through its **vestibular**

apparatus, which consists of the **semicircular canals** and the **otolithic organs** (the **utricle** and the **saccule**; Fig. 5.10). These structures detect angular and linear acceleration of the head and thus play an important role in proprioception. Specifically, the three semicircular canals are oriented perpendicularly to each other and are therefore able to detect angular acceleration in any plane, and the otolithic organs detect linear acceleration.

The sensory tissues of the semicircular canals are found in the **ampullary crest (crista ampullaris)**. Hair cells in this structure, like those of the organ of Corti, have apical stereocilia,

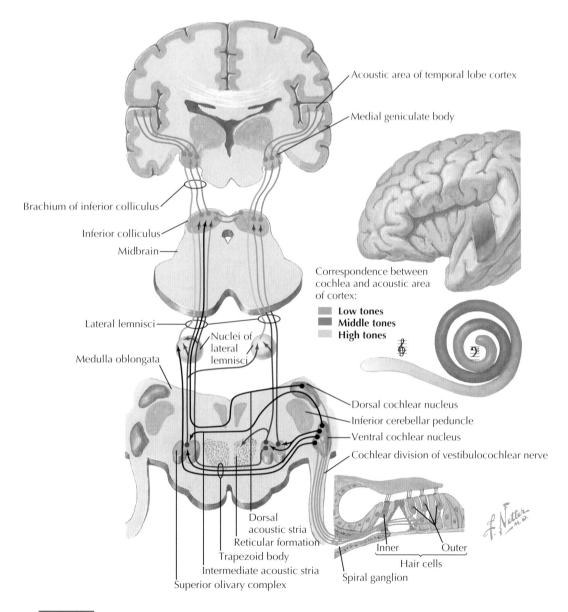

Acoustic area of temporal lobe cortex

Medial geniculate body

Brachium of inferior colliculus

Inferior colliculus

Midbrain

Correspondence between cochlea and acoustic area of cortex:

Low tones
Middle tones
High tones

Lateral lemnisci

Nuclei of lateral lemnisci

Medulla oblongata

Dorsal cochlear nucleus

Inferior cerebellar peduncle

Ventral cochlear nucleus

Cochlear division of vestibulocochlear nerve

Dorsal acoustic stria

Reticular formation

Trapezoid body

Intermediate acoustic stria

Superior olivary complex

Inner   Outer

Hair cells

Spiral ganglion

Figure 5.9 **Auditory Pathways** Electrical signals generated by hair cells in the organ of Corti are conveyed to the dorsal and ventral cochlear nuclei of the medulla, which project to the lateral lemniscus. After further relays, the pathway leads to the medial geniculate bodies of the thalamus, which sends projections to the primary auditory cortex.

but in this case, each hair cell also has a single large cilium known as the **kinocilium**. A gelatinous mass, the **cupula**, spans and occludes the ampulla at the ampullary crest. The semicircular canals are filled with endolymph, and during angular acceleration of the head, pressure against the cupula causes bending of the cilia. Because of the perpendicular orientation of the three semicircular canals, the pattern of pressure changes in the canals depends on the direction of movement of the head. When cilia bend toward the kinocilium, the membrane potential becomes depolarized (because of increased conductance to cations); bending in the opposite

direction causes hyperpolarization. Depolarization results in release of neurotransmitter, whereas hyperpolarization reduces neurotransmitter release by the hair cells. Changes in neurotransmitter release alter the firing rate of afferent nerve fibers ending on the hair cells. Impulses are carried by axons of these primary afferent nerves through cranial nerve VIII to the vestibular nuclei of the pons (Fig. 5.11). Subsequent processing involves both ascending and descending pathways that carry the information through secondary axons to the spinal cord, cerebellum, reticular formation, extraocular muscles, and cortex (by way of the thalamus).

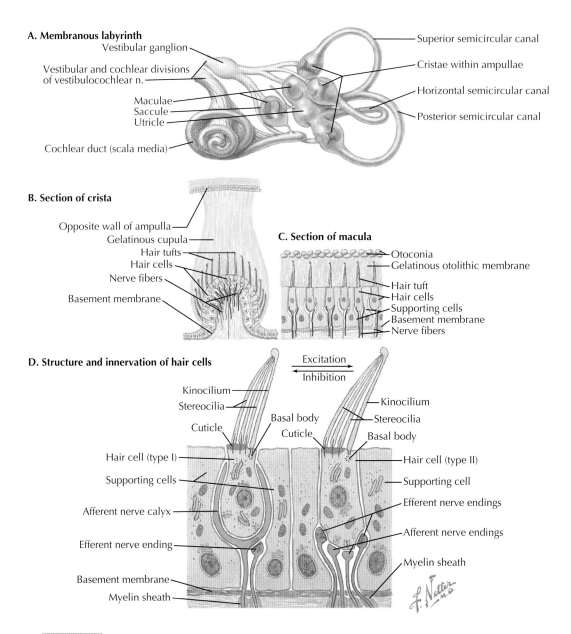

A. Membranous labyrinth
Vestibular ganglion
Vestibular and cochlear divisions of vestibulocochlear n.
Maculae
Saccule
Utricle
Cochlear duct (scala media)
Superior semicircular canal
Cristae within ampullae
Horizontal semicircular canal
Posterior semicircular canal

B. Section of crista
Opposite wall of ampulla
Gelatinous cupula
Hair tufts
Hair cells
Nerve fibers
Basement membrane

C. Section of macula
Otoconia
Gelatinous otolithic membrane
Hair tuft
Hair cells
Supporting cells
Basement membrane
Nerve fibers

D. Structure and innervation of hair cells
Excitation
Inhibition
Kinocilium
Stereocilia
Cuticle
Basal body
Cuticle
Kinocilium
Stereocilia
Basal body
Hair cell (type I)
Supporting cells
Afferent nerve calyx
Efferent nerve ending
Basement membrane
Myelin sheath
Hair cell (type II)
Supporting cell
Efferent nerve endings
Afferent nerve endings
Myelin sheath

Figure 5.10 **Vestibular Receptors** The semicircular canals and the otolithic organs (the utricle and saccule) constitute the vestibular apparatus of the inner ear, which has a critical role in balance and equilibrium **(A)**. The otolithic organs detect linear acceleration of the head, whereas the three perpendicular semicircular canals respond to angular acceleration. Sensory hair cells in the crista of the semicircular canals **(B)** and macula of the otolithic organs **(C)** respond to fluid (endolymph) movement during linear and angular acceleration **(D)**.

Umami (a Japanese word meaning "savory") has long been recognized as a basic taste classification in Japan. It is used to describe the savory taste of protein-rich dishes such as those containing meat. The discovery of a specific taste receptor that binds glutamates, which are prevalent in protein-rich food, led to the recognition of the "umami taste receptor," along with the traditional sweet, bitter, salty, and sour receptors on taste buds. The flavor-enhancing quality of monosodium glutamate can be ascribed to stimulation of the umami receptor. The molecular basis of taste reception is an area of active research today, and specific genes encoding various taste receptor proteins are being defined along with their functions.

In the standing position, the utricle senses horizontal acceleration, whereas the saccule is sensitive to vertical acceleration. This phenomenon is a result of the orientation of the sensory tissues of the otolithic organs, the **maculae** (see Fig. 5.10). The macula is oriented horizontally in the utricle and vertically in the saccule and contains hair cells like those in the crista of the semicircular canals. A gelatinous **otolithic membrane** lies above the macula. Linear acceleration produces pressure changes in endolymph-filled otolithic organs, shifting the macula and bending hair cells. Again, depolarization or hyperpolarization results, affecting transmitter release, afferent nerve firing, and transmission of signals to the brainstem (see Fig. 5.11). By virtue of the orientation of the maculae of the

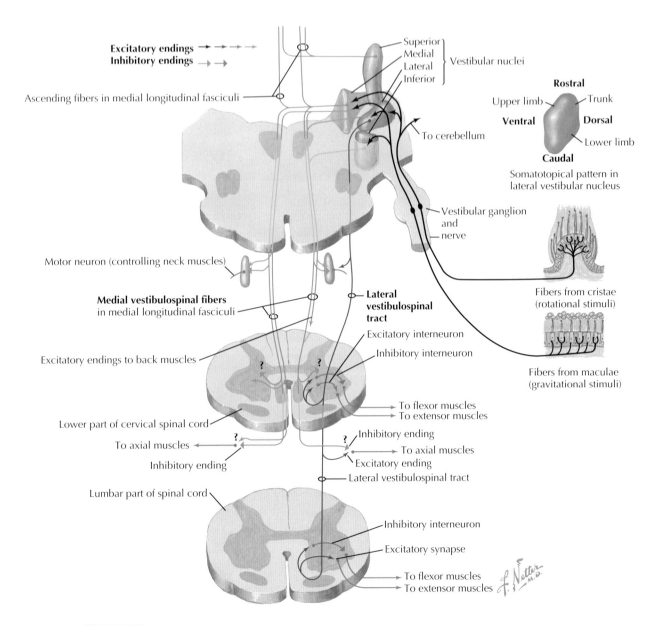

Excitatory endings → → → →
Inhibitory endings → →

Ascending fibers in medial longitudinal fasciculi

Superior
Medial
Lateral   Vestibular nuclei
Inferior

Rostral
Upper limb   Trunk
Ventral   **Dorsal**
Lower limb
Caudal
Somatotopical pattern in
lateral vestibular nucleus

To cerebellum

Vestibular ganglion
and
nerve

Motor neuron (controlling neck muscles)

Medial vestibulospinal fibers
in medial longitudinal fasciculi

**Lateral
vestibulospinal
tract**

Excitatory interneuron

Inhibitory interneuron

Excitatory endings to back muscles

To flexor muscles
To extensor muscles

Lower part of cervical spinal cord

To axial muscles
Inhibitory ending

Inhibitory ending
To axial muscles
Excitatory ending
Lateral vestibulospinal tract

Lumbar part of spinal cord

Fibers from cristae
(rotational stimuli)

Fibers from maculae
(gravitational stimuli)

Inhibitory interneuron

Excitatory synapse

To flexor muscles
To extensor muscles

Figure 5.11 **Vestibulospinal Tracts** Sensory signals from the vestibular apparatus are conveyed to the pontine vestibular nucleus and subsequently through secondary axons to the spinal cord, cerebellar vermis, reticular formation of the brainstem, extraocular muscles, and the cerebral cortex (by way of the thalamus). Vestibular sensory input is utilized to maintain balance and posture and in positioning of the head.

two otolithic organs, linear acceleration in any direction can be detected.

CHEMICAL SENSES

Taste Cells

Gustation (taste) and **olfaction** (smell) are two **chemical senses** involved in detecting chemical stimuli in the environment and the substances we ingest. Traditionally, tastes have been classified as sweet, bitter, salty, and sour. More

recently, a fifth classification, **umami** (savory), has become widely accepted (see the box regarding this topic). **Taste cells** (also called taste receptor cells or gustatory receptor cells) are found mainly in the **taste buds** of the papillae of the tongue but also in taste buds of the palate, larynx, and pharynx. Three types of papillae are recognized based on anatomic structure: the fungiform, foliate, and circumvallate papillae (Fig. 5.12). Taste buds within the papillae consist of groups of taste cells, along with supporting cells and basal cells; the latter differentiate to continuously replace taste cells. Individual taste cells respond through receptors to all of the fundamental taste types, although they may respond best to

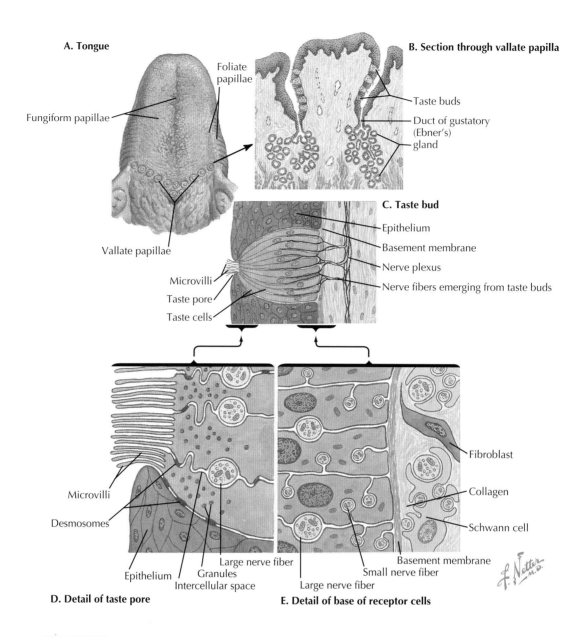

A. Tongue

Foliate papillae

Fungiform papillae

Vallate papillae

B. Section through vallate papilla

Taste buds

Duct of gustatory (Ebner's) gland

C. Taste bud

Epithelium

Basement membrane

Nerve plexus

Nerve fibers emerging from taste buds

Microvilli

Taste pore

Taste cells

Microvilli

Desmosomes

Epithelium

Granules

Intercellular space

Large nerve fiber

D. Detail of taste pore

Fibroblast

Collagen

Schwann cell

Basement membrane

Small nerve fiber

Large nerve fiber

E. Detail of base of receptor cells

Figure 5.12 **Taste Receptors** Taste receptors are found mainly in the taste buds of the three types of papillae of the tongue (**A** and **B**) but also in taste buds on the palate, larynx, and pharynx. Goblet-shaped taste buds (**C**) comprise taste cells with microvilli protruding through the taste pore (**D**). Nerve fibers are found in close association with taste cells (**D** and **E**). Taste cells, through receptors, respond to chemical signals (taste), ultimately producing depolarization and action potentials in afferent nerve fibers. At least five types of taste receptors exist: sweet, salty, sour, bitter, and umami (savory).

one type. Microvilli of taste cells recognize chemical signals in saliva, resulting in activation of G protein–coupled signal transduction mechanisms or changes in ion conductance, ultimately resulting in depolarization and action potentials in primary afferent nerve fibers. Various regions of the tongue are innervated by cranial nerves VII, IX, and X, which carry the signal to the medulla, from which it is transmitted to other regions of the brain and ultimately to the sensory cortex (Fig. 5.13).

At one time many texts provided a diagram of the tongue that depicted localization of perception for sweet, salty, bitter, and sour tastes to various regions on the surface of the tongue. Although it was thought for many years that different areas of the tongue are responsible for the detection of various tastes, it is now believed that taste perception is not exclusively differentiated in this manner. In other words, the "taste map" seen in some older texts is not well defined.

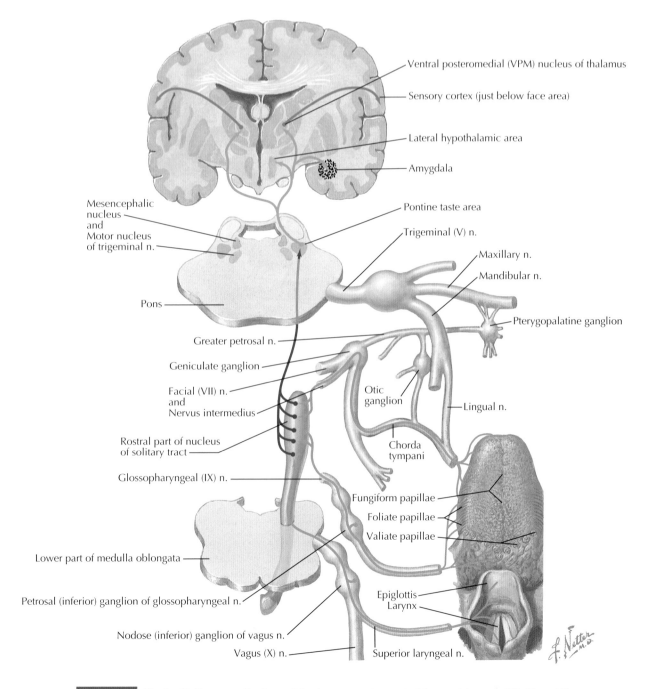

Figure 5.13 **Taste Pathways** Regions of the tongue are innervated by cranial nerves VII, IX, and X. These pathways transmit afferent signals to the medulla (specifically, the nucleus tractus solitarius), from which they are conveyed to other brain regions, eventually reaching the sensory cortex. *n,* nerve.

Olfactory Cells

The sense of smell has an important role in the stimulation of appetite in humans. However, compared with other mammals, in which olfaction can play a major role in reproductive and territorial behavior, this sense is poorly developed in humans. The six basic types of odors we sense are camphor, floral, ethereal, musky, putrid, and pungent. The sensory **olfactory cells** are found in the **olfactory epithelium** of the nasal cavity, along with supporting (sustentacular) and basal cells. As is the case with basal cells in taste buds, these basal cells differentiate to continuously replace sensory cells. The

cilia of olfactory cells extend into the nasal mucosa, where they bind odorant molecules on specific G protein–coupled receptors (Fig. 5.14). Binding of an odorant activates signal transduction mechanisms within the olfactory cells, resulting in opening of ion channels and depolarization. Olfactory cells are primary afferent neurons, and action potentials are carried by their axons to the **olfactory bulb**, where they synapse with **mitral cells** (Fig. 5.15). Projections from the mitral cells form the olfactory tract, which carries signals to higher centers, such as the piriform and orbitofrontal cortices, as well as the amygdala.

A. Distribution of olfactory epithelium (*blue area*)

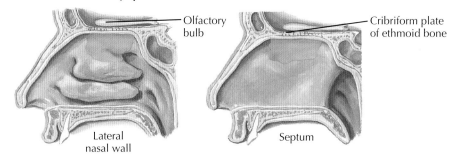

Olfactory bulb

Cribriform plate of ethmoid bone

Lateral nasal wall

Septum

B. Schema of section through olfactory mucosa

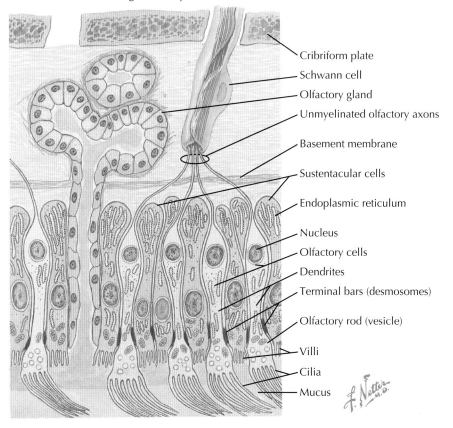

Cribriform plate

Schwann cell

Olfactory gland

Unmyelinated olfactory axons

Basement membrane

Sustentacular cells

Endoplasmic reticulum

Nucleus

Olfactory cells

Dendrites

Terminal bars (desmosomes)

Olfactory rod (vesicle)

Villi

Cilia

Mucus

Figure 5.14 **Olfactory Receptors** Chemical stimuli in the form of odorants result in depolarization of sensory olfactory cells of the olfactory epithelium, located in the upper nasal cavity **(A)**. An olfactory cell, through G protein–coupled receptors, is capable of responding to more than one odor type. Odorants are trapped and also removed by secretions of olfactory glands **(B)**.

In addition to detecting a multitude of odorants and discriminating between them, many animals can detect **pheromones** through their olfactory system. Pheromones are substances that constitute signals between members of the species, affecting behavior or reproductive physiology of the recipient. For instance, it is well known that pheromones can convey information on sex and reproductive status between deer, eliciting behavioral responses. Although human body odors have been shown to elicit some responses suggestive of pheromone activity (e.g., synchronization of sexual cycles between women living in close contact), definitive evidence for human pheromones has been elusive.

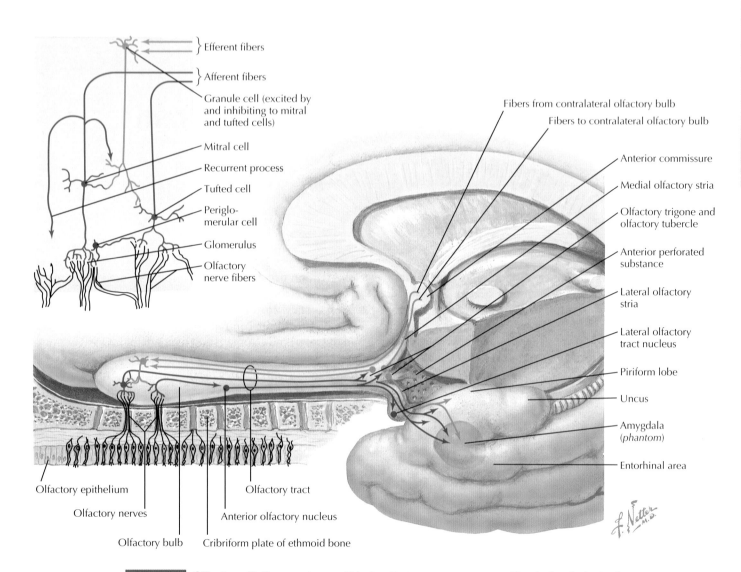

Efferent fibers

Afferent fibers

Granule cell (excited by
and inhibiting to mitral
and tufted cells)

Mitral cell

Recurrent process

Tufted cell

Periglo-
merular cell

Glomerulus

Olfactory
nerve fibers

Fibers from contralateral olfactory bulb

Fibers to contralateral olfactory bulb

Anterior commissure

Medial olfactory stria

Olfactory trigone and
olfactory tubercle

Anterior perforated
substance

Lateral olfactory
stria

Lateral olfactory
tract nucleus

Piriform lobe

Uncus

Amygdala
(phantom)

Entorhinal area

Olfactory epithelium

Olfactory nerves

Olfactory bulb

Olfactory tract

Anterior olfactory nucleus

Cribriform plate of ethmoid bone

Figure 5.15 **Olfactory Pathway** Axons of bipolar olfactory nerves synapse with tufted and mitral cell dendrites in glomeruli of the olfactory bulb (detail, including modulatory interneurons, in *inset diagram*). Mitral cell axons form the olfactory tract, which extends to the anterior commissure (where fibers project back to the contralateral olfactory bulb) and the olfactory trigone, ultimately projecting to the primary olfactory cortex, lateral entorhinal cortex, and amygdala.

Chapter 6

The Somatic Motor System

The motor system consists of two subdivisions, the **somatic motor system** and the **autonomic motor system**. The somatic nervous system controls **voluntary muscle activity** through motor neurons and skeletal muscle. The autonomic nervous system is responsible for the **involuntary muscle activity** involved in visceral organ function and maintenance of the internal environment.

MUSCLE SPINDLES

The somatic motor system, through the actions of skeletal muscle, motor neurons, the spinal cord, and the brain, is responsible for the complex task of controlling movement and posture. This task is accomplished through involuntary spinal reflexes and voluntary muscle activity and is dependent on coordinated contraction and relaxation of muscle. Most skeletal muscle fibers are the **extrafusal fibers** that are innervated by α-motor neurons and contract to generate the force necessary for movement and postural adjustment.

Intrafusal fibers are a second category of muscle fibers found specifically within **muscle spindles**; these spindles are distributed in parallel with extrafusal fibers and function as sensory organs that detect length changes in muscle (Fig. 6.1). The sensitivity of intrafusal fibers is regulated by the γ-motor neurons that innervate them. Muscle spindles are abundant in muscles that control fine movements—for example, in the eye. They have important roles in coordinating fine movements, in the return of muscles to their normal resting length, and in proprioception. Intrafusal fibers are further classified as **nuclear bag fibers** and **nuclear chain fibers**. Two types of sensory (afferent) nerves innervate intrafusal fibers:

- **Group Ia** afferent nerves innervate both nuclear bag fibers and nuclear chain fibers and transmit information regarding the rate of change in muscle length (contraction). Group Ia fibers form spiral endings around intrafusal fibers.
- **Group II** afferent nerves mainly innervate nuclear chain fibers and detect the length of muscle fibers; they form spraylike endings on the fibers.

Golgi tendon organs, like muscle spindles, are stretch receptors and are found at the insertion of skeletal muscles to tendons. They are innervated by group Ib afferent nerves. Excessive stretch or contraction of muscle will cause activation of Golgi tendon organs and their afferent nerves, producing a reflexive relaxation of the muscle. Golgi tendon organs are also important in proprioception; their role in this process, along with the role of muscle spindles and other receptors, is illustrated in Figures 6.2 and 6.3.

SPINAL REFLEXES

Stereotypic motor responses to sensory signals integrated completely within the spinal cord are called **spinal reflexes**. These reflexes are the simplest motor responses, but they may be modified by the central nervous system (CNS) during more complex motor activity. The **reflex arc** of a spinal reflex begins with stimulation of sensory receptors; the signal is carried to the spinal cord through sensory afferent nerves, integrated within the spinal cord, conducted to the muscle through motor neurons, and finally produces contraction or relaxation. There are three basic types of spinal cord reflexes (Fig. 6.4):

- The **stretch reflex** is the most basic, in which sensory afferent nerves synapse directly with α-motor neurons in the spinal cord. Stretch reflexes are therefore referred to as **monosynaptic reflexes**. In the **knee jerk reflex**, a type of stretch reflex, a tap on the patellar tendon of the knee stretches muscle spindles within the quadriceps muscle. This stretching initiates the reflex arc, in which type Ia afferent nerves depolarize and conduct the signal to the spinal cord, where they synapse directly with α-motor neurons. The α-motor neurons conduct the signal back to the quadriceps, producing contraction (the "knee jerk" response). Simultaneously, activation of interneurons results in relaxation of opposing muscles (see Fig. 6.4A).
- The **Golgi tendon reflex** (also known as the **inverse myotactic reflex**) is a **bisynaptic reflex**. It is a mechanism for prevention of muscle damage when muscle tension is excessive. Stretch of Golgi tendon organs activates type

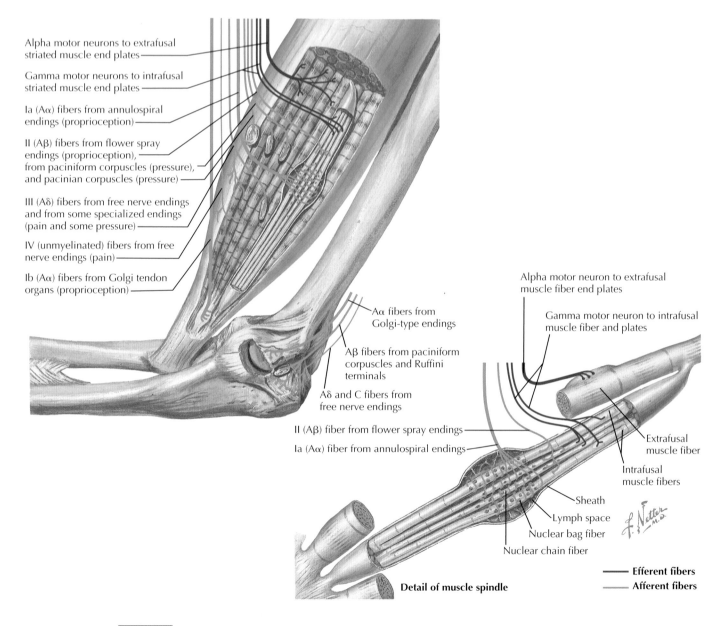

Alpha motor neurons to extrafusal
striated muscle end plates

Gamma motor neurons to intrafusal
striated muscle end plates

Ia (Aα) fibers from annulospiral
endings (proprioception)

II (Aβ) fibers from flower spray
endings (proprioception),
from paciniform corpuscles (pressure),
and pacinian corpuscles (pressure)

III (Aδ) fibers from free nerve endings
and from some specialized endings
(pain and some pressure)

IV (unmyelinated) fibers from free
nerve endings (pain)

Ib (Aα) fibers from Golgi tendon
organs (proprioception)

Aα fibers from
Golgi-type endings

Aβ fibers from paciniform
corpuscles and Ruffini
terminals

Aδ and C fibers from
free nerve endings

II (Aβ) fiber from flower spray endings

Ia (Aα) fiber from annulospiral endings

Alpha motor neuron to extrafusal
muscle fiber end plates

Gamma motor neuron to intrafusal
muscle fiber and plates

Extrafusal
muscle fiber

Intrafusal
muscle fibers

Sheath

Lymph space

Nuclear bag fiber

Nuclear chain fiber

Detail of muscle spindle

— Efferent fibers
— Afferent fibers

Figure 6.1 **Muscle and Joint Receptors** Muscle spindles are specialized sensory receptors composed of intrafusal fibers of two types, the nuclear chain fiber and the nuclear bag fiber. They function in coordinating fine movements, returning muscles to normal resting length, and proprioception. Intrafusal fibers are innervated by γ-motor neurons that regulate their sensitivity. Information about rate of change in muscle length is conveyed by group Ia afferent sensory nerves that innervate both nuclear chain fibers and nuclear bag fibers. These group Ia nerves form spiral endings around intrafusal fibers. Group II sensory afferent nerves convey information on the length of muscle. Golgi tendon organs are located at the insertion of skeletal muscles to tendons; when stretched, they produce a reflexive relaxation of the skeletal muscle. A number of other joint receptors are also illustrated.

Ib afferent sensory nerves that synapse in the spinal cord with interneurons that subsequently inhibit α-motor neurons to the muscle, causing relaxation. The reflex also causes antagonistic muscles to contract, coordinating the response (see Fig. 6.4*B*).

■ The **flexor withdrawal reflex** occurs in response to painful or otherwise noxious stimuli. When cutaneous pain receptors are stimulated—for example, by touching a hot object—afferent signals are conducted through sensory nerve fibers to the spinal cord, where activation of multiple interneurons produces simultaneous flexion and relaxation

of the appropriate muscles to withdraw the limb. This reaction is a **polysynaptic reflex** (see Fig. 6.4*C*).

ROLE OF HIGHER CENTERS IN MOTOR CONTROL

CNS areas involved in motor system control of balance, posture, and movement include the spinal cord, brainstem, cerebellum, basal ganglia, and cerebral motor cortex. For many types of voluntary movement, lower centers regulate

A. Spinal effector mechanisms

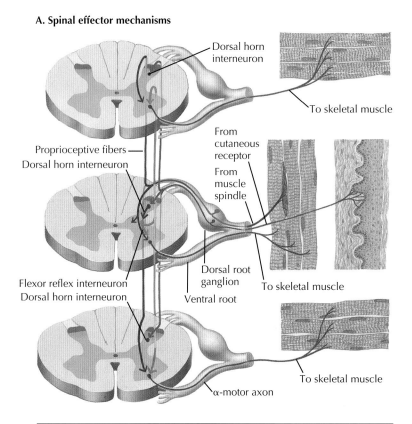

Dorsal horn interneuron

To skeletal muscle

Proprioceptive fibers
Dorsal horn interneuron

From cutaneous receptor
From muscle spindle

Flexor reflex interneuron
Dorsal horn interneuron

Dorsal root ganglion
Ventral root

To skeletal muscle

To skeletal muscle

α-motor axon

B. Schematic representation of motor neurons

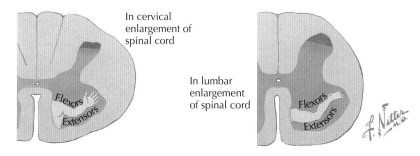

In cervical enlargement of spinal cord

In lumbar enlargement of spinal cord

Flexors
Extensors

Flexors
Extensors

Figure 6.2 **Proprioception: Spinal Effector Mechanisms A,** Spinal effector mechanisms for proprioception involve afferent input from cutaneous receptors and muscle spindles. Monosynaptic reflex pathways at one spinal cord segment and polysynaptic reflex pathways at multiple spinal cord segments activate α-motor neurons, producing muscle contraction. **B,** Motor neuron cell bodies producing flexion and extension of limb muscles are distributed somatopically in the ventral horn.

patterns of skeletal muscle activity when they are activated by the motor cortex. In the case of fine motor movements, particularly of the distal muscles of the limbs (e.g., in the fingers and hands), the cerebral cortex regulates muscle activity more directly.

Corticospinal Tract

The **corticospinal tract**, also known as the **pyramidal tract** (because it passes through the medullary pyramids), is the most important descending pathway for control of fine motor activity originating in the cortex (Fig. 6.5). Nerve fibers in this tract originate in the **primary motor cortex** as well as in the adjacent **premotor** and **supplementary**

motor areas and somatosensory areas posterior to the motor cortex. Most of the fibers cross in the lower medulla to form the **lateral corticospinal tract**; others descend through the **anterior corticospinal tract**. At various levels of the spinal cord, some fibers from these tracts synapse directly with second-order motor neurons (anterior horn cells). Secondary motor neurons innervated by axons of the lateral corticospinal tract are mainly those controlling distal limb muscles, whereas those innervated by the anterior corticospinal tract are mainly those controlling axial muscles. The **motor homunculus** is a graphic representation of the location and approximate relative size of areas of the motor cortex controlling motor activity of various regions of the body (see Fig. 6.5).

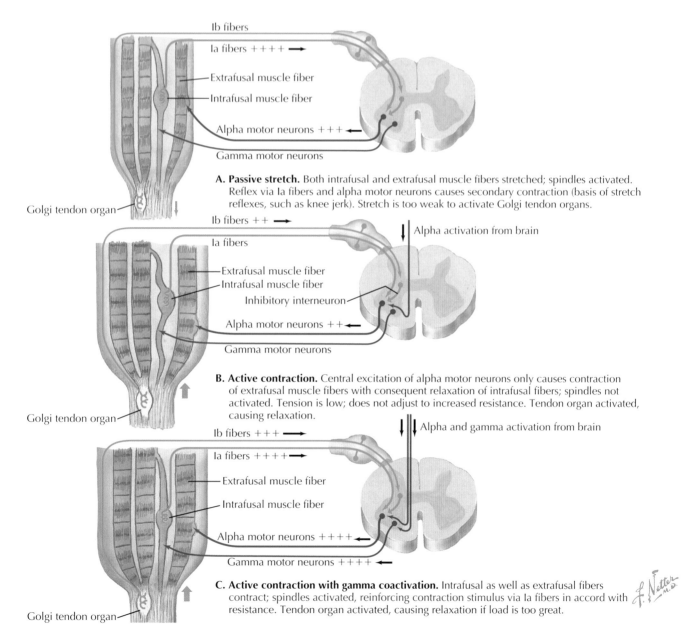

A. Passive stretch. Both intrafusal and extrafusal muscle fibers stretched; spindles activated. Reflex via Ia fibers and alpha motor neurons causes secondary contraction (basis of stretch reflexes, such as knee jerk). Stretch is too weak to activate Golgi tendon organs.

B. Active contraction. Central excitation of alpha motor neurons only causes contraction of extrafusal muscle fibers with consequent relaxation of intrafusal fibers; spindles not activated. Tension is low; does not adjust to increased resistance. Tendon organ activated, causing relaxation.

C. Active contraction with gamma coactivation. Intrafusal as well as extrafusal fibers contract; spindles activated, reinforcing contraction stimulus via Ia fibers in accord with resistance. Tendon organ activated, causing relaxation if load is too great.

Figure 6.3 **Proprioceptive Reflex Control of Muscle Tension**   **A,** During passive stretch of skeletal muscle, both intrafusal and extrafusal fibers are stretched, producing reflex contraction via Ia fibers and α-motor neurons, for example, during a knee jerk reflex. **B** and **C,** Reflex control of skeletal muscle tension in response to active contraction varies depending on the state of activation of γ-motor neurons (and its effects on muscle spindles) and degree of stretch of Golgi tendon organs.

Brainstem

The brainstem (medulla, pons, and mesencephalon) has an important role in the motor system, especially in the control of balance and posture:

- **Pontine reticular nuclei** send projections through the **pontine reticulospinal tract** of the anterior column of the spinal cord, with excitatory synapses on motor neurons innervating axial muscles involved in posture and support of the body against gravity.
- **Medullary reticular nuclei**, through projections descending in the **medullary reticulospinal tract** of the lateral column of the spinal cord, inhibit motor neurons that innervate axial muscles.
- **Vestibular nuclei** on the floor of the fourth ventricle have projections that descend by way of the anterior columns of the spinal cord in the **lateral** and **medial vestibulospinal tracts**. These nuclei receive input from the vestibular apparatus and have an important role in controlling muscles to maintain equilibrium, working in conjunction with the pontine reticular nuclei to regulate axial muscles.
- The **superior colliculus** of the mesencephalon receives input from visual nuclei; through the **tectospinal tract**, it is involved in control of head, neck, and eye muscles.

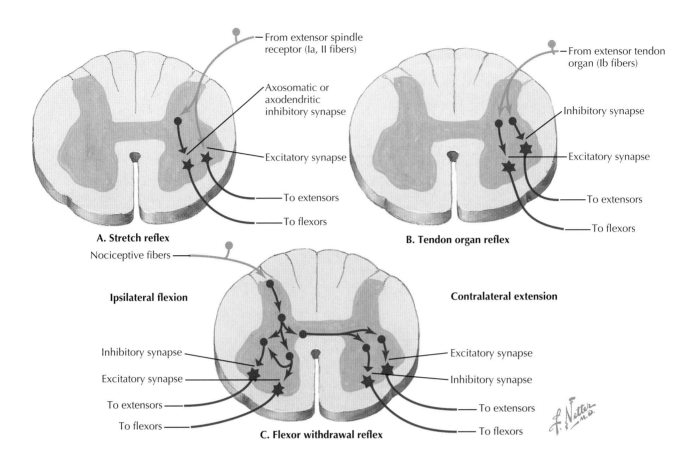

Figure 6.4 **Spinal Reflex Pathways for Stretch, Tendon Organ, and Flexor Withdrawal Reflexes** Reflexes elicited by stretch of muscle spindle receptor **(A)** or Golgi tendon organs **(B)** associated with an extensor muscle are illustrated. Opposite effects on flexors and extensors occur for reflexes initiated by stretch of a flexor muscle or tendon organ. In the flexor withdrawal reflex, nociceptive stimuli produce ipsilateral flexion and contralateral extension through the pathways illustrated **(C)**.

■ The **red nucleus** of the mesencephalon receives input from the motor cortex. It sends projections in the **rubrospinal tract** through the lateral columns of the spinal cord and is involved in voluntary control of large muscles, particularly in the limbs. It can be considered in some ways as an alternative pathway to the corticospinal tract for controlling voluntary muscle movement. Ordinarily this pathway is relatively minor in humans compared with other mammals, but it may gradually assume an expanded role when the corticospinal tract is injured.

Cerebellum

The **cerebellum** ("little brain" in Latin) has an important accessory role to the motor cortex with regard to coordination and control of posture, balance, and movement, and in planning and initiation of motion. These effects are accomplished through the activity of its three lobes (Fig. 6.6):

■ The **archicerebellum** (also known as the **vestibulocerebellum**) is involved in the regulation of posture and balance and in control of eye and head movement. It receives afferent signals from the vestibular apparatus and sends efferent signals through the relevant descending efferent pathways (Fig. 6.7).

■ The **paleocerebellum** (also called the **spinocerebellum**) has an important role in the regulation of proximal limb movement. Afferent sensory signals regarding the position and movement of limbs are used to "fine tune" limb motion through the relevant descending pathways (see Fig. 6.7).

■ The **neocerebellum** (also known as the **pontocerebellum**) has a coordinating role in the regulation of distal limb movement, based on input from the cerebral cortex (via the pontine nuclei). It aids in the planning and initiation of motor activity through its efferent fibers (see Fig. 6.7).

Alternatively, the lobes of the cerebellum can be conceptualized as anterior, middle, and flocculonodular lobes (see Fig. 6.6).

The role of the cerebellum in muscle control involves coordination and fine-tuning of motion. Deficits in cerebellar function are expressed as deficiency in fine motor activity, coordination, and equilibrium.

Cerebellar Cortex

All efferent signals of the cerebellar cortex are inhibitory and emanate from **Purkinje cells** that reside in the middle of three

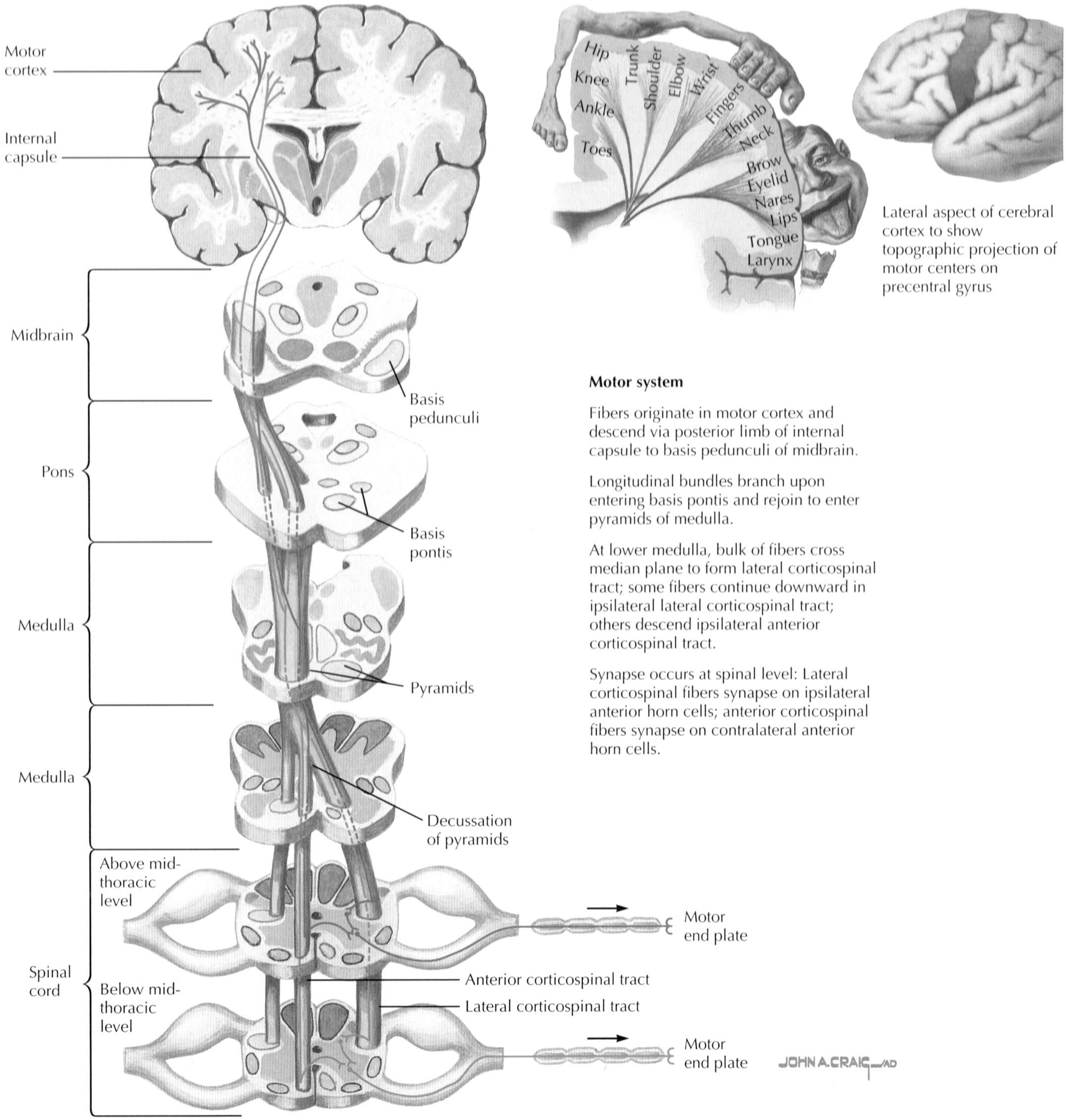
Motor cortex
Internal capsule
Midbrain
Pons
Medulla
Medulla
Spinal cord
Above mid-thoracic level
Below mid-thoracic level
Basis pedunculi
Basis pontis
Pyramids
Decussation of pyramids
Anterior corticospinal tract
Lateral corticospinal tract
Motor end plate
Motor end plate
JOHN A.CRAIG_AD
Hip
Knee
Ankle
Toes
Trunk
Shoulder
Elbow
Wrist
Fingers
Thumb
Neck
Brow
Eyelid
Nares
Lips
Tongue
Larynx
Lateral aspect of cerebral cortex to show topographic projection of motor centers on precentral gyrus
Motor system
Fibers originate in motor cortex and descend via posterior limb of internal capsule to basis pedunculi of midbrain.
Longitudinal bundles branch upon entering basis pontis and rejoin to enter pyramids of medulla.
At lower medulla, bulk of fibers cross median plane to form lateral corticospinal tract; some fibers continue downward in ipsilateral lateral corticospinal tract; others descend ipsilateral anterior corticospinal tract.
Synapse occurs at spinal level: Lateral corticospinal fibers synapse on ipsilateral anterior horn cells; anterior corticospinal fibers synapse on contralateral anterior horn cells.

Figure 6.5 Corticospinal Tract The corticospinal tract, also called the *pyramidal tract,* is the major descending pathway in voluntary motor activity and is particularly important in fine motor activity. The motor homunculus *(top center image)* illustrates the location and approximate relative size of areas in the motor cortex that control muscle activity in various regions of the body.

layers of the cortex. Signals from Purkinje cells are conducted to deep cerebellar nuclei (see Fig. 6.7). The three layers of the cortex contain various cell types that, through complex interactions between each other and with fibers projecting from other sites, regulate the inhibitory efferent output of Purkinje cells (Fig. 6.8):

- The inner **granular layer**, containing **granule cells, Golgi cells**, and **glomeruli.** Granule cells within this layer are the most numerous type of neuron in the CNS. The glomeruli are the sites of synapses between axons of **mossy fibers** from the spinocerebellar, vestibulocerebellar, and pontocerebellar tracts with the dendrites of the granule

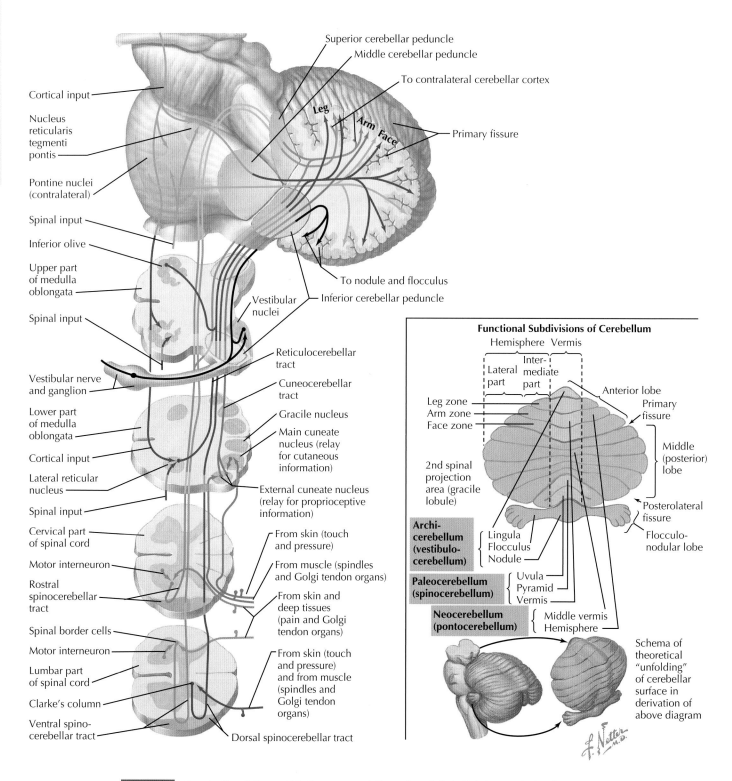

Figure 6.6 **Cerebellar Afferent Pathways and Functional Subdivisions of the Cerebellum**
The cerebellum acts as an accessory to the motor cortex in regulation of posture, balance, movement, and planning and initiation of motion, based on afferent sensory signals and cortical input. Specific functions ascribed to subdivisions of the cerebellum are illustrated in the right panel or discussed in the text.

and Golgi cells. Granule cells, which receive excitatory input from the mossy fibers, send axons through the Purkinje cell layer to the outer molecular layer (discussed later). They provide excitatory signals to Purkinje cell dendrites and dendrites of cells of the molecular layer. Golgi

cells are inhibitory interneurons that inhibit the effects of granule cells on Purkinje cells.

- The middle **Purkinje cell layer.** Purkinje cells are GABAergic (i.e., they use the neurotransmitter γ-aminobutyric acid) and are the only efferent output of the cerebellar

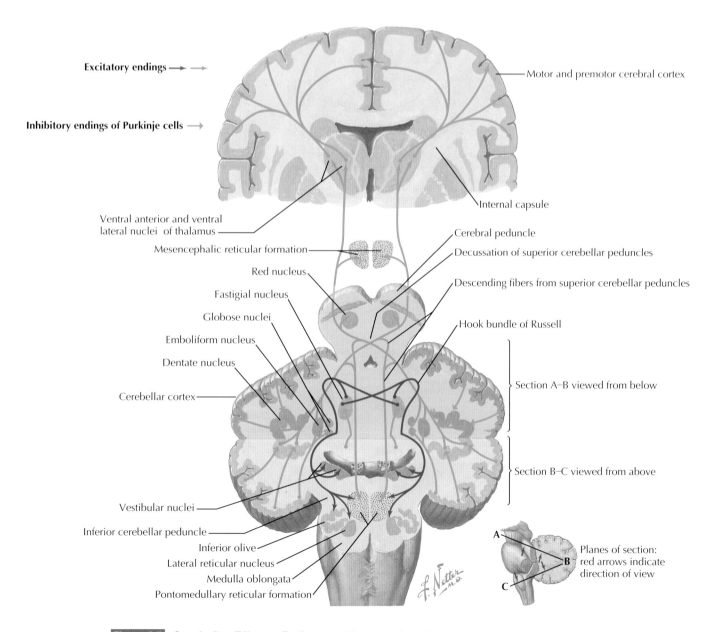

Excitatory endings ⟶ ⟶

Inhibitory endings of Purkinje cells ⟶

Motor and premotor cerebral cortex

Internal capsule

Ventral anterior and ventral lateral nuclei of thalamus

Cerebral peduncle

Mesencephalic reticular formation

Decussation of superior cerebellar peduncles

Red nucleus

Descending fibers from superior cerebellar peduncles

Fastigial nucleus

Globose nuclei

Hook bundle of Russell

Emboliform nucleus

Dentate nucleus

Cerebellar cortex

Section A–B viewed from below

Section B–C viewed from above

Vestibular nuclei

Inferior cerebellar peduncle

Inferior olive

Lateral reticular nucleus

Medulla oblongata

Pontomedullary reticular formation

A
B
C

Planes of section: red arrows indicate direction of view

Figure 6.7 **Cerebellar Efferent Pathways** The cerebellar efferent pathways originate from its deep nuclei, which receive inhibitory input from Purkinje cells of the cerebellar cortex. Axons project from the deep fastigial, globose, emboliform, and dentate nuclei to various nuclei in the brainstem, midbrain, and thalamus and modulate activity of descending motor pathways, resulting in coordination and fine-tuning of motion.

cortex. They receive input from various axons in the molecular layer and send inhibitory signals to the deep cerebellar nuclei and lateral vestibular nuclei.

■ The outer **molecular layer**, which contains **basket cells** and **stellate cells**, and the dendritic arbors of Purkinje cells (extending from the Purkinje cell layer). **Parallel fibers**, containing axons from the granule cells of the granular cell layer, form excitatory synapses with dendrites of Purkinje, basket, and stellate cells. Basket cells and stellate cells are interneurons that inhibit Purkinje cells through the parallel fibers. **Climbing fibers**, which project from the medullary inferior olive, form excitatory synapses with dendrites of Purkinje cells.

Thus the inhibitory cerebellar cortical output of the Purkinje cells is regulated by excitatory input from climbing fibers and mossy fibers (via the granule cells) and inhibitory input from various interneurons of the cortex.

Purkinje cells in the cerebellar cortex are stimulated by granule cells and climbing fibers; their activity is modulated by various cerebellar interneurons. When stimulated, Purkinje cells release the inhibitory neurotransmitter γ-aminobutyric acid (GABA) at synapses within deep cerebellar nuclei, thus regulating motion and posture.

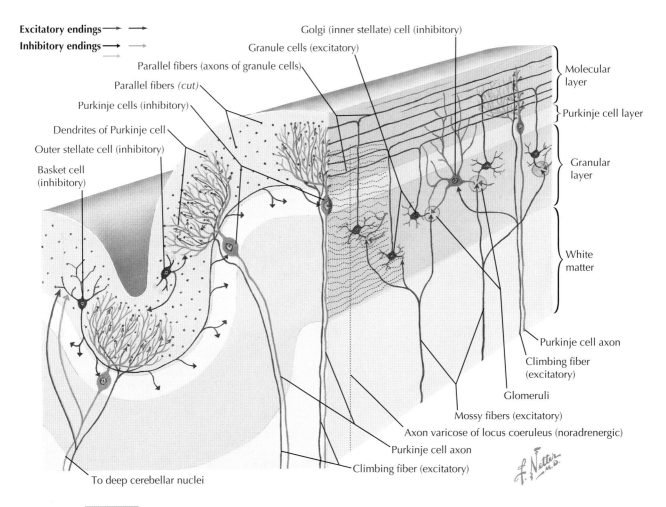

Excitatory endings → →
Inhibitory endings → →

Golgi (inner stellate) cell (inhibitory)
Granule cells (excitatory)
Parallel fibers (axons of granule cells)
Parallel fibers *(cut)*
Purkinje cells (inhibitory)
Dendrites of Purkinje cell
Outer stellate cell (inhibitory)
Basket cell (inhibitory)
Molecular layer
Purkinje cell layer
Granular layer
White matter
Purkinje cell axon
Climbing fiber (excitatory)
Glomeruli
Mossy fibers (excitatory)
Axon varicose of locus coeruleus (noradrenergic)
Purkinje cell axon
Climbing fiber (excitatory)
To deep cerebellar nuclei

Figure 6.8 **Cerebellar Neuronal Circuitry** The cerebellar cortex is composed of three layers. All output of the cortex is carried by inhibitory axons of the Purkinje cells. Their cell bodies are located in the Purkinje cell layer, and their dendritic arbors are in the outer molecular layer. The cortex receives input through climbing fibers, which project from the medullary inferior olive, and mossy fibers, which project from the spinocerebellar, vestibulocerebellar, and pontocerebellar tracts. Mossy fibers form excitatory synapses with dendrites of granule cells in the granular layer. The excitatory axons of granule cells reach the molecular layer, where they form parallel fibers that course through and synapse with dendritic arbors of many Purkinje cells. Climbing fibers form excitatory synapses directly with dendrites of Purkinje cells in the molecular layer. Various other interneurons and their actions are also illustrated and discussed in the text.

Basal Ganglia

The **basal ganglia** are deep nuclei of the cerebrum, diencephalon, and mesencephalon, and like the cerebellum, they have an accessory role to the motor cortex, regulating motor activity to produce smooth movement and maintain posture. The basal ganglia consist of the following nuclei:

- Striatum (caudate nucleus and putamen)
- Substantia nigra
- Globus pallidus
- Nucleus accumbens
- Subthalamic nucleus

These nuclei are associated with the cerebrum (**striatum, nucleus accumbens,** and **globus pallidus), diencephalon (subthalamic nucleus),** and midbrain (the **substantia nigra**). The input to the basal ganglia is from the motor cortex; the output to the cortex is through the thalamus. A complex series of interactions between components of the basal ganglia and associated structures leads to this output, which then regulates the level of excitation in the motor cortex. These interactions form an **indirect pathway** that inhibits cortical excitation and a **direct pathway** that is excitatory. The balance of these opposing pathways is responsible for coordinating smooth movement and maintenance of posture.

CLINICAL CORRELATE 6.1
Parkinson's Disease

Diseases of the basal ganglia produce a variety of deficits in motor activity and control of posture. Parkinson's disease is a progressive neurodegenerative disorder characterized by tremor of the hands and arms, rigidity, shuffling gait, and bradykinesia (slow movement). This disorder is caused by loss of dopaminergic cells of the substantia nigra that normally project to the striatum (caudate nucleus and putamen), affecting both the direct and indirect pathways of the basal ganglia. Parkinson's disease most commonly occurs after age 50 years. Although the most obvious early symptoms involve the patient's movement, depression and dementia may eventually occur. L-dopa, a dopamine precursor, is effective in treating some but not all Parkinsonian patients. When medications fail, deep brain stimulation may be used to alleviate some symptoms and improve quality of life. Deep brain stimulation is a major surgical procedure in which an electrical pacemaker is implanted to send high-frequency impulses that stimulate a specific area of the brain, often the subthalamic nucleus or the globus pallidus.

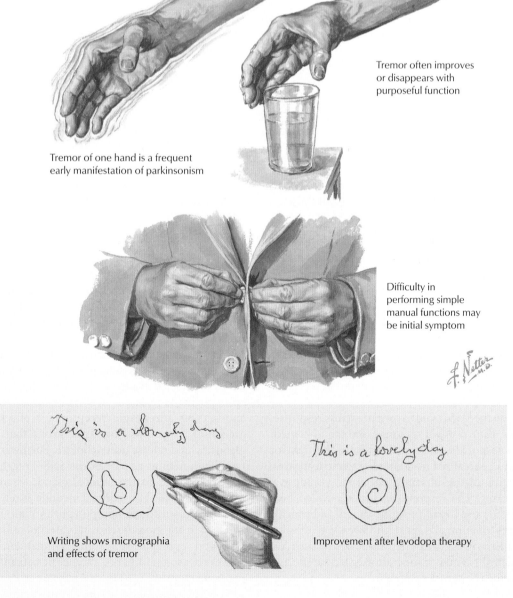

Tremor often improves or disappears with purposeful function

Tremor of one hand is a frequent early manifestation of parkinsonism

Difficulty in performing simple manual functions may be initial symptom

Writing shows micrographia and effects of tremor

Improvement after levodopa therapy

Parkinsonism: Early Manifestations

Chapter 7

The Autonomic Nervous System

Knowledge of the role of the autonomic nervous system is critical for understanding the function of any of the major organ systems. The autonomic nervous system is the primary mechanism for the involuntary control and coordinated activity of smooth muscle of visceral organs, cardiac muscle, and glands and is essential for most homeostatic processes. Within the brain, sensory information is integrated and activity of the autonomic nervous system is modulated to orchestrate this involuntary control of physiologic processes.

ORGANIZATION AND GENERAL FUNCTIONS OF THE AUTONOMIC NERVOUS SYSTEM

The two divisions of the autonomic nervous system are the **sympathetic nervous system (SNS)** and the **parasympathetic nervous system (PNS).** (The enteric nervous system of the gastrointestinal tract, which is discussed in Chapter 22, is also sometimes considered to be part of the autonomic nervous system). In many cases, the SNS and PNS have opposing actions on various organs and processes, and regulation of bodily functions often involves reciprocal actions of the two divisions. For example, heart rate is elevated by SNS activity and decreased by PNS activity.

As a generalization, the SNS is said to mediate **stress responses**, such as the classic *fight-or-flight* response, and the PNS mediates "vegetative" responses, such as digestion. The fight-or-flight response is a generalized reaction to extreme fear, stress, or physical activity and results in a patterned response in many organ systems. The response includes elevated heart rate, cardiac output, and blood pressure, as well as bronchial dilation, mydriasis (dilation of pupils), and sweating. Although the sympathetic nervous system often responds in such patterned manners, the PNS may produce more selective effects— for example, during the sexual response.

The autonomic nervous system has as its central components the hypothalamus, brainstem, and spinal cord; peripherally, it consists of sympathetic and parasympathetic nerves. Areas within the hypothalamus and brainstem regulate and coordinate various processes through the autonomic nervous system, including temperature regulation, responses to thirst and hunger, micturition, respiration, and cardiovascular function. This regulation is in response to sensory input and occurs through the reciprocal regulation of the SNS and PNS.

The fight-or-flight response was originally described in 1915 by Walter Canon, who also coined the term "homeostasis." The fight-or-flight response can be characterized as the physiologic response to acute stress in which generalized sympathetic activation occurs, resulting in effects such as tachycardia, bronchial dilation, mydriasis (dilation of the pupils), vasoconstriction in much of the body, piloerection, and inhibition of gastrointestinal motility. It has long been appreciated that the acute stress responses also involve activation of the hypothalamic-pituitary-adrenocortical endocrine axis (described in Section 7).

Peripherally, axons of **preganglionic neurons** of the SNS and the PNS emerge from the spinal cord and synapse with **postganglionic neurons** at sympathetic and parasympathetic ganglia, respectively; in both cases, **acetylcholine** is the neurotransmitter, acting at **nicotinic receptors** on postganglionic neurons (Fig. 7.1). Postganglionic neurons then send motor axons to effector organs and tissues. The catecholamine **norepinephrine** is released by postganglionic sympathetic axons and acts at **adrenergic receptors** of effector organs. One exception is the postganglionic axons that innervate sweat glands, which release acetylcholine. Furthermore, the adrenal medulla functions as part of the SNS. Preganglionic axons of the SNS extend to the adrenal gland, where they stimulate chromaffin cells of the adrenal medulla to release **epinephrine** (and to a lesser degree norepinephrine) into the bloodstream. Notably, in addition to releasing catecholamines (norepinephrine and epinephrine), some sympathetic postganglionic nerves release a number of adrenergic **cotransmitters**, including **neuropeptide Y, ATP**, and **substance P,** among others. In the PNS, acetylcholine, acting at **muscarinic receptors**, is the postganglionic neurotransmitter. These and other aspects of the two divisions of the autonomic nervous system are compared in Table 7.1 and illustrated in Figures 7.2 and 7.3. Actions of the autonomic nervous system in various organ systems and tissues are listed in Table 7.2, along with the receptor types involved.

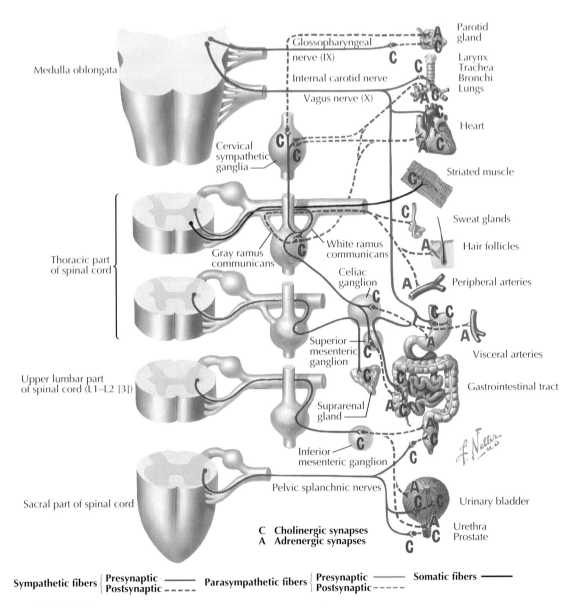

Parotid gland
Larynx
Trachea
Bronchi
Lungs
Heart
Striated muscle
Sweat glands
Hair follicles
Peripheral arteries
Visceral arteries
Gastrointestinal tract
Urinary bladder
Urethra
Prostate

Medulla oblongata
Glossopharyngeal nerve (IX)
Internal carotid nerve
Vagus nerve (X)
Cervical sympathetic ganglia
Thoracic part of spinal cord
Gray ramus communicans
White ramus communicans
Celiac ganglion
Superior mesenteric ganglion
Upper lumbar part of spinal cord (L1–L2 [3])
Suprarenal gland
Inferior mesenteric ganglion
Sacral part of spinal cord
Pelvic splanchnic nerves

C Cholinergic synapses
A Adrenergic synapses

Sympathetic fibers { Presynaptic ——— Postsynaptic - - - -   Parasympathetic fibers { Presynaptic ——— Postsynaptic - - - -   Somatic fibers ———

Figure 7.1 **Cholinergic and Adrenergic Synapses: Schema** Preganglionic neurons of the autonomic nervous system emerge from the spinal cord and synapse at autonomic ganglia. For both parasympathetic and sympathetic divisions, acetylcholine is the ganglionic neurotransmitter. Postganglionic fibers from parasympathetic ganglia release acetylcholine, which binds to muscarinic receptors at the effector organs. Sympathetic postganglionic axons release mainly norepinephrine, although acetylcholine is released at sweat glands. The adrenal gland (not illustrated) functions as part of the **sympathetic nervous system** (SNS). Chromaffin cells of the adrenal medulla act as postganglionic cells of the SNS, releasing epinephrine (and to some extent, norepinephrine) directly into the blood stream. Sympathetic postganglionic nerves, in addition to releasing catecholamines (norepinephrine and epinephrine), release several cotransmitters, including neuropeptide Y, ATP, and substance P.

Sweat gland secretion is stimulated by activation of the SNS. Most of the postganglionic sympathetic neurons innervating these glands are atypical, releasing the neurotransmitter acetylcholine instead of norepinephrine. Acetylcholine acts on muscarinic receptors, inducing sweat secretion. However, in some specific areas, such as the palms of the hands, adrenergic nerves stimulate sweat glands through the release of norepinephrine, which acts at α_1 receptors to stimulate secretion.

AUTONOMIC RECEPTORS

Autonomic receptors are coupled to specific G proteins within the cell membrane and produce their effects through various signal transduction systems in the effector cells. Acetylcholine receptors are classified as nicotinic or cholinergic, based on their pharmacology. Nicotinic receptors are activated by the drug **nicotine** and are blocked by **curare**, the active agent in South American dart poison. Muscarinic receptors, on the

Table 7.1 General Characteristics of the Parasympathetic and Sympathetic Nervous Systems

Characteristic	Parasympathetic Nervous System	Sympathetic Nervous System
Location of preganglionic nerve cell bodies	Brainstem (nuclei of cranial nerves II, VII, IX, and X) or sacral spinal cord (S2-S4; sacral parasympathetic nucleus)	Intermediolateral and intermediomedial cell columns of the thoracolumbar spinal cord (T1-L3)
Location of ganglia	In or adjacent to target organs	Paravertebral and prevertebral
Neurotransmitter of preganglionic neurons	Acetylcholine (acts at nicotinic receptors)	Acetylcholine (acts at nicotinic receptors)
Major neurotransmitter released by postganglionic neuron	Acetylcholine (acts at muscarinic receptors)	Norepinephrine (acts at α and β adrenergic receptors)

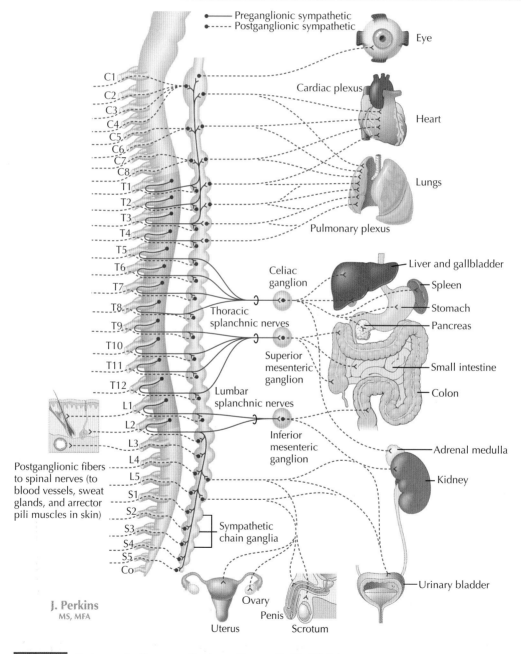

J. Perkins
MS, MFA

Figure 7.2 **Autonomic Nervous System: Sympathetic Division** The two divisions of the autonomic nervous system are the sympathetic nervous system (SNS) and the parasympathetic nervous system (PNS). Preganglionic fibers of the SNS emerge from the spinal cord at levels T1-L2. The SNS is involved in "fight or flight" responses, as well as responses to exercise, hemorrhage, and other challenges to homeostasis. Both the SNS and PNS innervate smooth and cardiac muscle and glands; in general, they work together in reciprocal fashion to regulate bodily function. *Co,* coccyx.

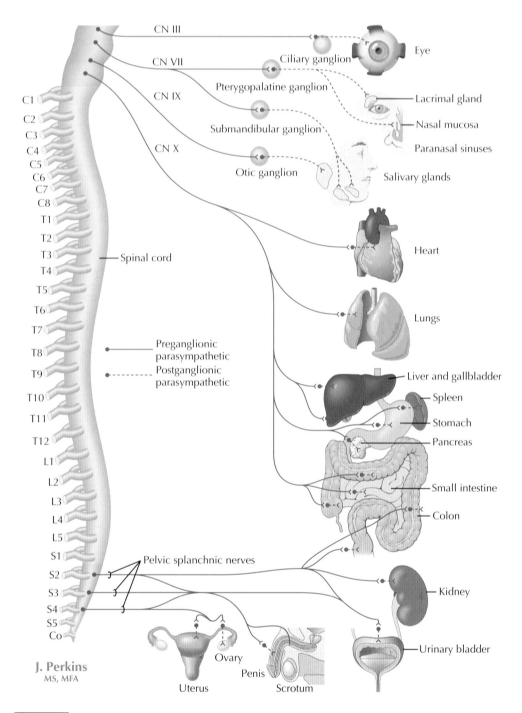

CN III
CN VII
Ciliary ganglion
Pterygopalatine ganglion
CN IX
Submandibular ganglion
CN X
Otic ganglion

Eye
Lacrimal gland
Nasal mucosa
Paranasal sinuses
Salivary glands

C1
C2
C3
C4
C5
C6
C7
C8
T1
T2
T3
T4
T5
T6
T7
T8
T9
T10
T11
T12
L1
L2
L3
L4
L5
S1
S2
S3
S4
S5
Co

Spinal cord

Preganglionic
parasympathetic
Postganglionic
parasympathetic

Pelvic splanchnic nerves

Heart
Lungs
Liver and gallbladder
Spleen
Stomach
Pancreas
Small intestine
Colon
Kidney
Urinary bladder

J. Perkins
MS, MFA

Ovary
Uterus
Penis
Scrotum

Figure 7.3 **Autonomic Nervous System: Parasympathetic Division**   Preganglionic fibers of the parasympathetic nervous system (PNS) are associated with cranial nerves (CN) III, VII, IX, and X and also emerge from the sacral spinal cord at levels S2-S4. The PNS is involved in "vegetative" processes such as digestion, as well as homeostatic functions. In general, the sympathetic nervous system and PNS work together in reciprocal fashion to regulate bodily function. *Co,* coccyx.

Table 7.2   **Actions of the Autonomic Nervous System**

Site of Action	Parasympathetic Nervous System		Sympathetic Nervous System	
	Action	Receptor Type	Action	Receptor Type
Cardiac pacemaker	Decreases heart rate	Muscarinic	Increases heart rate	β_1
Cardiac muscle	Decreases contractility of atria; limited effects on ventricles	Muscarinic	Increased contractility	β_1
Cardiac AV node	Decreases conduction velocity	Muscarinic	Increased conduction velocity	β_1
Vascular smooth muscle	Indirect vasodilation (genital organs and lower GI tract only) by nitric oxide released from endothelium	Muscarinic	Constriction (predominant effect in most vascular beds) Vasodilation	α_1 β_2
GI smooth muscle	Increases motility	Muscarinic	Reduces motility	α_2, β_2
	Relaxes sphincters	Muscarinic	Constricts sphincters	α_1
Gastric parietal cells	Acid secretion	Muscarinic		
Pancreas	Exocrine secretion	Muscarinic		
Lung, bronchial smooth muscle	Constricts	Muscarinic	Dilates	β_2
Sweat glands			Secretion (generalized) Secretion (specific areas such as palms)	Muscarinic* α_1
Male reproductive system	Erection	Muscarinic	Emission during orgasm	α
Female reproductive system	Vasocongestion, vaginal lubrication	Muscarinic	Orgasmic smooth muscle constriction	α
Pupil	Miosis (constriction)	Muscarinic	Mydriasis (dilation)	α_1

*Atypical sympathetic cholinergic nerve fibers secrete acetylcholine.
AV, atrioventricular; *GI,* gastrointestinal.

other hand, are activated by the mushroom toxin **muscarine** and are blocked by the drug **atropine**, the deadly nightshade toxin. Nicotinic receptors are membrane channels that conduct Na^+ and K^+ when acetylcholine is bound. There are several subtypes of muscarinic receptors; the most common are coupled to G proteins that activate phospholipase C and produce elevation of inositol trisphosphate (IP_3) and free intracellular Ca^{2+}.

Receptors for catecholamines are classified as α-adrenergic or β-adrenergic on the basis of their responses to specific pharmacologic agents (drugs). β-Receptors are blocked by the drug

propranolol, and α-receptors are blocked by **phentolamine**. Adrenergic receptors may be further subclassified as α_1, α_2, β_1, β_2, and β_3 subtypes on the basis of effects of more specific drugs. At α_1-receptors, binding of an agonist produces elevation of free intracellular Ca^{2+}, whereas binding of an agonist at α_2-receptors produces inhibition of adenyl cyclase and reduction of the levels of the second messenger cAMP. Activation of β-receptors results in activation of adenyl cyclase and elevation of cAMP. The manner in which an organ or tissue responds to sympathetic stimulation is largely a function of the type of adrenergic receptors present in the tissue (see Table 7.2).

CLINICAL CORRELATE 7.1
Pheochromocytoma

Tumors of the adrenal medulla that secrete epinephrine and nor-epinephrine are known as *pheochromocytomas;* catecholamine-secreting tumors in extraadrenal tissue may also occur. In either case, signs and symptoms are consistent with increased sympathetic activity. Catecholamines in plasma and 24-hour urine collections are elevated, and unlike catecholamines released by normal adrenal medulla, they are not suppressed by clonidine administration (clonidine acts centrally to suppress sympathetic activity). Treatment entails surgical resection of the tumor. Pheochromocytoma is a rare disorder.

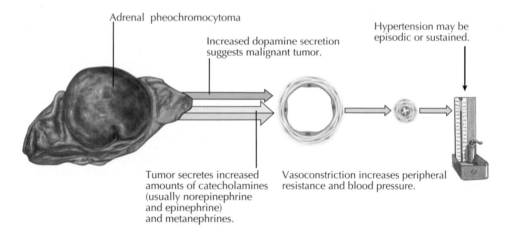

Adrenal pheochromocytoma

Increased dopamine secretion suggests malignant tumor.

Hypertension may be episodic or sustained.

Tumor secretes increased amounts of catecholamines (usually norepinephrine and epinephrine) and metanephrines.

Vasoconstriction increases peripheral resistance and blood pressure.

Pheochromocytoma is a chromaffin cell tumor secreting excessive catecholamines, resulting in increased peripheral vascular resistance and hypertension.

Clinical features of pheochromocytoma

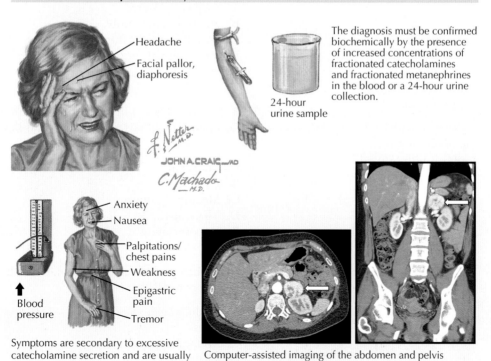

Headache

Facial pallor, diaphoresis

24-hour urine sample

The diagnosis must be confirmed biochemically by the presence of increased concentrations of fractionated catecholamines and fractionated metanephrines in the blood or a 24-hour urine collection.

Anxiety

Nausea

Palpitations/chest pains

Weakness

Epigastric pain

Tremor

Blood pressure

Symptoms are secondary to excessive catecholamine secretion and are usually paroxysmal. However, because of the increased use of CT imaging and familial testing, pheochromocytoma is diagnosed in up to 50% of patients before any symptoms develop.

Computer-assisted imaging of the abdomen and pelvis with CT or MRI should be the first localization test in a patient with a biochemically confirmed catecholamine-secteting tumor. A left adrenal pheochromocytoma (*arrow*) can be seen on the axial (*above left*) and coronal (*above right*) images of a contrast-enhanced CT scan of the abdomen.

Catecholamines Secreted by Tumors of the Adrenal Medulla or Other Sites Produce Excess Sympathetic-Like Activity

Review Questions

CHAPTER 3: NERVE AND MUSCLE PHYSIOLOGY

1. The resting membrane potential of a cell is measured to be −70 mV in a solution with ion concentrations resembling extracellular fluid. Which manipulation would result in a hyperpolarization of the cell?

A. Reduction in the membrane permeability to sodium ion
B. Reduction in the membrane permeability to potassium ion
C. Influx of calcium ion through the cell membrane
D. Increase in the extracellular sodium ion concentration
E. Increase in the extracellular potassium ion concentration

2. The rapid depolarization during phase 0 upstroke of the action potential in neurons is caused by opening of:

A. ligand-gated Ca^{2+} channels.
B. voltage-gated Ca^{2+} channels.
C. voltage-gated Na^+ channels.
D. voltage-gated K^+ channels.
E. voltage-gated Cl^- channels.

3. In the myelinated regions of an axon,

A. internal resistance is reduced.
B. membrane resistance is reduced.
C. the space constant is reduced.
D. membrane capacitance is reduced.
E. conduction velocity is reduced.

4. When depolarization of an axon reaches the synaptic bouton, release of neurotransmitter stored in presynaptic vesicles is most closely associated with:

A. influx of Ca^{2+}.
B. influx of Na^+.
C. efflux of Na^+.
D. influx of K^+.
E. efflux of K^+.

5. Which characteristic is primarily a characteristic of end plate potentials at the neuromuscular junction, but not of neuronal action potentials?

A. Rapid depolarization
B. Depolarization to a potential of +40 mV
C. Upstroke caused by opening of voltage-gated channels
D. Charge carried by a single, large channel for Na^+ and K^+
E. Repolarization associated with increased K^+ conductance

CHAPTER 4: ORGANIZATION AND GENERAL FUNCTIONS OF THE NERVOUS SYSTEM

6. The outermost tissue of the spinal cord is:

A. white matter.
B. gray matter.
C. pia mater.
D. the arachnoid membrane.
E. dura mater.

7. Cerebrospinal fluid circulating through the lateral ventricles, third and fourth ventricles, and the subarachnoid space is produced by:

A. the arachnoid membrane.
B. the choroid plexus.
C. the lymphatic system.
D. arachnoid granulations.
E. the pia mater.

8. Association pathways linking the two cerebral hemispheres are found in the:

A. corpus callosum.
B. anterior commissure.
C. posterior commissure.
D. hippocampal commissure.
E. all the above.

9. Centers involved in integration of cardiovascular and respiratory functions are mainly found in what region of the brainstem?

A. Midbrain
B. Pons
C. Medulla oblongata
D. Thalamus
E. Hypothalamus

10. Which part of the brain is anatomically continuous with the spinal cord?

A. Medulla oblongata
B. Pons
C. Midbrain
D. Thalamus
E. None of the above

CHAPTER 5: SENSORY PHYSIOLOGY

11. Which type of mechanoreceptor responds best to low-frequency stimuli such as flutter?

A. Meissner's corpuscles
B. Pacinian corpuscles
C. Merkel's disks
D. Hair follicle receptors
E. Ruffini's corpuscles

12. Light reception by photoreceptors results in reduction of permeability of what ion?

A. K^+
B. Na^+
C. Cl^-
D. HCO_3^-
E. Ca^{2+}

13. The endolymph of the scala media resembles which body fluid in its composition?

A. Interstitial fluid
B. Plasma
C. Cerebrospinal fluid
D. Lachrymal gland secretions
E. Intracellular fluid

14. During angular acceleration of the head, neurotransmitter release by hair cells within the ampullary crest of the semicircular canal alters the firing rate of afferent nerve fiber axons through cranial nerve:

A. IV.
B. V.
C. VI.
D. VII.
E. VIII.

15. In the auditory system, which structure is directly connected to the membranous oval window and thus directly transmits vibrations to that structure?

A. Tympanic membrane
B. Incus
C. Stapes
D. Malleus
E. None of the above

CHAPTER 6: THE SOMATIC MOTOR SYSTEM

16. The knee jerk reflex is an example of a:

A. stretch reflex.
B. Golgi tendon reflex.
C. flexor withdrawal reflex.
D. bisynaptic reflex.
E. polysynaptic reflex.

17. The most important descending pathway(s) for control of fine motor movements originating in the cortex is/are the:

A. pontine reticulospinal tract.
B. medullary reticulospinal tract.
C. lateral and medial vestibulospinal tracts.
D. tectospinal and rubrospinal tracts.
E. corticospinal (pyramidal) tract.

18. The inhibitory efferent output of the cerebellar cortex emanates from which type of cell?

A. Granule cells
B. Golgi cells
C. Purkinje cells
D. Basket cells
E. Stellate cells

19. Which of the following structures has as a primary function the regulation of the level of excitation in the motor cortex, resulting in smooth movement and maintenance of posture?

A. Limbic system
B. Basal ganglia
C. Medulla oblongata
D. Hypothalamus
E. Pons

20. Which area of the brainstem is most involved in control of head, neck, and eye muscles?

A. Red nucleus
B. Superior colliculus
C. Medullary reticular nuclei
D. Vestibular nuclei
E. Pontine reticular nuclei

CHAPTER 7: THE AUTONOMIC NERVOUS SYSTEM

21. Preganglionic nerve cell bodies in the sympathetic nervous system are located in the:

A. cervical spinal cord.
B. thoracolumbar spinal cord.
C. sacral spinal cord.
D. sympathetic chain ganglia.
E. brainstem.

22. Release of acetylcholine and subsequent binding to muscarinic receptors is the major mode of neurotransmission at:

A. parasympathetic ganglia.
B. neuroeffector junctions of the sympathetic nervous system.
C. sympathetic ganglia.
D. neuroeffector junctions of the parasympathetic nervous system.
E. all the above.

23. β_2-adrenergic receptors mediate which effect of the sympathetic nervous system?

A. Constriction of vascular smooth muscle
B. Increased cardiac contractility
C. Dilation of bronchial smooth muscle
D. Dilation of the pupil (mydriasis)
E. Sweat gland secretion

24. At which structure(s) do postganglionic neurons of the sympathetic nervous release acetylcholine?

A. Male reproductive organs
B. Pancreas
C. Gastric parietal cells
D. Cardiac atrioventricular node
E. Sweat glands

Section 3

CARDIOVASCULAR PHYSIOLOGY

The main function of the cardiovascular system is transport of gases, nutrients, and wastes. The cardiovascular system transports oxygen from the lungs to the rest of the body; carbon dioxide to the lungs from tissues throughout the body; nutrients between sites of absorption, utilization, and storage; and metabolic waste products from throughout the body to sites at which they are eliminated or recycled. The cardiovascular system also has important roles in thermoregulation, defense mechanisms, endocrine system function, and embryonic development. All of these roles are served by the continuous flow of blood, pumped by the heart, to capillary beds, where exchange between tissues and blood takes place. Nearly all living cells in our body are located no more than a few cell thicknesses from a capillary. In the following chapters, key concepts of cardiovascular physiology are presented, from basic electrophysiologic and biophysical principles to integrated mechanisms for control of cardiovascular system function.

Chapter 8

The Blood

Components of the blood are essential to the many functions of the cardiovascular system, and analysis of hematologic parameters is usually an important consideration in the diagnosis of disease. A brief overview of blood elements and their functions is presented in this chapter. More detailed information on some of the specific functions of blood, such as the role of red blood cells in gas transport, can be found in other chapters of this book.

COMPOSITION OF BLOOD

The 5 liters (L) of blood in an average, adult human is composed of approximately 55% plasma, with the balance being blood cells (Fig. 8.1). Plasma contains dissolved proteins that have many important functions, along with various other solutes, including electrolytes and organic nutrients and wastes.

Blood cells include **erythrocytes** (red blood cells), **leukocytes** (white blood cells), and **platelets**. In aggregate, these substances are often referred to as the "formed elements" of blood rather than blood cells because mature red blood cells are enucleate, having extruded their nuclei, and platelets (thrombocytes) are cell fragments formed by budding from bone marrow megakaryocytes. Figure 8.2 illustrates general functions and provides basic information about the formed elements. The leukocytes comprise several distinct cell types

and are classified as **granulocytes** or **agranulocytes** based on the presence or absence of cytoplasmic granules. They are cells of the immune system and are derived from bone marrow stem cells. Leukocytes also have important functions in inflammation—that is, the response to tissue injury.

HEMOSTASIS

In a healthy, intact vessel, free flow of blood is maintained by an intricate balance between factors associated with the endothelial lining of the vessel, anticoagulant and coagulant factors within the blood, and platelets. Endothelial cells lining the lumen of vessels produce the platelet inhibitors **prostacyclin** and **nitric oxide** and express anticoagulant molecules such as heparan sulfate and thrombomodulin on their surface. Although free flow is essential to normal physiological function, flow must be stopped in a vessel when damage to the vessel results in bleeding. This stoppage of bleeding, called **hemostasis**, occurs in three steps:

1. Vasospasm
2. Platelet plug formation
3. Blood coagulation (clotting)

The vasoconstriction response to vascular injury is rapid and is provoked by physical injury, reflexes, and substances released by platelets and endothelial cells. In platelet plug formation (Fig. 8.3), the initial adhesion of platelets is a passive process, but in a prothrombotic environment, platelets become activated, resulting in platelet shape change (from a disk to a sphere with pseudopods), activation of metabolic processes within platelets, release of granule contents, and ultimately aggregation of platelets to form a platelet plug to stanch blood flow (Fig. 8.3).

The third component of hemostasis, **blood coagulation**, is the process that results in the change in the state of blood from liquid to gel. The **coagulation cascade** refers to the cascade of enzymatic reactions that occurs to serially activate coagulation factors (zymogens) within the blood, ultimately resulting in the polymerization of fibrinogen to fibrin (Fig. 8.4). The platelet plug is stabilized and hemostasis is supported as polymerized fibrin is deposited on the plug, forming a clot.

CLINICAL CORRELATE 8.1
Plasma and Serum

Centrifugation of a blood sample anticoagulated with a substance such as ethylenediaminetetraacetic acid (EDTA), heparin, or sodium citrate produces the plasma and packed cell layers illustrated in Figure 8.1. The proportion of packed cell volume (mainly red blood cells) at the bottom of the sample to the total sample volume is the **hematocrit** and normally has a value between 40% and 45%. In contrast, to prepare serum (plasma minus the coagulation proteins), blood collected into a tube that does not contain anticoagulant is centrifuged. Thus plasma and serum concentrations of various substances are measured in clinical laboratories after collection and processing of blood samples with and without anticoagulant, respectively.

CLINICAL CORRELATE 8.2
HIV/AIDS

The **human immunodeficiency virus** (**HIV**) has infected tens of millions of people worldwide. This retrovirus enters a subclass of lymphocytes, the **helper T cells** (specifically the CD4+ T cells), as well as **macrophages** (differentiated monocytes in tissues). Viral RNA serves as a template for production of viral DNA through reverse transcription; viral DNA is then integrated into host DNA. After a variable latency period, the DNA is transcribed, the mRNA is processed and translated, and the viral RNA is then packaged with protein to produce new virus that is released to infect additional cells. Infected T cells are susceptible to apoptosis, direct killing by the virus, and killing by CD8 cytotoxic lymphocytes. The loss of CD4+ cells results in incompetence of the immune system, with susceptibility to opportunistic infections and cancers; this condition is known as **acquired immunodeficiency syndrome** (**AIDS**). HIV can be transmitted by an infected person through sexual contact, blood transfusion, and breastfeeding. HIV infection is a chronic disease with no cure; current therapy focuses on management of the disease using a combination of antiretroviral drugs to prevent progression to AIDS. These drug combinations may include entry inhibitors, reverse transcriptase inhibitors, integrase inhibitors, protease inhibitors, and drugs that modulate viral genes (see illustration).

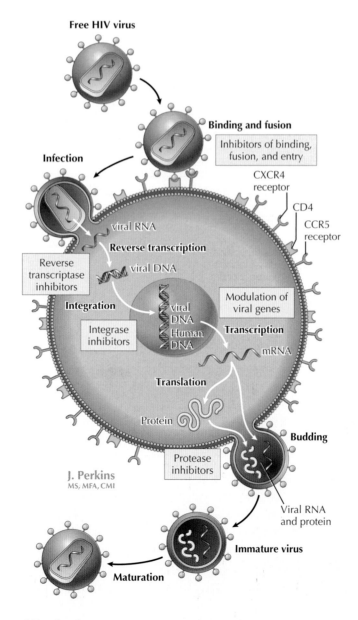

HIV Infection and Replication Sites of actions of drugs used to treat HIV infection are illustrated in the pink boxes.

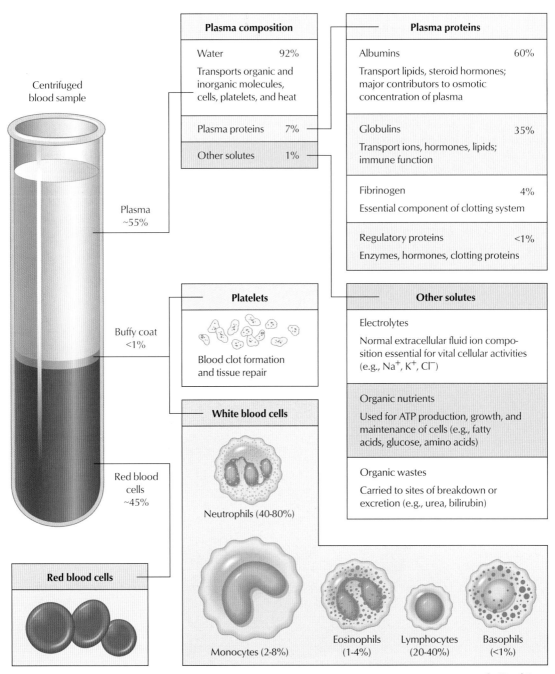

Centrifuged
blood sample

Plasma composition

Water	92%
Transports organic and inorganic molecules, cells, platelets, and heat	
Plasma proteins	7%
Other solutes	1%

Plasma proteins

Albumins	60%
Transport lipids, steroid hormones; major contributors to osmotic concentration of plasma	
Globulins	35%
Transport ions, hormones, lipids; immune function	
Fibrinogen	4%
Essential component of clotting system	
Regulatory proteins	<1%
Enzymes, hormones, clotting proteins	

Plasma
~55%

Platelets

Blood clot formation
and tissue repair

Other solutes

Electrolytes

Normal extracellular fluid ion composition essential for vital cellular activities (e.g., Na^+, K^+, Cl^-)

Organic nutrients

Used for ATP production, growth, and maintenance of cells (e.g., fatty acids, glucose, amino acids)

Organic wastes

Carried to sites of breakdown or excretion (e.g., urea, bilirubin)

Buffy coat
<1%

White blood cells

Neutrophils (40-80%)

Red blood
cells
~45%

Monocytes (2-8%)

Eosinophils
(1-4%)

Lymphocytes
(20-40%)

Basophils
(<1%)

Red blood cells

J. Perkins
MS, MFA

Figure 8.1 Components of Blood The composition of human blood is illustrated in the context of an anticoagulated, centrifuged sample.

Features of Erythrocytes and Platelets in Wright-Stained Blood Smears

Cells	Diameter (μm)	Life span (days)	No. of cells/ L of blood	Shape and nucleus type	Cytoplasm	Functions
Erythrocyte (red blood cell) 	7–10	120	5×10^{12} in males; 4.5×10^{12} in females	Biconcave disk, anucleate	Pink because of acidophilia of hemoglobin; halo in center	Transports hemoglobin that binds O_2 and CO_2
Platelet (thrombocyte) 	2–4	10	150 to 400×10^9	Oval biconvex disk, anucleate	Pale blue; central dark granulomere, peripheral less dense hyalomere	In hemostasis, promotes blood clotting; plugs endothelial damage

C. Machado, M.D.

Features of Leukocytes in Wright-Stained Blood Smears (Total Number: 5–10 x 10⁹/L Blood)

Cells	Diameter (μm)	Differential count (%)	Nucleus	Cytoplasm	Functions
Granulocytes					
Neutrophil 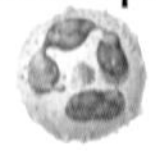	9–12	40–80	Segmented, 3–5 lobes, densely stained	Pale, finely granular, evenly dispersed specific granules	Phagocytoses bacteria; increases in number in acute bacterial infections
Eosinophil 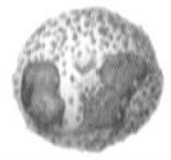	12–15	1–4	Bilobed, clumped chromatin pattern, densley stained	Large homogeneous red granules that are coarse and highly refractile	Phagocytoses antigen-antibody complexes and parasites
Basophil 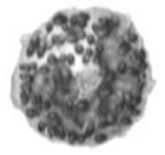	10–14	0–1	Bilobed or segmented	Large, blue, specific granules that stain with basic dyes and often obscure nucleus	Involved in anticoagulation; increases vascular permeability
Agranulocytes					
Monocytes 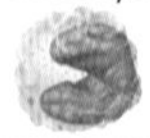	12–20	2–8	Indented, kidney shaped, lightly stained	Agranular pale-blue cytoplasm, with lysosomes	Is motile; gives rise to macrophages
Lymphocyte · Small · Medium to large	6–10 11–16	20–40	Small, round or slightly indented, darkly stained	Agranular, faintly basophilic, blue to gray	Acts in humoral (B cell) and cellular (T cell) immunity

Figure 8.2 **Formed Elements in Human Blood** From Ovalle WK, Nahirney PC: Netter's Essential Histology, ed 2, Philadelphia, 2013, Saunders.

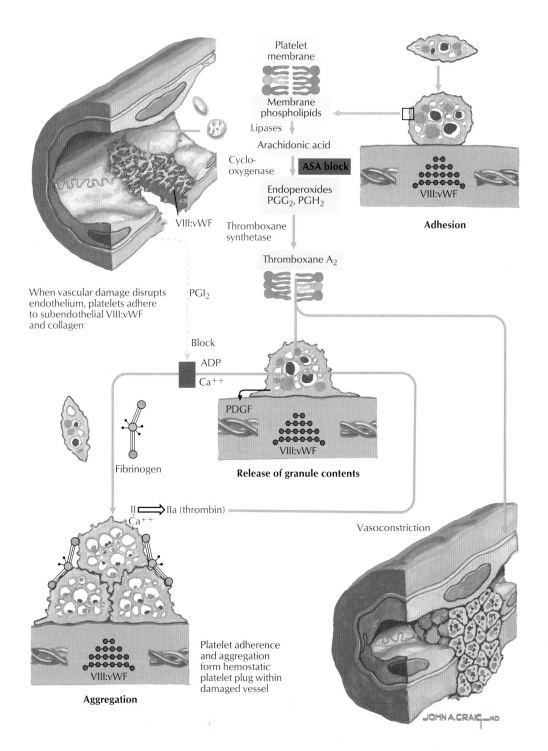

Platelet
membrane

Membrane
phospholipids

Lipases

Arachidonic acid

Cyclo-
oxygenase   ASA block

Endoperoxides
PGG$_2$, PGH$_2$

VIII:vWF

Adhesion

Thromboxane
synthetase

Thromboxane A$_2$

VIII:vWF

When vascular damage disrupts
endothelium, platelets adhere
to subendothelial VIII:vWF
and collagen

PGI$_2$

Block

ADP

Ca^{++}

PDGF

VIII:vWF

Release of granule contents

Fibrinogen

II ⟹ IIa (thrombin)
Ca^{++}

Vasoconstriction

VIII:vWF

Platelet adherence
and aggregation
form hemostatic
platelet plug within
damaged vessel

Aggregation

JOHN A. CRAIG—AD

Figure 8.3   Hemostasis: Platelet Adhesion, Release, and Aggregation With vascular injury,
platelets adhere to subendothelial collagen, von Willebrand factor (vWF; complexed with factor VIII), and
other molecules. Platelet activation results in synthesis of thromboxane A$_2$ from arachidonic acid; thromboxane A$_2$ amplifies the response, activates additional platelets, and also causes vasoconstriction. Platelet
release reaction results in extrusion of granule contents, including ADP, Ca^{2+}, and various growth factors.
Like thromboxane A$_2$, ADP amplifies the response and activates further platelets. Activation of platelets also
results in expression of adhesive glycoproteins on their surface. Platelets express binding sites for fibrinogen;
aggregation of platelets involves cross-linking on this plasma protein. Aspirin (ASA) blocks the enzyme
cyclooxygenase, and therefore inhibits synthesis of thromboxane A$_2$ in platelets (as well as synthesis of
other cyclooxygenase products such as prostaglandins and prostacyclin in various tissues). *PDGF,* platelet-
derived growth factor; *PGG$_2$,* prostaglandin G$_2$; *PGH$_2$,* prostaglandin H$_2$; *PGI$_2$,* prostaglandin I$_2$.

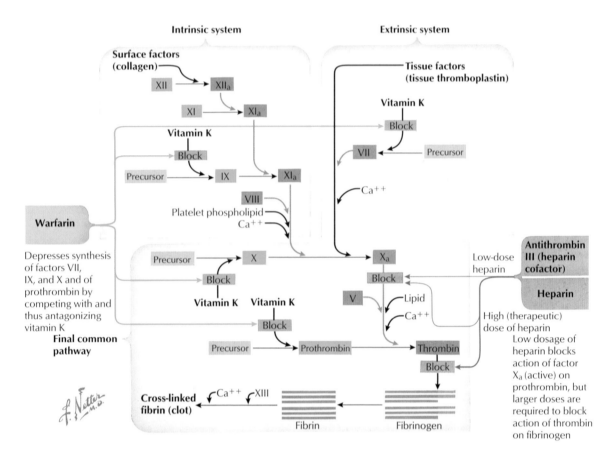

Figure 8.4   The Coagulation Cascade   Blood coagulation can be initiated through two pathways. In the intrinsic pathway, exposure of blood to surface factors such as collagen results in activation of factor XII, initiating a cascade that ultimately activates factor X, the first step in the final common pathway that leads to fibrinogen polymerization to fibrin. The extrinsic pathway is activated when blood is exposed to tissue factors during tissue injury; subsequent activation of factor VII leads to the final common pathway. Vitamin K is required for hepatic synthesis of several of the coagulation factors. Warfarin, the anticoagulant, inhibits vitamin K–dependent synthesis of these factors. Heparin acts as an anticoagulant through the illustrated mechanisms.

Hemostasis is obviously an important element in the early stage of wound repair. Ultimately, however, the fibrin clot formed during hemostasis is removed by a process known as **fibrinolysis**. When vessels are damaged, endothelial cells secrete tissue plasminogen activator, a protease that converts the plasma protein plasminogen to plasmin. Plasmin is an enzyme that breaks down and solubilizes fibrin.

CLINICAL CORRELATE 8.3
Hemophilia A and B

Hemophilia A and B are sex-linked genetic diseases that affect the coagulation system (both diseases are rare in females). The functional activity of clotting factor VIII and IX is reduced or lacking in hemophilia A and B, respectively, resulting in a potentially severe bleeding tendency and an impaired ability to achieve hemostasis. In persons with severe hemophilia, functional clotting factor is greatly reduced. Pathophysiological manifestations are the result of internal or external bleeding, and hemarthrosis (bleeding in joints), deep-muscle bleeding, and intracranial bleeds may occur with trauma. Treatment consists of periodic infusion of the factor that is lacking—factor VIII or IX.

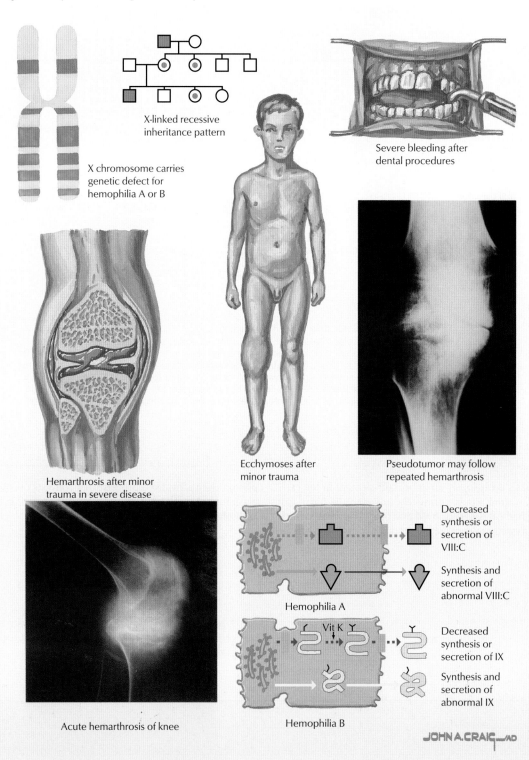

X-linked recessive inheritance pattern

X chromosome carries genetic defect for hemophilia A or B

Severe bleeding after dental procedures

Hemarthrosis after minor trauma in severe disease

Ecchymoses after minor trauma

Pseudotumor may follow repeated hemarthrosis

Acute hemarthrosis of knee

Decreased synthesis or secretion of VIII:C

Synthesis and secretion of abnormal VIII:C

Hemophilia A

Vit K

Decreased synthesis or secretion of IX

Synthesis and secretion of abnormal IX

Hemophilia B

JOHN A. CRAIG—AD

Pathophysiology of Hemophilia A and B

Chapter 9

Overview of the Heart and Circulation

The main functions of the cardiovascular system are gas transport, nutrient delivery, and waste removal. These functions depend on circulation of blood, which requires structural integrity of the heart and vessels. In this chapter, the major structural features of the human heart and circulation are presented.

GENERAL SCHEME OF THE CIRCULATION

The cardiovascular system consists of the heart, blood vessels, and blood. The overall circulation can be characterized as a series circuit (Fig. 9.1), in which:

- oxygenated blood is pumped by the left ventricle into the systemic arterial circulation,
- deoxygenated blood returns to the right atrium through the systemic veins,
- the right ventricle pumps this blood into the pulmonary circulation, and
- reoxygenated blood subsequently returns to the left atrium through the pulmonary veins.

Thus all of the oxygenated blood returning to the left side of the heart from the lungs flows through the systemic circulation, and all of the deoxygenated blood returning to the right side of the heart is pumped to the pulmonary circulation for reoxygenation. Within these circulations, the regional circulations are parallel to each other. For example, skeletal muscle vascular beds lie in parallel to the renal vasculature. In both the systemic and pulmonary circulations, arteries are the *distributing* vessels, supplying blood to capillary beds of the body and lungs, respectively. The veins can be considered *collecting* vessels for return of capillary blood to the heart; they also serve a reservoir function by which generalized venous constriction produces an overall rise in blood pressure, such as during compensation for dehydration or blood loss. Of the total blood volume of 5 liters (L), more than 60% is found in the systemic veins (see Fig. 9.1).

STRUCTURE OF THE HEART

The human heart consists of four chambers: the two muscular ventricles and the two atria. High-pressure ejection of blood into the systemic circulation is accomplished by the action of the muscular left ventricle (Fig. 9.2). The interventricular septum functions as part of the left ventricle in this pumping action. The right ventricle ejects blood at lower pressures into the pulmonary circulation. The contraction of the left ventricle consists of constriction of the chamber and shortening of the heart from base to apex; the less forceful contraction of the right ventricle consists of a lateral to medial bellows-like motion, with some shortening from base to apex. In the left heart, backflow of blood into the atrium is prevented by the **mitral valve** between the left ventricle and atrium. The right ventricle and atrium are separated by the **tricuspid valve**. The **chordae tendinae** are tendons connecting the **papillary muscles** to the mitral and tricuspid valves. The chordae tendinae and papillary muscles function to prevent eversion of the mitral and tricuspid valves into the atria during ventricular ejection, thereby preventing regurgitation of blood back into the atria. During ejection of blood, the aortic and pulmonic valves are open; they prevent backflow to the ventricles during cardiac filling. Valvular opening and closing are accomplished by passive responses of the valves to pressure gradients.

The four-chambered heart of mammals and birds is an evolutionary adaptation to prevent mixing of oxygenated blood returning to the heart from the lungs with deoxygenated blood returning from the systemic circulation. The efficiency of the series circulation created by this adaptation is consistent with the high metabolic rate and warm-blooded nature of these species. Some reptiles also have four-chambered hearts. Amphibians have a dual circulation with a three-chambered heart; the muscular ventricle is not separated into two pumps. In contrast to the dual circulation made possible by four chambers, fish have a single circulation and a two-chambered heart.

CONDUCTION SYSTEM OF THE HEART

Effective ejection of blood from the heart requires coordinated contraction and relaxation of the chambers. At rest, this cycle occurs approximately 70 times per minute. To achieve this rate, the sinoatrial node acts as the pacemaker, initiating the depolarization of the heart. Depolarization spreads from

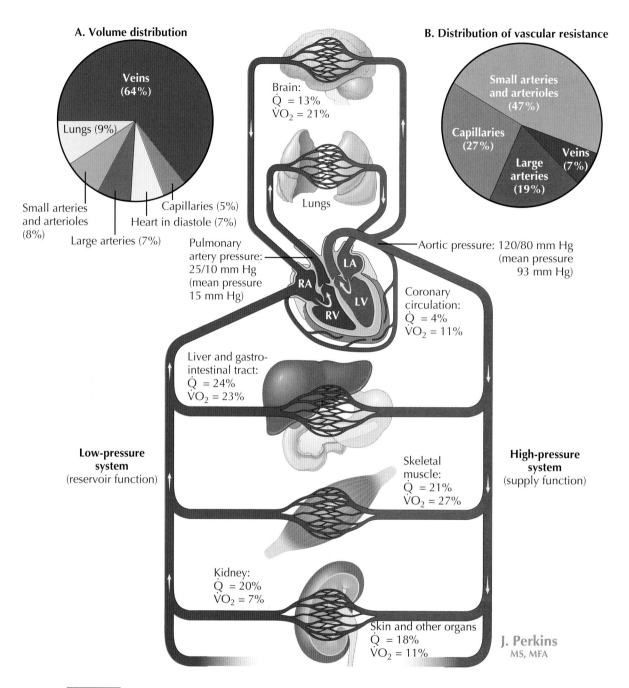

A. Volume distribution

Veins (64%)

Lungs (9%)

Small arteries and arterioles (8%)

Large arteries (7%)

Capillaries (5%)

Heart in diastole (7%)

B. Distribution of vascular resistance

Small arteries and arterioles (47%)

Capillaries (27%)

Large arteries (19%)

Veins (7%)

Brain:
$\dot{Q} = 13\%$
$\dot{V}O_2 = 21\%$

Lungs

Pulmonary artery pressure: 25/10 mm Hg (mean pressure 15 mm Hg)

Aortic pressure: 120/80 mm Hg (mean pressure 93 mm Hg)

LA
RA
LV
RV

Coronary circulation:
$\dot{Q} = 4\%$
$\dot{V}O_2 = 11\%$

Liver and gastro-intestinal tract:
$\dot{Q} = 24\%$
$\dot{V}O_2 = 23\%$

Low-pressure system (reservoir function)

Skeletal muscle:
$\dot{Q} = 21\%$
$\dot{V}O_2 = 27\%$

High-pressure system (supply function)

Kidney:
$\dot{Q} = 20\%$
$\dot{V}O_2 = 7\%$

Skin and other organs
$\dot{Q} = 18\%$
$\dot{V}O_2 = 11\%$

J. Perkins
MS, MFA

Figure 9.1 **General Scheme of the Circulation** The distribution of flow ($\dot{Q}$) to various organs as a percentage of cardiac output and the utilization of oxygen ($\dot{V}O_2$) relative to total O_2 consumption are illustrated. The pie charts show distribution of blood volume throughout the system **(A)** and the distribution of vascular resistance as a percentage of total resistance **(B)**. The greatest volume of blood in the cardiovascular system is in the systemic veins. The greatest resistance to flow occurs in small arteries and arterioles. All values in the figure are those observed at rest; these are adjusted to meet changes in physiological requirements, such as during exercise. *LA,* left atrium; *LV,* left ventricle; *RA,* right atrium; *RV,* right ventricle.

the sinoatrial node across the atria, depolarizing atrial muscle and initiating atrial contraction (Fig. 9.3). The atrioventricular node is the only pathway for spread of the depolarization to the ventricle. Electrical conduction through the atrioventricular node is slow, allowing atrial contraction to accomplish the final filling of the ventricles. Next, depolarization reaches the bundle of His, and depolarization proceeds rapidly through the left and right bundle branches to the Purkinje fibers and ventricular muscle, producing ventricular contraction.

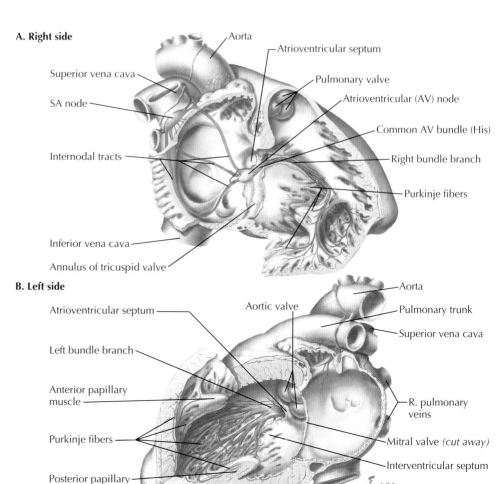

Figure 9.2 **Structure of the Heart** The muscular left (L) ventricle pumps blood into the high-pressure systemic circulation, and the less muscular right (R) ventricle pumps blood into the lower pressure pulmonary circulation. The mitral and tricuspid valves prevent backflow of blood into the atria, while the aortic and pulmonic valves prevent regurgitation of blood from the aorta and pulmonary artery to the ventricles.

Mitral valve
R. pulmonary vein
Ascending aorta
Aortic valve
Superior vena cava
Atrioventricular septum
R. atrium
Tricuspid valve
R. ventricle
Papillary muscles
Interventricular septum
L. ventricle
Mitral valve
L. pulmonary veins
L. atrium
Pulmonary trunk
Ascending aorta
Orifices of coronary arteries
R. atrium
Aortic valve
Outflow to pulmonary trunk
R. ventricle
Papillary muscle *(cut)*
L. ventricle
Plane of section

Figure 9.3 **Conduction System of the Heart** Contraction of cardiac muscle is initiated by electrical depolarization. The pacemaker for this electrical activity is in the sinoatrial (SA) node. Impulses travel from the SA node through internodal conducting pathways to the atrioventricular (AV) node and then spread through the His-Purkinje system of the ventricle. *R,* right.

A. Right side

Aorta
Superior vena cava
SA node
Internodal tracts
Inferior vena cava
Annulus of tricuspid valve
Atrioventricular septum
Pulmonary valve
Atrioventricular (AV) node
Common AV bundle (His)
Right bundle branch
Purkinje fibers

B. Left side

Atrioventricular septum
Left bundle branch
Anterior papillary muscle
Purkinje fibers
Posterior papillary muscle
Aortic valve
Aorta
Pulmonary trunk
Superior vena cava
R. pulmonary veins
Mitral valve *(cut away)*
Interventricular septum

CLINICAL CORRELATE 9.1
Cardiac Structural Changes in Disease

The normal structure of the heart may be altered in persons with valvular disease, systemic or pulmonary hypertension, heart failure, and other diseases. Common diseases affecting cardiac structure include the following:

■ Pulmonary hypertension: Right ventricular hypertrophy may occur in response to the chronically increased work of the right ventricle.

■ Aortic stenosis (narrowing of the aortic valve): Left ventricular hypertrophy occurs in response to the greater work performed by the ventricle.

■ Mitral incompetence: Left atrial dilation may develop as a result of the elevation of left atrial pressure and volume caused by mitral regurgitation (leakage of blood back into the atrium from the left ventricle).

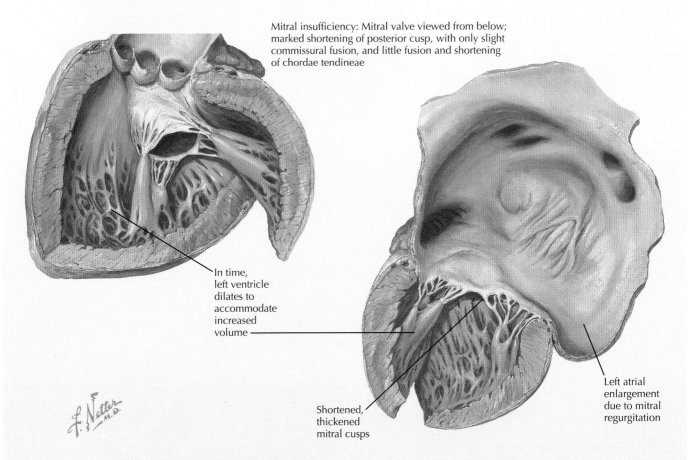

Mitral insufficiency: Mitral valve viewed from below; marked shortening of posterior cusp, with only slight commissural fusion, and little fusion and shortening of chordae tendineae

In time, left ventricle dilates to accommodate increased volume

Shortened, thickened mitral cusps

Left atrial enlargement due to mitral regurgitation

Cardiac Structural Changes in Mitral Regurgitation Incompetence of the mitral valve results in regurgitation of blood into the left atrium during systole. Chronic mitral regurgitation results in left atrial enlargement and left ventricular dilation.

Chapter 10

Cardiac Electrophysiology

Cells of the heart, like other excitable cells, have the ability to generate action potentials. These action potentials and their conduction throughout the heart are the basis of the rhythmic contraction and relaxation of the heart; the conduction of action potentials in the heart is recorded on the surface of the body as the electrocardiogram (ECG).

CARDIAC ACTION POTENTIALS

In a normal heart, the pacemaker activity of the **sinoatrial (SA) node** produces a resting heart rate of approximately 70 beats/min. The SA node is a group of specialized, noncontractile cardiomyocytes in the wall of the right atrium, adjacent to the opening of the superior vena cava. The resting membrane potential of SA node cells is initially around −60 millivolts (mV), but the cells undergo gradual, spontaneous depolarization as a result of an inward current, i_f, mainly carried by Na^+; an inward Ca^{2+} current, i_{Ca}; and reduced activity of an outward K^+ current. The *diastolic depolarization* resulting from the inward currents is responsible for the pacemaker activity of the SA node cells. Once threshold is reached, the upstroke of the action potential occurs as a result of opening of T-type and L-type Ca^{2+} channels (Fig. 10.1). Repolarization is caused by increased conductance of K^+ and closing of Ca^{2+} channels, completing the pacemaker cycle. Action potentials of **atrioventricular (AV) node cells**, found in the atrial septum at the junction of the atria and ventricles, are similar to those in the SA node, although the pacemaker activity of these cells is slower, overridden in normally functioning hearts by SA pacemaker activity.

In contrast to SA and AV nodal cells, depolarization of other cardiac cells largely depends on Na^+ influx rather than Ca^{2+} influx. Action potentials of ventricular and atrial myocytes and cells of the His-Purkinje system are characterized by five distinct phases (see Fig. 10.1):

- **Phase 4 (the resting membrane potential):** The resting membrane potential of these cells is mainly a function of K^+ efflux and is close to the Nernst potential for K^+. During phase 4, ion concentrations that were altered by the previous action potential are restored to resting levels by the Na^+/K^+ ATPase and an Na^+/Ca^{2+} exchanger, as well as an ATP-dependent Ca^{2+} pump.
- **Phase 0 (upstroke of the action potential):** Rapid depolarization occurs when the cells reach threshold and fast channels for Na^+ open. This mechanism is accompanied by reduced conductance of the inwardly rectified K^+ current (i_{K1}). The rapid upstroke results in rapid spread of depolarization through most of the conducting system of the heart.
- **Phase 1 (rapid repolarization to the plateau):** This phase is caused by inactivation of Na^+ channels and opening of voltage-sensitive K^+ channels, producing a transient outward K^+ current (i_{TO}).
- **Phase 2 (the plateau):** The membrane remains depolarized during the plateau because of opening of voltage-sensitive, slow L-type Ca^{2+} channels and inward current of Ca^{2+}. Simultaneously, an outward K^+ current occurs through a voltage-dependent K^+ channel (delayed rectifier K^+ channel).
- **Phase 3 (repolarization):** Gradual inactivation of the L-type Ca^{2+} channels leads eventually to activation of K^+ channels, causing rapid repolarization, substantially because of an inwardly rectified K^+ current (i_{K1}).

The plateau of the action potential is functionally important, because it normally prevents premature depolarization of cardiac cells and associated arrhythmias. There is an **effective refractory period**, beginning with phase 1 and extending through the plateau and much of phase 3, during which another action potential cannot be generated. This is followed by the **relative refractory period**, which lasts until the resting membrane potential is fully restored; during this period, it is more difficult to elicit another action potential than in a resting cell.

Chronotropic, dromotropic, and inotropic effects are those that alter heart rate, conduction velocity, and myocardial contractility, respectively. A positive chronotropic agent is a drug that increases heart rate. Stimulation of the sympathetic nervous system increases the chronotropic, dromotropic, and inotropic state of the heart, whereas the parasympathetic nervous system mainly reduces the chronotropic and dromotropic state.

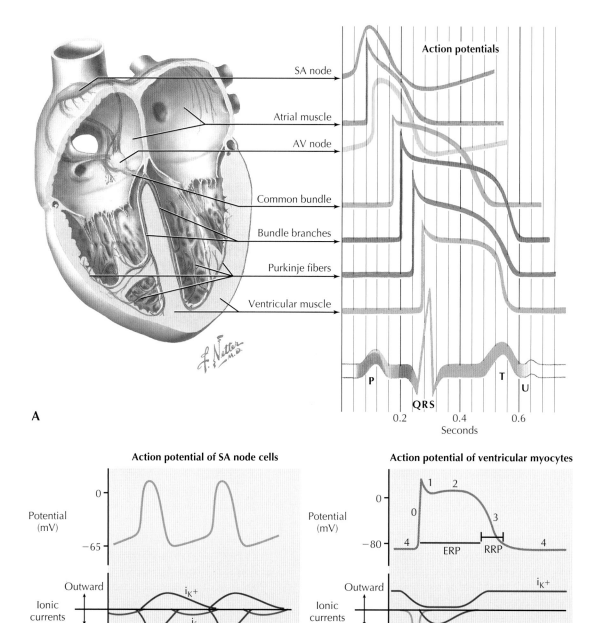

Figure 10.1 **Electrical Activity of the Heart** Action potentials vary as the wave of depolarization proceeds through the conducting pathway of the heart, beginning at the sinoatrial (SA) node and ending in the ventricular muscle. The electrocardiogram (ECG) records changes in surface potential on the body associated with depolarization and repolarization of the heart. P waves, QRS complexes, and T waves constitute the normal ECG. In 50% to 70% of persons, a U wave will also be recorded. Its origin is not well defined **(A)**. Differences in structure of the action potentials are based on differences in specific ion currents **(B)**. In atrioventricular (AV) nodal cells, note the spontaneous depolarization from the resting membrane potential that results in the action potential. In ventricular myocytes (and also the His-Purkinje system), note the sharp phase 0 upstroke that produces rapid conduction of the wave of depolarization. The phase 2 plateau is important in preventing premature depolarization. The effective refractory period (ERP) is the period during which another action potential cannot be elicited; during the relative refractory period (RRP), it is more difficult to elicit an action potential than during phase 4.

During the conduction of the wave of depolarization of the heart, forward propagation occurs as a result of the local currents created during action potentials, and unidirectional depolarization is maintained due to the refractory periods of cells once they have depolarized.

THE ELECTROCARDIOGRAM

An ECG is a recording of changes in surface potential on the body that is produced by the depolarization and repolarization of the heart. Electrocardiography may reveal the following:

- Abnormal cardiac rhythms and conduction
- Presence, location, and extent of ischemia or infarction
- Orientation of the heart in the thoracic cavity and size of chambers
- Effects of abnormal electrolyte levels and some drugs

These observations are facilitated by recording of the ECG with multiple lead configurations (Fig. 10.2). A 12-lead ECG includes standard leads I, II, and III; six unipolar leads (V_1 to V_6); and three augmented limb leads (aVL, aVR, and aVF). A normal ECG consists of:

- a P wave, caused by atrial depolarization
- a QRS complex, representing ventricular depolarization
- a T wave, representing ventricular repolarization

In a normal ECG, each P wave leads to a QRS complex and T wave. The heart is considered to be in "sinus rhythm," because the rhythm is being regulated by the SA node. **Bradycardia** refers to a resting heart rate of less than 60 beats per minute (beats/min), whereas **tachycardia** is a resting heart rate of greater than 100 beats/min. Some causes of bradyarrhythmias in the presence of normal sinus rhythm include high vagal (parasympathetic) tone; drugs; and various metabolic, endocrine, and neurologic disorders. Some causes of tachyarrhythmias with sinus rhythm include high sympathetic tone; drugs; low blood volume (dehydration); and metabolic, endocrine, and neurologic disorders. The specific electrical events leading to the components of the ECG are detailed in Figures 10.3 to 10.5.

CLINICAL CORRELATE 10.1
Atrioventricular Block

The normal conducting pathway in the heart is:

SA → AV → Bundle of His → Bundle branches → Purkinje fibers → Ventricular muscle

When conduction is disrupted in this pathway, arrhythmias result. AV block (heart block) is a group of cardiac arrhythmias associated with altered conduction through the AV node:

- First-degree AV block is abnormally delayed conduction through the AV node (PR interval on the ECG exceeding 200 milliseconds [msec]). Each P wave results in a QRS complex and thus ventricular contraction, as in normal sinus rhythm, but AV conduction is delayed. By itself, first-degree block does not produce bradycardia and is usually benign if it is not caused by underlying heart disease.
- Second-degree AV block is characterized by intermittent conduction through the AV node. As a result, some P waves on the ECG are followed by a QRS complex, whereas others are not.

Thus the ventricles contract less frequently than the atria. Myocardial ischemia or infarction may produce second-degree block.

- Third-degree AV block (complete block) occurs when the wave of depolarization is not conducted through the AV node (from atria to ventricles). On the ECG, both P-P intervals and R-R intervals are regular, but P waves are dissociated from the QRS complexes. When atrial pacemaker activity fails to be conducted to the ventricle, an escape pacemaker may arise in the AV node below the site of block or in the bundle of His, producing a ventricular rate of 40 to 55 beats/min (as determined by the R-R interval) that is only partially responsive to sympathetic stimulation. When the conduction block occurs in the bundle of His, the ventricular escape rhythm will produce an inadequate heart rate of 20 to 40 beats/min; cardiac output (flow rate from one ventricle) may be insufficient even at rest, and activity will be greatly restricted because of the inability to adequately adjust output. The usual therapy is implantation of a cardiac pacemaker.

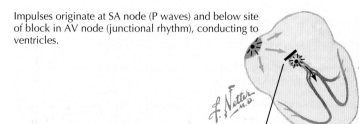

Impulses originate at SA node (P waves) and below site of block in AV node (junctional rhythm), conducting to ventricles.

Block

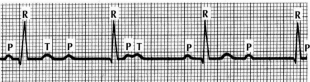

Atria and ventricles depolarize independently. QRS complexes less frequent; regular at 40 to 55 beats/min but normal in shape.

Third-Degree Heart Block

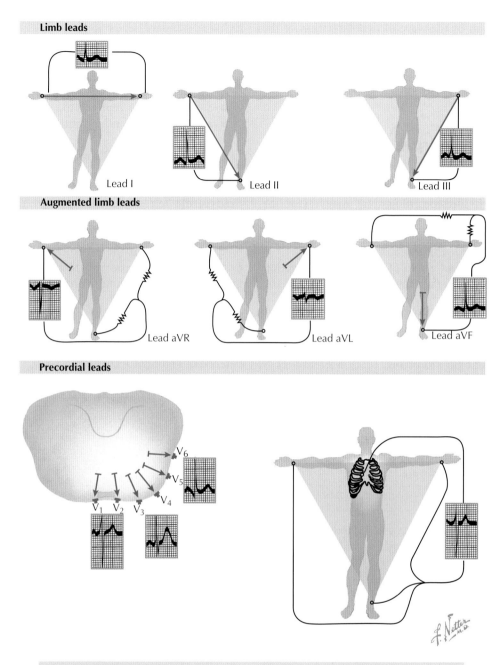

Limb leads

Lead I

Lead II

Lead III

Augmented limb leads

Lead aVR

Lead aVL

Lead aVF

Precordial leads

V₆

V₅

V₄

V₁ V₂ V₃

When current flows toward red arrowheads, upward deflection occurs in ECG

When current flows away from red arrowheads, downward deflection occurs in ECG

When current flows perpendicular to red arrows, no deflection or biphasic deflection occurs

Figure 10.2 **Electrocardiographic Leads** In electrocardiography, electrodes placed on the skin are used to record, on the surface of the body, electrical activity resulting from the depolarization and repolarization of the heart. In standard limb leads I, II, and III, voltage differences are recorded between the right arm and left arm, right arm and left leg, and left arm and left leg, respectively, with the first in each pair of electrodes being negative and the second being positive. For the three augmented leads, the negative electrode is a combination of two limb electrodes, and the third limb electrode is positive. For the six precordial leads, the three limb electrodes are combined as the negative electrode, and the positive electrodes are placed directly on the chest in specified locations. Analysis of specific types of disease is aided by comparison of tracings made with multiple leads.

Normal Sequence of Cardiac Depolarization and Repolarization and Derivation of ECG

A. Impulse origin and atrial depolarization
Impulse originates at SA node, and wave of depolarization spreads over atria, resulting in electrical vector directed downward and to left. This causes upward (positive) deflection in ECG tracing in leads I and aVF (P wave).

B. Septal depolarization
After brief delay at AV node, impulse traverses common bundle of His and right and left bundle branches and then enters interventricular septum, causing myocardial depolarization with electrical vector directed to right and downward. This results in small negative (downward) deflection in lead I (Q wave) and positive (upward) deflection in lead aVF (R wave).

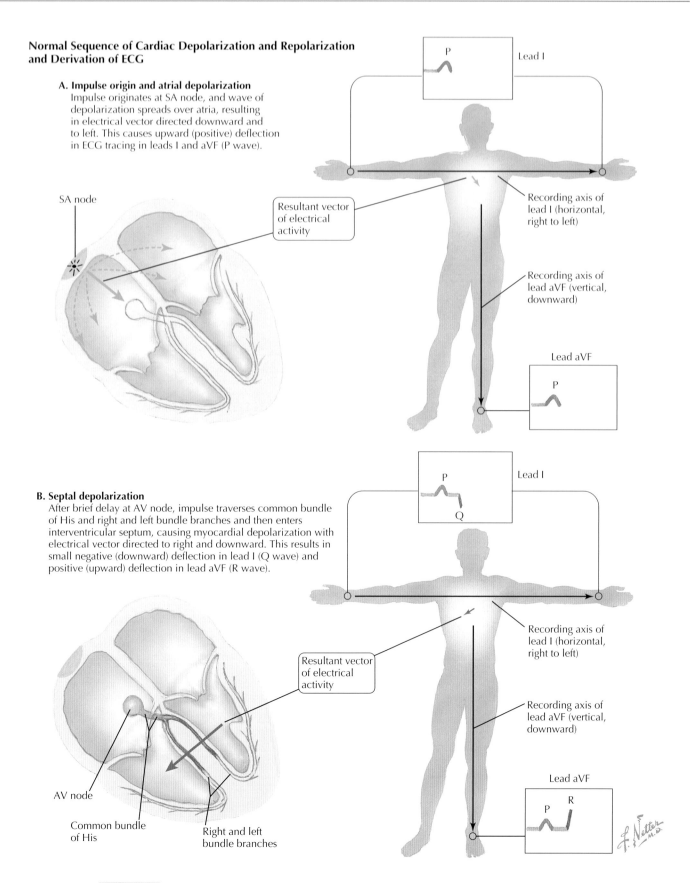

Figure 10.3 **Cardiac Depolarization and Repolarization Part I** Figures 10.3 to 10.5 depict the sequence of cardiac electrical events during the cardiac cycle and the surface recordings that result from these events and constitute the electrocardiogram (ECG). In Figure 10.3 the initial events are depicted, in which the firing of the sinoatrial (SA) node results in depolarization of the atria, which is recorded on the ECG as the P wave. This depolarization spreads to the atrioventricular (AV) node, the bundle of His, and bundle branches, with the resultant electrical activity vectors and surface ECG recordings. Note that the morphology of the ECG recording varies with lead placement.

Normal Sequence of Cardiac Depolarization and Repolarization and Derivation of ECG (continued)

C. Apical and early ventricular depolarization
Impulse continues along conduction system, causing depolarization of apical ventricular myocardium with electrical vector directed downward and to left. This results in large positive (upward) deflection (R wave) in lead I and extends R wave in lead aVF.

D. Late ventricular depolarization
As depolarization progresses over ventricles, vector shifts to become directed superiorly as well as to left, thus extending upward R wave in lead I and causing negative (downward) deflection (S wave) in lead aVF.

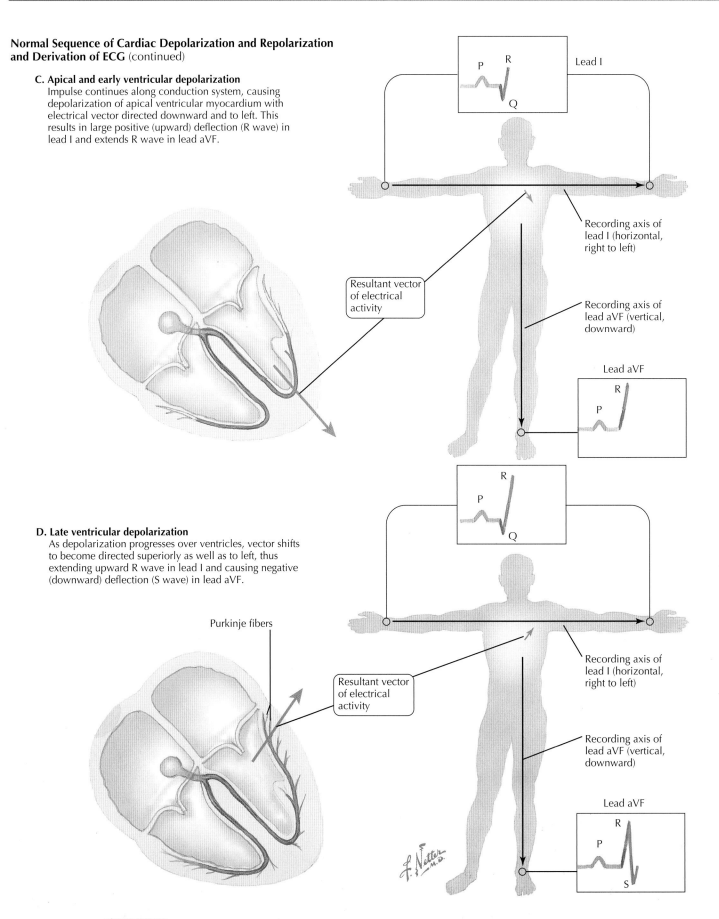

Figure 10.4 **Cardiac Depolarization and Repolarization Part II** Septal depolarization (see Fig. 10.3) is followed by apical and ventricular wall depolarization; resultant vectors of electrical activity produce the QRS complex of the electrocardiogram (ECG).

Normal Sequence of Cardiac Depolarization and Repolarization and Derivation of ECG (continued)

E. Repolarization

When heart is fully depolarized, there is no electrical activity for brief period (ST segment). Then repolarization begins from epicardium to endocardium, producing electrical vector directed downward and to left, causing upward (positive) deflection in both leads I and aVF (T waves). A period of no electrical activity follows, with tracing at baseline until next impulse originates at SA node.

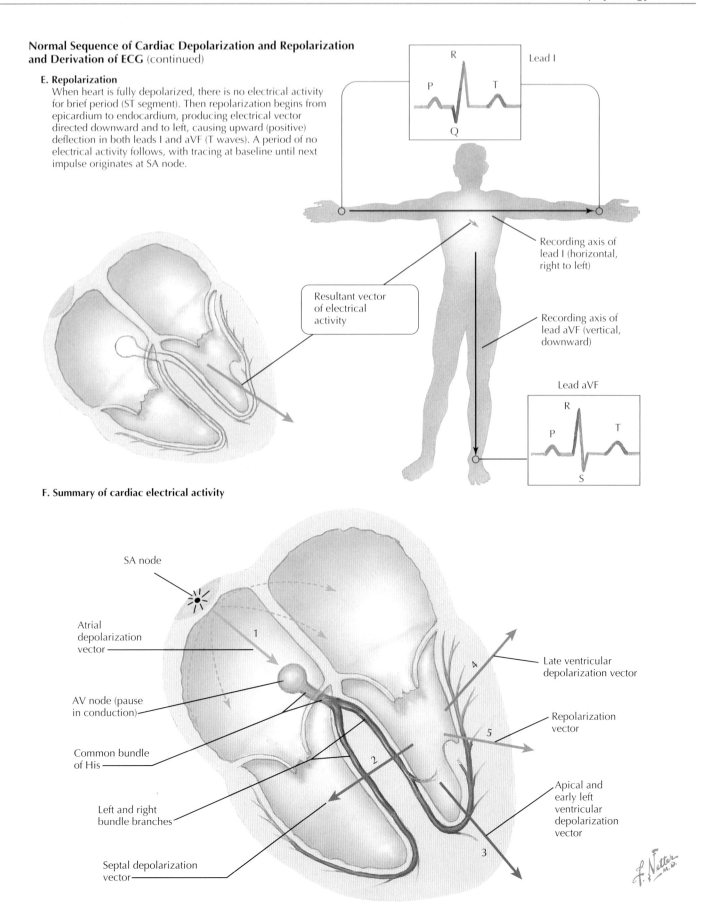

Resultant vector of electrical activity

Recording axis of lead I (horizontal, right to left)

Recording axis of lead aVF (vertical, downward)

F. Summary of cardiac electrical activity

SA node

Atrial depolarization vector

AV node (pause in conduction)

Common bundle of His

Left and right bundle branches

Septal depolarization vector

Late ventricular depolarization vector

Repolarization vector

Apical and early left ventricular depolarization vector

Figure 10.5 **Cardiac Depolarization and Repolarization Part III**   Ventricular depolarization (see Fig. 10.4) is followed by ventricular repolarization; the resultant vector of electrical activity produces the T wave of the electrocardiogram (ECG). The cardiac electrical activity and resultant vectors *(1-5)* through the cycle are summarized in part **F**. *AV*, atrioventricular; *SA*, sinoatrial.

Chapter 11

Flow, Pressure, and Resistance

Hemodynamics is the study of the forces involved in circulation of blood. Although arterial blood pressure is a convenient and readily measured parameter, understanding a person's broader hemodynamic state is essential when assessing cardiovascular disease.

BASIC HEMODYNAMICS

The rate of flow through the circulation (Q) is determined by the pressure gradient across the circulation (ΔP) and the resistance of the circulation (R):

$$Q = \Delta P/R \qquad \textbf{Eq. 11.1}$$

Pressure is defined as force per unit area. For example, in an auto tire, pressure is measured as pounds per square inch. The unit for measurement of blood pressure is most often millimeters of mercury (mm Hg; 1 mm Hg = 1 torr), and the pressure gradient (ΔP) for flow is arterial pressure minus venous pressure. Resistance is the impedance to flow and can be measured in units of mm Hg/milliliter (mL)/minute. Resistance can be quantified as the pressure rise associated with an incremental rise in flow. The greatest resistance in the circulation occurs in the smallest arteries and arterioles (see Fig. 9.1). Factors that determine resistance are considered later in this chapter (see "Biophysics of Circulation").

BLOOD PRESSURE

The blood pressures in the venous systems of both pulmonary and systemic circulations are considerably lower than are pressures in the respective arterial systems. In addition, blood pressures in the pulmonary circulation are lower than corresponding pressures in the systemic circulations. Systemic arterial pressure at rest is normally about 120/80 mm Hg (systolic/diastolic), compared with 25/10 mm Hg for pulmonary artery pressure. Some important definitions related to arterial pressure are as follows:

- **Systolic arterial pressure:** Peak arterial pressure reached during ejection of blood by the heart
- **Diastolic arterial pressure:** Lowest arterial pressure reached during diastole, while the heart is relaxed and filling (not ejecting blood)

 Flow, pressure, and resistance can be related in the formula:

$$Q = \Delta P/R$$

where Q is flow, ΔP is the pressure gradient, and R is resistance to flow.

This formula can be rearranged as:

$$\Delta P = QR$$

This equation is analogous to Ohm's law, which states that V = IR, where V is electrical potential (electrical gradient), I is current (electrical flow), and R is impedance (electrical resistance).

- **Arterial pulse pressure:** The difference between systolic and diastolic arterial pressures; dependent on stroke volume (volume ejected by one ventricle during one contraction), resistance, and arterial compliance
- **Mean arterial pressure (MAP):** The average pressure over a complete cardiac cycle of systole and diastole; dependent on peripheral resistance and cardiac output (volume ejected by one ventricle per unit time)

MAP is not the simple arithmetic mean of systolic and diastolic pressures because of the irregular shape of the arterial pressure curve (Fig. 11.1). MAP can be approximated by adding one third of the pulse pressure to the diastolic pressure.

Physiological pressures are usually given in units of mm Hg or cm H_2O. In other words, 1 mm Hg is the pressure that would support a column of mercury (Hg) at a height of 1 millimeter (mm), and 1 cm H_2O is the pressure that would support a column of water 1 centimeter (cm) high (1 mm Hg = 1.36 cm H_2O). Another way to conceptualize these units is that a column of water 1 cm high would exert 1 cm H_2O pressure at its base, and a column of mercury 1 mm high would exert 1 mm Hg pressure at its base. Although these may appear to be unconventional units for quantifying pressure, they are useful because water or mercury manometers are often used to measure pressure. In some texts, the unit *torr* is used; it is equal to 1 mm Hg.

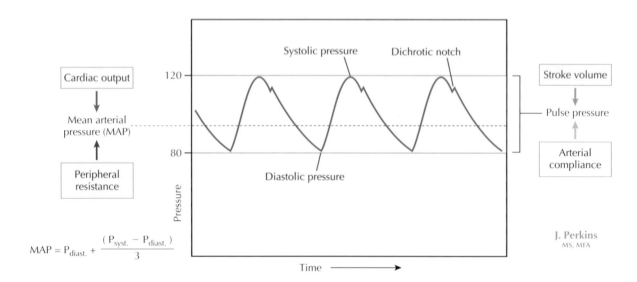

$$MAP = P_{diast.} + \frac{(P_{syst.} - P_{diast.})}{3}$$

J. Perkins
MS, MFA

Figure 11.1 **The Arterial Pressure Wave** The arterial pressure wave represents the changes in pressure in the arterial system over periods of systole (during which the stroke volume is ejected from the left ventricle) and diastole (during which the heart is refilling and blood in the arterial system continues to flow downstream). Arterial pressure is affected by cardiac output, stroke volume, arterial compliance, and peripheral resistance. Mean arterial pressure can be approximated based on the formula shown. The first upward deflection in the curve marks the beginning of systole, the period of cardiac ejection. The transient irregularity in the downward slope of the wave is known as the dicrotic notch, which is produced by closure of the aortic valve, marking the beginning of diastole. During diastole, the heart refills, while blood in the arterial system runs downstream, reducing arterial pressure.

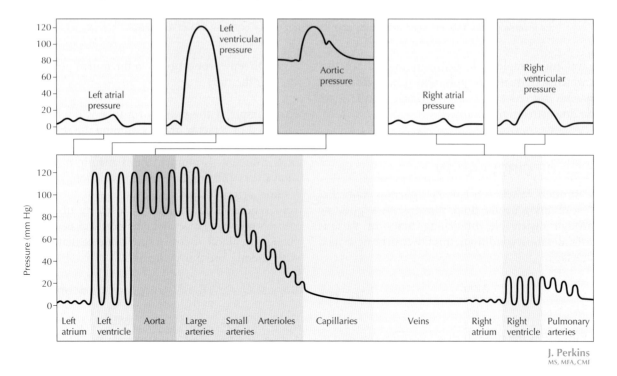

J. Perkins
MS, MFA, CMI

Figure 11.2 **Pulse Pressures Through the Circulation** Changes in pressure pulses are illustrated as blood flows from the left ventricle to the aorta, through the systemic circulation, and back to the right ventricle and pulmonary artery.

Pressure waves vary through the cardiovascular system (Fig. 11.2). Both high and low pressures are encountered in the ventricles (approximately 120/0 and 25/0 mm Hg at rest for left and right ventricles, respectively). High systolic pressure is necessary for pumping of blood through the circulation, whereas the low pressures are required for return of blood to the heart during diastole. Note that whereas mean pressures in the large arteries are somewhat lower than in the aorta, pressure pulsations are greater in larger arteries. This phenomenon is mainly attributable to two factors: changes in

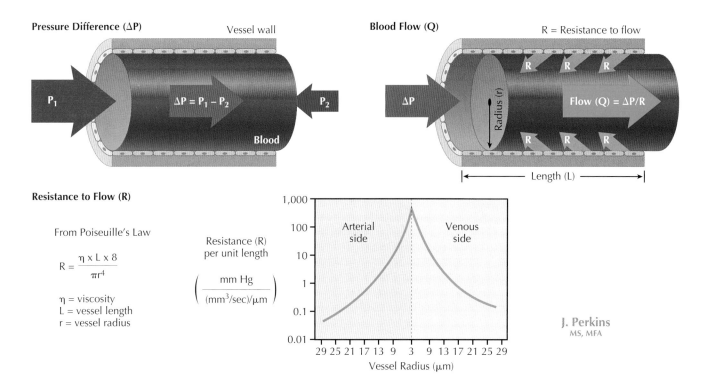

Figure 11.3 **Poiseuille's Law** Under normal physiological conditions, flow through vessels is governed by Poiseuille's law. By far, the greatest resistance to flow occurs in the smallest vessels that constitute the microcirculation, particularly the arterioles. This phenomenon is due to the small radius of these vessels and the inverse relationship between resistance and the fourth power of the radius of a vessel.

pressure travel more rapidly downstream than the actual blood flow, which accentuates the pulsatility downstream, and pressure changes are reflected back at branch points, again accentuating downstream pulsatility.

BIOPHYSICS OF CIRCULATION

Blood flow through vessels is a complex phenomenon that involves a nonhomogeneous fluid flowing in a pulsatile manner through distensible, branching tubes of various dimensions. Under most conditions, this flow can be described by Poiseuille's law:

$$Q = \frac{\Delta P \pi r^4}{\eta 8L}$$ **Eq. 11.2**

where Q is flow, ΔP is the pressure gradient from one end of a tube to the other, r^4 is the radius of the tube to the fourth power, η is the viscosity of the fluid, and L is the length of the tube. The effects of this relationship are illustrated in Figure 11.3.

Based on Poiseuille's law, flow (Q) through a tube will be:

- Directly proportional to the longitudinal pressure gradient (inflow pressure minus outflow pressure)
- Inversely proportional to the length of the tube
- Inversely proportional to the viscosity of the fluid

- Directly proportional to the fourth power of the radius of the tube

For example, if the radius of a tube is doubled, flow will be increased by a factor of 16, assuming the pressure gradient is maintained. Physiological regulation of regional blood flow on a moment-to-moment basis mainly involves changes in radius (vasodilation and vasoconstriction) of the small arteries and arterioles, taking advantage of this powerful factor. Under normal circumstances, viscosity of blood is not an issue; however, changes in hematocrit are associated with large changes in blood viscosity, such as those that occur in persons with anemia and polycythemia.

Because $Q = \Delta P/R$, resistance can be described as:

$$R = \frac{\eta 8L}{\pi r^4}$$ **Eq. 11.3**

💡 Of the factors affecting flow through a tube, the most important is the radius of the tube. Whereas flow is inversely proportional to the length of the tube and the viscosity of the fluid and directly proportional to the hydrostatic pressure gradient in the tube, it is directly proportional to the fourth power of the radius of the tube. Therefore, doubling the radius of the tube will cause a 16-fold increase in flow, if other factors are constant.

CLINICAL CORRELATE 11.1
Measurements of Blood Pressure

Arterial pressure is routinely measured by sphygmomanometry, in which a blood pressure cuff is inflated above the systolic arterial pressure, compressing vessels and stopping blood flow. As pressure in the cuff is gradually released, the practitioner listens for the **sounds of Korotkoff** through a stethoscope. These sounds are produced by the pulsatile flow of blood in arteries under the cuff when the cuff pressure is between the systolic and diastolic arterial pressures. Thus the sounds are first heard when cuff pressure falls below systolic pressure, and the sounds disappear when the cuff pressure falls below diastolic pressure. Arterial pressure can also be measured directly through an arterial catheter. Such a catheter can be passed in a retrograde direction (against the flow of blood) for measurement of pressures in the arteries, aorta, and left ventricle, as the catheter advances toward the heart.

Although arterial pressure monitoring is important for detection of hypertension, monitoring of additional pressures is often highly desirable in surgical and intensive care settings. Pulmonary capillary wedge pressure ("wedge pressure") is one such measurement. A venous catheter can be passed antegrade (in the direction of blood flow) for measurement of pressures in the veins, right atrium, and right ventricle. However, it is not possible to measure

pulmonary venous and left atrial pressures by these direct methods. Instead, we must rely on measurement of pulmonary capillary wedge pressure.

With the aid of a partially inflated balloon, a catheter is passed from a systemic vein, through the right heart, into the pulmonary artery, to a branch of the pulmonary artery. The lumen opens at the distal end of the catheter and is continuous; the balloon is on the outside of the catheter only. When the catheter can be advanced no further, the balloon is fully inflated (the catheter is actually a multilumen Swan-Ganz catheter; one lumen is continuous to the end of the catheter, another leads to the balloon, thus allowing inflation, and other lumens may open along the length of the catheter). Inflating the balloon has no effect on the patency of the catheter lumen through which pressure is measured. When the balloon is fully inflated, pressure falls downstream beyond that point. Vascular pressure beyond the point of occlusion equilibrates with downstream pressure, and wedge pressure (measured at the catheter tip) is thus an indicator of pulmonary venous and left atrial pressure. It also approximates left ventricular end-diastolic pressure (LVEDP), when left ventricular pressure has equilibrated with the pressure in the left atrium and pulmonary veins with filling of the ventricle. Wedge pressure is useful in hemodynamic assessment (e.g., in acute heart failure).

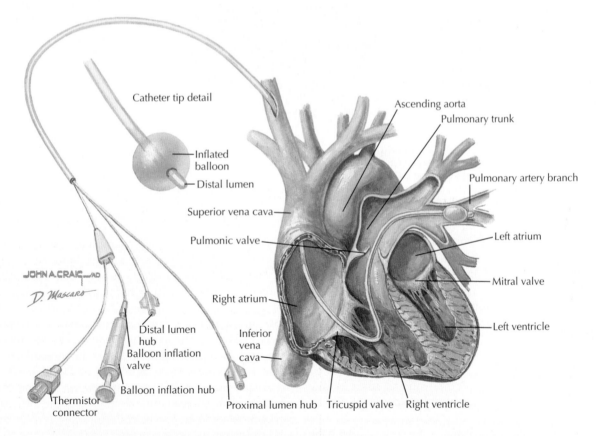

Pulmonary Artery Catheterization   Pulmonary artery catheters (Swan-Ganz catheters) are multilumen catheters that can be advanced into a branch of the pulmonary artery. With the tip thus wedged with the balloon inflated, flow is occluded and pressure can be measured beyond the balloon occlusion through one of the lumens. The measured *pulmonary capillary wedge pressure* is an approximation of left ventricular end-diastolic pressure.

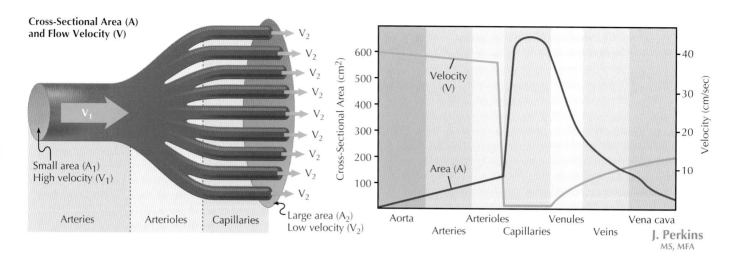

Figure 11.4 **Relationship Between Velocity of Blood Flow and Cross-Sectional Area** Proceeding downstream from the aorta, branching of arterial vessels increases total cross-sectional area and thus results in diminished velocity of blood flow from the aorta to the capillaries. Velocity increases from the capillaries to the large veins with the confluence of vessels and the resulting decrease in total cross-sectional area.

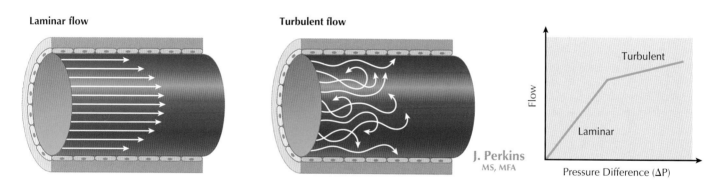

Figure 11.5 **Laminar and Turbulent Flow** Normally, laminar, or streamlined, flow occurs throughout most of the vascular system. Pathologic conditions such as coarctation (narrowing of a vessel), valvular abnormalities, and low blood viscosity (as in anemia) produce turbulence and are associated with murmurs in the heart or great vessels or vascular bruits heard by auscultation. Turbulence produces an increase in the pressure gradient required to produce flow.

The following key relationships must be remembered:

■ Resistance is directly proportional to the length of the tube.
■ Resistance is directly proportional to the viscosity of the fluid.
■ Resistance is inversely proportional to the fourth power of the radius of the tube.

Flow (Q) can also be related to cross-sectional area (A) and linear velocity of the flow (V) in this formula:

$$Q = VA \qquad \textbf{Eq. 11.4}$$

In the cardiovascular system, the overall flow rate (cardiac output) is the same at every level of the circulation, because the circulation is in series. Thus, based on the Q = VA relationship, the velocity of blood flow is slowest in the capillaries, which have the greatest aggregate cross-sectional area

(600 cm²) in the vascular system, and fastest in the aorta, where the cross-sectional area is only about 4 cm² (Fig. 11.4). The low velocity in capillaries is beneficial in terms of exchange of dissolved substances between blood and tissues.

Flow through vessels is generally laminar, with the greatest velocity of flow in the center of the vessel and the lowest velocity near the vascular wall (Fig. 11.5). This streamlining of flow is caused by shear stress produced as blood flows past the stationary vessel wall. Turbulent flow, on the other hand, is characterized by irregularities in flow patterns, such as whorls, vortices, and eddies. Vascular disease often is associated with sites of turbulence.

Reynolds number (R_e) relates the factors associated with turbulence:

$$R_e = \frac{VD\rho}{\eta} \qquad \textbf{Eq. 11.5}$$

where V is velocity of flow, D is the diameter of the tube, ρ is the density of the fluid, and η is the viscosity of the fluid. When R_e is below 2000, flow is usually laminar; values above 3000 result in turbulence. In the cardiovascular system, density of blood is always close to 1.0 and is not a factor; changes in the other variables are important, along with a number of other factors. Turbulence in blood vessels is associated with audible murmurs and is promoted by:

- High velocity of blood flow
- Large vessel diameter
- Low viscosity of blood (low hematocrit)
- Abrupt changes in vessel diameter, for example, in aneurysms (see Clinical Correlate 11.2) or coarctation (narrowing)
- Vascular branch points

Wall tension (T) is another important biophysical parameter in this system and can be conceptualized as the force necessary to hold together a theoretical slit occurring in a vessel wall. Wall tension is defined by Laplace's law:

$$T = P_t r \qquad \textbf{Eq. 11.6}$$

where P_t is the transmural pressure (i.e., the difference between pressure inside and outside the vessel; in other words, the pressure gradient across the vascular wall) and r is the vessel radius. Because of the small radii of capillaries and venules, they are protected against rupture despite the transmural pressure gradient. Wall tension is an important consideration in large arteries, where transmural pressures are high and radii are large, especially when the vascular wall is diseased (see Clinical Correlate 11.2).

CLINICAL CORRELATE 11.2
Aortic Aneurysm

An aortic aneurysm is a saclike enlargement in the wall of the aorta that is often associated with atherosclerosis. With bulging, wall stress is increased through increased radius of the vessel, according to Laplace's equation ($T = P_t r$). Increased radius is accompanied by decreased wall thickness, further reducing the ability of the wall to withstand stress, Risk factors for aortic aneurysm include hypertension, smoking, obesity, atherosclerosis, and hypercholesterolemia. Rupture of an aortic aneurysm is a medical emergency involving profuse internal hemorrhage and is often fatal. Dissection of an aneurysm is defined as bleeding into the vascular wall through a tear in the inner layer of the vessel (tunica intima), a common complication in thoracic aortic aneurysms. Aneurysms are treated by replacement of the affected aortic segment by a synthetic graft or by endovascular implantation of a graft in the area of the aneurysm.

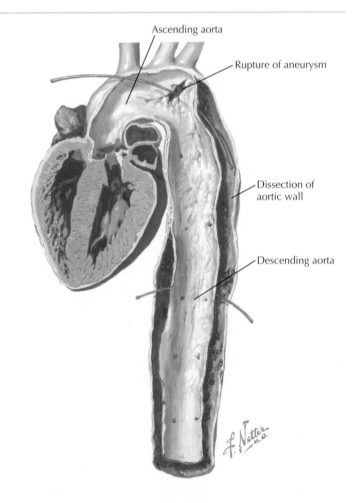

Ruptured and Dissecting Aortic Aneurysm

Chapter 12

The Cardiac Pump

Based on the biophysical principles described in Chapter 11, blood flows through the systemic and pulmonary circulations, with the energy for this flow being provided by the pumping action of the heart. Like early mechanical pumps, the heart functions in a cyclic manner, with periods of filling (diastole) and periods of contraction and ejection of blood (systole).

THE CARDIAC CYCLE

The cardiac cycle (or Wiggers diagram) consists of one cycle of ventricular systole and diastole. At rest the cycle is 0.86 seconds in duration if the heart rate is 70 beats per minute (beats/min). The changes in ventricular pressure and volume, aortic pressure and flow, atrial pressure, venous pulse, electrocardiogram (ECG), and phonocardiogram are all interdependent, and understanding the interrelationships among these variables is a key step in comprehending the complexities of hemodynamics (Fig. 12.1 and Video 12.1).

Left ventricular volume and pressures in the left heart and aorta are illustrated in Figure 12.1. Tracings for the right ventricular and pulmonary artery pressures are similar in shape, but the actual pressures are lower than those in the left ventricle and aorta. Within this cardiac cycle diagram are two short intervals known as periods of **isovolumetric contraction** and **isovolumetric relaxation**. During these periods, all of the valves of the heart are closed. The isovolumetric contraction period for the left heart begins with mitral valve closure (the aortic valve is already closed) and ends with opening of the aortic valve; the isovolumetric relaxation period begins with aortic valve closure and ends when the mitral valve reopens.

There is some slight asynchrony between the left heart and the right heart in terms of valve opening and closure, although the sequences are the same. The sequence of valve opening and closure during the cardiac cycle is as follows:

- *Left heart:* Mitral closure, aortic opening, aortic closing, mitral opening
- *Right heart:* Tricuspid closure, pulmonic opening, pulmonic closure, tricuspid opening

- *Aggregate sequence:* Mitral closure, tricuspid closure, pulmonic opening, aortic opening, aortic closure, pulmonic closure, tricuspid opening, mitral opening

The asynchrony between the left- and right-side valves is caused by differences in the pressure gradients between the sides of the circulation, and the degree of asynchrony varies during the respiratory cycle because of the effects of thoracic pressure on filling and pressures within the heart and central circulation.

Isovolumetric periods in the cardiac cycle refer to short intervals when all of the valves are closed. On the left side of the heart during isovolumetric contraction, ventricular pressure exceeds atrial pressure, and thus the mitral valve has closed. Ventricular pressure rises as the heart contracts, but it has not yet exceeded aortic pressure, and thus the aortic valve is still closed. During isovolumetric relaxation, ventricular pressure is less than aortic pressure, and thus the aortic valve is closed. However, ventricular pressure has not yet fallen below atrial pressure, and therefore the mitral valve is still closed. The overall sequence in the cardiac cycle in the ventricles consists of isovolumetric contraction, followed by ejection, isovolumetric relaxation, ventricular filling, isovolumetric contraction, and so forth.

Ventricular Pressure

The cardiac cycle is initiated by the P wave of the ECG. Depolarization of the atria results in atrial contraction and final filling of the ventricles, which accounts for the slight increase in **left ventricular pressure (LVP)**. When the wave of depolarization reaches the ventricle (QRS complex of the ECG) and ventricular contraction is initiated, LVP rises above **left atrial pressure (LAP)**, producing mitral valve closure and marking the beginning of isovolumetric contraction. **Left ventricular end-diastolic pressure** is normally about 5 mm Hg (right ventricular end-diastolic pressure is normally about 2 mm Hg). From this point, LVP rises rapidly as the ventricle attempts to contract against a fixed volume. When LVP rises above aortic pressure, the aortic valve opens, marking the end of isovolumetric contraction and the

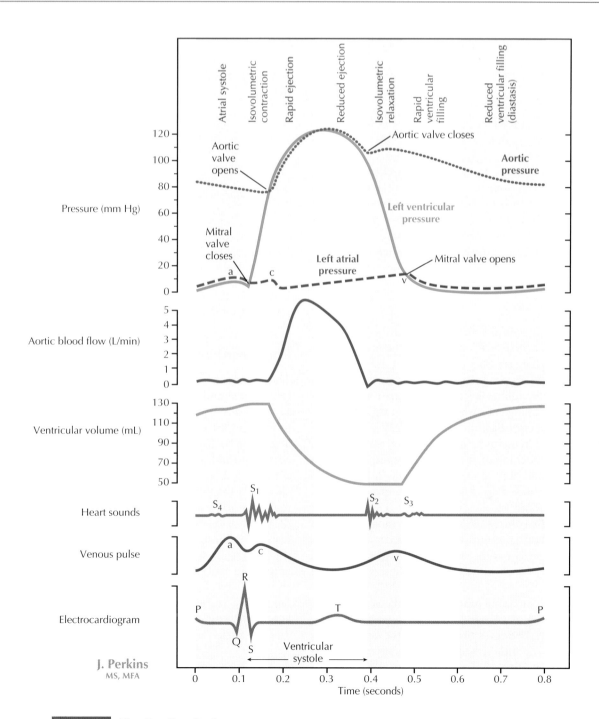

Figure 12.1 **The Cardiac Cycle** The cardiac cycle (or Wiggers diagram) is a simultaneous depiction of several parameters related to blood flow and volume, the electrocardiogram, and the echocardiogram through a cycle of cardiac systole and diastole. Ventricular systole begins with a short period of isovolumetric contraction, during which all of the heart valves are closed. This period is followed by the ejection phase of systole and then isovolumetric relaxation, when the heart valves are again all closed. Cardiac filling occurs after isovolumetric relaxation. The temporal relationships between features of these various curves or tracings are predictable, based on functional relationships within the cycle. The a, c, and v waves are identified on both the venous pulse and atrial pressure curves.

beginning of ventricular ejection. If aortic pressure is 120/80 mm Hg, this switch occurs at 80 mm Hg. LVP continues to rise for a period, but as the stroke volume is ejected, LVP reaches a peak (at approximately 120 mm Hg when aortic pressure is 120/80 mm Hg) and begins to fall. The T wave of the ECG occurs at this time, as ventricular repolarization takes place. LVP falls below aortic pressure, but aortic valve closure

is not immediate because of the momentum associated with ventricular ejection. Closure of the aortic valve marks the end of ejection and the beginning of isovolumetric relaxation. During this period, the ventricle is beginning to relax but the valves are closed, resulting in rapid fall of LVP. When LVP falls below LAP, the mitral valve opens, and ventricular filling begins. During the remainder of diastole, LVP remains slightly below LAP, consistent with the direction of blood flow during ventricular filling.

Aortic Pressure

The **aortic pressure** curve falls during diastole as blood flows to the peripheral circulation (see Fig. 12.1). The upstroke in this curve begins (at 80 mm Hg if aortic pressure is 120/80 mm Hg) with the opening of the aortic valve. As ejection proceeds, pressure rises to the systolic aortic pressure (120 mm Hg in this example) and then begins to fall. The dicrotic notch, a high-frequency deflection in the aortic pressure curve, occurs when the aortic valve closes. **Aortic flow** reaches a peak during rapid ejection and is lowest during diastole when no ejection is occurring.

Ventricular Volume

The **left ventricular volume** curve illustrates the filling and emptying of the ventricle (see Fig. 12.1). At the time of the P wave, ventricular filling is nearly complete, but the final filling (approximately 15%) is produced by atrial contraction. This phase of ventricular filling is known as **active ventricular filling**. Ventricular filling ends when the mitral valve closes, at the left ventricular **end-diastolic volume (EDV)**. Of course, ventricular volume remains constant during isovolumetric contraction. With opening of the aortic valve, rapid ejection of blood begins, and ventricular volume falls quickly. This phenomenon is followed by a period of reduced ejection, which ends with aortic valve closure. At this point, left ventricular **end-systolic volume (ESV)** has been reached, and this volume remains constant during isovolumetric relaxation. The stroke volume (SV) is the difference between EDV and ESV. Normal EDV is approximately 140 milliliters (mL), and normal SV is approximately 70 mL, although these values vary considerably. Opening of the mitral valve initiates a period of **rapid passive filling** that is followed by a period of **slow passive filling** (also known as **diastasis**), as the ventricle fills more slowly with the diminished pressure gradient between the atrium and ventricle.

Atrial Pressure

The shape of the LAP curve reflects both cardiac events and venous return (flow of blood back to the atrium); it consists of the a, c, and v waves. Following the P wave on the ECG, atrial contraction produces the first rise in atrial pressure, the a wave (see Fig. 12.1). During isovolumetric contraction, there is another upward wave in the LAP curve, the c wave, caused by bulging of the mitral valve back into the left atrium, as the ventricle attempts to contract against a fixed volume. During the ejection phase of the left ventricle, LAP rises slowly while venous return from the pulmonary circulation fills the atrium (the mitral valve is closed during this period), producing the v wave. When LVP falls below LAP, the mitral valve opens and LAP falls as blood flows into the left ventricle. For the remainder of diastole, LAP remains above LVP as the heart fills. Because there are no valves between the vena cavae and right atrium (or the pulmonary vein and left atrium), the **venous pulse** and the atrial pressure curve are similar in shape.

The events of the cardiac cycle are initiated when an action potential is generated in the sinoatrial node, producing depolarization and contraction of the atria, followed by contraction of the ventricles. When heart rate is increased, diastole is shortened, because the P wave (and thus, atrial and ventricular contraction) occurs earlier. As a result, the time for ventricular filling is diminished. However, at a resting heart rate of 70 beats/min, there is a significant period of diastasis (slow ventricular filling), and, up to a point, increasing heart rate will not limit the time for rapid passive filling, the period during which most ventricular filling occurs. Very rapid heart rates allow insufficient time for ventricular filling and can even compromise cardiac output.

Phonocardiogram

The **phonocardiogram** is an acoustical recording reflecting the **heart sounds** generated during the cardiac cycle (see Fig. 12.1). S_1, the first heart sound, is generated by closure of the mitral and tricuspid valves (the mitral valve closes slightly before the tricuspid valve). S_2 is associated with closure of the aortic and pulmonic valves; again, the valve on the left side closes just before the right-side valve. During inspiration, closure of the pulmonic valve is delayed more distinctly, normally resulting in audible **physiological splitting** of S_2. This delay is caused by increased filling of the right ventricle as a result of reduced intrathoracic pressure during inspiration. S_3, the third heart sound, is normal in children and is associated with rapid ventricular filling. It is not heard in healthy adults and may be a sign of volume overload, as in congestive heart failure. S_4, caused by active ventricular filling, also is not audible in healthy adults but may be present in a variety of disease states.

REGULATION OF CARDIAC OUTPUT

Cardiac output (CO) is 5 liters (L) per minute (min) in the average, resting adult. The **CO formula** is a key physiological concept, relating CO to heart rate (HR) and SV:

$$CO = HR \times SV \qquad \text{Eq. 12.1}$$

At rest, HR is approximately 70 beats/min and SV is approximately 70 mL, yielding a CO of approximately 5000 mL/min (5 L/min). To maintain adequate arterial blood pressure and blood flow at rest, during increased activity, or during challenges to homeostasis such as exercise, CO must be adjusted through regulation of HR and SV. Many of the mechanisms that regulate HR or SV affect both parameters, and furthermore, they are closely linked to mechanisms involved in blood pressure regulation.

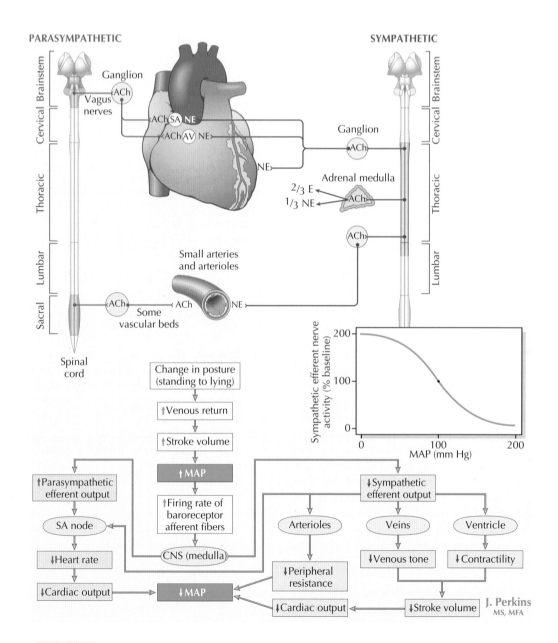

Figure 12.2 **Autonomic Nervous System and Arterial Baroreceptor Reflexes** Both sympathetic and parasympathetic nerves innervate the sinoatrial (SA) and atrioventricular (AV) nodes. The myocardium is innervated by sympathetic nerves. Arterial and venous vessels throughout most of the body are innervated by sympathetic nerves, whereas the parasympathetic nervous system innervates vessels in the genital organs and gastrointestinal tract. Autonomic efferent activity is regulated by the baroreceptor reflex, in response to changes in arterial pressure detected at the carotid sinus and aortic arch baroreceptors. The response to a change in posture is illustrated. *ACh,* acetylcholine; *AV,* atrioventricular; *CNS,* central nervous system; *E,* epinephrine; *MAP,* mean arterial pressure; *NE,* norepinephrine; *SA,* sinoatrial.

The **autonomic nervous system** plays an important role in regulation of CO through modulation of the sinoatrial (SA) nodal pacemaker rate, myocardial contractility, and vascular smooth muscle tone (Fig. 12.2). Sympathetic nerve endings at the heart and vessels release norepinephrine, whereas parasympathetic nerves release acetylcholine at the heart (primarily the SA node) and in a few vascular beds (vessels in most regions do not have parasympathetic innervation). In addition to direct release of catecholamines by sympathetic nerves in the heart and vessels, circulating catecholamines (predominantly epinephrine) released by the adrenal gland contribute to cardiovascular regulation when the sympathetic nervous system is strongly activated.

Heart Rate Regulation

The arterial **baroreceptor reflex** is a key mechanism involved in autonomic regulation of cardiovascular function (see Fig. 12.2). The baroreceptors consist of specialized cells in the vascular wall of the carotid sinuses and aortic arch.

Baroreceptors respond to the stretch associated with elevation of arterial pressure by increasing afferent impulses to the medullary cardiovascular center, where autonomic nervous system activity is regulated. In response to increased afferent input, sympathetic nerve activity is reduced and parasympathetic outflow is increased. The effects of these autonomic responses on the heart and vessels result in a return of arterial blood pressure toward its original level; specific aspects of this regulation are discussed in the following text.

The following mechanisms contribute to **regulation of HR**:

- The autonomic nervous system and baroreceptors
- The Bainbridge reflex response to atrial stretch
- Effects of thoracic pressure changes during respiration on venous return

Under normal conditions, HR regulation is accomplished primarily by the autonomic nervous system and baroreceptor reflexes. Sympathetic nerves release norepinephrine at the SA node, which acts at β_1 receptors, increasing cAMP production and ultimately raising the activity of the pacemaker and elevating HR. Parasympathetic activation reduces HR through release of the neurotransmitter acetylcholine. Thus, when arterial pressure rises, the baroreceptor reflex results in a fall in the SA node pacemaker rate. The resulting reduction in HR contributes to a fall in CO and a return to normal arterial pressure (see Fig. 12.2). Conversely, reduced arterial pressure will result in less stretch of baroreceptors, reduced afferent impulses to the cardiovascular center, greater sympathetic activity, and reduced parasympathetic activity, raising HR and returning arterial blood pressure to the normal level.

Another reflex potentially regulating HR is the **Bainbridge reflex**. When right atrial volume is increased, low-pressure stretch receptors (that is, receptors responding to stretch at the low pressures typical in the atria) initiate a neural reflex that increases HR through sympathetic nerves. Note that arterial baroreceptors respond to stretch by decreasing HR, whereas atrial baroreceptors respond to stretch by increasing HR. In the former case, the response is part of a mechanism to regulate arterial pressure, and in the latter case, the response is to increased blood volume. When atrial stretch results in increased HR, blood volume will be redistributed. In addition, a number of other mechanisms for regulating blood volume and pressure will be activated; these mechanisms will be considered later in this text (see "Regulation of Arterial Blood Pressure" in Chapter 13).

HR may also be affected by the respiratory cycle, particularly in infants and children. In so-called **respiratory sinus arrhythmia**, HR is increased during inspiration and decreased during expiration. The low-pressure baroreceptors within the atria are stretched by increased venous return during inspiration, producing the cyclic variation in HR.

The Bainbridge reflex and the arterial baroreceptor reflex produce opposite responses to an intravenous infusion. The Bainbridge reflex is usually observed when a rapid intravenous infusion is administered to a subject with a slow HR. Stretch of atrial receptors produces a reflexive increase in HR under these circumstances. On the other hand, when HR is high to begin with, such as after hemorrhage, intravenous infusion usually decreases the HR. In this circumstance, the increased ventricular filling associated with the infusion produces an increase in CO and therefore an increase in arterial pressure. The rise in arterial pressure causes stretch of arterial baroreceptors, resulting in diminished sympathetic nervous system activity and enhanced parasympathetic activity, along with a slowing of the HR.

Stroke Volume Regulation

SV regulation is dependent on several parameters:

- **Preload:** The degree of stretch of myocardial fibers prior to contraction. Preload in the intact heart is closely related to EDV of the ventricle. Increased preload is associated with increased force of contraction, and hence, increased SV.
- **Afterload:** The force against which the heart contracts. In the intact heart, afterload is closely related to arterial pressure or LVP during systole. Increased afterload opposes ejection by the heart and therefore will tend to reduce SV.
- **Contractility (inotropism):** The intrinsic ability of cardiac muscle to generate force at a given fiber length. Specifically, contractility is *not* synonymous with force of contraction, because force of contraction is dependent on the degree of initial stretch of the muscle fibers (preload).

The Frank-Starling Relationship

The **Frank-Starling relationship** is named for Otto Frank, who described the relationship between pressure generated in the frog heart and the diastolic volume, and for Ernest Starling, who demonstrated, in an intact, canine heart-lung preparation, the relationship between SV and EDV. The **cardiac function curve** illustrates this relationship (Fig. 12.3A). When preload is increased by raising atrial pressure, SV (and therefore, CO) rises, up to the optimum level of preload. This intrinsic mechanism for regulation of SV is not associated with a change in inotropism. The molecular basis for this relationship involves stretch of sarcomeres; at optimal preload, conditions for subsequent crossbridge formation and recycling are best.

Sympathetic Regulation of Stroke Volume

SV is also regulated by the **sympathetic nervous system** (Fig. 12.3B). Cardiac muscle is directly innervated by

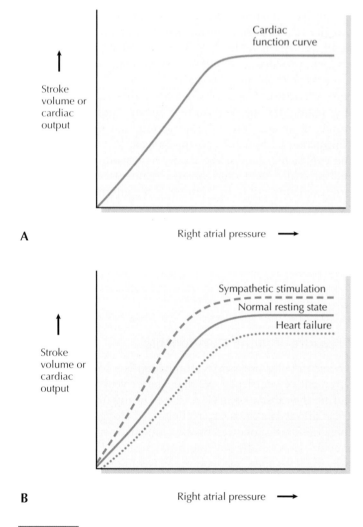

A

Right atrial pressure ➞

B

Right atrial pressure ➞

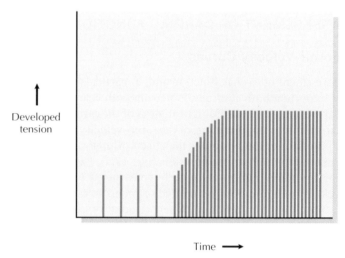

Time ➞

Figure 12.4 **Effect of Heart Rate on Force of Contraction (Treppe or Staircase Effect)** When the intervals between cardiac muscle contractions are long, the tension developed is low. Note the staircase-like increase in the force of contraction when frequency (heart rate) is increased. Because this effect does not depend on changes in resting fiber length, contractility (inotropism) is increased.

Figure 12.3 **The Frank-Starling Relationship** **A,** An increase in preload of the ventricle results in an increase in stroke volume and thus cardiac output. **B,** Sympathetic stimulation of the heart shifts this curve upward and to the left; heart failure results in a lower slope of the curve. The slope of the cardiac function curve is one measure of contractility of the heart. In this graph, it is assumed that afterload (arterial pressure) is held constant.

The Frank-Starling relationship is an important mechanism for matching CO and venous return and left- and right-side CO. If output from the right ventricle is suddenly increased (for example, because of a change in posture), return to the left ventricle will be rapidly elevated. The resulting stretch of the left ventricle will cause an increase in its output, matching the rise in output that occurred on the right side. Thus, the Frank-Starling relationship maintains balance between right- and left-side output. It is also an important mechanism in the adjustment of SV and CO during the complex physiological changes associated with exercise or volume depletion (for example, in hemorrhage).

sympathetic nerves, which release norepinephrine. Norepinephrine, through binding to β-adrenergic receptors (β₁), produces elevation of intracellular free Ca^{2+} levels, resulting in increased contractility of the heart. This is a true increase in inotropism or contractility because it is independent of any change in preload. Release of epinephrine by the adrenal medulla may also produce greater contractility of the heart, although levels of circulating epinephrine rarely reach levels that significantly affect inotropic activity. The parasympathetic nervous system has only limited effects on contractility of human ventricles. Clinically, a number of drugs are used to promote inotropism in persons with heart failure, including digitalis, dopamine, and dobutamine. Although the basic cardiac function curve simply describes the effects of preload on CO or SV, the slope of this curve is a measure of inotropism. With sympathetic stimulation, the slope is steeper, indicating enhanced contractility.

Treppe, also known as the **staircase effect**, is an intrinsic mechanism for regulation of SV. When cardiac contractions are infrequent, force of contraction is reduced, whereas as the HR increases, force of contraction is elevated (Fig. 12.4). Because these changes occur independently of changes in preload, they reflect a change in contractility. The increased contractility at higher HRs is associated with elevated free intracellular Ca^{2+} in myocardial fibers.

During a baroreceptor-mediated response to a fall in arterial pressure (see Fig. 12.2), many mechanisms contribute to the change in CO. Sympathetic activation elevates HR, as discussed previously, but unless mechanisms are available for maintaining or elevating SV simultaneously, elevation of HR is not an effective mechanism for raising CO, because filling time for the heart is reduced. By raising contractility of the heart, sympathetic activation enhances SV as well, and, simultaneously, sympathetic constriction of the systemic venous system elevates preload of the heart, allowing the Frank-Starling relationship to further augment SV and CO.

ASSESSMENT OF CARDIAC FUNCTION

Force–Velocity Curves

Assessment of cardiac function and, in particular, inotropism is important both clinically, in persons with heart failure, and experimentally. In addition to analysis of the **cardiac function curve** and its slope, analysis of the **force–velocity relationship** in cardiac muscle illustrates the effects of preload and altered contractility on cardiac function (Fig. 12.5). Experimentally, force–velocity curves are usually constructed by measuring velocity of contraction of isolated segments of cardiac muscle against various afterloads. When velocity of contraction is plotted against afterload, an inverse relationship is revealed: when the force of contraction is higher (i.e., when the muscle contracts against higher afterload), velocity of shortening

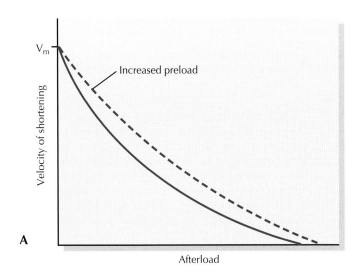

A

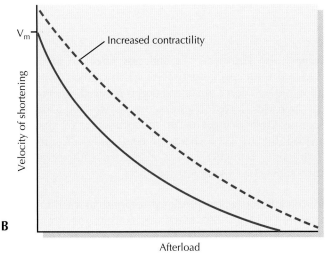

B

Figure 12.5 **Force–Velocity Relationship in Cardiac Muscle** Force–velocity curves depict the effect of afterload on velocity of contraction of cardiac muscle. **A,** When the force of contraction is higher (i.e., when the muscle contracts against higher afterload), velocity of shortening is reduced. Velocity is highest (V$_m$) at zero afterload. Increased preload shifts the curve upward, but V$_m$, a measure of contractility, is unchanged. **B,** Positive inotropic influences such as sympathetic stimulation change V$_m$. The force–velocity curve is shifted upward.

is reduced. Maximum force of contraction occurs at zero velocity, that is, during an isometric contraction. On the other hand, velocity is highest (V$_m$) at zero afterload. Changes in preload produce a family of curves, all with the same y-intercept (the same maximum velocity of contraction at zero afterload). These curves are another manifestation of the Frank-Starling relationship: greater preload generally results in greater force of contraction, and thus velocity of contraction, although V$_m$ is unchanged. V$_m$ is a measure of the contractile state of the tissue, representing the maximal rate at which actin and myosin filaments can interact to produce contraction. In contrast to the effects of changes in preload, positive inotropic influences such as sympathetic stimulation change V$_m$. The force–velocity curve is shifted upward and to the right, with an increase in the maximal force that can be developed by the muscle (during an isometric contraction, at the x-intercept) and an increase in V$_m$ (at the y-intercept). An increase in V$_m$ is indicative of a positive inotropic effect.

Pressure–Volume Relationship

The ventricular **pressure–volume loop** is a continuous measurement of ventricular volume against ventricular pressure during the cardiac cycle (Fig. 12.6). During diastole, the ventricle fills at low pressure, as illustrated along the bottom of the loop. This is followed by isovolumetric contraction, when pressure rises rapidly but volume is constant. During the ejection phase of the cycle, volume falls while pressure remains high. Subsequently, during isovolumetric relaxation, volume is constant and pressure falls rapidly. This cycle is then repeated over and over, with each beat. The shape and the area within this loop are affected by changes in preload, afterload, and contractility.

Ejection Fraction

A simple and useful measurement for assessment of myocardial function is the **ejection fraction**. Ejection fraction is the ratio of SV to EDV. In a healthy person at rest, this ratio should be greater than 50%. If a positive inotropic drug is administered, the ejection fraction will rise; in persons with myocardial ischemia or heart failure, the ejection fraction is diminished. The ejection fraction can be estimated noninvasively by echocardiography.

VASCULAR FUNCTION AND CARDIAC OUTPUT

Regulation of CO depends not only on adjustment of HR and SV but also on regulation of the function of veins and venous return. Venous return is defined as the flow returning to one side of the heart. Thus, because CO is defined as flow from one ventricle, the normal value for venous return will be identical to CO when averaged over time. If CO is 5 L/min, venous return will also be 5 L/min. Matching of CO and venous return is an important aspect of cardiovascular function.

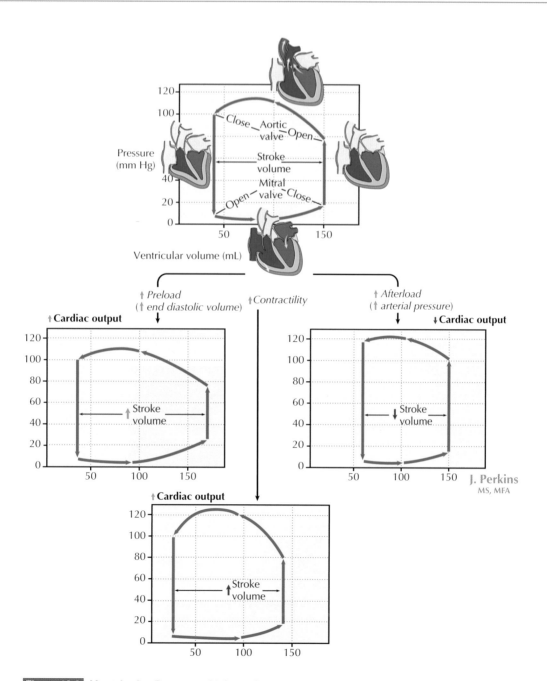

Figure 12.6 **Ventricular Pressure–Volume Loop**   A continuous plot of ventricular volume and ventricular pressure during the cardiac cycle yields a closed loop. Following the red arrows from the bottom left of the top graph, during diastole, filling of the ventricle occurs with little change in pressure; during isovolumetric contraction, volume is constant, but pressure rises rapidly. During the ejection phase of the cycle, volume falls, while pressure remains high. During isovolumetric relaxation, volume is constant and pressure falls rapidly, to the starting point on the diagram. Changes in preload, afterload, and contractility affect stroke volume as illustrated.

CLINICAL CORRELATE 12.1
Echocardiography

Echocardiography is a technique in which inaudible, high-frequency sound is transmitted through the thorax (transthoracic echocardiography) or esophageal wall (transesophageal echocardiography) into the heart. The echoes of the ultrasound, reflected at the interfaces of tissue and fluid, are recorded to produce a graphic image of the heart. In standard echocardiography, the resulting echocardiogram is a two-dimensional image of a slice of the heart. In Doppler echocardiography, the use of continuous wave Doppler ultrasound allows assessment of the velocity of blood flow within the heart. Using these techniques, it is possible to (1) determine the size and structure of valves and chambers; (2) determine ejection fraction and other parameters; and (3) detect abnormal wall motion, valvular regurgitation (leakage) or narrowing, bacterial growths (vegetation) within the heart, and other conditions.

Transducer Positions in Echocardiographic Examination

Parasternal position

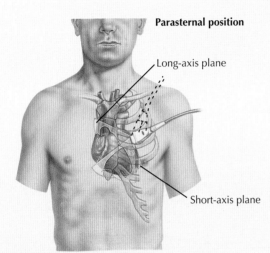

Long-axis plane

Short-axis plane

Left parasternal position allows views in long- and short-axis planes. Tilting transducer allows multiple sections.

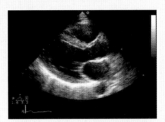

Normal long-axis view during systole

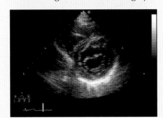

Normal short-axis view at mitral valve level

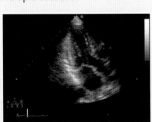

Normal apical long-axis view

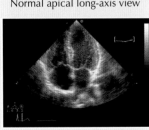

Normal apical four-chamber view

Apical position

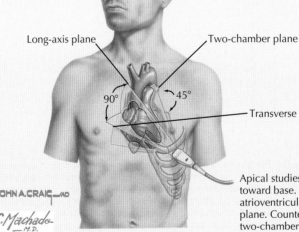

Long-axis plane

Two-chamber plane

90°   45°

Transverse (four-chamber, five-chamber) plane

JOHN A. CRAIG_MD
C. Machado _M.D.

Apical studies imaged from point of maximal impulse toward base. Four-chamber plane passes through atrioventricular valves; upward tilt gives five-chamber plane. Counterclockwise rotation of 45 degrees gives two-chamber plane; 90-degree rotation gives long-axis plane.

Just as CO is affected by various factors, so is venous return. Flow from the capillaries to the right atrium occurs despite a small pressure gradient. Because this pressure gradient is small, venous return is affected by factors that are often different from those affecting the arterial system.

CO and venous return can be conceptualized as two sides of the same coin, with CO being the flow from one side of the heart and venous return being the flow back to one side of the heart. Measured over time, CO and venous return must be equal, because CO is completely dependent upon return of blood to the heart (venous return). A change in one of these parameters will produce an equivalent change in the other.

Venous Compliance and Gravitational Effects

An important factor affecting veins is **compliance** (i.e., the change in volume associated with a change in pressure). Compliance of veins is approximately 20-fold greater than that of arteries. Thus, when a person stands from a recumbent position, blood pools in the lower extremities as veins distend because of the hydrostatic pressure associated with the column of blood between the lower extremities and the right atrium. CO is transiently reduced upon standing as pooling occurs. To prevent **orthostatic hypotension** (i.e., a decrease in blood pressure upon standing) and to restore venous return under these conditions, the most important compensatory mechanism is the **baroreceptor** reflex (see Fig. 12.2). Increased sympathetic outflow results not only in constriction of the arteries and augmentation of rate and contractile force of the heart, but it also constricts veins, reducing the pooling effects of gravity. Veins, in fact, act as a reservoir of blood, which can be mobilized by sympathetic nervous system activation and venoconstriction to increase venous return and, thus, CO. This mechanism is important in maintaining CO during volume depletion as well. Movement of skeletal muscles is also important in preventing orthostatic hypotension. During walking, for example, movement of leg muscles compresses veins and augments venous return, because veins outside the central venous system contain one-way valves that prevent backflow of blood.

Effect of Respiration

Another factor affecting venous return is the **effect of respiration**. During inspiration, as the rib cage expands and the diaphragm moves down, negative pressure is created in the thorax. Simultaneously, pressure in the abdominal cavity rises as a result of the downward movement of the diaphragm. Thus veins in the abdomen are subjected to a positive pressure, which augments venous return toward the negative pressure in the thorax. During expiration, the gradient is reduced. With increased depth and frequency of respiration, such as during exercise, a pulsatile increase in venous return occurs.

Cardiac Function and Vascular Function Curves

The interactions between vascular function and cardiac function can be illustrated by the simultaneous consideration of two relationships: the **cardiac function curve** (previously considered) and the **vascular function curve** (Fig. 12.7). The cardiac function curve illustrates the Frank-Starling relationship, in which right atrial pressure is the independent variable and CO is the dependent variable. A rise in right atrial pressure (preload) produces a rise in CO. The vascular function curve as illustrated is an unconventional graph, in which the independent variable (CO) is plotted on the y-axis and the dependent variable (right atrial pressure) is on the x-axis. This is an inverse relationship: a rise in CO produces a fall in right atrial pressure (or preload). In other words, greater CO will result in redistribution of blood volume, with reduction of preload. Note that the x intercept of the vascular function curve is the **mean circulatory pressure**, which is the pressure in the system when CO is zero. It is dependent on blood volume and the compliance of the vascular system as a whole. Thus, if the heart is stopped, pressure equilibrates across the entire cardiovascular system. A positive mean circulatory pressure is necessary for the heart to effectively pump blood.

Mean circulatory pressure is the residual pressure that would exist throughout the cardiovascular system (after equilibration) if the heart were suddenly stopped and vascular tone throughout the system were to remain unchanged. The normal value for mean circulatory pressure is about 7 mm Hg and is a function of vascular tone (particularly venous tone) and blood volume. Without positive mean circulatory pressure, efficient circulation of blood would not be possible, because preload would be compromised at even a very low CO.

Integration of the Cardiac and Vascular Function Curves

When the cardiac function and vascular function curves are plotted on the same graph, the two curves intersect at one point: the resting CO and right atrial pressure (normally about 5 L/min and 2 mm Hg, respectively). In other words, this point is the steady-state or equilibrium point for normal resting cardiovascular function. When one of the curves is altered, a new equilibrium point is reached. For example, an increase in blood volume (hypervolemia) will shift the vascular function curve upward and to the right (Fig. 12.7B). At the new equilibrium point for the two curves (point B), CO is increased because of higher preload. Venous constriction would have the same effect as increased volume. On the other hand, hypovolemia will shift the vascular function curve to the left and downward, and the new equilibrium point A represents lower CO at rest. The cardiac mechanism resulting in these changes in CO with changes in the vascular function curve is the Frank-Starling relationship.

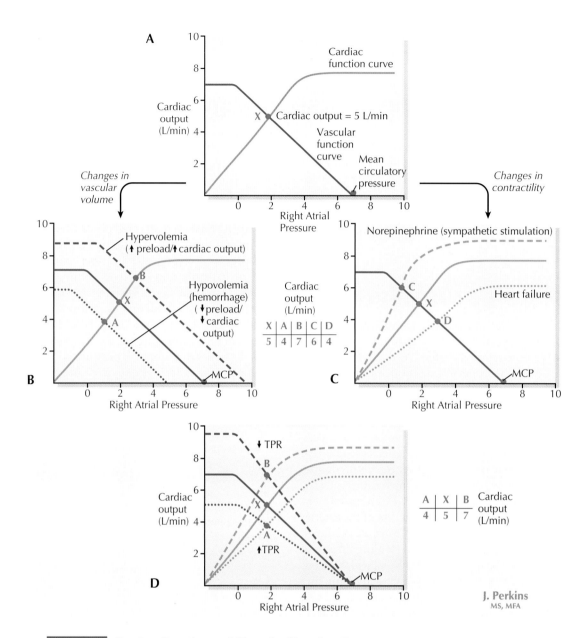

Figure 12.7 **Cardiac Function and Vascular Function Curves** **A,** In the cardiac function curve, right atrial pressure is the independent variable and cardiac output (CO) is the dependent variable. A rise in right atrial pressure produces a rise in CO, as predicted by the Frank-Starling relationship. In the vascular function curve, the independent variable is CO, plotted unconventionally on the y-axis, and the dependent variable is right atrial pressure. This is an inverse relationship; a rise in CO produces a fall in right atrial pressure. The x-intercept of the vascular function curve is the mean circulatory pressure (MCP), which is the pressure throughout the system when CO is zero. The two curves intersect at point X, the resting CO and right atrial pressure, which is the normal resting steady state for cardiovascular function. When one of the curves is displaced—for example, by a change in **(B)** blood volume, **(C)** contractility, or **(D)** total peripheral resistance (TPR)—a new steady state is reached.

Similarly, changes in the cardiac function curve also produce new equilibrium points between the two curves (Fig. 12.7C). An increase in contractility—for example, as a result of sympathetic stimulation of the heart—will increase the slope of the cardiac function curve, with higher CO at the new equilibrium. Heart failure is associated with a lower slope in the cardiac function curve, and CO is reduced as a result.

Changes in total peripheral resistance (TPR) affect both the vascular function and cardiac function curves (Fig. 12.7D). In this analysis, it is assumed that arterial resistance (specifically, resistance in small arteries and arterioles) is altered, while venous tone is unchanged. Thus, if TPR is diminished, the vascular function curve rotates up and to the right, because venous and right atrial pressure will be raised as a result of the lower arterial resistance. Note that mean circulatory pressure is unchanged because total compliance of the vascular system is not significantly affected. The cardiac function curve shifts upward as SV is ejected against lower arterial pressure (reduced afterload) as a result of reduced TPR. Thus the equilibrium point for the system moves from point X to point B, at which point CO is elevated from the original steady state. Opposite changes occur when TPR is elevated: the vascular function curve rotates downward and to the left and the cardiac function curve shifts downward because of the effects of increased afterload. The equilibrium shifts from point X to point A, at which point CO is reduced from the original steady state.

CLINICAL CORRELATE 12.2
Venous Pressure in Heart Failure

In right-heart failure, contractility of the right ventricle is reduced, often because of myocardial infarction. As a result, CO will be reduced and venous pressure will be elevated. This knowledge can be useful diagnostically. If a healthy person is lying in bed with his upper body tilted up 30 to 60 degrees on a pillow, the external jugular vein is ordinarily collapsed just above the level of the clavicle. The height of the column of blood in the external jugular vein is a measure of central venous pressure or right atrial pressure. If a patient is experiencing heart failure and the central venous pressure is elevated, the vein will be distended significantly above the clavicle.

Chapter 13

The Peripheral Circulation

The peripheral circulation consists of the systemic arteries, veins, and microcirculation. The structures of arteries and veins include the following three tissue layers (Fig. 13.1):

- **Tunica intima:** This innermost layer consists of a single endothelial cell layer forming the inner lining of the vessel; these cells rest on a basement membrane that separates the intima from the media.
- **Tunica media:** The media consists mainly of smooth muscle and is the contractile portion of the vascular wall.
- **Tunica adventitia:** The adventitia consists mainly of connective tissue.

Variation occurs in the absolute and relative thickness of media and adventitia between arteries and veins, as well as between small and large vessels; differences are also observed between vessel types in the connective tissue and cellular constituents of these layers. For example, the walls of large arterial vessels are rich in elastic tissue and have a relatively thick adventitia compared with smaller arteries. On the other hand, smaller arteries have a relatively more dominant, muscular medial layer. Capillaries, unlike other vessels, have no media or adventitia. Their vascular walls consist simply of endothelial cells and basement membrane.

THE MICROCIRCULATION

The microcirculation consists of vessels less than 100 micrometers (μm) in diameter and includes arterioles, metarterioles, capillaries, and venules (Fig. 13.2). Muscular arterioles may directly feed into capillaries or into metarterioles that structurally are between capillaries and arterioles. Precapillary sphincters are bands of smooth muscle found at the point at which blood enters capillaries. Constriction and relaxation of the smallest arteries, arterioles, and precapillary sphincters regulate flow into capillary beds (see the section "Regulation of Blood Flow").

The thin wall of capillaries consists of a single layer of endothelial cells and the associated basement membrane. This simple structure is well adapted for the diffusion of gases, nutrients, and wastes between blood and the interstitial fluid of tissues, which occurs only in capillaries. The exchange of fluid between the vascular space and interstitial space also occurs across the capillary wall by simple diffusion (see Fig. 2.1). The net filtration pressure for diffusion of fluid out of capillaries is governed by the Starling equation. When entering the capillary at the arteriolar end, hydrostatic pressure is approximately 30 millimeters of mercury (mm Hg); it falls to approximately 10 mm Hg at the venular end of the capillary. Interstitial hydrostatic pressure is uniform within a local region, whereas plasma oncotic pressure is high in plasma and substantially lower in the interstitial space, and thus net filtration of fluid out of the capillary usually occurs at the arteriolar end. Toward the venular end of the capillary, where capillary hydrostatic pressure is lower, net reabsorption of fluid into the capillary usually occurs.

The histologic differences between various types of vessels reflect their functional roles. The thin wall of the capillaries, consisting of only an intimal layer, allows the efficient exchange of nutrients, waste, and dissolved gases between blood and tissues. On a relative basis, the muscular tunica media is more prominent in small arteries and arterioles than in large arteries or veins, reflecting their role in regulation of blood flow. Larger arteries have thick adventitia with significant elastic tissue, consistent with their role as "distributing vessels," whereas veins have more compliant adventitia. The tunica media of veins contains smooth muscle arranged both circularly and longitudinally, reflecting the function of veins as "capacitance vessels." Similarly, elastic tissue in the adventitia of veins is less prominent than in arteries.

Edema is the swelling associated with an increase in interstitial fluid in a tissue or body cavity. The volume of fluid in the interstitial space depends on the rate at which interstitial fluid is produced and the rate at which it is removed. When high capillary hydrostatic pressure, low capillary oncotic pressure, or leakage of proteins from the vascular space (producing high interstitial oncotic pressure) results in excessive fluid in the interstitial space, it is ordinarily removed by the lymphatic system. Edema may occur with blockage of the lymphatic system or excessive flux of fluid from the vascular space.

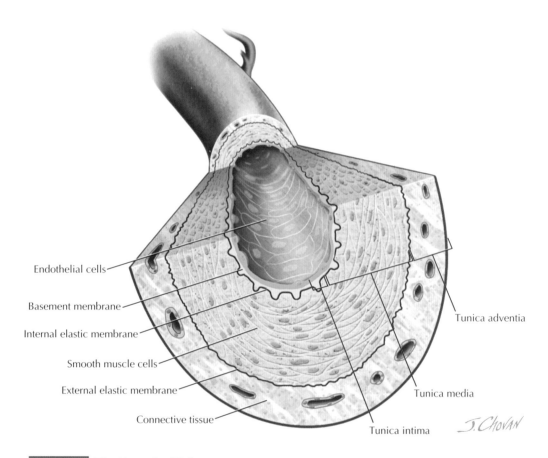

Endothelial cells

Basement membrane

Internal elastic membrane

Smooth muscle cells

External elastic membrane

Connective tissue

Tunica adventia

Tunica media

Tunica intima

J. Chovan

Figure 13.1 **The Vascular Wall**   The vascular walls of arteries and veins have three tissue layers, with the relative thickness of medial and adventitial layers varying among various vessel types. The capillary wall consists of only the intimal layer of endothelial cells and basement membrane. The vessel illustrated in this figure is a large artery, with prominent medial smooth muscle. The walls of large arteries and veins have their own vascular supply (vasa vasora). The tunica media of arteries is bounded by an internal elastic membrane and an external elastic membrane.

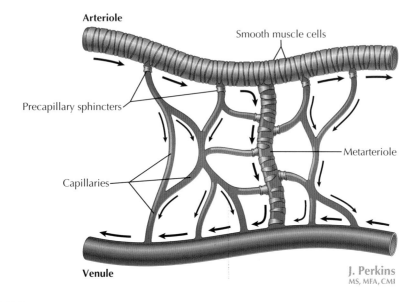

Arteriole

Smooth muscle cells

Precapillary sphincters

Metarteriole

Capillaries

Venule

J. Perkins
MS, MFA, CMI

Figure 13.2 **Components of the Microcirculation**   Arterioles, metarterioles, capillaries, and venules constitute the microcirculation. Direction of blood flow is indicated by arrows. Blood flow is regulated by constriction and dilation of smooth muscle of the arterioles, metarterioles, and precapillary sphincters. Sympathetic nerves extend to arterioles, metarterioles, and venules in the microcirculation.

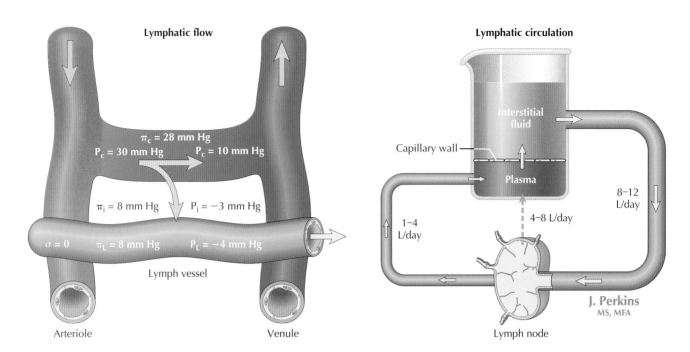

Figure 13.3 **The Lymphatic Circulation** Diffusion of fluid from the vascular space to the interstitial space is governed by the Starling equation (see Chapter 1). Excess fluid in the interstitial space is transported by the lymphatic system back to the central venous circulation after filtration through lymph nodes. Movement of fluid from interstitium to the lymphatic system is driven solely by the hydrostatic pressure gradient, because there is no oncotic pressure gradient (protein flows freely through the lymphatic vessel wall). (P_c = HP_c, hydrostatic capillary pressure.)

THE LYMPHATIC SYSTEM

Excess interstitial fluid is returned to the blood by the lymphatic circulation (Fig. 13.3). Movement of fluid from the interstitium into the lymphatic capillaries is driven by the hydrostatic pressure gradient between the interstitial space and the lymph. The edges of lymphatic capillary endothelial cells overlap, acting as one-way valves; one-way flow of lymph is also maintained in larger lymphatic vessels (Fig. 13.4). Flow is augmented by smooth muscle filaments of the collecting lymphatics that contract when stretched. Lymph is ultimately returned into the central venous circulation through the thoracic duct and lymphatic duct (Fig. 13.5).

The lymphatic system is closely associated with lymphoid organs and tissues, including bone marrow, thymus, lymph nodes, and spleen, and has other functions in addition to its primary function of returning lymph to the circulation, including:

- Circulation of immune cells
- Transport of antigen-presenting cells to lymph nodes
- Filtration of debris, pathogens, and cancerous cells from lymph through lymph nodes
- Absorption of lipids from the digestive system and transport to the blood (see Section 6)

REGULATION OF BLOOD FLOW

Flow through a specific tissue can be altered either by changes in perfusion pressure or changes in resistance of the arterial vessels perfusing the tissue. Obviously, when blood flow requirements of a specific tissue are altered, the most efficient mechanisms for adjusting blood flow will involve a change in regional resistance. On the other hand, when systemic metabolic needs vary, such as during exercise, regulation of flow is accomplished through changes in both systemic hemodynamics (blood pressure and cardiac output [CO]) and regional resistances, resulting in changes in both overall flow (CO) and flow distribution between various tissues.

At the tissue level, both intrinsic and extrinsic factors may influence smooth muscle tone in the resistance vessels (arterioles, precapillary sphincters, and small arteries) in response to local and systemic events. Both smooth muscle and endothelial cells participate in this regulation. Basic aspects of smooth muscle structure and function are discussed in Section 2.

Regulation of Vascular Tone by Endothelial Cells

The role of the endothelium in regulation of smooth muscle contraction and relaxation is a relatively recent discovery. The

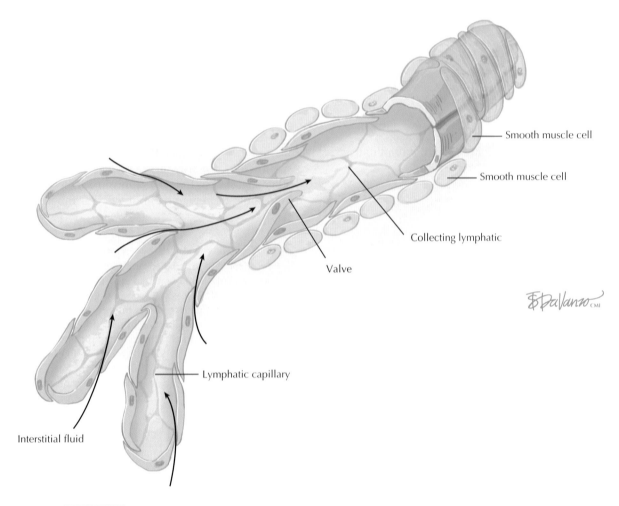

Smooth muscle cell

Smooth muscle cell

Collecting lymphatic

Valve

Lymphatic capillary

Interstitial fluid

Figure 13.4 **Lymphatic Vessels** Interstitial fluid flows freely into lymphatic capillaries when hydrostatic pressure in the interstitium exceeds pressure in the lymphatic capillary. Flow occurs through pores between lymphatic endothelial cells, which act as valves; in the larger collecting lymphatics, one-way flow is maintained by valves. This flow is augmented by contraction of smooth muscle associated with collecting vessels, as well as by external forces acting on lymphatics, such as movement of adjacent skeletal muscles.

endothelium participates in this regulation through the following processes (Fig. 13.6):

- Release of vasodilators such as **nitric oxide** (NO) and prostacyclin
- Release of the vasoconstrictor endothelin
- Conversion of angiotensin I to angiotensin II

Of particular importance is the endothelium's role in vascular regulation through production of NO. When endothelial cells are exposed to **endothelium-dependent vasodilators** such as acetylcholine, histamine, and bradykinin, the enzyme NO synthase (NOS) is activated, resulting in the production of the short-lived but powerful vasodilator NO from the amino acid arginine:

$$\text{L-arginine} \xrightarrow{\text{NOS}} \text{NO} + \text{L-citrulline} \qquad \textbf{Eq. 13.1}$$

NO diffuses to adjacent smooth muscle cells and acts on smooth muscle guanylyl cyclase, elevating cGMP. cGMP causes reduction in free intracellular Ca^{2+}, producing relaxation of smooth muscle and thus vasodilation. This mechanism is particularly important in vascular responses to inflammatory stimuli, as well as in parasympathetic-mediated vasodilation in the genital system and lower gastrointestinal tract (vessels in other regions are not innervated by the parasympathetic nervous system). In addition, many substances that directly act on smooth muscle cells to produce constriction (for example, norepinephrine) also release endothelial NO. This simultaneous release and action of NO results in a dampened response to vasoconstrictors in healthy vessels, as opposed to vessels in which the endothelium has been damaged. Many of the same stimuli that release NO also release **prostacyclin (PGI$_2$)**, a vasodilator metabolite of arachidonic acid, from endothelial cells. Although PGI$_2$ is less important than NO in regulation of vascular tone, both mediators are important in preventing activation of platelets and their adherence to the vascular wall. The actions of PGI$_2$ are opposed by **thromboxane A$_2$**, which is a major product of arachidonate in activated platelets. Thus thromboxane A$_2$ is a potent prothrombotic substance and

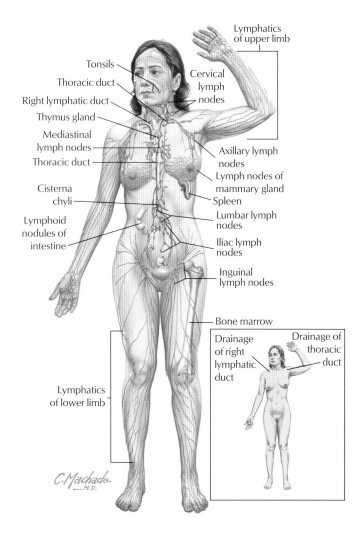

Figure 13.5 **The Lymphatic System and Associated Organs**
The primary role of the lymphatic system is to return excess interstitial fluid to the cardiovascular system after filtration through lymph nodes, as illustrated. Areas of drainage served by the thoracic duct and the right lymphatic duct are illustrated in the inset figure. The lymphatic system is also part of the immune system, serving in the maturation and circulation of immune cells and filtration of lymph. Associated lymphatic organs include the lymph nodes, thymus, spleen, and tonsils, which are involved in the development, storage, and function of immune cells (primarily lymphocytes) produced in the bone marrow, which is itself a lymphatic organ.

vasoconstrictor that is important in inflammatory states and hemostasis.

The release of NO by **shear stress** is believed to participate in vascular control as follows: When blood flow to a region is increased (e.g., by vasodilation in the arterial microcirculation), the higher flow causes elevated shear stress in the endothelial cells of vessels supplying the region, which respond by increasing NO production. Adjacent smooth muscle cells relax when exposed to NO, resulting in vasodilation and further augmentation of flow.

Endothelin is a powerful vasoconstrictor protein released by endothelial cells when vessels are damaged. Its role in vascular

regulation in normal tissues is controversial, although it is an important pathophysiologic mediator, for example, in pulmonary hypertension and preeclampsia. **Angiotensin-converting enzyme** is an enzyme found on the surface of endothelial cells. This enzyme cleaves angiotensin I to **angiotensin II**, a potent vasoconstrictor. Like the posterior pituitary hormone **vasopressin**, angiotensin II is important in long-term regulation of blood pressure through its effects on renal sodium and water retention (see Section 5) and plays a role in acute responses to hypotensive crises such as hemorrhage. However, neither angiotensin II nor vasopressin is believed to be important in short-term regulation of vascular function.

The NO produced by endothelial cells is sometimes referred to as "EDRF," or endothelium-dependent relaxing factor. This name is based on the original discovery by Robert Furchgott that vessel segments experimentally denuded of endothelium did not display smooth muscle relaxation when stimulated by acetylcholine and other "endothelium-dependent" vasodilators. Subsequent studies during the 1980s by several investigators led to the identification of EDRF as NO. The journal *Science* named NO "Molecule of the Year" in 1993, and the Nobel Prize in Medicine or Physiology went to Furchgott, Louis Ignarro, and Ferid Murad for their work in this area.

Until the second half of the twentieth century, the endothelium was generally believed to function mainly as a lining or barrier between blood and tissues. In the past few decades, appreciation of its role in many physiological processes has blossomed. The endothelium is important in regulating vascular tone, angiogenesis (formation of blood vessels), and the hemostatic process (healthy endothelium is "antithrombogenic" because of its formation of NO and PGI_2). In addition, the endothelium has a role in metabolism. An example is conversion of angiotensin I to angiotensin II by its angiotensin-converting enzyme. The endothelium has an important role in disease processes, such as in atherosclerosis, in which diminished endothelial NO production and alterations in other endothelial functions are among the earliest changes observed.

Local Control of Blood Flow

Local regulation of blood flow occurs through mechanisms that involve responses to local metabolic products and transmural pressure (i.e., pressure gradient across the vascular wall; Fig. 13.7A). If blood flow to a region is occluded temporarily, when flow is reestablished, **reactive hyperemia** occurs. In other words, blood flow is elevated above the original level. Reactive hyperemia occurs as a result of **local metabolites** that accumulate during the occlusion. These metabolites, which include CO_2, H^+, K^+, lactic acid, and adenosine, act directly on smooth muscle of arterioles and precapillary sphincters in the region to produce vasodilation. Thus flow is elevated until the tissue levels of O_2 are restored and accumulated metabolites are removed. In contrast to reactive hyperemia, **active**

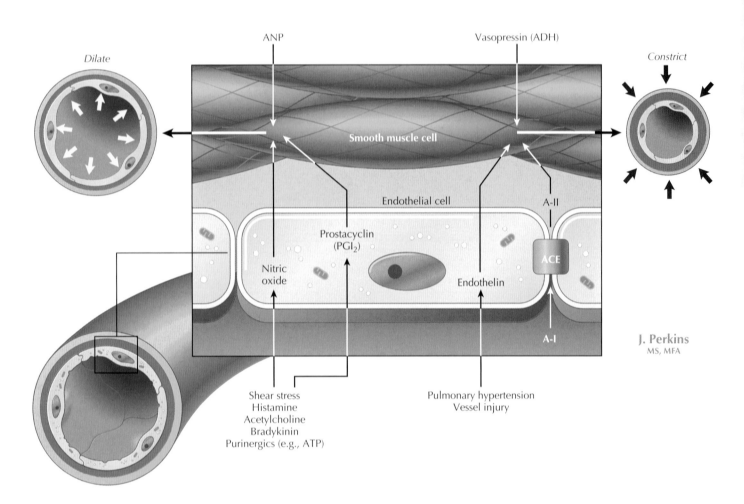

Figure 13.6 **Control of Arteriolar Tone** The greatest resistance to flow occurs in the small arteries and arterioles. The state of constriction or relaxation of these vessels is regulated in part by the sympathetic nervous system and the release of norepinephrine. Circulating hormones, including vasopressin (ADH) and angiotensin II (A-II), may contribute to constriction through their actions on vascular smooth muscle; atrial natriuretic peptide (ANP) is a smooth muscle dilator. The endothelium plays an important role in regulating vascular tone by its release of nitric oxide and prostacyclin (PGI$_2$) in response to many factors, including shear stress, acetylcholine, and bradykinin. Endothelin is a potent, endothelium-derived vasoconstrictor important in some pathophysiological states. The endothelial cell surface also has angiotensin-converting enzyme (ACE), which forms angiotensin II by cleavage of circulating angiotensin I (A-I, an inactive precursor).

hyperemia refers to the increase in flow that occurs in tissues when metabolism is elevated (Fig. 13.7B), such as the greatly increased flow to skeletal muscles during exercise. Ongoing production of local metabolites in the working skeletal muscles causes vasodilation, thus enhancing flow to the muscles.

Autoregulation of local blood flow can occur without changes in local metabolism (Fig. 13.7C). In many tissues and organs, if blood flow is artificially increased by raising the perfusion pressure to the vascular bed, flow will be elevated immediately, as expected; however, it soon returns toward the basal rate. According to the **myogenic hypothesis**, smooth muscle cells constrict in response to elevated transmural pressure—in other words, in response to stretch. With this mechanism flow to a tissue is kept fairly constant, despite changes in pressure, when metabolic needs of the tissue are not changing. Myogenic regulation does not involve the vascular endothelium but is a direct response of smooth muscle cells.

Extrinsic Regulation of Peripheral Blood Flow

Extrinsic regulation of peripheral blood flow involves vasoconstriction or vasodilation in response to neural mechanisms and circulating vasoactive substances. Smooth muscle cells have several types of adrenergic receptors, including α-receptors and β$_2$-receptors:

- α-Receptors mediate the constrictor responses to catecholamines. Receptors of the subtype α$_1$ are the predominant subtype in vascular tissue; α$_2$-receptors also produce vasoconstriction by inhibiting reuptake of norepinephrine. α$_1$-Adrenergic vasoconstriction is mediated by the second messenger IP$_3$, and α$_2$ vasoconstriction is mediated by reduced cAMP levels.
- β$_2$-Receptors mediate dilator responses to catecholamines. β$_2$-Adrenergic vasodilation is mediated by the second-messenger cAMP.

Local Regulation of Blood Flow

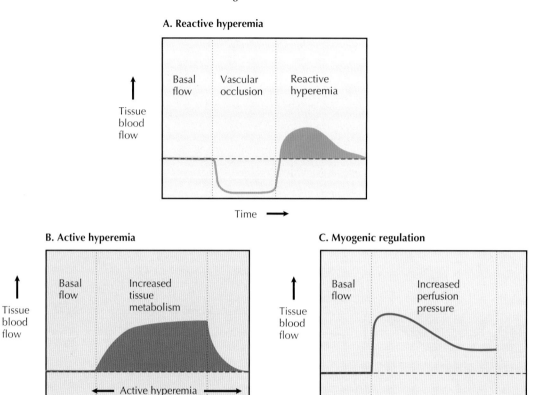

A. Reactive hyperemia

Tissue blood flow

Basal flow | Vascular occlusion | Reactive hyperemia

Time →

B. Active hyperemia

Tissue blood flow

Basal flow | Increased tissue metabolism

← Active hyperemia →

Time →

C. Myogenic regulation

Tissue blood flow

Basal flow | Increased perfusion pressure

Time →

Figure 13.7 **Local Regulation of Blood Flow** Regional blood flow is controlled by local, neural, and humoral factors. **A,** Reactive hyperemia occurs when blood flow is reestablished after occlusion. Buildup of vasodilatory metabolites such as CO_2, H^+, K^+, lactic acid, and adenosine results in relaxation of arterioles and precapillary sphincters directly exposed to these substances. Hence a period of increased blood flow (hyperemia) occurs upon reperfusion. **B,** Active hyperemia refers to the increase in blood flow to a tissue when metabolism in the tissue is elevated; this hyperemia is a result of increased production of vasodilatory metabolites. **C,** Myogenic regulation refers to the autoregulation of blood flow that occurs when perfusion pressure is increased (with no change in metabolic activity in the tissue). Initially, when perfusion pressure is raised, flow to the tissue rises as expected but returns toward the baseline. Smooth muscle of the arterial microcirculation constricts in response to a rise in transmural pressure, thus autoregulating blood flow.

When the sympathetic nerves are stimulated, the responses of vessels will be dependent on the type of adrenergic receptors activated (a function of receptor density and agonist concentration), with the predominant systemic response being vasoconstriction. The major arterial response to sympathetic nervous system activation is α-receptor–mediated vasoconstriction, such as in baroreceptor-mediated responses (see Fig. 12.2). Note that generalized arterial constriction produces elevation of arterial blood pressure because arterioles and small arteries are the major site of action, and constriction of these vessels produces increased peripheral resistance. Sympathetic activation results in widespread constriction of veins, elevating venous pressure and enhancing preload on the heart. β2-Receptors are present in some arterial vessels, such as in skeletal muscle, where sympathetic outflow in anticipation of exercise produces increased blood flow to muscle.

In addition to release of catecholamines at sympathetic nerve endings in the vasculature, epinephrine may be released into the bloodstream by the adrenal medulla during sympathetic nervous system activation. In this case, epinephrine acts as a circulating hormone; other circulating hormones that affect vascular tone include angiotensin II and vasopressin.

Whereas the sympathetic nervous system innervates vessels in most parts of the circulation, parasympathetic innervation of blood vessels is absent in most tissues. Exceptions are the genital organs, salivary glands, and lower gastrointestinal tract. In circulations innervated by the parasympathetic nervous system, release of acetylcholine stimulates endothelial cells to release NO, producing NO-dependent vasodilation of underlying smooth muscle. Because acetylcholine is short-lived in the circulation, it affects only the vessels that are directly innervated by parasympathetic nerves.

Chemoreceptors also participate in the extrinsic regulation of vascular tone. The aortic and carotid bodies are densely vascularized groups of specialized cells found in the proximity of the arterial baroreceptors. Although they are more important in regulation of respiration (see Section 4), they respond to reduced arterial P_{O_2}, and to a lesser extent to elevated P_{CO_2} and H^+ concentration, by initiating a neural reflex that produces arterial and venous constriction.

REGULATION OF ARTERIAL BLOOD PRESSURE

From the preceding information, it should be clear that local, intrinsic mechanisms for vascular regulation are mainly aimed at regulation of regional blood flow, whereas neural and humoral mechanisms are often aimed at regulation of arterial blood pressure. The maintenance of mean arterial pressure (MAP) near its normal level is necessary for adequate perfusion of tissues throughout the systemic circulation. By rearranging the flow equation ($Q = \Delta P/R$), the relationship $\Delta P = R \times Q$ results, where ΔP is the pressure gradient, R is resistance, and Q is flow. The pressure gradient is the difference between arterial and central venous pressure. For the overall systemic circulation, Q is equal to CO and R is total peripheral resistance (TPR), yielding

$$\Delta P = CO \times TPR \qquad \text{Eq. 13.2}$$

Because venous pressure is very low, this equation can be essentially reduced to:

$$MAP = CO \times TPR \qquad \text{Eq. 13.3}$$

Thus regulation of arterial blood pressure involves regulation of CO and systemic resistance, through the various mechanisms already described. Blood pressure is monitored at several points in the system (Fig. 13.8):

- Aortic arch and carotid sinus baroreceptors
- Renal juxtaglomerular apparatus
- Low-pressure (cardiopulmonary) baroreceptors

Short-Term Regulation of Blood Pressure by Arterial Baroreceptors

The **high-pressure arterial baroreceptors** of the aortic arch and carotid sinus and the associated baroreceptor reflexes are most important for moment-to-moment regulation of arterial pressure. During normal, quiet daily activities, an inverse relationship is usually observed between changes in blood pressure and heart rate, reflecting this role of the baroreceptor reflex in maintaining pressure: when pressure falls, heart rate rises; when pressures rises, heart rate falls. These fluctuations in heart rate reflect changes in sympathetic and parasympathetic outflow from the medullary cardiovascular centers in response to the degree of baroreceptor stretch.

Efferent sympathetic nerve activity is inversely related to MAP between the range of 60 to 160 mm Hg in a healthy individual (see Fig. 12.2). At pressures below 60 mm Hg, sympathetic nerve activity is maximal. In addition to MAP, baroreceptors are also sensitive to pulse pressure. If pulse pressure is dampened while MAP is held constant, afferent nerve impulses from the baroreceptors to the cardiovascular center will be less frequent, and sympathetic efferent activity and thus arterial blood pressure will be elevated.

The arterioles of the **juxtaglomerular apparatus** of the kidneys also contain high-pressure baroreceptors; in this case, reduced stretch (low blood volume) results in release of the enzyme renin by the kidney. Renin enzymatically cleaves the plasma protein angiotensinogen (a liver product) to form angiotensin I, which is converted to angiotensin II by endothelial angiotensin-converting enzyme. This mechanism is important in short-term regulation of blood pressure only during pathophysiological states such as hemorrhage. The role of the renin-angiotensin-aldosterone system in regulation of extracellular fluid volume is discussed in Chapter 20; its role in long-term regulation of blood pressure is discussed later in this chapter.

"Resetting" of arterial baroreceptor sensitivity occurs when blood pressure is chronically elevated. In a healthy person with normal blood pressure, baroreceptors function to return arterial pressure to the normal level when pressure falls or rises. With chronic hypertension, a new set point is established, whereby baroreceptor activity is aimed at maintaining the higher resting blood pressure. Although on the surface this change appears to be maladaptive, it allows adequate short-term regulation of blood pressure despite higher basal blood pressure.

Role of Low-Pressure Baroreceptors and Atrial Stretch

Low-pressure baroreceptors are found in low-pressure sites of the circulation, specifically the atria and large vessels of the pulmonary circulation, and respond to changes in blood volume. The Bainbridge reflex was discussed earlier; increased atrial stretch initiates a reflexive increase in heart rate. Reduction of blood volume, such as during hemorrhage, is detected by low-pressure baroreceptors in the left atrium, as well as by arterial baroreceptors. Neural afferent signals through the vagus nerve to the hypothalamus result in vasopressin release by the posterior pituitary. Like angiotensin II, vasopressin participates in short-term responses to hemorrhage, but not in acute regulation of blood pressure under normal circumstances.

Increased blood volume and stretch of atria stimulate atrial myocytes to secrete stored **atrial natriuretic peptide (ANP)**. ANP dilates some vessels (although it has little role in acute blood pressure regulation) and has important effects on

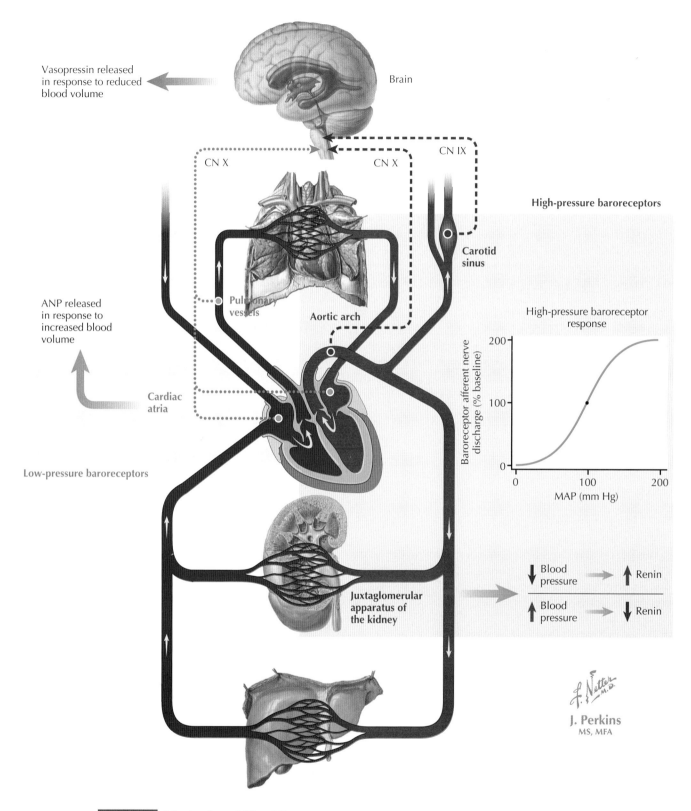

Vasopressin released in response to reduced blood volume

Brain

CN X

CN X

CN IX

High-pressure baroreceptors

Carotid sinus

ANP released in response to increased blood volume

Pulmonary vessels

Aortic arch

High-pressure baroreceptor response

Cardiac atria

Low-pressure baroreceptors

Baroreceptor afferent nerve discharge (% baseline)

200

100

0

0   100   200

MAP (mm Hg)

Juxtaglomerular apparatus of the kidney

↓ Blood pressure → ↑ Renin

↑ Blood pressure → ↓ Renin

J. Perkins
MS, MFA

Figure 13.8 **Monitoring of Blood Pressure**   To maintain adequate blood flow to tissues, the body has a complicated system for monitoring and regulating blood pressure. High-pressure baroreceptors in the aortic arch and carotid sinus are extremely important in acute regulation of blood pressure through their effects on the autonomic nervous system. Afferent arterioles in the renal juxtaglomerular apparatus also contain high-pressure baroreceptors that are involved in regulation of renin release and, consequently, regulation of sodium and water balance, which is important in long-term regulation of blood pressure. Low-pressure baroreceptors in the heart and pulmonary circulation respond to changes in blood volume and modulate sympathetic activity and vasopressin release. The cardiac atria also release atrial natriuretic peptide (ANP) in response to elevated blood volume. *CN,* cranial nerve; *MAP,* mean arterial pressure.

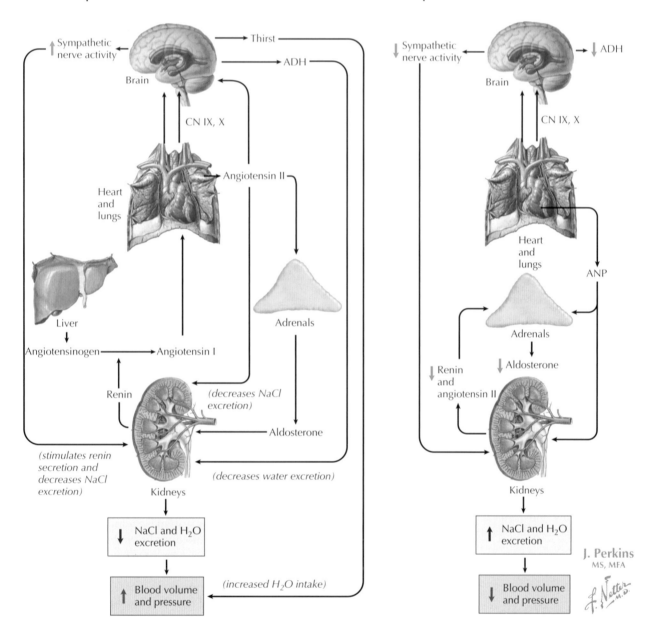

Response to Decreased Blood Volume and Pressure

Response to Increased Blood Volume and Pressure

Figure 13.9 **Long-term Response to Changes in Blood Volume and Pressure**   In addition to evoking mechanisms for acute adjustment of blood pressure, changes in blood volume and pressure will also activate renal mechanisms for adjusting blood volume. Reduced blood volume (and therefore arterial pressure) will stimulate the renin-angiotensin-aldosterone system, with the end result of sodium and water retention. Reduced blood pressure will also activate the sympathetic nervous system, which will stimulate renin secretion and have direct effects on the kidneys. On the other hand, increased volume will stimulate atrial natriuretic peptide (ANP) release by the heart. ANP has direct renal effects (natriuresis and diuresis) and also inhibits aldosterone release by the adrenal cortex. *ADH,* antidiuretic hormone; *CN,* cranial nerve.

sodium and water balance (see Section 5) and long-term regulation of blood pressure (see the next section).

Long-Term Regulation of Blood Pressure

In contrast to moment-to-moment regulation of blood pressure, which relies heavily on baroreceptor reflexes and adjustments to cardiac and vascular function, **long-term regulation**

of blood pressure is accomplished mainly by mechanisms that control blood volume through neural and humoral pathways (Fig. 13.9). These mechanisms are covered in further detail in Sections 5 and 7. Briefly, when blood volume is depleted and blood pressure is consequently reduced, direct effects of reduced renal perfusion, as well as increased sympathetic nerve activity, stimulate renin production by the kidneys (sympathetic activation occurs in part as a result of baroreceptor responses). Renin action produces elevation of angiotensin I, which is subsequently cleaved to angiotensin II by endothelial cell angiotensin-converting enzyme (much of this enzymatic cleavage occurs in the pulmonary circulation). **Angiotensin II** has direct effects on sodium retention by the kidney and has the important effect of stimulating aldosterone release by the zona glomerulosa of the adrenal cortex. **Aldosterone** also promotes renal sodium (and hence fluid) retention. Low blood volume, through reflexes discussed earlier, stimulates posterior pituitary antidiuretic hormone (ADH; also known as vasopressin). ADH release, as well as thirst, is also stimulated by elevated plasma osmolality associated with volume depletion. ADH promotes water retention by the kidney. Thus the effects of multiple hormones (ADH, angiotensin II, and aldosterone) on sodium and fluid retention and water intake result in an increase in blood volume, which helps to maintain blood pressure. These mechanisms are important in the physiological responses to hemorrhage or dehydration; various aspects are also implicated in elevation of blood pressure in some forms of hypertension.

The **diving reflex** is a specialized mechanism for regulation of blood pressure and heart rate. This reflex is an adaptation for conservation of oxygen in diving mammals, allowing protracted stays underwater without breathing. During diving, the heart rate is slowed by increased vagal activity, while pressure is maintained by arterial vasoconstriction. Thus blood flow to vital organs is maintained, while blood flow to much of the body besides the coronary and cerebral circulations is reduced; the work of the heart is also reduced. Although the diving reflex is weaker in humans, its effects can still be observed when the face is submerged in cold water and breath is held. Receptors in the face and nasal cavities are stimulated, and reflexive bradycardia and peripheral vasoconstriction occur. The diving reflex is thought to be responsible for the survival of some children after long periods of accidental submersion in cold water.

SPECIAL CIRCULATIONS

Blood flow to various tissues is regulated by local and extrinsic mechanisms, and the importance of these varies among tissues (Fig. 13.10). In addition, some circulations have unique aspects. A number of these "special circulations" merit individual consideration.

Cerebral Circulation

The **cerebral circulation** is supplied by the arterial **circle of Willis**, which is derived from the internal carotid and the vertebral arteries. Because the brain is a vital tissue that is housed in a rigid case (the cranium), it is necessary to closely regulate its blood flow. Flow below 35 milliliters per 100 g of brain per minute (mL/100 g/min) causes neuronal dysfunction; high flow may result in cerebral edema and high intracranial pressure, which also produce dysfunction. In a healthy person, cerebral blood flow is maintained at a constant level of 50 mL/100 g/min for MAPs between 50 mm Hg and 150 mm Hg. This level is maintained by autoregulation (myogenic regulation).

Cerebral blood flow is also under the control of arterial P_{CO_2}, although this blood gas does not vary significantly in a healthy person except during extreme exercise or hyperventilation or hypoventilation. Decreased arterial P_{CO_2} will produce vasoconstriction and reduced cerebral blood flow; high arterial P_{CO_2} levels will cause vasodilation and increased flow. Note that this response is to *arterial* P_{CO_2}. Global cerebral blood flow is not significantly regulated by local metabolic factors (such as CO_2, O_2, K^+, and H^+), but blood flow to various regions of the brain is adjusted according to neuronal activity in those areas. Effects of neurotransmitters on smooth muscle and endothelium, as well as vasoactive effects of substances released by astrocytes, apparently contribute to this regulation.

The **circle of Willis** provides a high degree of collateralization between the large arteries that supply oxygenated blood to the brain. The redundant pathways for blood flow help to ensure adequate blood flow to the brain in case of injury or disease—for example, as a result of stenosis of vessels or blockage due to thrombosis. Considerable variation exists in the exact anatomic structure of the circle of Willis between individuals, and less than half of the population has the most common structure illustrated in many textbooks.

In addition, the following two reflexes may affect cerebral blood flow:

- **Central nervous system ischemic reflex:** If the vasomotor center in the medulla becomes ischemic, strong sympathetic outflow acts on the heart and peripheral circulation to increase arterial blood pressure above the autoregulatory range for cerebral blood flow, thus increasing blood flow to the brain in a "last-ditch" effort to reverse the ischemia. The cerebral circulation is itself not significantly innervated by sympathetic nerves.
- **Cushing reflex:** An increase in intracranial pressure, which is usually associated with traumatic head injury, may impede cerebral blood flow. A strong sympathetic outflow will result, raising arterial pressure in an attempt to overcome the high intracranial pressure that is impeding flow. This mechanism is also a last-ditch effort to preserve cerebral perfusion. High arterial pressure is accompanied by bradycardia because of activation of arterial baroreceptors.

Coronary Circulation

Like any metabolically active tissue, the myocardium receives its own arterial supply. The **coronary circulation** is fed by the right and left coronary arteries that originate at the base of the aorta (Fig. 13.11). These arteries are also called epicardial

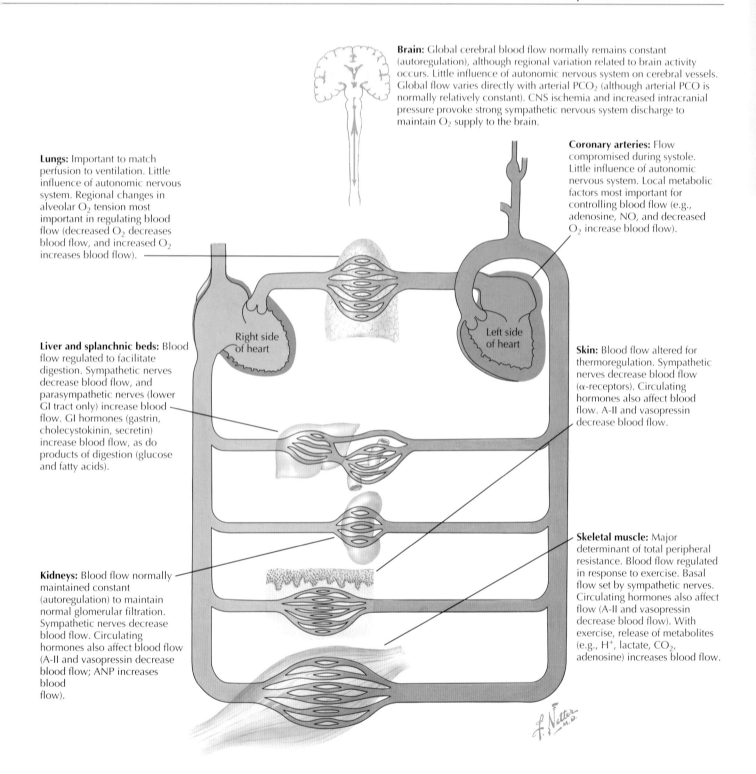

Brain: Global cerebral blood flow normally remains constant (autoregulation), although regional variation related to brain activity occurs. Little influence of autonomic nervous system on cerebral vessels. Global flow varies directly with arterial PCO_2 (although arterial PCO is normally relatively constant). CNS ischemia and increased intracranial pressure provoke strong sympathetic nervous system discharge to maintain O_2 supply to the brain.

Coronary arteries: Flow compromised during systole. Little influence of autonomic nervous system. Local metabolic factors most important for controlling blood flow (e.g., adenosine, NO, and decreased O_2 increase blood flow).

Lungs: Important to match perfusion to ventilation. Little influence of autonomic nervous system. Regional changes in alveolar O_2 tension most important in regulating blood flow (decreased O_2 decreases blood flow, and increased O_2 increases blood flow).

Liver and splanchnic beds: Blood flow regulated to facilitate digestion. Sympathetic nerves decrease blood flow, and parasympathetic nerves (lower GI tract only) increase blood flow. GI hormones (gastrin, cholecystokinin, secretin) increase blood flow, as do products of digestion (glucose and fatty acids).

Skin: Blood flow altered for thermoregulation. Sympathetic nerves decrease blood flow (α-receptors). Circulating hormones also affect blood flow. A-II and vasopressin decrease blood flow.

Right side of heart

Left side of heart

Kidneys: Blood flow normally maintained constant (autoregulation) to maintain normal glomerular filtration. Sympathetic nerves decrease blood flow. Circulating hormones also affect blood flow (A-II and vasopressin decrease blood flow; ANP increases blood flow).

Skeletal muscle: Major determinant of total peripheral resistance. Blood flow regulated in response to exercise. Basal flow set by sympathetic nerves. Circulating hormones also affect flow (A-II and vasopressin decrease blood flow). With exercise, release of metabolites (e.g., H^+, lactate, CO_2, adenosine) increases blood flow.

Figure 13.10 **Circulation to Special Regions** Mechanisms regulating blood flow vary among regions in the systemic circulation, reflecting the physiological functions and needs of the tissues. *A-II,* angiotensin II; *ANP,* atrial natriuretic peptide; *GI,* gastrointestinal; *NO,* nitric oxide.

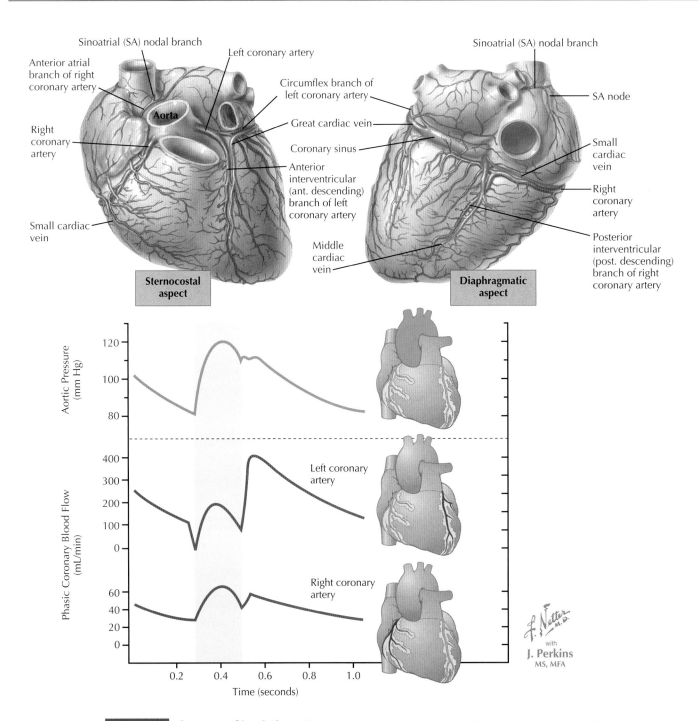

Figure 13.11 **Coronary Circulation** The coronary arteries supply arterial blood to the muscular wall of the heart. The pressure gradient for flow through the arteries is affected by tissue pressure in the wall of the heart during systole, particularly in the left coronary circulation. Thus, while flow through both right and left coronary arteries is related to aortic pressure, these compressive forces reduce left coronary flow during systole. During diastole, flow is greatly enhanced in the left coronary circulation by the actions of vasodilator metabolites that build up during systole. The metabolite adenosine is believed to be particularly important in this enhancement of coronary flow.

arteries because they lie on the surface of the heart (epicardium) and supply branches into the myocardium. These branches form a very extensive microcirculation, supplying oxygen and nutrients and removing metabolites from the highly metabolically active myocardium. The heart has a very large arteriovenous O_2 gradient, even at rest.

The left coronary artery gives rise to the left anterior descending artery and the circumflex artery; likewise, the right coronary artery gives rise to several branches. The venous drainage of the heart returns blood to the coronary sinus, which empties into the right atrium. The layers of the heart, from inside to outside, are as follows:

- **Endocardium:** The innermost layer is connective tissue, lined by endothelial cells.
- **Myocardium:** The muscular layer that forms the bulk of the heart wall. The subendocardium is the inner $\frac{1}{3}$ of the myocardium, and the subepicardium is the outer $\frac{1}{3}$ of the myocardium.
- **Epicardium:** The outer, connective tissue layer of the heart. It forms the visceral pericardium. The parietal pericardium is a sac enclosing the heart; the pericardial cavity is the space between the visceral and parietal pericardium.

Flow to the myocardium is affected by factors that are in some ways different from those regulating flow in other circulations. In particular, coronary blood flow is affected greatly by the following mechanisms:

- Compression of the coronary circulation caused by contraction of myocardium.
- Powerful metabolic vasodilation during diastole.

The pumping action of the left ventricle generates the normal resting arterial pressure of 120/80 mm Hg to supply blood to the systemic circulation. In doing so, the myocardium of the left ventricle, and in particular, the subendocardium, must generate an extravascular tissue pressure that is higher than left ventricular and arterial pressures. This intramyocardial pressure impedes left coronary flow during systole, when the net coronary perfusion pressure is arterial pressure minus the intramyocardial pressure. Thus left coronary artery flow is low during systole (see Fig. 13.11). During isovolumetric contraction, flow rapidly decreases because tissue pressure has exceeded arterial pressure. During the remainder of systole, the shape of the left coronary flow curve is similar to the aortic pressure curve, but flow remains low because of high intramyocardial pressure.

With the onset of isovolumetric relaxation, left coronary artery flow increases markedly as a result of two factors. During systole, when flow is low, metabolites (H^+, CO_2, K^+, prostaglandins, lactic acid, adenosine, and others) accumulate and O_2 tension falls. These changes produce coronary vasodilation; adenosine is believed to be the major factor involved.

The vasodilation, combined with the fall in intramyocardial pressure, results in the large increase in left coronary flow during diastole.

Flow through the right coronary artery is less affected by intramyocardial pressure, because tissue pressure in the right ventricle does not nearly reach that of the left ventricle and does not exceed arterial pressure. Thus the shape of the right coronary artery blood flow curve (see Fig. 13.11) resembles the aortic pressure curve, with the highest flow occurring during systole.

Sympathetic nerves innervate coronary arteries and can affect blood flow, but to a lesser extent than in most other circulations. Activation of the sympathetic nervous system increases the work of the heart, and thus metabolic vasodilation largely overrides sympathetic coronary vasoconstriction. However, it is believed that sympathetic vasoconstriction may limit the extent of metabolic vasodilation, and thus, in a compromised heart (e.g., in coronary heart disease), sympathetic stimulation may contribute to ischemia.

Autoregulation of coronary blood flow can be demonstrated experimentally but is a less important factor in normal regulation of coronary flow than metabolic factors or the effects of intramyocardial pressure.

Blood Flow in Exercise and Effects on Specific Circulations

The importance of regional differences in regulatory mechanisms governing blood flow can be gleaned by considering the changes in flow that occur during exercise. Regulation of arterial blood pressure and flow during **dynamic exercise** is a complex process, involving large changes in CO and regional resistances (Fig. 13.12). Dynamic exercise, also called aerobic exercise, involves rhythmic contraction/relaxation of large skeletal muscle groups, such as during jogging, swimming, or aerobics. Young, fit persons are able to exercise at levels associated with COs of 20 to 30 L/min, by increasing both heart rate and stroke volume. Sympathetic nervous system activation results in vasoconstriction and reduced flow (as a proportion of CO) in many but not all regional circulations (see Fig. 13.12):

- Coronary blood flow increases, mainly as a result of metabolic vasodilation.
- Skin blood flow is reduced at first but will eventually rise as a result of thermoregulatory mechanisms.
- Skeletal muscle flow increases dramatically, as constrictor effects of catecholamines are overridden by metabolic vasodilation (and to some extent, activation of β-adrenergic receptors).

Total peripheral resistance is reduced as a result of these changes. Thus aerobic exercise is associated with high CO and low resistance. Constriction of veins results in increased

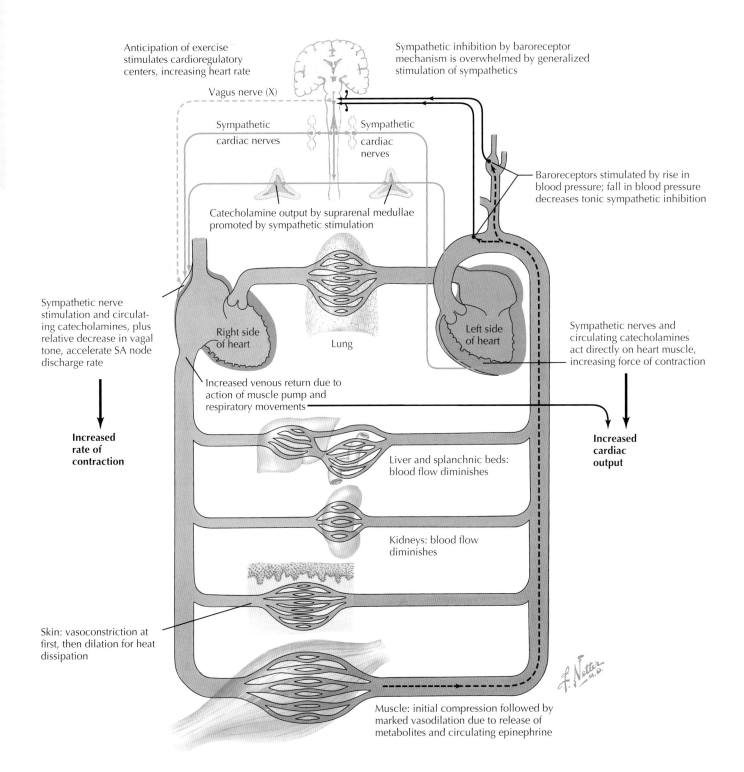

Anticipation of exercise stimulates cardioregulatory centers, increasing heart rate

Sympathetic inhibition by baroreceptor mechanism is overwhelmed by generalized stimulation of sympathetics

Vagus nerve (X)

Sympathetic cardiac nerves

Sympathetic cardiac nerves

Baroreceptors stimulated by rise in blood pressure; fall in blood pressure decreases tonic sympathetic inhibition

Catecholamine output by suprarenal medullae promoted by sympathetic stimulation

Sympathetic nerve stimulation and circulating catecholamines, plus relative decrease in vagal tone, accelerate SA node discharge rate

Right side of heart

Lung

Left side of heart

Sympathetic nerves and circulating catecholamines act directly on heart muscle, increasing force of contraction

Increased venous return due to action of muscle pump and respiratory movements

Increased rate of contraction

Increased cardiac output

Liver and splanchnic beds: blood flow diminishes

Kidneys: blood flow diminishes

Skin: vasoconstriction at first, then dilation for heat dissipation

Muscle: initial compression followed by marked vasodilation due to release of metabolites and circulating epinephrine

Figure 13.12 **Circulatory Response to Exercise** Dynamic (aerobic) exercise elicits an integrated circulatory response. The autonomic nervous system is important in the neural regulation of this response, affecting heart rate, contractility, and vascular tone. During exercise, skeletal muscle blood flow is greatly elevated mainly because of the production of vasodilatory metabolites. *SA*, sinoatrial.

central venous pressure, augmenting cardiac filling pressure and stroke volume. Arterial blood pressure is elevated during dynamic exercise, but it is mainly systolic pressure (and thus pulse pressure) that rises because of the large stroke volume. Diastolic pressure may be slightly elevated or slightly reduced, because low peripheral resistance results in a rapid fall in arterial pressure during diastole.

Fetal Circulation

Although the **fetal circulation** is not a special circulation in the same sense as the cerebral and coronary circulations, several significant differences between the prenatal and postnatal circulations merit discussion. The fetus, enveloped in the amniotic sac and floating in the amniotic fluid, is dependent on the placental circulation for exchange of gases and nutrients. To adapt the circulation to prenatal life, the fetus has six structures that are not normally seen in adults (Fig. 13.13):

- Two large umbilical arteries that branch from the systemic arterial circulation and supply the placental circulation, where gas exchange and nutrient and waste exchange occur.
- The umbilical vein, which returns placental blood to the systemic venous circulation, supplying O_2 and nutrients to the fetus.
- The ductus venosus, a shunt between the umbilical vein and the inferior vena cava. Although most of the placental blood passes through the liver, a fraction of the placental circulation passes directly into the vena cava through the ductus venosus.
- Foramen ovale, a "right-to-left" shunt through which most of the blood from the inferior vena cava flows into the left atrium, bypassing the right ventricle and the pulmonary circulation.
- Ductus arteriosus, another right-to-left shunt between the pulmonary artery and the aorta. Pulmonary vascular resistance is high in the fetus because of the collapsed state of the lungs and low oxygen tension in the lungs; 90% of pulmonary arterial blood flows through the ductus arteriosus to the aorta.

These structures all normally close shortly after birth, although anatomic closure, as opposed to functional closure, requires more time. Inflation of the lungs and inhalation of oxygen-rich air reduces pulmonary vascular resistance, reversing flow through the ductus arteriosus. High oxygen tension initiates closure of the ductus. With increased venous return from the lungs to the left atrium and reduced flow to the right atrium because of occlusion of the placental circulation, left atrial pressure rises above right atrial pressure, closing a valve over the foramen ovale. With closure of the ductus arteriosus and foramen ovale, the "adult" pattern of circulation of blood through the right heart, lungs, left heart, and systemic circulation is established.

Unlike the adult circulation, the fetal circulation can be considered a parallel circulation. Much of the right ventricular output is shunted past the lungs to the aorta through the ductus arteriosus, mixing with blood ejected by the left ventricle. Flow through the pulmonary circulation is not critical, because oxygenation of blood does not take place in the lungs in utero. With closure of the foramen ovale and ductus arteriosus shortly after birth, the fetal circulation is transformed to a series circulation.

CLINICAL CORRELATE 13.1
Myocardial Ischemia

Myocardial ischemia is reduced blood flow to heart muscle, resulting in poor oxygenation. Angina, arrhythmias, myocardial infarction, and sudden death may result. In patients with coronary heart disease, buildup of atherosclerotic plaque in epicardial coronary arteries results in reduced perfusion of the affected vessels. Functionally, ischemia and infarction may result in reduced contractility of the heart, arrhythmias, electrocardiogram changes, wall motion abnormalities, and even ventricular aneurysm. Myocardial infarction is one of the major causes of congestive heart failure. Severe cases of coronary occlusion are treated by angioplasty or coronary artery bypass graft surgery. In bypass surgery, a section of vessel, typically a piece of saphenous vein, is grafted to bypass the blockage. In angioplasty, a balloon catheter is used to open the occluded artery. To reduce the occurrence of restenosis (reocclusion), a stent is often deployed.

Continued

CLINICAL CORRELATE 13.1
Myocardial Ischemia—cont'd

A. Coronary angioplasty and stent deployment

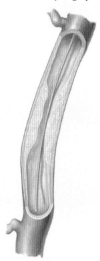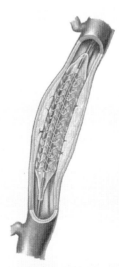

As the first step, a coronary guide wire is advanced across the stenotic atherosclerotic plaque.

A double-lumen catheter with a balloon is slid over the guide wire; the balloon is inflated to compress the plaque and open the obstruction.

A balloon catheter containing the stent is placed in the dilated area.

The balloon is expanded, deploying the stent.

Once the stent has been deployed, the catheter and the guide wire are removed.

B. Manifestations of myocardial infarction

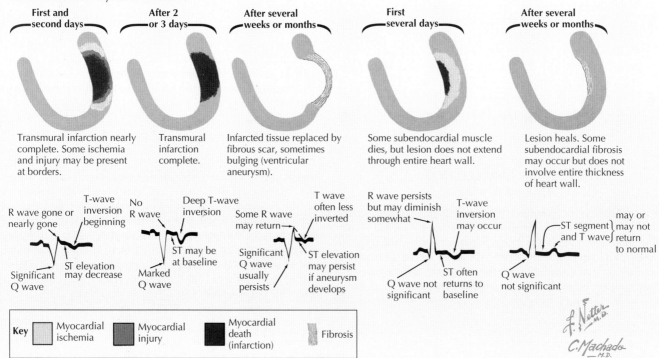

First and second days

Transmural infarction nearly complete. Some ischemia and injury may be present at borders.

R wave gone or nearly gone
T-wave inversion beginning
Significant Q wave
ST elevation may decrease

After 2 or 3 days

Transmural infarction complete.

No R wave
Deep T-wave inversion
Marked Q wave
ST may be at baseline

After several weeks or months

Infarcted tissue replaced by fibrous scar, sometimes bulging (ventricular aneurysm).

Some R wave may return
T wave often less inverted
Significant Q wave usually persists
ST elevation may persist if aneurysm develops

First several days

Some subendocardial muscle dies, but lesion does not extend through entire heart wall.

R wave persists but may diminish somewhat
T-wave inversion may occur
Q wave not significant
ST often returns to baseline

After several weeks or months

Lesion heals. Some subendocardial fibrosis may occur but does not involve entire thickness of heart wall.

ST segment and T wave } may or may not return to normal
Q wave not significant

Key
Myocardial ischemia
Myocardial injury
Myocardial death (infarction)
Fibrosis

Prenatal circulation

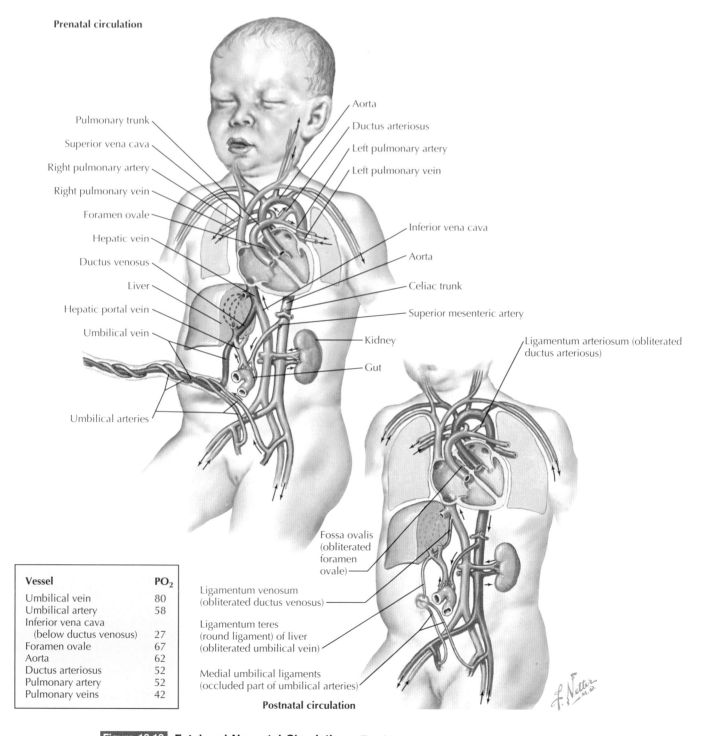

Pulmonary trunk

Superior vena cava

Right pulmonary artery

Right pulmonary vein

Foramen ovale

Hepatic vein

Ductus venosus

Liver

Hepatic portal vein

Umbilical vein

Umbilical arteries

Aorta

Ductus arteriosus

Left pulmonary artery

Left pulmonary vein

Inferior vena cava

Aorta

Celiac trunk

Superior mesenteric artery

Kidney

Gut

Ligamentum arteriosum (obliterated ductus arteriosus)

Fossa ovalis (obliterated foramen ovale)

Ligamentum venosum (obliterated ductus venosus)

Ligamentum teres (round ligament) of liver (obliterated umbilical vein)

Medial umbilical ligaments (occluded part of umbilical arteries)

Postnatal circulation

Vessel	PO$_2$
Umbilical vein	80
Umbilical artery	58
Inferior vena cava (below ductus venosus)	27
Foramen ovale	67
Aorta	62
Ductus arteriosus	52
Pulmonary artery	52
Pulmonary veins	42

Figure 13.13 **Fetal and Neonatal Circulation**   The fetal circulation is specifically adapted to efficiently exchange gases, nutrients, and wastes through placental circulation. Upon birth, the shunts (foramen ovale, ductus arteriosus, and ductus venosus) close and the placental circulation is disrupted, producing the series circulation of blood through the lungs, left atrium, left ventricle, systemic circulation, right heart, and back to the lungs.

Review Questions

CHAPTER 8: THE BLOOD

1. Effective hemostatic function requires:

A. normally functioning platelets.
B. fibrinogen.
C. vitamin K.
D. factor VIII.
E. all the above.

2. Blood volume in a child is determined to be 4.0 L; red blood cell volume is measured to be 1.6 L. Hematocrit in this child is:

A. 25%.
B. 33%.
C. 40%.
D. 50%.
E. 68%.

3. Tissue macrophages are formed by differentiation of:

A. lymphocytes.
B. monocytes.
C. basophils.
D. eosinophils.
E. neutrophils.

4. Coagulation proteins are removed in the preparation of:

A. serum.
B. whole blood.
C. plasma.
D. platelet-rich plasma.
E. All the above.

CHAPTER 9: OVERVIEW OF THE HEART AND CIRCULATION

5. The greatest volume of blood is found in the:

A. systemic arteries.
B. systemic veins.
C. pulmonary arteries.
D. pulmonary veins.
E. chambers of the heart.

6. Contraction of which chamber of the heart is best described as a constriction of the chamber, resulting in shortening from base to apex?

A. Left atrium
B. Right atrium
C. Left ventricle
D. Right ventricle
E. All the above

7. Of the following parts of the circulation, which receives the smallest percentage of cardiac output at rest?

A. Coronary circulation
B. Liver and gastrointestinal tract
C. Skeletal muscle
D. Kidneys
E. Cerebral circulation

8. Of the following parts of the circulation, which receives the largest percentage of cardiac output at rest?

A. Kidney
B. Liver and gastrointestinal tract
C. Lung
D. Brain
E. Skin

CHAPTER 10: CARDIAC ELECTROPHYSIOLOGY

9. Rapid phase 3 repolarization of cardiac myocytes during the action potential is mainly associated with:

A. opening of Na^+ channels.
B. reduced conductance of the inwardly rectified K^+ current (i_{K1}).
C. opening of voltage-sensitive slow L-type Ca^{2+} channels.
D. inactivation of Na^+ channels.
E. activation of K^+ channels.

10. An electrocardiogram may reveal:

A. abnormalities in cardiac rhythm and conduction.
B. location and extent of myocardial ischemia.
C. orientation of the heart and size of chambers.
D. effects of abnormal electrolyte levels.
E. all the above.

11. Conduction velocity is slowest in the:

A. atrial myocytes.
B. atrioventricular node.
C. bundle of His.
D. Purkinje fibers.
E. ventricular myocytes.

12. The phase 2 plateau of the cardiac action potential is associated with:

A. open voltage-sensitive, slow L-type Ca^{2+} channels.
B. open voltage-dependent K^+ channels.
C. outward K^+ current and inward Ca^{2+} current.
D. refractoriness to another depolarization.
E. all the above.

CHAPTER 11: FLOW, PRESSURE, AND RESISTANCE

13. Given a resting heart rate of 75 beats/min and arterial blood pressure of 130/80 mm Hg, what is the approximate mean arterial pressure?

A. 90 mm Hg
B. 97 mm Hg
C. 105 mm Hg
D. 110 mm Hg
E. 115 mm Hg

14. The highest pulse pressure in the circulation is observed in the:

A. left ventricle.
B. right ventricle.
C. aorta.
D. pulmonary artery.
E. systemic arterioles.

15. Which of the following changes would result in the greatest increase in flow through a tube, assuming that other factors are held constant?

A. Doubling of the pressure gradient between ends of the tube
B. Reducing the viscosity of the fluid by half
C. Doubling the viscosity of the fluid
D. Reducing the length of the tube by half
E. Doubling the radius of the tube

16. Which of the following conditions favors laminar flow in a tube?

A. Wide tube diameter
B. Rapid velocity of fluid flow
C. High viscosity of fluid
D. High density of fluid
E. Pulsatility of flow

17. Wall tension in a blood vessel would be increased by:

A. reduction of vessel radius.
B. reduction of hydrostatic pressure within the vessel lumen.
C. reduction of hydrostatic pressure in the interstitial space.
D. reduction of transmural pressure.
E. none of the above.

CHAPTER 12: THE CARDIAC PUMP

18. The beginning of the isovolumetric contraction period for the right ventricle is marked by:

A. closure of the aortic valve.
B. closure of the pulmonic valve.
C. closure of the mitral valve.
D. closure of the tricuspid valve.
E. opening of the mitral valve.

19. The correct order of valve opening and closure during a cardiac cycle is:

A. mitral closure, tricuspid closure, pulmonic opening, aortic opening, aortic closure, pulmonic closure, tricuspid opening, mitral opening.
B. tricuspid closure, mitral closure, pulmonic opening, aortic opening, aortic closure, pulmonic closure, tricuspid opening, mitral opening.
C. mitral closure, tricuspid closure, aortic opening, pulmonic opening, aortic closure, pulmonic closure, tricuspid opening, mitral opening.
D. mitral closure, tricuspid closure, pulmonic opening, aortic opening, pulmonic closure, aortic closure, tricuspid opening, mitral opening.
E. mitral closure, tricuspid closure, pulmonic opening, aortic opening, aortic closure, pulmonic closure, mitral opening, tricuspid opening.

20. In respiratory sinus arrhythmia, the increase in heart rate during inspiration is initiated by:

A. reduced stretch of arterial baroreceptors.
B. increased stretch of arterial baroreceptors.
C. release of atrial natriuretic peptide.
D. reduced stretch of atrial stretch receptors.
E. increased stretch of atrial stretch receptors.

21. Activation of the sympathetic nervous system will produce:

A. increased heart rate.
B. arterial vasoconstriction.
C. constriction of veins.
D. increased myocardial contractility.
E. all the above.

22. An increase in myocardial contractility will result in which of the following changes in the force–velocity relationship?

A. An increase in the maximum force of contraction without change in V_m
B. An increase in the maximum force of contraction and an increase in V_m
C. A decrease in the maximum force of contraction without change in V_m
D. A decrease in the maximum force of contraction and an increase in V_m
E. A decrease in the maximum force of contraction and a decrease in V_m

23. When the vascular function and cardiac function curves are plotted on the same graph, the point of intersection represents the:

A. point at which baroreceptor activation occurs.
B. resting cardiac output and central venous pressure.
C. normal mean circulatory pressure.
D. normal mean arterial pressure.
E. point of maximum contractility.

CHAPTER 13: THE PERIPHERAL CIRCULATION

24. Release of nitric oxide by the endothelial cells in a small artery would be expected to produce all of the following effects **except:**

A. elevation of cyclic guanosine monophosphate in smooth muscle underlying the endothelium.
B. inhibition of platelet adhesion to the vascular wall.
C. elevation of free intracellular Ca^{2+} in the underlying smooth muscle.
D. vasodilation of the vessel.
E. increased hydrostatic pressure in downstream capillaries.

25. In an experimental model, the perfusion pressure for an artery is increased. A rise in blood flow through the artery is observed, but despite continued high perfusion pressure, flow falls over a 2-minute period toward the baseline level. This reduction in flow is most likely the result of:

A. autoregulation of blood flow.
B. reactive hyperemia.
C. active hyperemia.
D. shear stress–induced nitric oxide release.
E. increased sympathetic nerve activity.

26. Vasodilation is evoked by binding of adrenergic transmitter by which type of vascular adrenergic receptor?

A. α_1
B. α_2
C. β_1
D. β_2
E. All the above

27. Afferent nerve activity from the arterial baroreceptors will be greatest when:

A. mean arterial pressure (MAP) and pulse pressure are both high.
B. MAP is high and pulse pressure is low.
C. MAP and pulse pressure are both low.
D. MAP is low and pulse pressure is high.
E. MAP and pulse pressure are in the normal resting range.

28. Left coronary artery blood flow is highest during:

A. early systole.
B. late systole.
C. early diastole.
D. mid diastole.
E. late diastole.

Section 4

RESPIRATORY PHYSIOLOGY

The components of the respiratory system are the lungs, airways, and the muscles of breathing. The primary function of this system is to facilitate the exchange of oxygen and carbon dioxide between the body and the environment. The respiratory system participates in additional, important physiological processes, including acid-base regulation, temperature regulation, immune function, and metabolic functions, among others. In the following chapters, the basic physiology of ventilation, pulmonary perfusion, and gas transport are covered, along with the critical role of the respiratory system in acid-base balance.

Chapter 14

Pulmonary Ventilation and Perfusion and Diffusion of Gases

Respiration is a complex process that begins with ventilation of the lungs and diffusion of gases between the lungs and blood. Meanwhile, the lungs are perfused with blood, which transports gases between lungs and tissues, where the biochemical processes involved in cellular respiration occur.

BLOOD FLOW IN THE LUNGS

The general principles of blood flow are discussed in Section 3, but the systemic and pulmonary circulations are different in some important aspects. Although right ventricular output is the same as left ventricular output (normally about 5 L/min at rest), pressures in the pulmonary circulation are lower than pressures at equivalent points in the systemic circulation (Fig. 14.1). To accommodate high flow rate at lower pressure, resistance is also low in the pulmonary circulation. This low resistance is a result of the dense microcirculation and shorter vessels in the lungs compared with the systemic circulation.

Passive Control of Pulmonary Resistance

Pulmonary resistance is subject to **passive control** by pulmonary artery pressure and lung volume (Fig. 14.2). Ordinarily, some pulmonary capillaries are collapsed. When **pulmonary artery pressure** increases, **recruitment** and **distension** of pulmonary capillaries occurs. In other words, collapsed capillaries are recruited to the circulation by the higher perfusion pressure, and individual capillaries distend, further reducing resistance. Such recruitment and distension are important in reducing pulmonary vascular resistance during high cardiac output, such as during exercise. **Lung volume** has different effects on extraalveolar vessels and alveolar capillaries. As lung volume increases, the associated traction causes distension of extraalveolar vessels, reducing resistance within them. In contrast, with increased lung volume, alveolar vessels are compressed by the inflation of alveoli. Thus as lungs are inflated from a very low volume, resistance first falls as a result of the effects on extraalveolar vessels, but as inflation continues, resistance begins to rise as alveolar vessels are compressed.

Active Control of Pulmonary Resistance

Pulmonary vascular resistance is also subject to **active control** by **chemical and humoral substances** (Fig. 14.2). Hypoxia affects pulmonary vessels differently than it affects systemic vessels. In the systemic circulation, tissue hypoxia results in metabolic vasodilation of the arterial microcirculation. In contrast, in the pulmonary system, when **alveolar oxygen concentration** (PA_{O_2}) falls, vasoconstriction of arterioles in the hypoxic region occurs. This is the most important mechanism for short-term regulation of regional pulmonary blood flow and results in better perfusion of well-oxygenated portions of the lungs.

FUNCTIONAL ANATOMY OF THE LUNGS AND AIRWAYS

The human lungs consist of the three lobes of the right lung and two lobes of the left lung (Fig. 14.3). The bronchi, blood vessels, lymphatic vessels, and nerves leave and enter at the hilum of each lung. The airways consist of the trachea (windpipe), right and left main bronchi, smaller bronchi, and bronchioles (Figs. 14.4 and 14.5). The trachea branches into the right and left main bronchi, which enter the lungs, further divide, and become smaller in diameter. Up to 23 generations of branching airways are present in the system beginning at the trachea and leading to the alveoli.

Conducting Zone of the Lung

Airways from the trachea to the terminal bronchioles constitute the **conducting zone** of the lungs and are not capable of gas exchange, which takes place only in the respiratory bronchioles and alveoli. The conducting zone is also known as **anatomic dead space** because of the lack of gas exchange in this area. In an adult, the anatomic dead space contains approximately 150 mL of air. The trachea has cartilaginous rings that extend around three quarters of its circumference, providing the structural support necessary to hold airways open but also allowing coughing. Cartilaginous plates are present in the walls of bronchi but are lost at the level of the bronchioles.

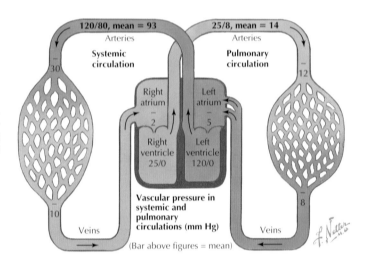

Figure 14.1 **Pulmonary and Systemic Circulations** The pulmonary circulation exists in series with the systemic circulation, and for this reason, equal cardiac output and stroke volumes are ejected from the right and left ventricles. Pressures and resistances are lower in the pulmonary circulation than in the systemic circulation.

A. Effects of increases in pulmonary blood flow and vascular pressures

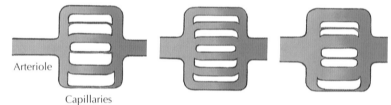

Arteriole

Capillaries

Normally some pulmonary capilliaries are closed and conduct no blood

Recruitment: More capillaries open as pulmonary vascular pressure or blood flow increases

Distention: At high vascular pressures individual capillaries widen and acquire a larger cross-sectional area

B. Effects of lung volume

Figure 14.2 **Pulmonary Vascular Resistance** Resistance in the pulmonary circulation is affected by pulmonary artery pressure and blood flow, lung volume, and chemical and humoral substances. Elevation of pulmonary artery pressure causes distension of vessels and recruitment of capillaries that are otherwise collapsed **(A)**, resulting in reduced resistance. An increase in lung volume causes traction and thus distension of extraalveolar vessels but results in compression of alveolar capillaries **(B)**; the combination of these two effects results in lowest pulmonary vascular resistance at intermediate lung volumes. Pulmonary vessels also constrict and dilate in response to a number of chemical and humoral mediators **(C)**.

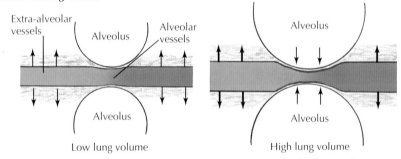

Extra-alveolar vessels

Alveolar vessels

Alveolus

Alveolus

Alveolus

Alveolus

Low lung volume

High lung volume

As lung volume increases, increasing traction on extra-alveolar capillaries produces distension, and their resistance falls. Alveolar vessels, in contrast, are compressed by enlarging alveoli, and their resistance increases.

C. Effects of chemical and humoral substances

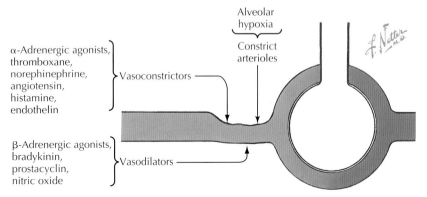

α-Adrenergic agonists, thromboxane, norephinephrine, angiotensin, histamine, endothelin
} Vasoconstrictors

β-Adrenergic agonists, bradykinin, prostacyclin, nitric oxide
} Vasodilators

Alveolar hypoxia

Constrict arterioles

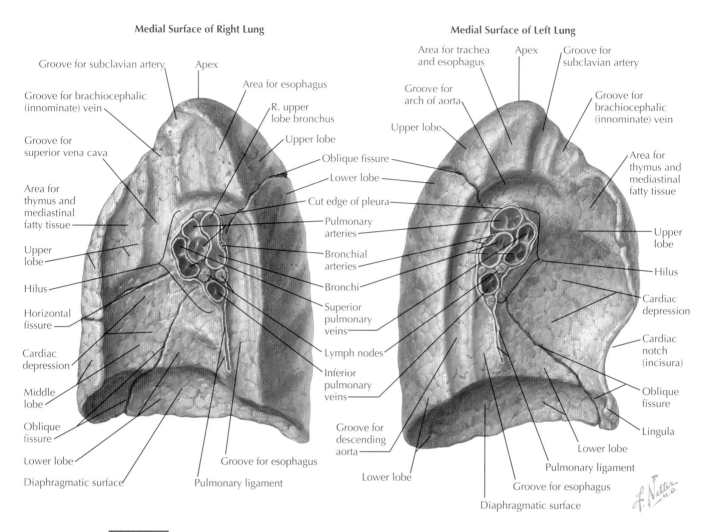

Figure 14.3 **Gross Anatomy of the Lung**   The three lobes of the right (R) lung and the two lobes of the left lung are depicted in this medial view of the lungs. Blood vessels, nerves, right and left main bronchi, and lymphatics enter and exit the lungs at the hilum.

Most of the conducting system is lined by ciliated, **pseudostratified columnar epithelial cells**, mucus-secreting **goblet cells**, and several other cell types, with ciliated cells and goblet cells constituting the majority of cells lining the larger airways (Fig. 14.6). Mucus prevents desiccation of the epithelium and traps particulate matter in inspired air. Particles are carried upward (out of the lung) by ciliary action. This process is referred to as **mucociliary transport**. The epithelium becomes **cuboidal epithelium** in bronchioles, where ciliated cells are predominant. Goblet cells diminish and are absent in terminal bronchioles. **Club cells** (formerly known as *Clara cells*) in bronchioles secrete substances that line the bronchioles and play a role in the defense system of the airways; they also act as stem cells for regeneration of bronchiolar epithelium.

The walls of the conducting airways also contain smooth muscle, which is regulated by the autonomic nervous system. Sympathetic stimulation dilates airways through the effects of norepinephrine (or circulating epinephrine) on β_2 receptors

on smooth muscle cells. Parasympathetic stimulation constricts airways through muscarinic receptor activation by acetylcholine. Airway dilation is characteristic of the "fight-or-flight" response (see Chapter 7).

Respiratory Zone of the Lung

Inspired air passes through the trachea, bronchi, and bronchioles, eventually entering the terminal bronchioles, respiratory bronchioles, and alveoli. The respiratory bronchioles and alveoli are the **respiratory zone of the lungs**, where gas exchange occurs. There are an estimated 300 million alveoli in the human lungs, with an exchange area of 50 to 100 m². This enormous surface area, along with the thinness of the air-blood interface, allows for efficient gas exchange. This interface, known as the **alveolar-capillary membrane**, consists of the capillary endothelium, the alveolar epithelium, and the basement membrane, upon which these two single-cell layers are formed. The alveolar-capillary membrane is approximately 500 nm thick. Most of the alveolar surface is surrounded in

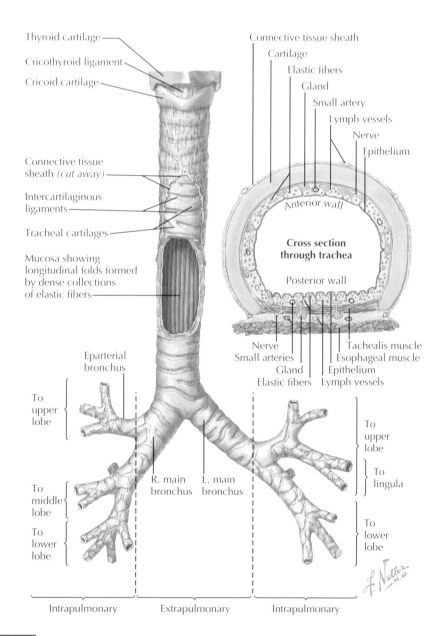

Thyroid cartilage

Cricothyroid ligament

Cricoid cartilage

Connective tissue sheath *(cut away)*

Intercartilaginous ligaments

Tracheal cartilages

Mucosa showing longitudinal folds formed by dense collections of elastic fibers

Connective tissue sheath

Cartilage

Elastic fibers

Gland

Small artery

Lymph vessels

Nerve

Epithelium

Anterior wall

Cross section through trachea

Posterior wall

Nerve

Small arteries

Gland

Elastic fibers

Tachealis muscle

Esophageal muscle

Epithelium

Lymph vessels

Eparterial bronchus

To upper lobe

To middle lobe

To lower lobe

R. main bronchus

L. main bronchus

To upper lobe

To lingula

To lower lobe

Intrapulmonary

Extrapulmonary

Intrapulmonary

Figure 14.4 **Structure of the Trachea and Major Bronchi** The trachea divides into the right (R) and left (L) main bronchi before entering the lungs. The airways branch up to 23 times, leading to the alveoli. The airways from the trachea to the terminal bronchioles are the conducting zone of the lung. With branching, the diameter of the airways diminishes. Collagen is less prevalent in the walls of smaller airways and is lost in the bronchioles.

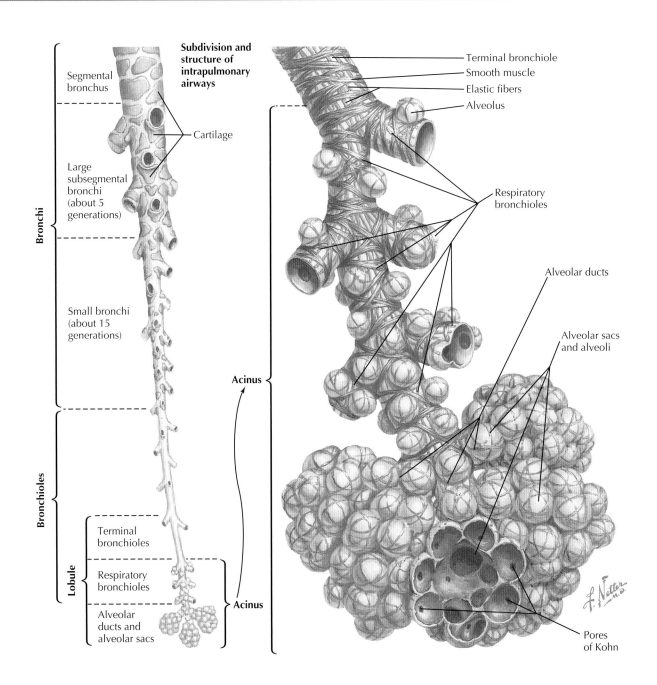

Subdivision and structure of intrapulmonary airways

Segmental bronchus

Cartilage

Large subsegmental bronchi (about 5 generations)

Small bronchi (about 15 generations)

Bronchi

Bronchioles

Terminal bronchioles

Respiratory bronchioles

Alveolar ducts and alveolar sacs

Lobule

Acinus

Acinus

Terminal bronchiole

Smooth muscle

Elastic fibers

Alveolus

Respiratory bronchioles

Alveolar ducts

Alveolar sacs and alveoli

Pores of Kohn

Figure 14.5   Intrapulmonary Airways   After branching up to 23 times, the conducting system leads to the terminal bronchioles in the pulmonary lobules. Respiratory bronchioles and alveolar ducts give rise to alveolar sacs, the major site of gas exchange. The respiratory zone (gas exchange region) consists of structures distal to the terminal bronchioles (alveolar sacs, ducts, and respiratory bronchioles).

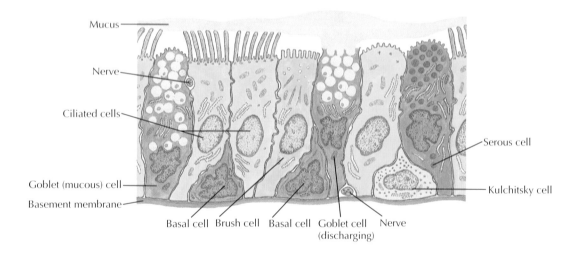

Trachea and large bronchi. Ciliated and goblet cells predominant, with some serous cells and occasional brush cells and club cells. Numerous basal cells and occasional Kulchitsky cells are present.

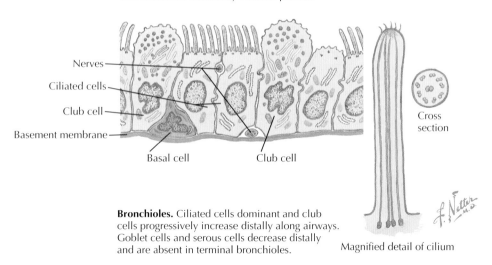

Bronchioles. Ciliated cells dominant and club cells progressively increase distally along airways. Goblet cells and serous cells decrease distally and are absent in terminal bronchioles.

Magnified detail of cilium

Figure 14.6 **Ultrastructure of Tracheal, Bronchial, and Bronchiolar Epithelium**   The trachea and bronchi are lined mainly by pseudostratified columnar epithelium and goblet cells, along with several other, less common cell types. The goblet cells have the important function of mucus secretion. Kulchitsky cells are neuroendocrine-like cells that secrete paracrine factors and are part of the "diffuse neuroendocrine system." The functions of brush and serous cells are not well defined. Basal cells are pulmonary epithelial stem cells. The epithelial lining of bronchioles contains columnar epithelial cells; goblet cells are lost in terminal bronchioles. Club cells (formerly known as *Clara cells*) are secretory cells of the bronchioles.

this manner by pulmonary capillaries, allowing for "sheet flow" of blood around alveolar sacs, with a broad surface area for gas exchange (Figs. 14.7 and 14.8).

Alveoli are lined by type I and type II epithelial cells:

- **Type I epithelial cells** constitute more than 90% of the surface area. Their squamous structure is adapted for gas diffusion across the alveolar-capillary membrane.
- **Type II epithelial cells** are cuboidal cells that secrete surfactant, a complex lipoprotein that lines the surface of alveoli and reduces their surface tension, increasing compliance of the lung (see "Surfactant and Surface Tension" in Chapter 15).

PULMONARY VOLUMES AND CAPACITIES

To describe pulmonary function in health and disease, it is necessary to understand the volumes and capacities associated with the lungs and breathing. There are four basic pulmonary volumes:

- **Tidal volume (V_T):** The volume of air inhaled and exhaled during breathing. Resting V_T is approximately 500 mL.
- **Residual volume (RV):** The volume remaining in the lungs after a maximal exhalation.
- **Expiratory reserve volume (ERV):** The additional volume a subject is *capable* of exhaling after a normal, quiet expiration.

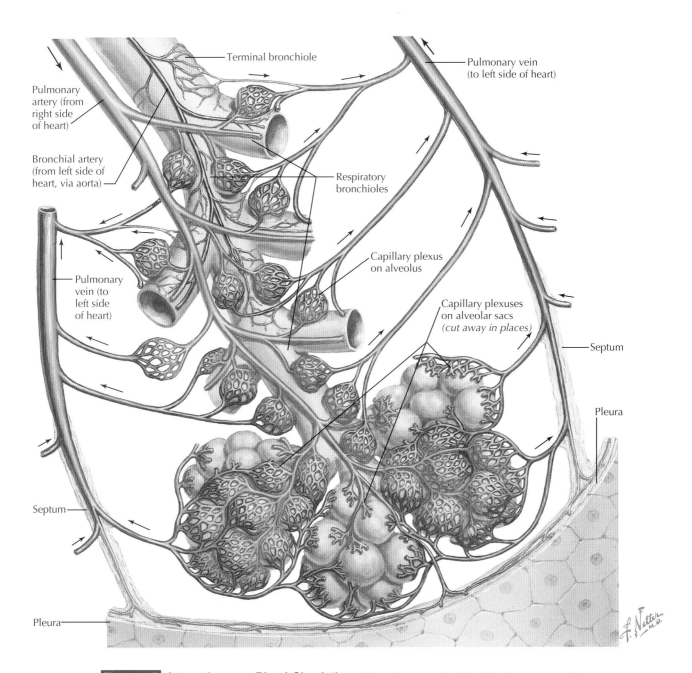

Terminal bronchiole

Pulmonary vein
(to left side of heart)

Pulmonary
artery (from
right side
of heart)

Bronchial artery
(from left side of
heart, via aorta)

Respiratory
bronchioles

Pulmonary
vein (to
left side
of heart)

Capillary plexus
on alveolus

Capillary plexuses
on alveolar sacs
(cut away in places)

Septum

Pleura

Septum

Pleura

Figure 14.7 **Intrapulmonary Blood Circulation** The pulmonary circulation is a low-pressure, low-resistance circulation. Blood from the right ventricle and pulmonary artery is distributed to the pulmonary capillaries, where gas exchange takes place. The interface between the alveolar lumen and pulmonary capillary blood consists of a single layer of alveolar epithelium, basement membrane, and the one-cell-layer-thick capillary endothelium. Capillaries cover alveoli in this manner, providing for efficient gas exchange.

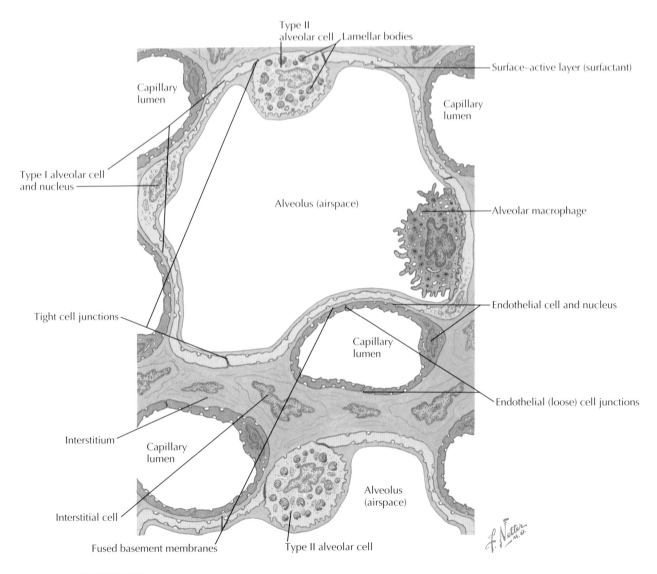

Figure 14.8 **Ultrastructure of Pulmonary Alveoli and Capillaries**   The alveolar epithelium consists of type I and type II alveolar epithelial cells. Type I cells constitute the largest surface area; type II epithelial cells secrete surfactant. Gas diffusion takes place across the thin alveolar-capillary membrane composed of alveolar epithelium, basement membrane, and capillary endothelium. Interstitial tissue is minimal in most of the interface between the alveolar epithelium and capillary endothelium in healthy lungs.

- **Inspiratory reserve volume (IRV):** The additional volume a subject is *capable* of inhaling after a normal, quiet inspiration.

The following four capacities are associated with pulmonary function:

- **Total lung capacity (TLC):** The volume of gas in the lungs after maximal inspiration. TLC is approximately 7 L in healthy adults.
- **Vital capacity (VC):** The maximum volume of air that a subject can exhale after maximal inspiration. The normal value is approximately 5 L. **Forced vital capacity** is vital capacity measured during expiration at maximum force.
- **Functional residual capacity (FRC):** The volume remaining in the lungs after expiration during normal, quiet breathing.

- **Inspiratory capacity (IC):** The maximum volume that can be inspired after expiration during normal, quiet breathing.

Spirometry

These capacities and volumes are measured by **spirometry** and related techniques. With spirometry, the subject breathes in and out of a device known as a *spirometer*. Essentially, a spirometer is composed of two vessels: one contains water and the other floats upside down in the first vessel (Fig. 14.9). As the subject breathes through the attached tube, air flows in and out of the inner vessel, which consequently moves up and down. This up-and-down motion is recorded as a spirogram (Fig. 14.10), which is calibrated to reflect changes in volume of the inner vessel.

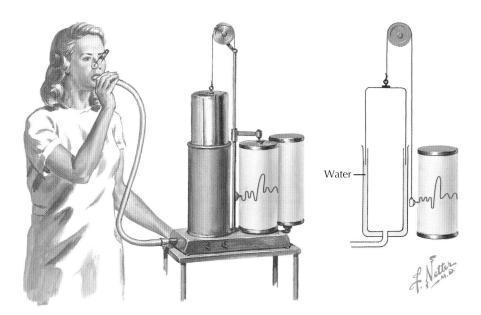

Figure 14.9 **The Spirometer** Pulmonary function testing involves multiple instruments and techniques. One basic instrument is the spirometer. The subject breathes in and out of a tube, resulting in vertical displacement of the inner canister, which floats upside down in water. The recorded tracing is analyzed to obtain several basic lung volumes and capacities (see Fig. 14.10).

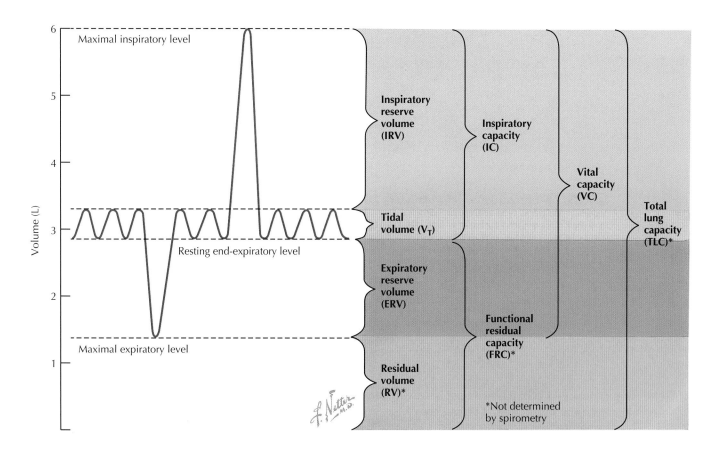

Figure 14.10 **Measurement of Lung Volumes and Capacities by Spirometry** In this spirogram, beginning on the left, the subject breathed quietly for a few breaths, exhaled maximally, breathed quietly for a few breaths, inhaled maximally, and then breathed normally again. Note that residual volume, functional residual capacity (FRC), and total lung capacity cannot be measured by spirometry alone. FRC is measured by another technique (often helium dilution); when FRC is known, total lung capacity and residual volume can be calculated from the spirometry tracing. The volumes represented in the tracing are those of a typical healthy adult.

Full appreciation of pulmonary volumes and capacities requires understanding the relationships between these parameters. For example, TLC is the sum of four of the described volumes:

$$TLC = RV + ERV + V_T + IRV$$

Similarly, VC, FRC, and IC can be equated to the sums of particular lung volumes:

$$VC = IRV + V_T + ERV$$

$$FRC = ERV + RV$$

$$IC = V_T + IRV$$

V_T is measured by spirometry during normal, quiet breathing and is the difference between end inspiratory and end expiratory levels (see Fig. 14.10). Note that the actual end inspiratory and end expiratory volumes are not known because the spirometer records *changes* in volume rather than actual volume inside the lung. By having the subject expire maximally, ERV can be determined, because ERV is the difference between the resting end expiratory level and maximal expiratory level. Likewise, IRV can be measured by having the subject inspire maximally and comparing the maximal inspiratory level to the end inspiratory level during normal, quiet breathing. VC and IC can be determined by similar comparisons: VC is the difference between the maximal inspiratory and expiratory levels, and IC is the difference between maximal inspiratory level and resting end-expiratory level.

Measurement of Capacities (FRC, RV, and TLC)

To measure TLC, RV, and FRC, one of these parameters must be measured indirectly, such as by nitrogen washout, helium dilution, or body plethysmography. In the **helium dilution technique**, a small volume of helium is added to the spirometer before the subject begins the test, resulting in a known initial helium concentration within the volume of the system. When the subject begins breathing (beginning at FRC), this initial helium concentration becomes diluted as the gas equilibrates between the lungs of the subject and the spirometer. Because helium does not diffuse through the alveolar capillary membrane and is inert, a stable equilibrium is reached quickly. This final concentration is dependent only on the initial concentration of helium within the spirometer and the volume of the spirometry system plus FRC. After measuring the final concentration of helium, FRC can be determined from the following formula:

$$C_1 V_S = C_2 \times (V_S + FRC) \qquad \text{Eq. 14.1}$$

where C_1 is the initial, measured concentration of helium in the spirometer, V_S is the known volume of the spirometry

system, and C_2 is the measured concentration of helium during breathing (after equilibrium is reached).

Once FRC is known, TLC and RV can be readily calculated from spirometry measurements (see Fig. 14.10):

$$TLC = FRC + IC$$
$$RV = TLC - VC \qquad \text{Eq. 14.2}$$

Whole-body plethysmography is based on **Boyle's law**, which states that the product of pressure and volume for a gas is constant:

$$P_1 V_1 = P_2 V_2 \qquad \text{Eq. 14.3}$$

To use whole-body plethysmography to measure FRC, the subject is put into an airtight box equipped with a mouthpiece through which he or she breathes outside air. During normal, quiet breathing, the subject is asked to stop and relax after a normal, quiet exhalation. At that point, the mouthpiece is closed, and the subject attempts to inhale through the closed mouthpiece. This attempt causes the air in the lungs (FRC) to expand as a result of the negative pressures created by the expansion of the chest wall and results in an equal reduction of the volume of the air in the box outside the patient's body. The pressure of the air in the box increases as a result of this reduction in volume. Changes in pressure in the box are used to calculate the change in volume in the box outside the patient's body using Boyle's law. This change in volume is the same as the change in volume of the lungs from the resting value (FRC). Based on the change in pressure in the respiratory system (measured at the mouthpiece during the attempt to inhale) and the change in volume of the lungs, the initial volume (FRC) can be calculated using Boyle's law.

VENTILATION AND ALVEOLAR GAS COMPOSITION

Ventilation is the movement of air in and out of the respiratory system. During normal, quiet breathing, the volume of air inhaled and exhaled with each breath is approximately 500 mL. This is known as the resting V_T. The normal, resting **respiratory rate** is 12 to 20 breaths/min. When taking an average rate of 15 breaths/min, the **minute ventilation** ($\dot{V}_E$, the volume exhaled per minute) is calculated by using the following formula:

$$\dot{V}_E = R \times V_T \qquad \text{Eq. 14.4}$$

where R is the respiratory rate and V_T is the tidal volume. Thus 15 breaths/min $\times$ 500 mL yields a $\dot{V}_E$ of 7500 mL/min (7.5 L/min).

$\dot{V}_E$ is greater than **alveolar ventilation** ($\dot{V}_A$, ventilation of the respiratory zone of the lungs) because a portion of the V_T remains in the anatomic dead space (the conducting zone) and is not involved in gas exchange. Because anatomic dead space is approximately 150 mL, this **dead space ventilation** ($\dot{V}_D$) is roughly 150 mL × 15/min, or 2250 mL/min. $\dot{V}_A$ is calculated by using the following formula:

$$\dot{V}_A = R\,(V_T - V_D) \qquad \textbf{Eq. 14.5}$$

where R is respiratory rate, V_T is tidal volume, and V_D is dead space volume. Stated another way,

$$\dot{V}_A = \dot{V}_E - \dot{V}_D \qquad \textbf{Eq. 14.6}$$

The symbols used in respiratory physiology can be confusing. The uppercase letter V is used to designate volume, whereas airflow rates are designated by the symbol $\dot{V}$. Uppercase letter A used as a subscript, as in $\dot{V}_A$, specifies that the parameter refers to the alveolar space; thus $\dot{V}_A$ is alveolar ventilation. Lowercase letters a and v are used to specify arterial and venous measurements, respectively, as in Pa_{CO_2}, or Pv_{CO_2} (partial pressure of carbon dioxide in arterial and venous blood, respectively). PA_{CO_2} refers to partial pressure of carbon dioxide in alveolar gas.

Of the 500 mL V_T, only 350 mL is entering the alveoli with each breath, and of the 7500 mL/min $\dot{V}_E$, only the $\dot{V}_A$ of 5250 mL/min (7500 mL/min − 2250 mL/min) is available for gas exchange.

Composition of Alveolar Air

The composition of alveolar air is dependent on several factors, including the composition of inspired air, $\dot{V}_A$, and the concentration of dissolved gases in mixed venous blood. Our atmosphere is composed of 21% oxygen, 79% nitrogen, and less than 1% other gases, including carbon dioxide, with a total atmospheric pressure of 760 mm Hg at sea level. Gas concentrations may also be expressed as fractional concentrations in an inspired gas mixture; in our atmosphere, fractional concentration of inspired oxygen (FiO_2) = 0.21 and FiN_2 = 0.79. According to **Dalton's law**, the sum of the partial pressures of gases in a mixture is equal to the total pressure (P_{tot}), and thus for dry air at sea level:

$$
\begin{aligned}
PO_2 &= 0.21 \times 760 \text{ mm Hg} = 160 \text{ mm Hg} \\
PN_2 &= 0.79 \times 760 \text{ mm Hg} = 600 \text{ mm Hg} \qquad \textbf{Eq. 14.7} \\
P_{tot} &= 760 \text{ mg Hg}
\end{aligned}
$$

As air is inspired, it warms rapidly to body temperature and becomes saturated with water vapor. At 37°C, the vapor pressure of water is 47 mm Hg and must be accounted for when determining the gas composition of inspired air. In **inspired air:**

$$
\begin{aligned}
P_{H_2O} &= 47 \text{ mm Hg} \\
PO_2 &= 0.21 \times (760 - 47) \text{ mm Hg} = 150 \text{ mm Hg} \\
PN_2 &= 0.79 \times (760 - 47) \text{ mm Hg} = 563 \text{ mm Hg} \\
P_{tot} &= (150 + 563 + 47) \text{ mm Hg} = 760 \text{ mm Hg}
\end{aligned} \qquad \textbf{Eq. 14.8}
$$

This composition of inspired air is constant throughout the conducting zone of the lung, where no gas exchange occurs. Within the respiratory zone, oxygen diffuses from alveolar air to blood, whereas carbon dioxide diffuses from blood to alveolar air, resulting in alveolar air composition that is different from composition of inspired air. The relationship between partial pressure of oxygen and carbon dioxide in alveolar air is described by the important **alveolar gas equation:**

$$PA_{O_2} = PI_{O_2} - PA_{CO_2}/R \qquad \textbf{Eq. 14.9}$$

where PA_{O_2} and PI_{O_2} are the partial pressures of oxygen in alveolar air and inspired air, respectively; PA_{CO_2} is partial pressure of carbon dioxide in alveolar air; and R is the **respiratory quotient**, which usually has a value of 0.8. This equation can be used to predict the PA_{O_2} based on measurement of carbon dioxide in systemic arterial blood in a healthy person, because PCO_2 is normally fully equilibrated between blood and alveolar air in the alveolar capillaries. For example, at 760 mm Hg atmospheric pressure (sea level), when partial pressure of oxygen in arterial blood (Pa_{O_2}) in an arterial blood gas determination is 40 mm Hg,

$$PA_{O_2} = 150 \text{ mm Hg} - 40 \text{ mm Hg}/0.8 = 100 \text{ mm Hg} \qquad \textbf{Eq. 14.10}$$

Because PO_2 is also fully equilibrated between alveolar air and blood as it passes through alveolar capillaries, and nearly all of pulmonary blood flow is exposed to alveolar air, the normal value for Pa_{O_2} is nearly 100 mm Hg. Partial pressures of oxygen and carbon dioxide in inspired and alveolar air and mixed venous and arterial blood are illustrated in Figure 14.11.

At the same $\dot{V}_E$, deep, slow breathing yields greater $\dot{V}_A$ than rapid, shallow breathing. A comparison of $\dot{V}_A$ at a respiratory rate of 15/min and V_T of 500 mL with $\dot{V}_A$ at a respiratory rate of 30/min and V_T of 250 mL yields the following values:

$$\dot{V}_A = 15/\text{min}\,(500 \text{ mL} - 150 \text{ mL}) = 5250 \text{ mL/min}$$

$$\dot{V}_A = 30/\text{min}\,(250 \text{ mL} - 150 \text{ mL}) = 3000 \text{ mL/min}$$

Thus slower, deeper ventilation produces greater $\dot{V}_A$ than more rapid ventilation at proportionally less V_T.

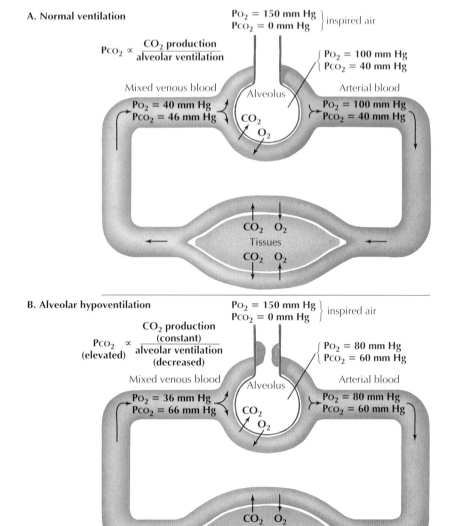

A. Normal ventilation

$$P_{O_2} = 150 \text{ mm Hg}$$
$$P_{CO_2} = 0 \text{ mm Hg}$$ inspired air

$$P_{CO_2} \propto \frac{CO_2 \text{ production}}{\text{alveolar ventilation}}$$

$$P_{O_2} = 100 \text{ mm Hg}$$
$$P_{CO_2} = 40 \text{ mm Hg}$$

Mixed venous blood

$$P_{O_2} = 40 \text{ mm Hg}$$
$$P_{CO_2} = 46 \text{ mm Hg}$$

Alveolus

CO_2

O_2

Arterial blood

$$P_{O_2} = 100 \text{ mm Hg}$$
$$P_{CO_2} = 40 \text{ mm Hg}$$

CO_2 O_2

Tissues

CO_2 O_2

B. Alveolar hypoventilation

$$P_{O_2} = 150 \text{ mm Hg}$$
$$P_{CO_2} = 0 \text{ mm Hg}$$ inspired air

$$P_{CO_2} \atop \text{(elevated)} \propto \frac{\begin{array}{c}CO_2 \text{ production}\\ \text{(constant)}\end{array}}{\begin{array}{c}\text{alveolar ventilation}\\ \text{(decreased)}\end{array}}$$

$$P_{O_2} = 80 \text{ mm Hg}$$
$$P_{CO_2} = 60 \text{ mm Hg}$$

Mixed venous blood

$$P_{O_2} = 36 \text{ mm Hg}$$
$$P_{CO_2} = 66 \text{ mm Hg}$$

Alveolus

CO_2

O_2

Arterial blood

$$P_{O_2} = 80 \text{ mm Hg}$$
$$P_{CO_2} = 60 \text{ mm Hg}$$

CO_2 O_2

Tissues

CO_2 O_2

Figure 14.11 **Partial Pressures of Oxygen and Carbon Dioxide in Blood and Alveolar Air**
The normal partial pressures of gases in mixed venous blood entering the alveolar capillaries are indicated
on the left side of part **A**. As blood flows through the alveolar capillaries in healthy lungs, oxygen and carbon
dioxide levels equilibrate between alveolar air and blood. Thus partial pressures of oxygen and carbon
dioxide determined by arterial blood gas measurement are approximately equal to the partial pressures in
alveolar air in healthy subjects. During hypoventilation (**B**, indicated by partial blockage of the airway), P_{O_2}
is reduced and P_{CO_2} is elevated in arterial blood (the subject is hypoxic and hypercapnic), as well as in the
alveolar gas.

DIFFUSION OF GASES

Diffusion of gas ($\dot{V}_{gas}$) between alveolar gas and capillary
blood follows Fick's law,

$$\dot{V}_{gas} = \frac{A \times D(P_1 - P_2)}{T}$$ **Eq. 14.11**

where A is the area of a membrane separating two compartments, T is the thickness of the membrane, D is the diffusion

constant, and P_1 and P_2 are the gas concentrations in the two
compartments. Thus diffusion of a gas is:

- **Directly** related to the surface area for diffusion.
- **Directly** related to the difference in partial pressure of
  the gas on each side of the membrane.
- **Directly** related to the diffusion constant of the gas.
- **Inversely** related to the thickness of the membrane.

The diffusion constant of a gas is directly related to solubility
and inversely related to the square root of its molecular weight.

The respiratory quotient (R) is the ratio of carbon dioxide production to oxygen consumption and is dependent on metabolism. For pure carbohydrate metabolism, R has a value of 1.0, because oxidative metabolism of one mole of glucose requires six moles of oxygen and produces six moles of carbon dioxide. During pure lipid metabolism, the approximate value of R would be 0.7, whereas the respiratory quotient for amino acids varies. Under metabolic conditions involving oxidation of a typical mixture of substrates, the respiratory quotient is approximately 0.8 (0.8 moles of carbon dioxide are produced for each mole of oxygen consumed).

Perfusion-Limited Gas Transport

The concentrations of dissolved gases in blood equilibrate with the gas concentrations of alveolar air as blood flows through pulmonary capillaries (Fig. 14.12). Under resting conditions, the transit time for blood through the alveolar capillary is only approximately 0.75 seconds. In normal, healthy lungs at rest, oxygen and carbon dioxide are equilibrated between blood and alveolar gas by the time blood has passed one third of the way through an alveolar capillary. Transport of these gases is **perfusion limited** under these conditions, because the only way to increase gas transfer is to increase perfusion (blood flow, rather than diffusion, is the limiting factor for exchange). As a consequence, cardiac output can be increased substantially without compromising oxygenation of blood despite reduced transit time through the alveolar capillaries, such as during exercise. Nitrous oxide (laughing gas; N_2O) is a textbook example of perfusion limitation. N_2O, unlike many other gases, including O_2 and CO_2, is not bound by blood components but exists only as dissolved gas in blood. As a result, in a subject breathing a gas mixture containing N_2O, blood and alveolar P_{N_2O} equilibrate rapidly in blood flowing through the alveolar capillary, reaching this equilibrium one fifth of the way through the capillary.

Transit time is the time required for blood or a formed element to pass through a portion of the circulation. Because blood volume in the entire pulmonary circulation is approximately 500 mL and cardiac output at rest is about 5000 L/min (83 mL/sec), transit time through the entire pulmonary circulation is approximately 6 seconds. Of this, about 0.75 second is transit time through the alveolar capillaries, during which all exchange of gases takes place. In healthy lungs, oxygen is equilibrated between blood and alveolar gas by the time blood has passed one third of the way through an alveolar capillary.

Diffusion-Limited Gas Transport

During very strenuous aerobic exercise in which cardiac output is greatly increased, or in some disease states such as interstitial fibrosis, oxygen transport may be **diffusion limited**. In other words, P_{O_2} is not completely equilibrated between alveolar air and blood leaving the alveolar capillary.

The classic example of a diffusion-limited gas is carbon monoxide (CO). When a subject breathes a gas mixture containing CO, CO diffuses from the alveolar air to blood. However, CO is bound to hemoglobin with very high affinity, and as a result, large amounts of CO can be transferred to blood with little change in blood P_{CO}. Thus the partial pressures of CO in alveolar air and blood do not fully equilibrate during the transit of blood through the alveolar capillary, and its transport is limited only by the diffusion capacity of the alveolar membrane.

Diffusion in the lung can be assessed using a measurement known as **diffusion capacity of the lung for CO (DLCO)**. CO is useful for this purpose because its transport is diffusion limited. According to Fick's law,

$$\dot{V}_{CO} = \frac{A \times D_{CO} \times (P_{CO_1} - P_{CO_2})}{T}$$ **Eq. 14.12**

DLCO is defined as a transfer factor equal to $(A/T) \times D_{CO}$ and is substituted in equation 14.12, yielding

$$\dot{V}_{CO} = DLCO \times (P_{CO_1} - P_{CO_2})$$ **Eq. 14.13**

Rearranging and substituting $P_{A_{CO}}$ for $P_{CO_1} - P_{CO_2}$ (because the concentration of CO in blood is initially zero),

$$DLCO = \frac{\dot{V}_{CO}}{P_{A_{CO}}}$$ **Eq. 14.14**

VENTILATION AND PERFUSION GRADIENTS

Optimal function of the lung requires proper balance between ventilation and perfusion. In other words, ventilation should be matched to blood flow for efficient exchange of gases between the environment and the blood. However, neither ventilation nor perfusion of the lung is uniform from apex to base.

In the standing position, the weight of the lungs stretches the top portion of the organ, such that the alveoli near the apex have larger volume than those near the base. Thus the smaller alveoli toward the base of the lung have greater compliance than do those that are already stretched by gravitational force, and as a result, ventilation is greater toward the bottom of the lung. In other words, there is an increasing ventilation gradient from the top to the bottom of the lung.

Zones of the Lung

An even larger gradient than the one for ventilation exists for pulmonary blood flow (perfusion) in the standing position (Fig. 14.13), whereby greater flow occurs toward the bottom of the lungs. This phenomenon is the result of gravitational effects on vascular pressures and the relationship between vascular pressure and alveolar pressure as blood flows through alveolar capillaries. Perfusion pressure in most

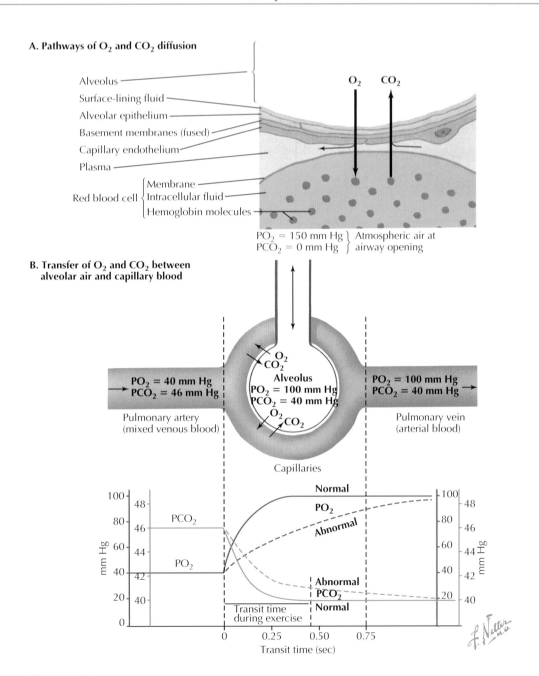

A. Pathways of O₂ and CO₂ diffusion

Alveolus
Surface-lining fluid
Alveolar epithelium
Basement membranes (fused)
Capillary endothelium
Plasma
Red blood cell { Membrane / Intracellular fluid / Hemoglobin molecules

O₂   CO₂

$PO_2 = 150$ mm Hg | Atmospheric air at
$PCO_2 = 0$ mm Hg } airway opening

B. Transfer of O₂ and CO₂ between alveolar air and capillary blood

O₂ / CO₂

$PO_2 = 40$ mm Hg
$PCO_2 = 46$ mm Hg

Alveolus
$PO_2 = 100$ mm Hg
$PCO_2 = 40$ mm Hg

$PO_2 = 100$ mm Hg
$PCO_2 = 40$ mm Hg

Pulmonary artery
(mixed venous blood)

Pulmonary vein
(arterial blood)

O₂ / CO₂

Capillaries

Normal
PO₂
Abnormal

PCO₂

PO₂

Abnormal
PCO₂
Normal

Transit time
during exercise

mm Hg

Transit time (sec)

Figure 14.12 O₂ and CO₂ Exchange Diffusion of carbon dioxide and oxygen between blood and alveolar air occurs by diffusion through the thin alveolar-capillary membrane **(A)**. Partial pressures of gases in inspired air, alveolar air, and blood are illustrated in the center panel. At resting cardiac output, blood normally traverses the length of the alveolar capillary in 0.75 sec, and O₂ and CO₂ are equilibrated between blood and alveolar air as blood passes through the first third of the capillary **(B)**. When the alveolar-capillary membrane is thickened by interstitial fibrosis, the resulting diffusion barrier impedes gas exchange *(dashed lines)*.

CLINICAL CORRELATE 14.1
DLCO and Interstitial Pulmonary Fibrosis

Interstitial fibrosis is a type of restrictive lung disease in which functional lung volume is reduced. It is an inflammatory disease in which deposition of connective tissue and subsequent scar formation in the interstitial space of the alveolar-capillary membrane results in thickening of the membrane. As a consequence, diffusion across the alveolar-capillary membrane is impeded and mechanical properties of the lung are altered, resulting in "restriction" (discussed in context in Chapter 15). Changes in the diffusing capacity of the lung in persons with interstitial fibrosis are typically assessed by measurement of DLCO. Beginning at RV, the subject takes in a single, large breath of air containing 0.3% CO and 10% helium (inhaling to TLC), holds this breath for 10 seconds, and then exhales. A sample of exhaled air is collected after exhalation of dead space volume for measurement of the final concentration of gases in alveolar air. DLCO is then determined by the formula $DLCO = \dot{V}CO/P_{ACO}$. P_{ACO} in this formula is the initial value upon inspiration of the gas mixture and is calculated based on CO concentration in inspired air and helium dilution; $\dot{V}CO$ is calculated based on alveolar volume, breath-hold time, and the change in CO concentration in alveolar air.

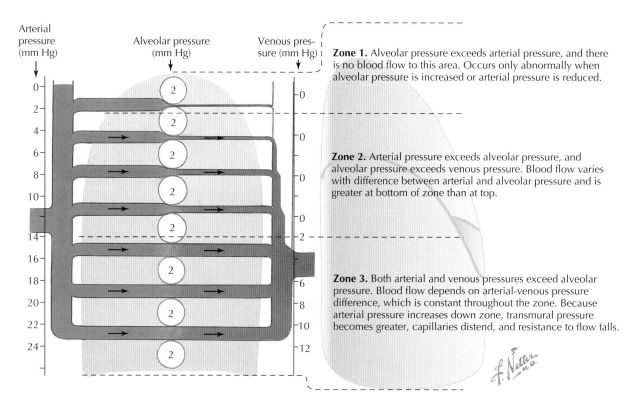

Arterial pressure (mm Hg)

Alveolar pressure (mm Hg)

Venous pressure (mm Hg)

Zone 1. Alveolar pressure exceeds arterial pressure, and there is no blood flow to this area. Occurs only abnormally when alveolar pressure is increased or arterial pressure is reduced.

Zone 2. Arterial pressure exceeds alveolar pressure, and alveolar pressure exceeds venous pressure. Blood flow varies with difference between arterial and alveolar pressure and is greater at bottom of zone than at top.

Zone 3. Both arterial and venous pressures exceed alveolar pressure. Blood flow depends on arterial-venous pressure difference, which is constant throughout the zone. Because arterial pressure increases down zone, transmural pressure becomes greater, capillaries distend, and resistance to flow falls.

Figure 14.13 **Distribution of Pulmonary Blood Flow** As a result of gravity, blood flow through the lungs is not uniformly distributed in the standing position. At the apex of the lungs, hydrostatic pressure is reduced, potentially resulting in collapse of capillaries by alveolar pressure; as a result, flow is reduced toward the apex and may cease if alveolar pressure is increased or blood pressure is reduced (zone 1). In the middle of the lungs (zone 2), arterial pressure exceeds alveolar pressure, but alveolar pressure is higher than venous pressure. Flow in zone 2 is more than that in zone 1 but less than flow in zone 3, where higher hydrostatic pressure causes distension of vessels and therefore reduced vascular resistance.

regional circulations is the difference between arterial and venous pressure. However, in the pulmonary system, as blood flows through a capillary, the rate of flow is potentially affected by air pressure in the alveoli on either side of the capillary. Blood enters the lung at the hilum, approximately midway between apex and base; above this point of entry, hydrostatic pressure in the pulmonary vessels is reduced by the effect of gravity. At the very top of the lung, alveolar pressure may even exceed arterial pressure under some conditions, producing a region of no blood flow, known as **zone 1**. In regions of the lung in which alveolar pressure is higher than venous pressure but less than arterial pressure, perfusion pressure is equal to arterial pressure minus alveolar pressure, which occurs in **zone 2** of the lung (see Fig. 14.13). Thus as blood passes through alveolar capillaries in zone 2, the pressure gradient for blood flow is the difference between pulmonary arterial pressure and the pressure within the alveoli, which exceeds venous pressure. In contrast, in **zone 3** (in the lower portions of the lung), pulmonary vascular pressures are elevated by the effect of gravity, and both arterial and venous pressures exceed alveolar pressure. The higher hydrostatic pressure causes distension of vessels and therefore reduced resistance; as a result, blood flow is greatest in zone 3 of the lung.

The Ventilation-to-Perfusion Ratio

As a result of these gradients in ventilation and perfusion, the ratio $\dot{V}_A/\dot{Q}_C$ (where $\dot{Q}_C$ is pulmonary capillary blood flow) is greatest at the top of the lung and lowest at the bottom (Fig. 14.14), and ventilation and perfusion are best matched in the middle portion. Because resting pulmonary blood flow is approximately 5 L/min (equal to cardiac output of the systemic circulation) and $\dot{V}_A$ is also approximately 5 L/min, $\dot{V}_A/\dot{Q}_C$ is approximately 1 in this middle region, less than 1 at the base of the lung, and more than 1 toward the apex.

Dead Space and Shunt: Extremes of $\dot{V}_A/\dot{Q}_C$ Imbalance

Dead space and **shunt flow** are extremes of imbalance. Alveoli that are ventilated but not perfused constitute **physiological dead space** (as opposed to anatomic dead space). In dead space, $\dot{V}_A/\dot{Q}_C = \infty$. In contrast to dead space (ventilation without perfusion), **shunt flow** represents perfusion without ventilation. In other words, shunted blood returning to the left heart has not been exposed to ventilated alveoli. If alveolar capillaries are perfused in an area of the lung that is not

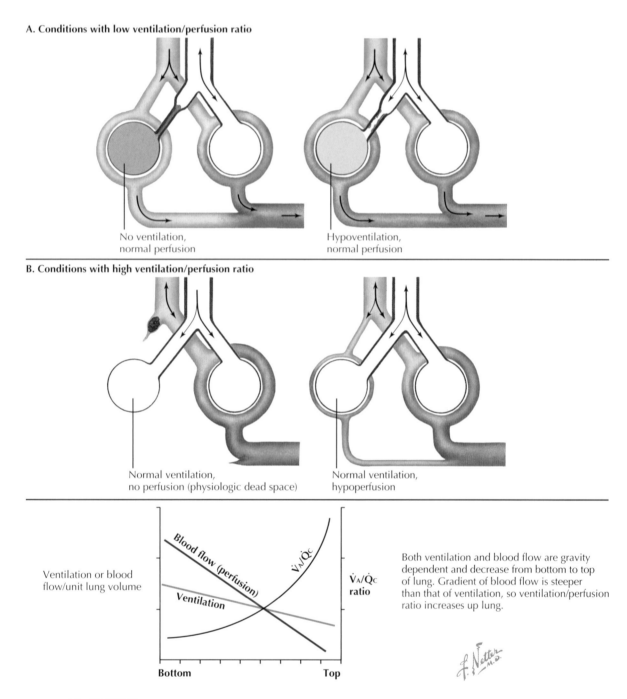

A. Conditions with low ventilation/perfusion ratio

No ventilation,
normal perfusion

Hypoventilation,
normal perfusion

B. Conditions with high ventilation/perfusion ratio

Normal ventilation,
no perfusion (physiologic dead space)

Normal ventilation,
hypoperfusion

Ventilation or blood
flow/unit lung volume

Blood flow (perfusion)

Ventilation

$\dot{V}_A/\dot{Q}_C$

$\dot{V}_A/\dot{Q}_C$
ratio

Both ventilation and blood flow are gravity
dependent and decrease from bottom to top
of lung. Gradient of blood flow is steeper
than that of ventilation, so ventilation/perfusion
ratio increases up lung.

Bottom   Top

Figure 14.14 **Ventilation-Perfusion ($\dot{V}_A/\dot{Q}_C$) Relationships**   In the standing position, the effects of gravity result in gradients in both perfusion and ventilation of the lung from base to apex. Because the perfusion gradient is steeper than the ventilation gradient, the ratio of ventilation to perfusion, $\dot{V}_A/\dot{Q}_C$, is lowest at the bottom of the lung and greatest at the top of the lung **(B)**. $\dot{V}_A/\dot{Q}_C$ is also affected by various other conditions affecting ventilation and perfusion **(A** and **B)**.

ventilated (e.g., when an airway is blocked), **physiological shunt** is present ($\dot{V}_A/\dot{Q}_C = 0$). Shunt can also be **anatomic** (bypassing the alveoli). In addition to blood flow originating in the pulmonary arteries, the pulmonary veins receive some flow from the bronchopulmonary (bronchial) circulation, which originates from the systemic circulation and perfuses tissues in the conducting zone of the lungs. Thus blood from the bronchopulmonary circulation returning to the left

heart is not oxygenated. The venous admixture of oxygenated and deoxygenated blood resulting from anatomic shunt accounts for most of the small **alveolar-to-systemic arterial PO$_2$ gradient (A-a PO$_2$ gradient)** in healthy subjects (6 to 9 mm Hg). While $\dot{V}_A/\dot{Q}_C$ varies even within normal lungs, ventilation-perfusion imbalances are a central issue in pulmonary pathophysiology (discussed in various contexts in the following pages).

HYPOXEMIA

Hypoxemia (low Pa_{O_2}) may result from one of five causes, all of which reflect problems related to ventilation, perfusion, or diffusion. In the simplest case, ventilation (breathing) in an atmosphere with low P_{O_2} (e.g., at high altitude) will cause hypoxemia that can be corrected by breathing air with higher P_{O_2}. A second cause of hypoxemia is **hypoventilation**. With insufficient ventilation (e.g., with shallow breathing), PA_{CO_2} rises and PA_{O_2} falls because of inefficient exchange of alveolar air with the atmosphere. Consequently, hypoxemia and **hypercapnea** (high Pa_{CO_2}) result. Hypoxemia of this type can be corrected by adjusting respiratory rate and V_T, or breathing air containing a high concentration of oxygen.

Diffusion abnormalities are a third cause of hypoxemia. Poor diffusion (e.g., as a result of thickening of the alveolar capillary membrane) will produce hypoxemia with an elevated A-a P_{O_2} gradient; this abnormal gradient occurs because P_{O_2} is not fully equilibrated between alveolar air and the blood leaving alveolar capillaries. Although hypoxemia will be corrected by administration of 100% oxygen, the A-a P_{O_2} gradient will remain elevated.

Shunt flow also produces hypoxemia. In the presence of shunt flow, venous admixture of oxygenated and deoxygenated blood produces an elevated A-a P_{O_2} gradient. In congenital abnormalities of the heart in which right-to-left shunting of blood occurs through atrial or ventricular septal defects, large A-a P_{O_2} gradients may be seen. Airway obstruction also results in shunt flow and increased A-a P_{O_2} gradient. Hypoxemia (low arterial P_{O_2}) caused by shunt flow can be differentiated from hypoxemia caused by other defects by measurement of PaO_2 before and after several minutes of breathing of 100% oxygen. Neither the hypoxemia nor the A-a P_{O_2} gradient associated with shunt flow can be fully corrected by breathing 100% oxygen because of venous admixture of shunt flow to oxygenated blood.

Ventilation-perfusion imbalance is a fifth cause of low Pa_{O_2}. As discussed earlier, gradients in ventilation and perfusion produce some degree of ventilation-perfusion imbalance even in a healthy subject in the standing position, and regional ventilation-perfusion imbalances contribute to the normal, small A-a gradient in P_{O_2}. If the imbalance is abnormally pronounced, mixing of blood from areas of the lung that are well perfused but underventilated with blood from other areas (where ventilation and perfusion are better balanced) causes an exaggerated A-a P_{O_2} gradient and hypoxemia. This type of hypoxemia can be corrected by administration of 100% oxygen.

Chapter 15

The Mechanics of Breathing

The physical forces resulting in ventilation of the lungs are analogous to those creating blood flow in the cardiovascular system. A pressure gradient is required, and in the case of the lung, the gradient is created by movement of the chest wall and diaphragm. Flow of air occurs against the resistance of the airways, analogous to the resistance of blood vessels. However, factors affecting pressure, flow, and resistance are complex and are often different in the two systems. This chapter details the physical forces involved in ventilation of the lungs and some of the changes that take place in the presence of disease.

BASIC MECHANICS OF THE VENTILATORY APPARATUS

Ventilation occurs as a result of mechanical forces associated with the chest wall and lungs. Both the lungs and the chest wall are **elastic**—that is, they passively recoil after being distended. **Elastic recoil pressure** is the pressure caused by distension. Functionally, the chest wall includes the diaphragm and abdominal muscles in addition to the rib cage. The **visceral pleura** (outer lining) of the lungs apposes the **parietal pleura** of the chest wall, and the small, fluid-filled space between the pleurae, the **pleural cavity**, contains only a few milliliters of fluid. The muscles of breathing are illustrated in Figure 15.1. The **diaphragm** is the major muscle for inspiration during normal, quiet breathing. As it contracts and its domes descend, the thoracic space is enlarged, decreasing alveolar pressure and resulting in inward flow of air through the airways. During more active ventilation, such as during exercise, the **intercostal muscles** have greater involvement in inspiration, elevating the ribs and expanding the chest as they contract. Expiration is a passive process during normal, quiet breathing and results from passive recoil of the lungs. During active breathing, various muscles of the abdominal wall, along with some of the intercostal muscles, contribute to the force, resulting in expiration.

Elastic Recoil of the Chest Wall and Lungs

The interactions between the forces of the chest wall and lungs during normal, quiet breathing are illustrated in Figure 15.2. In the context of elastic properties of the system at all lung volumes, from residual volume (RV; the volume remaining in the lungs after maximal exhalation) to total lung capacity (TLC), chest wall and lung forces are illustrated in Figure 15.3. At functional residual capacity (FRC), the mechanical system is at rest, and the pressure within the airways and alveolar space is equal to atmospheric pressure (see Fig. 15.2A). In this state, muscles of the chest wall are relaxed, and the outward elastic recoil pressure of the chest wall is equal and opposite to the inward elastic recoil pressure of the lungs. In other words, at FRC, although the tendency of the chest wall is to expand and the tendency of the lungs is to collapse, these two forces are in balance, keeping the volume of the lungs at FRC. The algebraic sum of the negative recoil pressure created by the chest wall and the positive recoil pressure of the lungs is zero; pleural pressure (pressure in the pleural cavity) is negative (subatmospheric). Note the balance between chest wall recoil and lung recoil pressures at FRC in Figure 15.3.

Mechanically, the lungs and chest wall can be conceptualized as a respiratory pump, acting in unison to produce pressures required to generate airflow. Flow, which is driven by the pressure gradient between the alveolar space and the mouth opening, is produced by the elastic properties of the system and the activity of respiratory muscles. An additional characteristic of this pump is airway resistance (R_{aw}), defined by the following formula,

$$R_{aw} = \frac{(P_A - P_{ATM})}{\text{Rate of airflow}}$$

where P_A is alveolar pressure and P_{ATM} is atmospheric pressure. This formula can be rearranged as follows:

$$\text{Rate of airflow} = \frac{(P_A - P_{ATM})}{R_{aw}}$$

This formula is analogous to the formula for blood flow ($Q = \Delta P/R$, where ΔP is the pressure gradient along which flow occurs and R is the resistance to flow).

Forces During Inspiration and Expiration

Inspiration occurs when the rib cage expands and the diaphragm moves downward, creating negative pressure

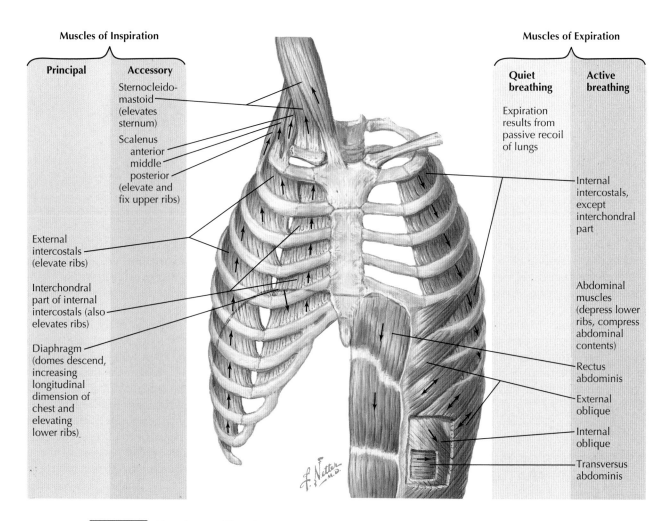

Muscles of Inspiration

Principal | **Accessory**

Sternocleido-mastoid (elevates sternum)

Scalenus anterior middle posterior (elevate and fix upper ribs)

External intercostals (elevate ribs)

Interchondral part of internal intercostals (also elevates ribs)

Diaphragm (domes descend, increasing longitudinal dimension of chest and elevating lower ribs)

Muscles of Expiration

Quiet breathing | **Active breathing**

Expiration results from passive recoil of lungs

Internal intercostals, except interchondral part

Abdominal muscles (depress lower ribs, compress abdominal contents)

Rectus abdominis

External oblique

Internal oblique

Transversus abdominis

Figure 15.1 **Respiratory Muscles** Contraction of the diaphragm is the main factor producing inspiration during normal, quiet breathing; expiration is a passive process in this type of breathing, caused by passive recoil of the lungs. Active breathing requires the activity of additional muscles and involves energy expenditure for both inspiration and expiration.

within the lungs (see Fig. 15.2*B*). Pleural pressure is more negative than in the resting state, as the chest wall exerts greater outward pressure. Air flows into the lung until alveolar pressure reaches atmospheric pressure. At the end of inspiration, during normal, quiet breathing (tidal volume of 500 mL), chest wall recoil pressure is still negative, but recoil pressure of the total respiratory system is positive as a result of increased recoil pressure of the lungs (see Fig. 15.3). At TLC, both chest wall and lung recoil pressures are positive.

During **expiration**, with the relaxation of inspiratory muscles, the elastic recoil pressure of the respiratory system (increased because of its expansion) results in elevation of alveolar pressure above atmospheric pressure, causing outward airflow until alveolar pressure has fallen to atmospheric pressure (see Fig. 15.2*C*). The system returns to FRC, unless air is actively expired beyond that level; expiration is eventually limited by the large negative elastic recoil pressure of the chest wall as residual volume (RV) is approached.

COMPLIANCE, ELASTANCE, AND PRESSURE-VOLUME RELATIONSHIPS

Elastance is defined as the tendency of a hollow organ to return to its original size when distended; it can be quantified as elastic recoil pressure. The **elastic recoil** of the lungs and chest wall can be related to pleural pressure:

- Elastic recoil of the lungs is equal to P_A minus pleural pressure.
- Elastic recoil of the chest wall is equal to pleural pressure minus P_{ATM}.

Lung compliance (C_L) is a measure of the distensibility of the lung and is the inverse of lung elastance. Thus C_L is measured as the change in lung volume resulting from a change in transpulmonary pressure, where transpulmonary pressure is the pressure across the lung, or the difference between alveolar and pleural pressure (Fig. 15.4). Experimentally, compliance can be measured in an isolated lung (Fig. 15.5). Referring to the experimental preparation in Figure 15.5 in which the lung

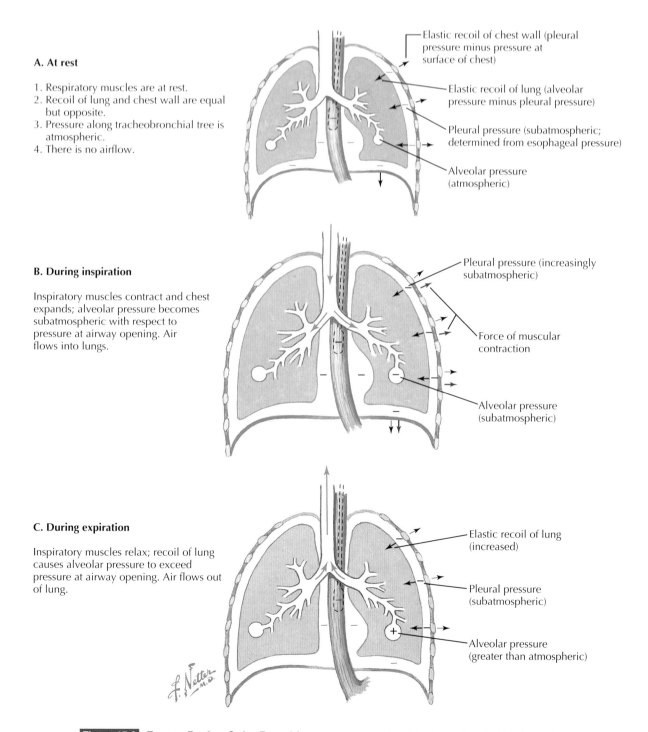

A. At rest

1. Respiratory muscles are at rest.
2. Recoil of lung and chest wall are equal but opposite.
3. Pressure along tracheobronchial tree is atmospheric.
4. There is no airflow.

Elastic recoil of chest wall (pleural pressure minus pressure at surface of chest)

Elastic recoil of lung (alveolar pressure minus pleural pressure)

Pleural pressure (subatmospheric; determined from esophageal pressure)

Alveolar pressure (atmospheric)

B. During inspiration

Inspiratory muscles contract and chest expands; alveolar pressure becomes subatmospheric with respect to pressure at airway opening. Air flows into lungs.

Pleural pressure (increasingly subatmospheric)

Force of muscular contraction

Alveolar pressure (subatmospheric)

C. During expiration

Inspiratory muscles relax; recoil of lung causes alveolar pressure to exceed pressure at airway opening. Air flows out of lung.

Elastic recoil of lung (increased)

Pleural pressure (subatmospheric)

Alveolar pressure (greater than atmospheric)

Figure 15.2 **Forces During Quiet Breathing**   Normal, quiet breathing is produced mainly by contraction of the diaphragm, resulting in inspiration as alveolar pressure falls below atmospheric pressure; expiration occurs when the diaphragm relaxes and recoil pressure of the lungs elevates alveolar pressure above atmospheric pressure. The dynamic interactions of elastic recoil pressure of the lungs and chest wall and contraction and relaxation of the diaphragm producing airflow are illustrated **(A** to **C)**. Pleural pressure is always negative during quiet breathing.

During normal, quiet breathing, inspiration is mainly driven by the contraction of the muscular diaphragm, whereas expiration is a passive process, driven by the positive elastic recoil pressure of the respiratory system achieved during inspiration. In this type of breathing, pleural pressure is always negative.

During active expiration, contraction of abdominal muscles compresses the viscera, forcing the diaphragm upward and producing positive pleural pressure. Thus, in contrast to quiet breathing, in which work is performed only for inspiration, energy is expended for both inspiration and expiration in active breathing.

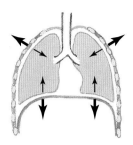
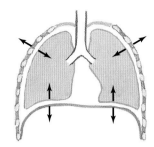
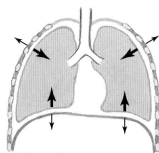

A. At residual volume
Elastic recoil of chest wall directed outward is large; recoil of lung directed inward is very small

B. At functional residual capacity
Elastic recoils of lung and chest wall are equal but opposite

C. At larger lung volume
Elastic recoil of chest wall becomes smaller, and recoil of lung increases

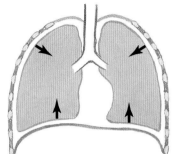
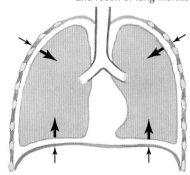

D. At approximately 70% of total lung capacity
Equilibrium position of chest wall (its recoil equals zero)

E. At total lung capacity
Elastic recoil of both lung and chest wall directed inward, favoring decrease in lung volume

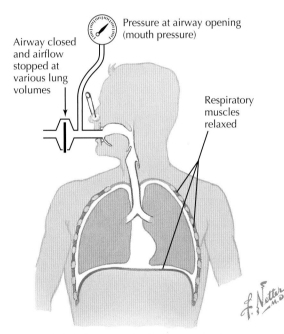

F. Pressure recorded at airway opening is same as alveolar pressure when airflow is stopped; provides a measure of elastic recoil of respiratory system when respiratory muscles are relaxed

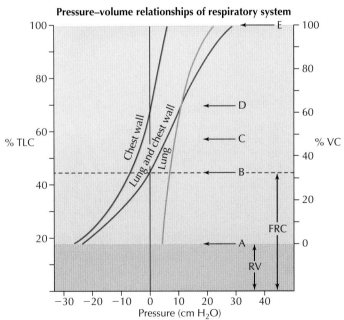

G. Elastic recoil pressure of respiratory system is algebraic sum of recoil pressures of lung and chest wall

Figure 15.3 **Elastic Properties of Respiratory System: Lung and Chest Wall** Elastic recoil pressure of the respiratory system is the algebraic sum of recoil pressures of the lungs and chest wall **(G)**. These pressures can be measured at various lung volumes from residual volume (RV) to total lung capacity (TLC) **(A** to **E)** and represent the forces resulting in inspiration and expiration. Above functional residual volume (FRC), the net recoil pressure of the respiratory system is positive, and relaxation of inspiratory muscles will result in expiration; below FRC, net recoil pressure is negative and relaxation of expiratory muscles will cause inspiration. At FRC, the system is in equilibrium. Elastic recoil when respiratory muscles are relaxed can be determined by stopping airflow **(F)**. *VC*, vital capacity.

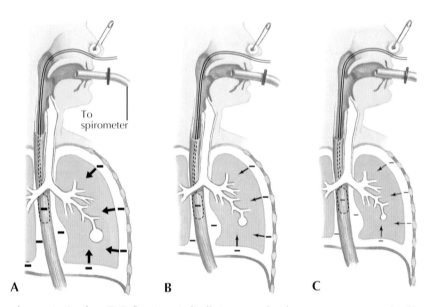

During a slow expiration from TLC, flow is periodically interrupted and measurements are made of lung volume and of transpulmonary pressure. Transpulmonary pressure is the difference between alveolar and pleural pressures. Pleural pressure is determined from pressure in the esophagus. Because there is no airflow, alveolar pressure is the same as pressure at the airway opening.

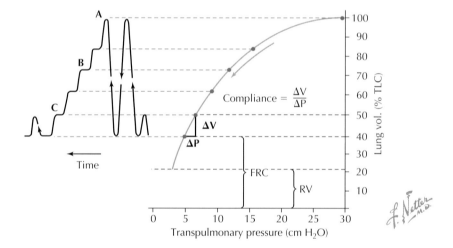

$$Compliance = \frac{\Delta V}{\Delta P}$$

Figure 15.4 **Measurement of Elastic Properties of the Lung** Lung compliance ($\Delta V/\Delta P$) is a measure of lung distensibility and can be measured by determining transpulmonary pressure at various lung volumes (**A** to **C**). Transpulmonary pressure is the difference between alveolar and pleural pressure. Alveolar pressure is the same as pressure measured at the airway opening when flow is stopped; pleural pressure is measured by esophageal balloon catheter. Note that lung compliance is highest at residual volume (RV) and is reduced at greater lung volumes. *FRC*, functional residual volume; *TLC*, total lung capacity.

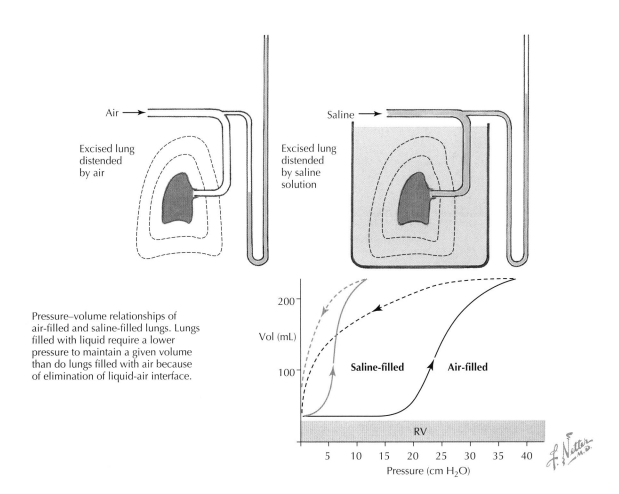

Pressure–volume relationships of air-filled and saline-filled lungs. Lungs filled with liquid require a lower pressure to maintain a given volume than do lungs filled with air because of elimination of liquid-air interface.

Figure 15.5 **Compliance and Surface Forces of the Lung** Measurement of the pressure-volume relationship in an isolated lung results in the illustrated pressure-volume loops. Greater pressure is required to inflate an air-filled lung compared with a saline solution–filled lung because of the surface tension at the air-alveolar interface in the air-filled lung. Surfactant is produced by type II alveolar epithelial cells and reduces surface tension of the air-filled lung; without it, even greater force would be required to inflate the lungs with air. *RV,* residual volume.

is filled with air, as pressure applied to the lung is changed, the volume of the lung changes. The slope of this curve is the compliance of the lung. Note that the slope differs between inflation and deflation, with the volume at a given pressure being lower during inflation than during deflation, a phenomenon known as **hysteresis**. The difference in the inflation and deflation curves is mainly associated with the effects of surface tension during inflation, when surface tension at the liquid-air interface of the lung must be overcome as volume rises. When a fluid-filled lung (in which surface tension is not an issue) is exposed to the same manipulations, compliance is greater and hysteresis is less pronounced.

As with the lungs, the chest wall also is characterized by compliance and elastance. These properties become apparent when the chest wall is punctured, creating a **pneumothorax** (see Clinical Correlate Box 15.1).

SURFACTANT AND SURFACE TENSION

Surface tension is the elastic-like force at the surface of a liquid (at the gas-liquid interface) caused by intermolecular attraction of the liquid molecules at that surface. In the lung, surface tension reduces lung compliance and has the potential for causing collapse of small airways. The potential problems of surface tension and low pulmonary compliance are overcome by the production of **surfactant** by type II alveolar epithelial cells. Surfactant is a complex lipoprotein containing the phospholipid **dipalmitoyl phosphatidyl choline**. It is amphipathic and lines the surface of the alveolar epithelium and small airways (with hydrophilic regions of the phospholipid oriented toward the epithelial surface and hydrophobic regions facing the lumen). Surfactant reduces surface tension of airways and alveoli and increases lung compliance, reducing the work of breathing.

CLINICAL CORRELATE 15.1
Pneumothorax

Pneumothorax is a condition in which air accumulates in the pleural cavity as a result of injury to the chest wall or lungs or as a result of pulmonary disease. Ordinarily, the pleural cavity contains only a few milliliters of fluid. Introduction of air into the pleural space through an open (sucking) pneumothorax will uncouple the mechanical forces associated with the chest walls and lung, causing collapse of the lung on the damaged side, thereby affecting ventilation in that lung. In a tension pneumothorax, air is able to enter the pleural space but cannot leave (a piece of tissue acts as a one-way valve). Air accumulates with each breath, raising intrathoracic pressure and thereby causing severe dyspnea (shortness of breath) and circulatory collapse. Tension pneumothorax is an unstable condition requiring immediate medical intervention.

Pathophysiology

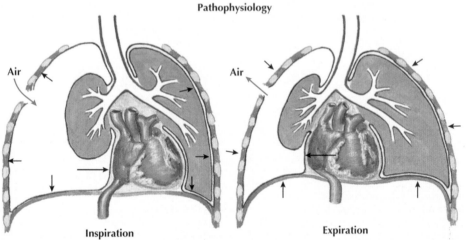

Inspiration

Air enters pleural cavity through open, sucking chest wound. Negative pleural pressure is lost, permitting collapse of ipsilateral lung and reducing venous return to heart. Mediastinum shifts, compressing opposite lung.

Expiration

As chest wall contracts and diaphragm rises, air is expelled from pleural cavity via wound. Mediastinum shifts to affected side and mediastinal flutter further impairs venous return by distortion of venae cavae.

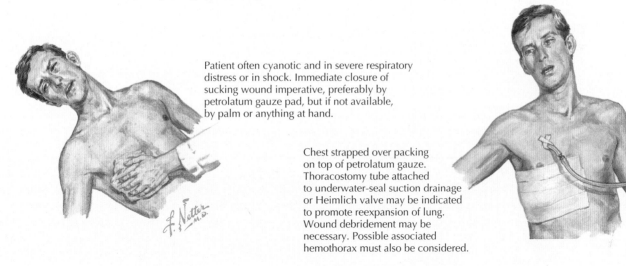

Patient often cyanotic and in severe respiratory distress or in shock. Immediate closure of sucking wound imperative, preferably by petrolatum gauze pad, but if not available, by palm or anything at hand.

Chest strapped over packing on top of petrolatum gauze. Thoracostomy tube attached to underwater-seal suction drainage or Heimlich valve may be indicated to promote reexpansion of lung. Wound debridement may be necessary. Possible associated hemothorax must also be considered.

Open Pneumothorax

CLINICAL CORRELATE 15.2
Compliance in Pulmonary Diseases

Lung diseases are often classified as restrictive and obstructive diseases. Restrictive diseases are characterized by reduced functional volume of the lung, whereas obstructive diseases are characterized by reduced flow rate. Restrictive diseases include interstitial lung diseases (e.g., idiopathic pulmonary fibrosis, sarcoidosis, and asbestosis); examples of obstructive lung diseases include chronic obstructive pulmonary disease (COPD) and asthma. In restrictive diseases, compliance of the respiratory system is reduced, resulting in lower FRC and TLC and reduced

slope of the pressure-volume relationship. Obstructive lung disease does not directly affect lung compliance but may ultimately change compliances. For example, in persons with emphysema, which is often associated with tobacco smoking and COPD, elastic fibers in the lungs are destroyed, alveolar architecture is compromised, and lung compliance is increased (i.e., elasticity is reduced). FRC and TLC are increased, and the slope of the pressure-volume relationship is greater. Despite increased compliance and greater tidal volume, persons with emphysema have reduced ability to exchange gas, based on the destruction of alveoli.

Work of Breathing

A. Normal

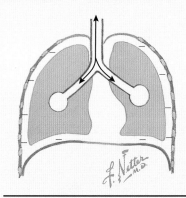

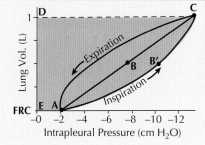

Work performed on lung during breathing can be determined from dynamic pressure–volume loop. Work to overcome elastic forces is represented by area of trapezoid EABCD. Additional work required to overcome flow resistance during inspiration is represented by area of right half of loop AB'CBA.

B. Obstructive disease

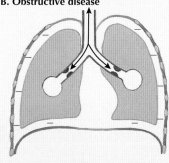

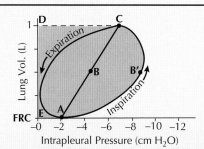

In disorders characterized by airway obstruction, work to overcome flow resistance is increased; elastic work of breathing remains unchanged.

C. Restrictive disease

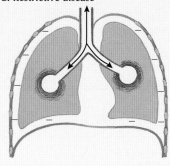

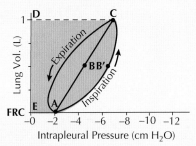

Restrictive lung diseases result in increase of elastic work of breathing; work to overcome flow resistance is normal.

Work of Breathing in Obstructive and Restrictive Lung Disease Compared with work performed in normal breathing **(A)**, in obstructive disease, although compliance may be unchanged, the work of breathing is increased by the elevated airway resistance **(B)**. In restrictive disease, lung compliance is low, and the elastic work of breathing is increased **(C)**. *FRC,* functional residual capacity.

AIRWAY RESISTANCE

Flow of air in and out of the lungs is dependent on the pressure gradient from the opening of the mouth to the alveoli. At the end of inspiration or expiration, the gradient is zero. **Poiseuille's law**, presented in the context of blood flow in Section 3, also applies to flow of air (Q) through tubes (see Fig. 15.5),

$$Q = \frac{\Delta P \pi r^4}{\eta 8 L} \qquad \textbf{Eq. 15.1}$$

where ΔP is the pressure gradient from one end of a tube to the other, r^4 is the radius of the tube to the fourth power, η is the viscosity of air, and L is the length of the tube. Thus, based on this relationship, airflow (Q) through a tube should be:

- Directly proportional to the longitudinal pressure gradient (inflow pressure minus outflow pressure).
- Inversely proportional to the length of the tube.
- Inversely proportional to the viscosity of air.
- Directly proportional to the fourth power of the radius of the tube.

In the respiratory system as a whole, the *greatest resistance to flow actually occurs in medium-sized airways* (fourth to eighth generation). Recall from Section 3 that in parallel tubes, total resistance is less than the resistance of the individual tubes. Considering both the radius and number of tubes at this level, resistance is higher in the medium-sized bronchi (in aggregate) than in the larger or smaller airways.

Poiseuille's law applies to laminar flow of air but becomes less precise with turbulent flow. Laminar flow occurs in small airways, whereas flow in the largest airways is turbulent; flow in the remainder of the system tends to be transitional, with some turbulence (Fig. 15.6). Factors producing **turbulent and laminar flow** are discussed in Section 3, in the context of blood flow. In the largest airways, high velocity and airway diameter contribute to producing turbulent flow. In small, peripheral airways, smaller diameter and lower velocity result in laminar flow of air.

The effects of surface tension in the lung have often been incorrectly presented in textbooks in the context of **Laplace's law**, which states that in a sphere, surface tension creates pressure (P),

$$P = \frac{2T}{r}$$

where T is surface tension and r is radius of the sphere. According to this construct, in an isolated alveolus, this pressure is the pressure tending to cause its collapse (or the equivalent pressure required to keep the alveolus open). Laplace's law would lead to the prediction that when multiple spherical units are subjected to the same inflation pressure through a branching tube, larger units would expand and smaller units would collapse, as the higher surface tension in smaller units causes air to flow toward larger units in which surface tension is lower. Textbooks have typically illustrated this situation as a Y-tube with two alveoli attached, or as alveoli in the configuration of a "bunch of grapes." According to this line of argument, a major role of surfactant is to overcome the effects of surface tension in alveoli of different sizes and promote more uniform inflation of the lungs.

This argument is incorrect for several reasons: alveoli are actually prismatic in shape (appearing polygonal in histologic sections) and exist in a honeycomb-like geometry with common walls, sometimes even with pores between them. Thus alveoli do not act as independent units, and inflation of an alveolus affects inflation of adjacent alveoli. Rather than viewing surfactant's role in terms of alveolar surface tension, its effects on the overall compliance of the lung and on small airway patency should be emphasized.

(From Prange H: Laplace's law and the alveolus: a misconception of anatomy and a misapplication of physics, *Adv Physiol Educ* 27:34-40, 2003.)

Effects of Autonomic Nerves on Airway Resistance

Airway resistance is also affected by the autonomic nervous system:

- **Activation of parasympathetic nerves** innervating smooth muscle of conducting airways causes bronchoconstriction and promotes glandular secretions in the lungs.
- **Activation of sympathetic nerves** innervating smooth muscle of conducting airways results in bronchodilation and reduced airway resistance (through activation of β_2-adrenergic receptor–linked pathways) in mammalian species, although human lungs have little sympathetic innervation. Release of epinephrine by the adrenal medulla during sympathetic activation will also reduce airway resistance through activation of the pulmonary β_2-receptor mechanism.

Lung Volume and Airway Resistance

In addition, a relationship exists between **lung volume and airway resistance**. At high lung volume, the diameter of airways tends to be increased by the radial traction placed on the airways by the expanded lungs. At low lung volumes, in the absence of traction, small airways have a greater tendency to collapse.

DYNAMIC COMPRESSION OF AIRWAYS DURING EXPIRATION

Airway resistance is also affected by **dynamic compression**, which is the compression of airways during *forced expiration*.

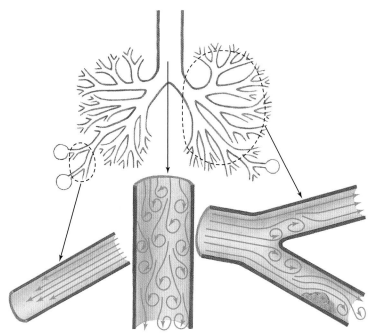

Laminar flow occurs mainly in small peripheral airways where rate of airflow through any airway is low. Flow is proportional to driving pressure.

Turbulent flow occurs at high flow rates in trachea and larger airways. Flow is proportional to the square root of the driving pressure.

Transitional flow occurs in larger airways, particularly at branches and at sites of narrowing.

Poiseuille's law. Resistance to laminar flow is inversely proportional to tube radius to the fourth power and directly proportional to length of tube. When radius is halved, resistance is increased 16-fold. If driving pressure is constant, flow will fall to one sixteenth. Doubling length only doubles resistance. If driving pressure is constant, flow will fall to one half.

Resistance ~1

Resistance ~16

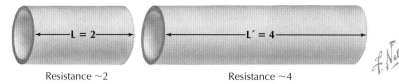

Resistance ~2   Resistance ~4

Figure 15.6 **Flow** Airflow through large airways tends to be turbulent, whereas flow in smaller airways tends to be laminar. Laminar flow is described by Poiseuille's law. Airway diameter, according to this law, is the greatest factor in determining resistance and flow (flow is proportional to the fourth power of the radius of a tube).

CLINICAL CORRELATE 15.3
Treatment of Asthma with Sympathetic Adrenergic Drugs

Asthma is a chronic respiratory disease that is characterized by episodic bronchoconstriction and mucus secretion, resulting in increased airway resistance and dyspnea (labored breathing). Asthma can be life-threatening in severe cases. Asthma attacks may be initiated by a variety of factors, including inhaled allergens and irritants, cold air, stress, and exercise, which exacerbate the ongoing airway inflammation in persons with asthma. Drugs used to prevent asthma attacks are mainly aimed at reducing inflammation, whereas drugs used to relieve acute attacks are generally bronchodilators (e.g., inhaled β_2-adrenergic agonists such as salbutamol or terbutaline). These drugs reduce airway resistance by relaxing bronchial smooth muscle.

An **expiratory flow–volume curve** is illustrated in Figure 15.7 (*A*, solid line), in which a subject performs a **forced vital capacity (FVC) maneuver,** inspiring to TLC and then exhaling as forcibly as possible to RV. Note that in the expiratory flow–volume curve generated:

■ The peak of this curve represents the **peak expiratory flow rate**.
■ The downward slope (expiratory phase) of the flow-volume curve is **effort-independent**; during this phase of the curve, flow is limited by **dynamic compression** of the airways.

When less effort is exerted, lower peak flow is reached, but the associated flow-volume curve converges with the maximum

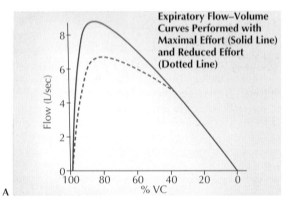

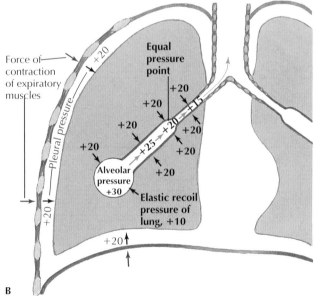

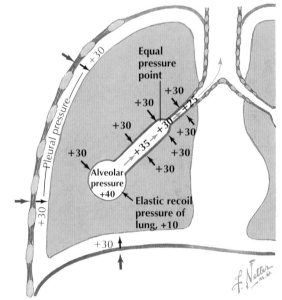

Determinants of Maximal Expiratory Flow

At onset of maximal airflow, contraction of expiratory muscles at a given lung volume raises pleural pressure above atmospheric level (+20 cm H₂O). Alveolar pressure (sum of pleural pressure and lung recoil pressure) is yet higher (+30 cm H₂O). Airway pressure falls progressively from alveolus to airway opening in overcoming resistance. At equal pressure point of airway, pressure within airway equals pressure surrounding it (pleural pressure). Beyond this point, as intraluminal pressure drops further, below pleural pressure, airway will be compressed.

With further increases in expiratory effort, at same lung volume, pleural pressure is greater and alveolar pressure is correspondingly higher. Fall in airway pressure and location of equal pressure point are unchanged, but beyond equal pressure point, intrathoracic airways will be compressed to a greater degree by higher pleural pressure. Once maximal airflow is achieved, further increases in pleural pressure produce proportional increases in resistance of segment downstream from equal pressure point, so rate of airflow does not change.

Figure 15.7 **Expiratory Flow–Volume Relationship**   The expiratory flow–volume curve generated during a forced vital capacity (VC) maneuver *(solid line in graph)* is characterized by a peak expiratory flow rate at the top of the curve and a downward slope during the remainder of expiration **(A)**. The maximal expiratory flow rate at various lung volumes along this downward slope is effort-independent because of dynamic compression of the airways **(B)**. When effort is reduced *(dotted line in graph)*, peak expiratory flow rate is lower, but note the overlap in the downward slope of the two lines, consistent with the effort independence of maximal expiratory flow rate.

effort curve and the downslope is the same, consistent with the effort independence of expiratory flow during active expiration (see Fig. 15.7A, *dotted line*).

During normal, quiet breathing, pleural pressure is always negative. However, during active expiration, contraction of expiratory muscles results in elevation of pleural pressure above atmospheric pressure (see Fig. 15.7). Under these circumstances, during expiration, alveolar pressure is the sum of the positive pleural pressure and the elastic recoil pressure of

the lungs. Pressure in the airways falls between the alveoli and the opening of the mouth, reaching atmospheric pressure at the mouth. Thus, at some point downstream from the alveoli, an **equal pressure point** is reached, at which airway pressure is equal to pleural pressure. Beyond this point, airways will be compressed. This **dynamic compression** of airways limits expiratory flow rate and accounts for the effort-independent nature of airflow in active expiration. When greater effort is exerted, greater compression occurs, so that airflow remains the same.

CLINICAL CORRELATE 15.4
Respiratory Distress Syndrome of the Newborn

Immediately after delivery, a newborn takes its first breath. Negative pressure of 40 to 100 cm H_2O is required to draw air into the collapsed airways and inflate the alveoli. During this first breath, in a healthy, normal, term infant, surfactant stored in type II alveolar epithelial cells is released and forms a monomolecular layer at the air-fluid interface of the small airways and alveoli. By the third breath, only a small negative pressure is required to inflate the lung. Respiratory distress syndrome (previously known as *hyaline membrane disease*) is the most common cause of death in premature infants and is caused by a lack of surfactant production. Without surfactant, the negative pressure required to inflate the lungs remains high (CL is low) and portions of the lung collapse, leading to respiratory distress and potentially respiratory failure and death. Treatment consists of ventilatory support and surfactant therapy through the breathing tube.

A. Risk factors for development of respiratory distress syndrome (RDS) of newborn

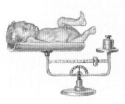

Prematurity
Birth wt. >2.5 kg; RDS not likely
Birth wt. <2.5 kg; likelihood of RDS increases in relation to lower wt. (if viable)

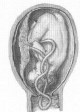

Perinatal asphyxia
(second born of twins therefore more susceptible)

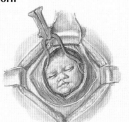

Cesarean birth

Diabetes mellitus
(maternal)

B. Surfactant effects during lung inflation in the neonate

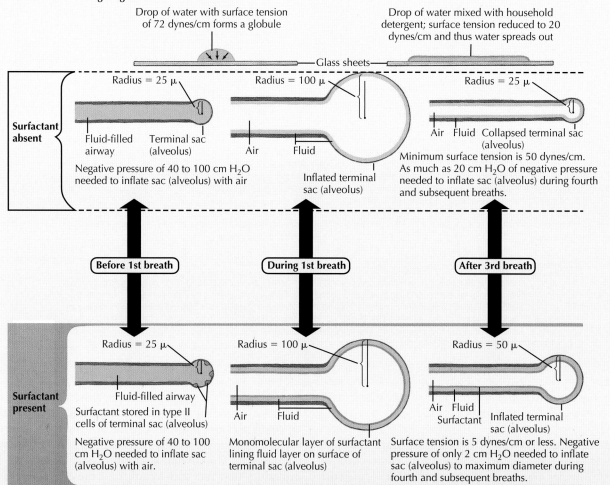

The initial inflation of the collapsed lungs of a neonate requires large negative pressure (40 to 100 cm H_2O), but surface tension is reduced in subsequent breaths as surfactant lines the alveoli and small airways, reducing the work required to inflate the lungs. Premature birth is often associated with surfactant deficiency and respiratory distress.

OBSTRUCTIVE AND RESTRICTIVE PULMONARY DISEASES AND PULMONARY FUNCTION TESTS

Measurements of flow-volume relationships are important in assessing **obstructive and restrictive pulmonary diseases** (see Clinical Correlate 15.2). Severe COPD is characterized by emphysema, in which inflammation leads to destruction of alveolar walls and capillaries, resulting in high C_L. Reduced

elastic recoil of the alveoli and airways also results in early formation of the equal pressure point (at a point closer to the alveolus) during expiration, and, as a result of this dynamic compression, "trapping" of air takes place in the lung. These changes ultimately produce an increase in TLC, FRC, and RV. The changes in flow are clinically assessed by spirometry and related tests (Fig. 15.8). Notably, expiratory flow rate and the forced expiratory volume in the first second of a vital capacity

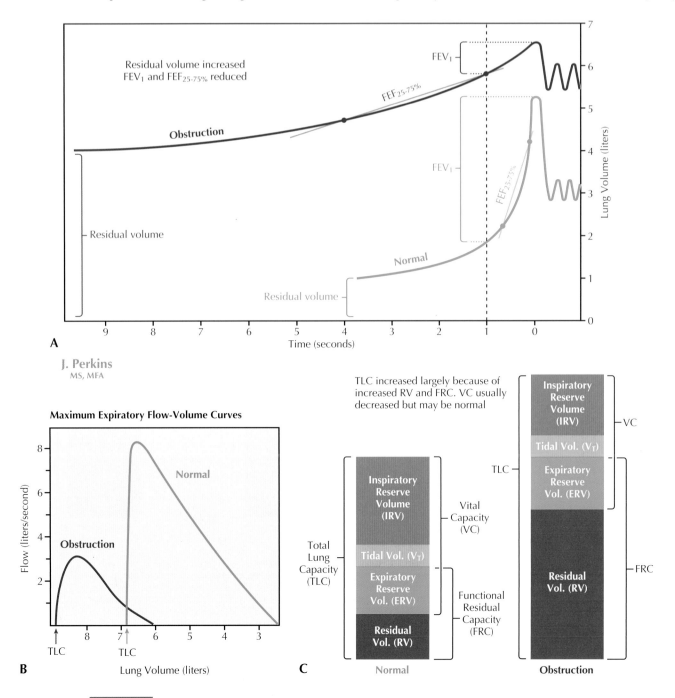

Figure 15.8 **Pulmonary Function in Obstructive Lung Disease** In emphysema, a chronic obstructive lung disease often associated with smoking, inflammatory destruction of elastic tissues in the lung occurs, resulting in reduced elastic recoil of the lung. Changes in lung volumes (**A**), flow-volume curves (**B**), and spirometric measurements (**C**) associated with emphysema are illustrated. Notably, forced expiratory volume in one second (FEV_1) is reduced in persons with obstructive lung disease, as is the ratio of FEV_1 to forced vital capacity (**A**). The forced expiratory flow rate during the middle portion of a forced expiration ($FEF_{25\%-75\%}$) is also reduced.

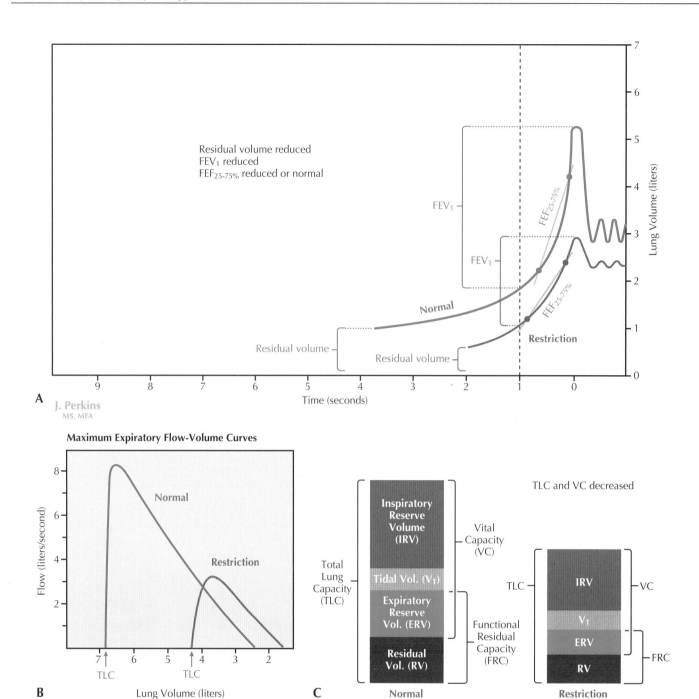

Residual volume reduced
FEV$_1$ reduced
FEF$_{25-75\%}$ reduced or normal

A

J. Perkins
MS, MFA

Maximum Expiratory Flow-Volume Curves

B

C

TLC and VC decreased

Figure 15.9 **Pulmonary Function in Restrictive Lung Disease** Lung compliance is reduced in restrictive lung diseases such as interstitial fibrosis, resulting in diminished lung volumes. Changes in spirometric measurements **(A)**, flow-volume curves **(B)**, and lung volumes **(C)** associated with restrictive lung disease are illustrated. Because both forced expiratory volume in one second (FEV$_1$) and forced vital capacity (FVC) are reduced in restrictive lung disease **(A)**, the ratio of FEV$_1$ to FVC is usually normal but may even be increased when FVC is greatly reduced. The forced expiratory flow rate during the middle portion of a forced expiration (FEF$_{25\%-75\%}$) is normal or reduced in persons with restrictive disease.

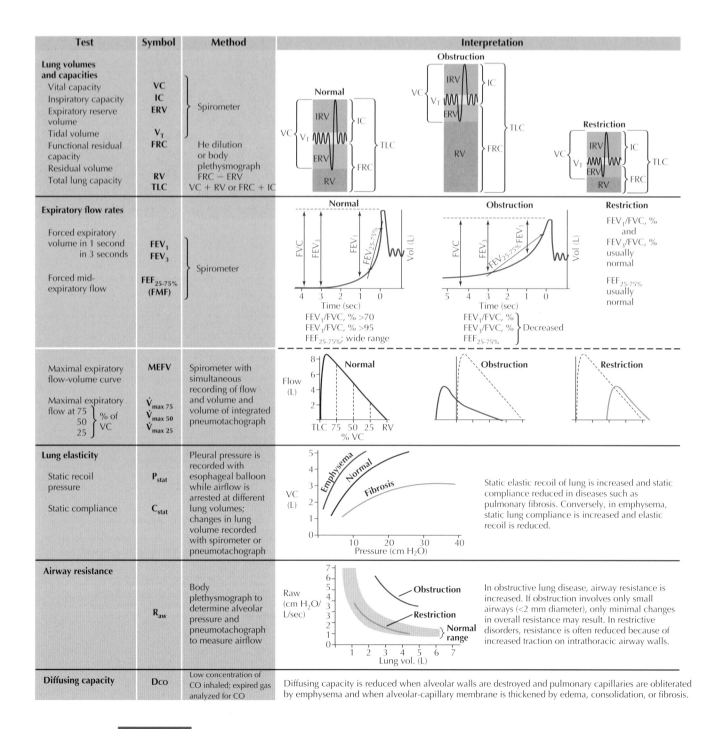

Test	Symbol	Method	Interpretation
Lung volumes and capacities Vital capacity Inspiratory capacity Expiratory reserve volume Tidal volume Functional residual capacity Residual volume Total lung capacity	**VC** **IC** **ERV** V_T **FRC** **RV** **TLC**	Spirometer He dilution or body plethysmograph FRC − ERV VC + RV or FRC + IC	Normal / Obstruction / Restriction (illustrated)
Expiratory flow rates Forced expiratory volume in 1 second in 3 seconds Forced mid-expiratory flow	FEV_1 FEV_3 $FEF_{25-75\%}$ (FMF)	Spirometer	Normal: FEV_1/FVC, % >70; FEV_3/FVC, % >95; $FEF_{25-75\%}$ wide range. Obstruction: FEV_1/FVC, %, FEV_3/FVC, %, $FEF_{25-75\%}$ Decreased. Restriction: FEV_1/FVC, % and FEV_3/FVC, % usually normal; $FEF_{25-75\%}$ usually normal
Maximal expiratory flow-volume curve Maximal expiratory flow at 75, 50, 25 % of VC	**MEFV** $\dot{V}_{max\ 75}$ $\dot{V}_{max\ 50}$ $\dot{V}_{max\ 25}$	Spirometer with simultaneous recording of flow and volume of integrated pneumotachograph	Normal / Obstruction / Restriction (illustrated)
Lung elasticity Static recoil pressure Static compliance	P_{stat} C_{stat}	Pleural pressure is recorded with esophageal balloon while airflow is arrested at different lung volumes; changes in lung volume recorded with spirometer or pneumotachograph	Static elastic recoil of lung is increased and static compliance reduced in diseases such as pulmonary fibrosis. Conversely, in emphysema, static lung compliance is increased and elastic recoil is reduced.
Airway resistance	R_{aw}	Body plethysmograph to determine alveolar pressure and pneumotachograph to measure airflow	In obstructive lung disease, airway resistance is increased. If obstruction involves only small airways (<2 mm diameter), only minimal changes in overall resistance may result. In restrictive disorders, resistance is often reduced because of increased traction on intrathoracic airway walls.
Diffusing capacity	Dco	Low concentration of CO inhaled; expired gas analyzed for CO	Diffusing capacity is reduced when alveolar walls are destroyed and pulmonary capillaries are obliterated by emphysema and when alveolar-capillary membrane is thickened by edema, consolidation, or fibrosis.

Figure 15.10A **Pulmonary Function Tests** Tests of pulmonary function are defined and illustrated, with comparisons between values observed in normal lungs and in restrictive and obstructive disease.

maneuver (FEV_1) are reduced and the FEV_1/**FVC ratio** is less than the normal value of 75% (although FEV_1 is reduced, FVC is only slightly lower than normal). In contrast, in restrictive diseases such as interstitial fibrosis, thickening of alveolar walls results in decreased CL. Lung volumes are reduced as a result (Fig. 15.9). Although FEV_1 is reduced, the FEV_1-to-FVC ratio is usually normal or elevated, because FVC is also diminished in restrictive disease. Various **pulmonary function tests** and typical findings in obstructive and restrictive disease are summarized in Figure 15.10.

Because of **dynamic compression** of airways, maximum expiratory flow is effort-independent. For any given lung volume, the associated maximum expiratory flow cannot be exceeded by increasing expiratory effort. The explanation for this phenomenon is that increased expiratory effort physically compresses airways as well as alveoli, increasing resistance and limiting airflow.

Test	Symbol	Method	Interpretation
Tests for small airway disease Closing volume Closing capacity	**CV** **CC**	Following a full inspiration of O_2 the expired lung volume from TLC to RV is plotted against the N_2 concentration	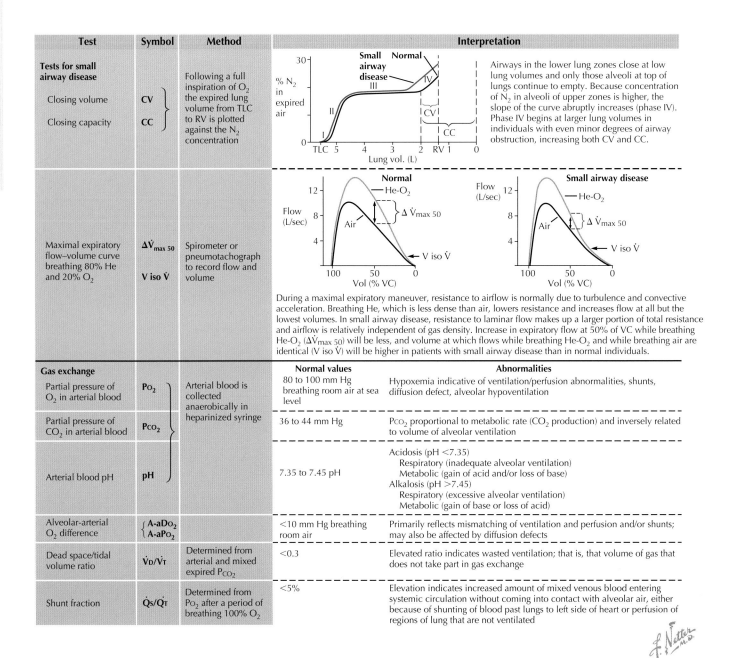 Airways in the lower lung zones close at low lung volumes and only those alveoli at top of lungs continue to empty. Because concentration of N_2 in alveoli of upper zones is higher, the slope of the curve abruptly increases (phase IV). Phase IV begins at larger lung volumes in individuals with even minor degrees of airway obstruction, increasing both CV and CC.
Maximal expiratory flow–volume curve breathing 80% He and 20% O_2	$\Delta \dot{V}_{max\ 50}$ V iso $\dot{V}$	Spirometer or pneumotachograph to record flow and volume	During a maximal expiratory maneuver, resistance to airflow is normally due to turbulence and convective acceleration. Breathing He, which is less dense than air, lowers resistance and increases flow at all but the lowest volumes. In small airway disease, resistance to laminar flow makes up a larger portion of total resistance and airflow is relatively independent of gas density. Increase in expiratory flow at 50% of VC while breathing He-O_2 ($\Delta \dot{V}_{max\ 50}$) will be less, and volume at which flows while breathing He-O_2 and while breathing air are identical (V iso $\dot{V}$) will be higher in patients with small airway disease than in normal individuals.

Test	Symbol	Method	Normal values	Abnormalities
Gas exchange Partial pressure of O_2 in arterial blood	P_{O_2}	Arterial blood is collected anaerobically in heparinized syringe	80 to 100 mm Hg breathing room air at sea level	Hypoxemia indicative of ventilation/perfusion abnormalities, shunts, diffusion defect, alveolar hypoventilation
Partial pressure of CO_2 in arterial blood	P_{CO_2}		36 to 44 mm Hg	P_{CO_2} proportional to metabolic rate (CO_2 production) and inversely related to volume of alveolar ventilation

Arterial blood pH	**pH**		7.35 to 7.45 pH	Acidosis (pH <7.35)
				   Respiratory (inadequate alveolar ventilation)
				   Metabolic (gain of acid and/or loss of base)
				Alkalosis (pH >7.45)
				   Respiratory (excessive alveolar ventilation)
				   Metabolic (gain of base or loss of acid)

Alveolar-arterial O_2 difference	A-aD_{O_2} A-aP_{O_2}		<10 mm Hg breathing room air	Primarily reflects mismatching of ventilation and perfusion and/or shunts; may also be affected by diffusion defects
Dead space/tidal volume ratio	$\dot{V}_D/\dot{V}_T$	Determined from arterial and mixed expired P_{CO_2}	<0.3	Elevated ratio indicates wasted ventilation; that is, that volume of gas that does not take part in gas exchange
Shunt fraction	$\dot{Q}_S/\dot{Q}_T$	Determined from P_{O_2} after a period of breathing 100% O_2	<5%	Elevation indicates increased amount of mixed venous blood entering systemic circulation without coming into contact with alveolar air, either because of shunting of blood past lungs to left side of heart or perfusion of regions of lung that are not ventilated

Figure 15.10B Pulmonary Function Tests Tests of pulmonary function are defined and illustrated, with comparisons between values observed in normal lungs and in restrictive and obstructive disease.

Chapter 16 Oxygen and Carbon Dioxide Transport and Control of Respiration

While Fick's law describes the diffusion of gases across the alveolar-capillary membrane, many other factors are important in determining the actual content of gases in blood and in the process of gas transport. The contribution of other factors can be gleaned by considering that average oxygen consumption in humans at rest is 250 mL O_2/min, a rate that cannot be supported simply by diffusion and delivery of dissolved gas to tissues. In fact, transport of gases is a complex process that affects other processes beyond respiration, including acid-base balance, which will also be discussed in this chapter.

TRANSPORT OF OXYGEN

The concentration of dissolved gas in a liquid is directly proportional to the partial pressure of the gas in the atmosphere to which the liquid is exposed (Henry's law) and the solubility of the gas in the solvent. For each millimeter of mercury of Po_2, only 0.003 mL O_2 will dissolve in 100 mL of blood at body temperature. Because Pa_{O_2} (Po_2 in arterial blood) is normally approximately 100 mm Hg (equilibrated with PA_{O_2}, i.e., Po_2 in alveolar air), the amount of **dissolved oxygen** in arterial blood is normally only 0.3 mL O_2/100 mL blood (Fig. 16.1). The actual measured concentration of oxygen in normal arterial blood is approximately 20.4 mL O_2/100 mL blood. What accounts for this large difference? The answer is binding of O_2 to the **hemoglobin** (Hb) in red blood cells. The average concentration of hemoglobin is 15 g/100 mL blood, and each gram of hemoglobin binds 1.34 mL O_2 when fully saturated. Therefore, by far, the bulk of oxygen transport is accomplished by hemoglobin.

Oxygen-Binding Capacity and Oxygen Content of Blood

The oxygen-binding capacity of blood is determined by using the following formula:

$$O_2 \text{ binding capacity} = (1.34 \text{ mL } O_2/\text{g Hb}) \times (\text{g Hb}/100 \text{ mL blood}) \quad \textbf{Eq. 16.1}$$

Thus, for the normal hemoglobin concentration of 15 g/100 mL blood,

$$O_2 \text{ binding capacity} = (1.34 \text{ mL } O_2/\text{g Hb}) \times (15 \text{ g Hb}/100 \text{ mL blood}) \quad \textbf{Eq. 16.2}$$
$$= 20.1 \text{ mL } O_2/100 \text{ mL blood}$$

Oxygen content of blood can be calculated by using the following formula:

$$O_2 \text{ content} = \% \text{ saturation} \times O_2 \text{ binding capacity} + \text{dissolved oxygen} \quad \textbf{Eq. 16.3}$$

Given normal values of 15 g hemoglobin/100 mL blood and approximately 100% oxygen saturation of arterial blood at Po_2 of 100 mm Hg, oxygen content in arterial blood can be calculated as shown below:

$$\text{Arterial } O_2 \text{ content} = 100\% \times (1.34 \text{ mL } O_2/\text{g Hb}) \times (15 \text{ g Hb}/100 \text{ mL blood}) + (0.003 \text{ mL } O_2/100 \text{ mL blood/mm Hg}) \times (100 \text{ mm Hg}) = 20.4 \text{ mL } O_2/100 \text{ mL blood}$$
$$\textbf{Eq. 16.4}$$

Arteriovenous Oxygen Gradient and Oxygen Consumption

Compared with this arterial oxygen content (20.4 mL O_2/100 mL blood), in normal venous blood in which Po_2 is approximately 40 mm Hg and hemoglobin is 75% saturated,

$$\text{Venous } O_2 \text{ content} = 75\% \times (1.34 \text{ mL } O_2/\text{g Hb}) \times (15 \text{ g Hb}/100 \text{ mL blood}) + (0.003 \text{ mL } O_2/100 \text{ mL blood/mm Hg}) \times (40 \text{ mm Hg}) = 15.2 \text{ mL } O_2/100 \text{ mL blood}$$
$$\textbf{Eq. 16.5}$$

Each deciliter of blood delivers approximately 5 mL of oxygen to the tissues as it passes through the systemic circulation, because the arteriovenous difference in the oxygen content of blood is 5 mL. Based on this arteriovenous difference in oxygen content of blood and cardiac output, oxygen consumption can be estimated. Thus, at rest,

$$O_2 \text{ consumption} = [a-v]_{o_2} \times \text{cardiac output} = (5 \text{ mL } O_2/100 \text{ mL}) \times 5000 \text{ mL/min} = 250 \text{ mL } O_2/\text{min}$$
$$\textbf{Eq. 16.6}$$

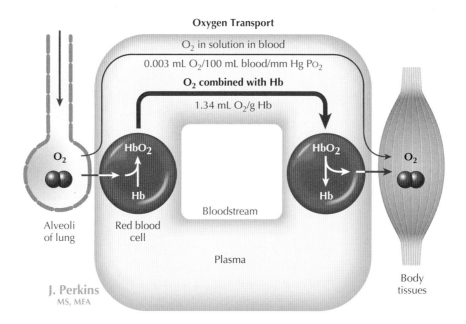

Figure 16.1 **Oxygen Transport** Oxygen diffuses into the blood flowing through alveolar capillaries and is transported to tissues, where it diffuses out of the blood along its concentration gradient. Transport of oxygen in the blood is mainly in the form of oxygen combined with hemoglobin (Hb), with only a minor portion carried in the form of dissolved oxygen.

At the usual respiratory quotient (CO_2 production/O_2 consumption) of 0.8, CO_2 production is 200 mL/min.

Oxygen-binding capacity is the maximum amount of oxygen that can be bound to hemoglobin in blood (at 100% saturation). The actual oxygen content of blood depends on the degree of saturation of hemoglobin. Dissolved oxygen is a minor portion of the oxygen content of blood; in a person breathing normal air, it is only 0.3 mL/100 mL blood, compared with the total content of oxygen of 20.4 mL/100 mL blood at 100% saturation.

THE OXYHEMOGLOBIN DISSOCIATION CURVE

As discussed, at P_{O_2} of 100 mm Hg, hemoglobin is nearly 100% saturated with oxygen (actually, 97.5%), and at 40 mm Hg, hemoglobin is 75% saturated. The **oxyhemoglobin dissociation curve** (Fig. 16.2) describes the relationship between oxygen saturation of blood (S_{O_2}) and P_{O_2}. The sigmoidal shape of the curve is due to cooperative binding of hemoglobin. Each molecule of hemoglobin is capable of binding four oxygen molecules; when one molecule binds to an oxygen binding site, the other three sites bind oxygen more readily, causing the middle portion of the curve to be steep. Thus exposure of blood to the high P_{O_2} in the respiratory zone of the lung results in binding of a substantial amount of oxygen. Note also that because of the sigmoidal shape, $P_{A_{O_2}}$ can fall substantially without greatly affecting the degree of saturation of hemoglobin (at P_{O_2} of 80 mm Hg, S_{O_2} is still

above 95%). On the other hand, as P_{O_2} falls in blood coursing through systemic capillaries, typically to 40 mm Hg, blood P_{O_2} is on the steep portion of the curve, which facilitates delivery of oxygen to the tissues. Dissociation of one molecule of oxygen from a molecule of hemoglobin makes dissociation of the other bound oxygen molecules from that molecule more likely.

To summarize:

- Cooperative binding of oxygen to four binding sites per hemoglobin molecule is responsible for the sigmoidal shape of the oxyhemoglobin dissociation curve.
- In the lungs, blood will become fully saturated with oxygen over a wide range of P_{O_2} because of the flat, upper portion of the oxyhemoglobin dissociation curve.
- At the lower P_{O_2} levels in tissue capillaries, small changes in P_{O_2} result in dissociation of relatively large amounts of oxygen because of the steep middle portion of the curve, facilitating oxygen delivery to the tissues.

Factors Affecting the Oxyhemoglobin Dissociation Curve

In addition to these qualities, another important characteristic of the oxyhemoglobin dissociation curve is that it is shifted to the right under conditions of increased P_{CO_2}, low pH, and high temperature (see Fig. 16.2). These conditions occur locally during tissue hypoxia and increased metabolism (for example, during exercise), and the rightward shift of the curve

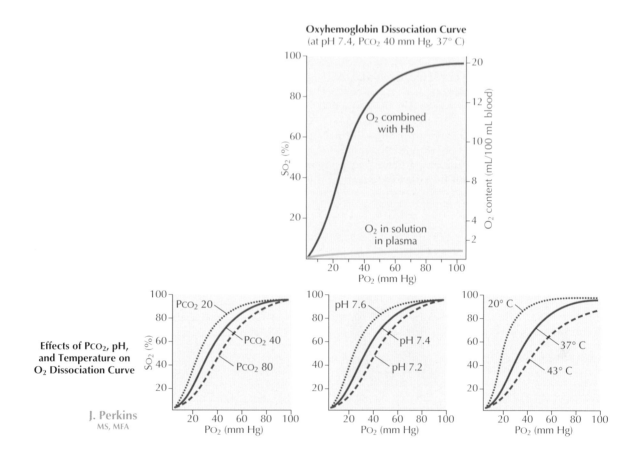

Figure 16.2 **Oxyhemoglobin Dissociation Curves** The oxyhemoglobin binding curve describes the relationship between P_{O_2} and the degree of oxygen saturation of hemoglobin (Hb). Saturation of hemoglobin is nearly 100% (97.5%) when P_{O_2} is 100 mm Hg. The sigmoidal shape of this curve results in a high degree of oxygen saturation of blood after passing through alveolar capillaries and significant dissociation of oxygen from hemoglobin at the P_{O_2} levels to which blood is exposed as it perfuses systemic capillaries. The values given for oxygen content on the graph are those expected at the normal blood hemoglobin concentration of 15 g/100 mL blood. Note that the amount of dissolved hemoglobin in blood is very low over a wide range of P_{O_2}. High P_{CO_2}, low pH, and high temperature shift the oxyhemoglobin dissociation curve to the right, which promotes oxygen dissociation from hemoglobin in capillaries supplying actively metabolizing tissues.

results in decreased hemoglobin affinity for oxygen and thus enhances delivery of oxygen to tissues. The metabolite of red blood cell glycolysis, **2,3-diphosphoglycerate** (2,3-DPG; also known as 2,3-bisphophoglycerate), also shifts the curve to the right and is elevated during hypoxia.

Oxygen delivery to the fetus involves transfer of oxygen from maternal blood to fetal blood across the placenta. Fetal blood contains a form of hemoglobin known as *fetal hemoglobin* (hemoglobin F) that has higher affinity for oxygen than adult hemoglobin (hemoglobin A), facilitating transfer of oxygen from maternal to fetal blood. Thus the oxyhemoglobin dissociation curve for hemoglobin F is shifted to the left compared with the dissociation curve for hemoglobin A. At a given P_{O_2}, oxygen saturation of hemoglobin F will be higher than that of hemoglobin A. Hemoglobin F is replaced by hemoglobin A within the first 3 months of life.

TRANSPORT OF CARBON DIOXIDE

The concentration of carbon dioxide is highest in the mitochondria, where it is produced during cellular respiration. From there, it diffuses to the interstitium and eventually into the blood, which transports it to the alveoli. Within the blood, carbon dioxide is transported in three forms (Fig. 16.3):

- Approximately 7% of CO_2 in blood is **dissolved CO_2**. Because solubility of CO_2 in plasma is relatively high (20 times the solubility of O_2), the dissolved form of CO_2 has a significant role in its transport.
- Up to 23% of CO_2 may be combined with protein, including hemoglobin (as **carbaminohemoglobin**, which gives venous blood its bluish tinge). CO_2 binds to terminal amino groups of blood proteins.
- About 70% of CO_2 in blood is carried in the form of **bicarbonate anion (HCO_3^-)**.

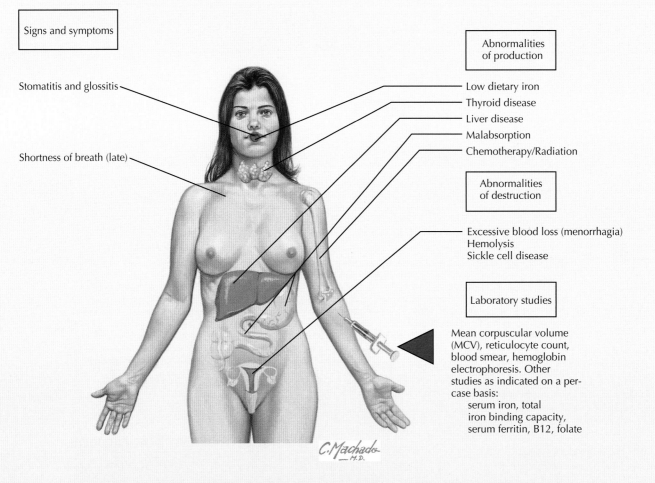

Anemia Destruction of red blood cells, bleeding, or abnormalities of red blood cell production are causes of anemia. Abnormalities in red blood cell production may be related to malabsorption of iron and/or vitamin B12, or liver disease (iron and vitamin B12 are stored in the liver; see Chapter 26). Laboratory studies are useful in determining a differential diagnosis of the cause of anemia.

Carbon Dioxide Transport in the Form of Bicarbonate Ion

The bulk of carbon dioxide is transported in the form of HCO_3^- within red blood cells (see Fig. 16.3). Carbon dioxide dissolved in blood reacts with H_2O to form carbonic acid (H_2CO_3), which dissociates to form H^+ and HCO_3^-:

$$CO_2 + H_2O \leftrightarrow H_2CO_3 \leftrightarrow H^+ + HCO_3^- \qquad \textbf{Eq. 16.7}$$

This reaction, which is normally slow, is catalyzed by the enzyme **carbonic anhydrase** in red blood cells. As HCO_3^- is formed, it diffuses out of the red blood cell while Cl^- diffuses into the cell to maintain electrochemical equilibrium. This process is known as the **chloride shift**. Most of the H^+ formed is buffered within the red blood cell by binding to hemoglobin. The reaction forming H_2CO_3 is driven forward in capillary blood, as carbon dioxide diffuses from tissues into the

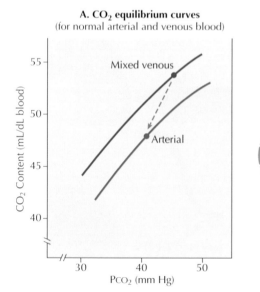

A. CO$_2$ equilibrium curves
(for normal arterial and venous blood)

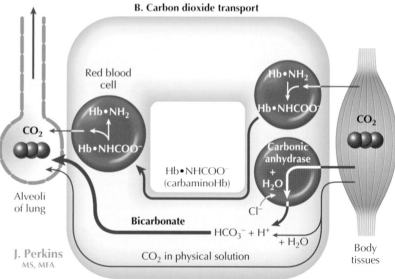

B. Carbon dioxide transport

Figure 16.3 **Carbon Dioxide Transport**   Carbon dioxide is transported in blood as bicarbonate anion (approximately 70%), carbaminohemoglobin (approximately 23%), and dissolved CO$_2$ (approximately 7%) **(B)**. The CO$_2$ equilibrium (dissociation) curve **(A)** is steep and linear, unlike the oxyhemoglobin dissociation curve, accounting for the relatively small difference in PCO$_2$ between arterial and venous blood (40 mm Hg vs. 45 mm Hg). Note also that the dissociation curve is shifted to the left when hemoglobin (Hb) is in the form of deoxyhemoglobin, as in venous blood (the Haldane effect).

CLINICAL CORRELATE 16.2
Sickle Cell Disease

Patients with sickle cell disease have a variant form of hemoglobin known as *hemoglobin S*. The allele causing this disease is recessive and is most common in people of sub-Saharan African origin. Hemoglobin S has the tendency to polymerize when it is deoxygenated, causing red blood cells to assume a sickle-like shape. In a "sickle cell crisis," red blood cells lodge in the microcirculation, causing painful ischemia and infarction of tissue. Sickle cell disease confers resistance to malaria, a parasite attacking red blood cells, and is believed to have evolved and persisted in sub-Saharan Africa as a result of the evolutionary advantage associated with malaria resistance.

Sickled Red Blood Cells

blood. At the lungs, the reverse reaction occurs, as carbon dioxide is breathed off.

In normal venous blood, the partial pressures of carbon dioxide and oxygen are similar (approximately 45 mm Hg and 40 mm Hg, respectively). According to Henry's law, the concentration of dissolved gas in a solution is directly proportional to its partial pressure. Although this law applies to both oxygen and carbon dioxide, because the solubility of carbon dioxide is 20 times that of oxygen, at similar partial pressures, much more carbon dioxide is present in blood in the form of dissolved gas than is the case for oxygen.

Although the approximate values given earlier for carbon dioxide transport as dissolved gas, carbaminohemoglobin, and bicarbonate anion are prevalent in the literature and textbooks, the value for carbaminohemoglobin may be significantly overstated, because the original studies of carbon dioxide carriage in blood were performed in the absence of 2,3-DPG, which binds with higher affinity to hemoglobin than carbon dioxide.

The Haldane Effect

Unlike the sigmoidal oxyhemoglobin dissociation curve, the dissociation curve for carbon dioxide in blood is linear (see Fig. 16.3). However, it is shifted to the left when hemoglobin is in the form of deoxyhemoglobin, as in venous blood. This occurrence is known as the **Haldane effect**. As a result of the Haldane effect, as hemoglobin is deoxygenated in systemic capillaries, its affinity for carbon dioxide is increased, facilitating carbon dioxide transport. Binding affinity for H$^+$ (generated in red blood cells along with HCO$_3^-$) is also increased. In

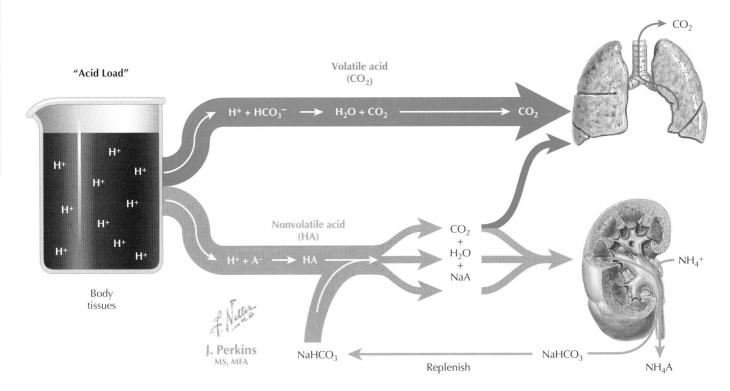

Figure 16.4 **Role of Lungs and Kidneys in Acid-Base Balance** The lungs and kidneys have a critical role in maintaining proper acid-base balance in the face of the acid load created by cellular metabolism of nutrients. Carbon dioxide ("volatile acid") produced by oxidative metabolism of carbohydrates and fats is efficiently eliminated by respiration in the lungs to maintain pH balance, whereas nonvolatile acids are primarily buffered by bicarbonate anion in extracellular fluid and intracellular proteins, with the kidneys replenishing the bicarbonate anion and excreting acid. H^+ is also eliminated in the urine as bicarbonate is regenerated. When a metabolic acid-base disturbance occurs, intracellular and extracellular buffering systems (involving primarily proteins and bicarbonate, respectively) are the first line of defense, along with rapid compensation by the lungs, which adjust the rate of CO_2 (volatile acid) elimination. Over a longer period (hours to days), renal mechanisms compensate by adjusting acid secretion and bicarbonate regeneration. When a respiratory acid-base disturbance occurs, the compensation is primarily renal.

the pulmonary circulation, as hemoglobin is oxygenated, its affinity for carbon dioxide is reduced, and as a result, transfer of carbon dioxide from blood to alveolar air is facilitated.

CARBON DIOXIDE TRANSPORT AND ACID-BASE BALANCE

Normal blood pH is approximately 7.4 and is regulated within a tight range (7.35 to 7.45) by renal and respiratory mechanisms and by various other buffering systems (Fig. 16.4; see Chapter 21). Significant deviation of pH from the normal range is incompatible with life, because protein structure is affected and enzymatic function is disturbed. Carbon dioxide transport plays an important role in maintaining **acid-base equilibrium**. The **"acid load"** of the body consists of volatile acid, which is carbon dioxide in its various forms, and nonvolatile acids such as lactic acid and amino acids. Nonvolatile acids are buffered by intracellular and extracellular mechanisms; the bicarbonate buffering system of extracellular fluids, including blood, are important in this regard. Acid balance is also maintained by renal excretion of acid (renal mechanisms

are covered in Chapter 21). Normally, carbon dioxide formed during metabolism of lipids and carbohydrates is readily eliminated by the respiratory system, but disturbances in respiration can result in acid-base imbalance, because changes in carbon dioxide elimination will directly affect carbonic acid levels. In addition, by adjusting respiration, the respiratory system can compensate for pH imbalances produced by metabolic disturbances.

The Henderson-Hasselbalch Equation

The pH of buffer systems can be calculated by the **Henderson-Hasselbalch equation,**

$$pH = pK + \log \frac{[A^-]}{[HA]} \qquad \textbf{Eq. 16.8}$$

where K is the acid dissociation constant. For the bicarbonate buffer system,

$$pH = pK + \log \frac{[HCO_3^-]}{[H_2CO_3]} \qquad \textbf{Eq. 16.9}$$

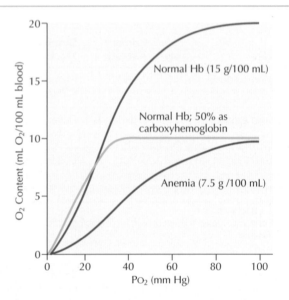
The pK for this system is 6.1. H_2CO_3 exists in low concentrations in extracellular fluids but is in equilibrium with dissolved CO_2, and thus, $0.03 \times P_{CO_2}$ (0.03 mmol/L/mm Hg is the solubility coefficient for P_{CO_2}) can be substituted for $[H_2CO_3]$ and the equation becomes:

$$pH = 6.1 + \log \frac{[HCO_3^-]}{0.03 \times P_{CO_2}} \qquad \textbf{Eq. 16.10}$$

Substituting normal values for arterial $[HCO_3^-]$ and P_{CO_2} yields the normal pH of 7.4:

$$pH = 6.1 + \log \frac{[24]}{0.03 \times 40} = 7.4 \qquad \textbf{Eq. 16.11}$$

Based on these equations, it should be apparent that changes in P_{CO_2} will result in alteration of pH. If hypoventilation results in a rise in Pa_{CO_2}, for example, the accompanying rise in Pa_{CO_2} will cause a fall in blood pH.

ACID-BASE DISTURBANCES

Acid-base disturbances are covered briefly in the following discussion and in detail in Chapter 21. **Acidemia** is defined as increased acidity of blood (pH below 7.35), whereas **alkalemia** is increased alkalinity of blood (pH above 7.45). Technically, **acidosis** and **alkalosis** are more general terms referring to low pH and high pH, respectively, in body fluids and tissues but are usually used synonymously with acidemia and alkalemia. Alterations in pH caused by respiratory abnormality are referred to as **respiratory acidosis** or **respiratory alkalosis**. Respiratory alkalosis is caused by hyperventilation, whereas respiratory acidosis is caused by hypoventilation (either because of an acute cause, such as airway obstruction or central nervous system [CNS] depression, or a chronic lung disease). In contrast, when the primary cause of the acid-base imbalance is metabolic—for example, due to a metabolic disease or abnormal renal function—it is described as **metabolic acidosis** or **metabolic alkalosis**. Compensation for respiratory acidosis or alkalosis occurs by renal mechanisms, whereas respiratory adjustments compensate for metabolic acidosis or alkalosis (Table 16.1). The general role of the kidney and lungs in acid-base balance is illustrated in Figure 16.4.

CONTROL OF RESPIRATION

Although breathing can be controlled voluntarily (for example, during breath holding or hyperventilation), it is ultimately an

Table 16.1 **Acid-Base Disorders**

Disorder	pH	Primary Alteration	Defense Mechanisms
Metabolic acidosis	↓	↓ [HCO_3^-]	Buffers, ↓P_{CO_2}, ↑NAE
Metabolic alkalosis	↑	↑ [HCO_3^-]	Buffers, ↑P_{CO_2}, ↓NAE
Respiratory acidosis	↓	↑ P_{CO_2}	Buffers and ↑NAE
Respiratory alkalosis	↑	↓ P_{CO_2}	Buffers and ↓NAE

NAE, net acid excretion.
From Hansen J: *Netter's Atlas of Human Physiology,* Philadelphia, 2002, Elsevier.

Differentiating between metabolic and respiratory causes of acidosis and alkalosis is usually a relatively simple process. To identify the condition:

1. First examine the pH. Values below 7.35 are defined as acidosis; pH above 7.45 is by definition alkalosis.
2. Second, examine the Pa_{CO_2}. If the pH disturbance is respiratory in origin, the Pa_{CO_2} level will be abnormal and predictive of the pH change. In other words, in respiratory acidosis, Pa_{CO_2} will be elevated above 40 mm Hg, whereas in respiratory alkalosis (caused by hyperventilation), Pa_{CO_2} will be lower than 40 mm Hg. If the condition is acute, HCO_3^- level will be normal; if the respiratory condition is chronic, HCO_3^- will be elevated in acidosis and depressed in alkalosis, reflecting renal compensation. Because of the compensation, pH will be closer to normal in chronic acidosis or alkalosis than would be expected based on a change in Pa_{CO_2} alone.
3. If the Pa_{CO_2} level is abnormal in the opposite direction predicted for it to be the primary alteration, the disturbance is metabolic. Examination of HCO_3^- should reveal that its level is consistent with a primary metabolic disturbance: HCO_3^- will be high in metabolic alkalosis and low in metabolic acidosis. Pa_{CO_2} levels will be depressed in metabolic acidosis and elevated in metabolic alkalosis, reflecting respiratory compensation. Because of this compensation, the pH will be closer to normal than would be predicted based on a change in HCO_3^- alone.

involuntary process that closely controls Pa_{O_2} and Pa_{CO_2}. Changes in both depth and rate of respiration are involved in this process. Three essential components of the involuntary control system are as follows:

- Brainstem respiratory centers
- Peripheral and central chemoreceptors
- Mechanoreceptors in lungs and joints

Ultimately, signals involved in control of breathing are integrated in the **medullary respiratory center**, resulting in regulation of the activity of respiratory muscles (see Fig. 15.1) and affecting tidal volume and respiratory rate and pattern. Within

the medulla, respiratory control is accomplished by the following areas:

- The **ventral respiratory group**, which includes the nucleus retroambiguus, nucleus ambiguus, and nucleus retrofacialis and innervates both inspiratory and expiratory muscles. It is involved in regulation of inspiratory force and in voluntary expiration.
- The **dorsal respiratory group** within the nucleus tractus solitarius, which innervates inspiratory muscles.

The medullary respiratory center receives input from the following two important pontine areas:

- The **pneumotaxic center**, which regulates rate and depth of respiration by cyclical inhibition of inspiration. This center has input from the cerebral cortex.
- The **apneustic center**, which stimulates inspiration. It is antagonized by the pneumotaxic center.

Damage to the pons or upper medulla may result in **apneusis** (i.e., breathing characterized by prolonged inspiratory efforts and brief, intermittent exhalations). Experimentally, apneusis can be produced by ablation of the pneumotaxic center and transection of the vagus nerve.

Role of Central and Peripheral Chemoreceptors

Sensory information from central and peripheral chemoreceptors is important in this regulation of respiration by the brainstem. **Central chemoreceptors** located at the ventrolateral surface of the medulla respond indirectly to changes in arterial P_{CO_2} and play a critical role in acute regulation of Pa_{CO_2}. The blood-brain barrier is largely impermeable to HCO_3^- and H^+, but carbon dioxide readily diffuses across the barrier and into the cerebrospinal fluid (CSF), where it affects CSF pH (by mechanisms discussed earlier). Thus, when Pa_{CO_2} is altered, respiration is affected:

- A rise in Pa_{CO_2} will cause a fall in CSF pH, which is detected by central chemoreceptors, resulting in an increase in respiratory rate.
- A fall in Pa_{CO_2} will cause a rise in CSF pH, which is detected by central chemoreceptors, resulting in decreased ventilation.

Peripheral chemoreceptors, located in the carotid bodies and aortic bodies (see Section 3), also convey information concerning the quality of arterial blood to the respiratory center in the brainstem, thereby affecting ventilation. Unlike the central chemoreceptors, these receptors respond directly to changes in Pa_{O_2} and Pa_{CO_2}, as well as pH. Through the peripheral chemoreceptor mechanism, ventilation is stimulated by the following mechanisms:

- A fall in Pa_{O_2}: The ventilatory effects of changes in Pa_{O_2} are relatively small when Pa_{O_2} is above 60 mm Hg, but

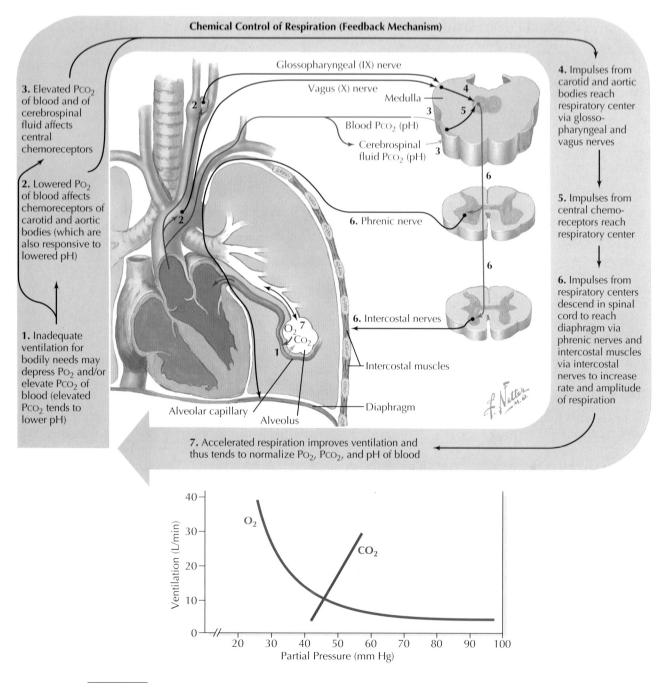

Chemical Control of Respiration (Feedback Mechanism)

Glossopharyngeal (IX) nerve

Vagus (X) nerve

Medulla

Blood Pco_2 (pH)

Cerebrospinal fluid Pco_2 (pH)

6. Phrenic nerve

6. Intercostal nerves

Intercostal muscles

Alveolar capillary

Alveolus

Diaphragm

3. Elevated Pco_2 of blood and of cerebrospinal fluid affects central chemoreceptors

2. Lowered Po_2 of blood affects chemoreceptors of carotid and aortic bodies (which are also responsive to lowered pH)

1. Inadequate ventilation for bodily needs may depress Po_2 and/or elevate Pco_2 of blood (elevated Pco_2 tends to lower pH)

4. Impulses from carotid and aortic bodies reach respiratory center via glosso-pharyngeal and vagus nerves

5. Impulses from central chemo-receptors reach respiratory center

6. Impulses from respiratory centers descend in spinal cord to reach diaphragm via phrenic nerves and intercostal muscles via intercostal nerves to increase rate and amplitude of respiration

7. Accelerated respiration improves ventilation and thus tends to normalize Po_2, Pco_2, and pH of blood

Figure 16.5 **Control of Respiration**   Central and peripheral chemoreceptors regulate respiration by responding to arterial blood gas levels. Central chemoreceptors respond primarily to changes in arterial Pco_2, which diffuses into the cerebrospinal fluid (CSF) and alters the pH of the CSF (the blood-brain barrier is largely impermeable to HCO_3^- and H^+), while peripheral chemoreceptors in the carotid bodies and aortic bodies respond to changes in Pa_{O_2} and to changes in Pa_{CO_2} and pH. Brainstem respiratory centers adjust the rate and depth of respiration, producing changes in Pa_{O_2} and Pa_{CO_2} (and thus pH). In the bottom graph, the effects of Pa_{CO_2} and Pa_{O_2} on minute ventilation are illustrated.

peripheral chemoreceptors are very responsive when Pa_{O_2} falls below this level.

■ A rise in Pa_{CO_2}: Changes in Pa_{CO_2} affect respiration through both central and peripheral chemoreceptors, although the central chemoreceptor mechanism is more responsive to the changes.

■ A fall in pH: Changes in H^+ concentration in arterial blood affect peripheral chemoreceptors directly, independently of the effects of Pa_{CO_2}.

Chemical control of respiration by Pa_{O_2} and Pa_{CO_2} is illustrated in Figure 16.5.

Sleep apnea is a disorder characterized by the periodic interruption of normal breathing during sleep. The pauses in breathing can be the length of two to three breaths, and thus a significant reduction in gas transport can occur during these episodes. Sleep apnea can be central, obstructive, or complex (both central and obstructive). In central sleep apnea, brain centers are dysregulated, resulting in a lack of effort to breathe; in obstructive sleep apnea, although respiratory effort is normal, obstruction prevents airflow. Disrupted sleep and fatigue are common symptoms. Chronic sleep apnea is associated with increased incidence of heart disease and stroke.

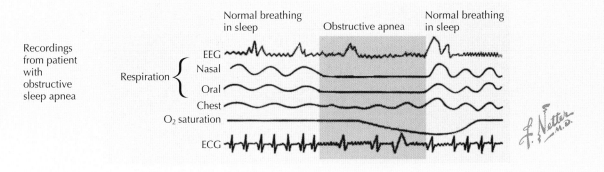

Additional Mechanisms Controlling Respiration

Respiration is also controlled by a number of additional peripheral mechanisms:

- **Pulmonary mechanoreceptors** respond to inflation of the lung and result in termination of inspiration. The afferent signals from these receptors in the smooth muscle of airway walls are transmitted through the vagus nerve to the medulla, where they inhibit the apneustic center, thereby terminating inspiration and avoiding overinflation. This response to lung inflation is known as the **Hering-Breuer reflex** (specifically, the Hering-Breuer inspiratory-inhibitory reflex).
- **Irritant receptors** in the large airways respond to noxious gases and particulate matter, such as in cigarette smoke. Activation of these receptors results in afferent signals to the CNS mainly through the vagus nerve and causes reflexive bronchoconstriction and coughing.
- **Juxtacapillary receptors** (J receptors) in the alveoli are stimulated by hyperinflation of the lungs and various chemical stimuli; reflexive rapid, shallow breathing occurs as a result.
- **Joint and muscle mechanoreceptors** are stimulated during movement of joints and muscles, producing an increase in respiratory rate.

Respiratory Control in Exercise

Control of respiration is a critical component of the **integrated response to exercise** (Fig. 16.6). During dynamic (aerobic) exercise, oxygen consumption rises from the average resting rate of 250 mL O_2/min to as high as 4 L O_2/min, without substantial change in either Pa_{O_2} or Pa_{CO_2}. At the onset of dynamic exercise, a rapid increase in respiration occurs through neural and reflexive mechanisms, although the control mechanisms are not fully understood. Activation of motor pathways results in collateral activation of the respiratory center, and respiration is further stimulated by afferent signals from muscle and joint mechanoreceptors and other, unknown factors. With continuing exercise, feedback mechanisms become important. The rise in core body temperature and elevation of lactic acid production (and plasma H^+ concentration) contribute to the further, more gradual rise in ventilation. Although Pa_{O_2} and Pa_{CO_2} change only modestly (except at high levels of exercise), respiratory control systems may be more sensitive to such changes during exercise. When exercise is terminated, ventilation diminishes rapidly at first but requires some time to fall to the resting level because of the continued activation of feedback mechanisms until the metabolic alterations associated with exercise (including the elevation of lactic acid) are reversed.

CLINICAL CORRELATE 16.5
Respiratory Control in Chronic
Obstructive Pulmonary Disease

Chronic obstructive pulmonary disease (COPD) is most often associated with tobacco smoking and may be exacerbated by occupational exposure to certain pollutants. It is characterized by chronic bronchitis (coughing with sputum production) and emphysema (destruction of alveolar walls resulting in fewer, enlarged "alveoli" with a reduction in total surface area for gas diffusion). As a result, patients with this disease become hypoxic and hypercapnic (Pa_{O_2} is below normal and Pa_{CO_2} is elevated); pH is somewhat lower than normal, although significant compensation occurs by elevation of HCO_3^- through renal mechanisms. In other words, persons with COPD experience chronic respiratory acidosis, with metabolic compensation. In this state of chronic hypercapnia, normal CSF pH is maintained by elevation of HCO_3^- in the CSF. The CNS is chronically exposed to high P_{CO_2}, and the central chemoreceptors become unresponsive to carbon dioxide. Thus "hypoxic respiratory drive" develops in persons with COPD, in which respiratory drive is mainly mediated by peripheral chemoreceptor responses to low Pa_{O_2}. When an acute exacerbation arises in such a patient, if supplemental oxygen is administered carelessly, results can be catastrophic and even fatal: Pa_{O_2} will rise, but respiratory drive may fail as a result, causing a fall in minute ventilation and a further rise in Pa_{CO_2}. Ventilatory support with some supplemental oxygen may be beneficial, but suppression of ventilatory drive must be avoided.

COPD is characterized by chronic bronchitis and emphysema. Persons with COPD usually experience both conditions to some extent and are classified on the basis of the dominant symptomatology. "Pink puffers," who mainly have emphysema, have ruddy complexions and an elevated respiratory rate. "Blue bloaters" mainly have chronic bronchitis, resulting in hypoxemia, cyanosis (bluish lips and skin), and, often, symptoms of right heart failure, including swelling of feet and ankles ("bloating").

Mixed Chronic Bronchitis and Emphysema

The typical patient with COPD has clinical, physiological, and radiographic features of both chronic bronchitis and emphysema. She may have chronic cough and sputum production and need accessory muscles and pursed lips to help her breathe. Pulmonary function testing may reveal variable degrees of airflow limitation, hyperinflation, and reduction in the diffusing capacity, and arterial blood gases may show variable decreases in P_{O_2} and increases in P_{CO_2}. Radiographic imaging often shows components of airway wall thickening, excessive mucus, and emphysema.

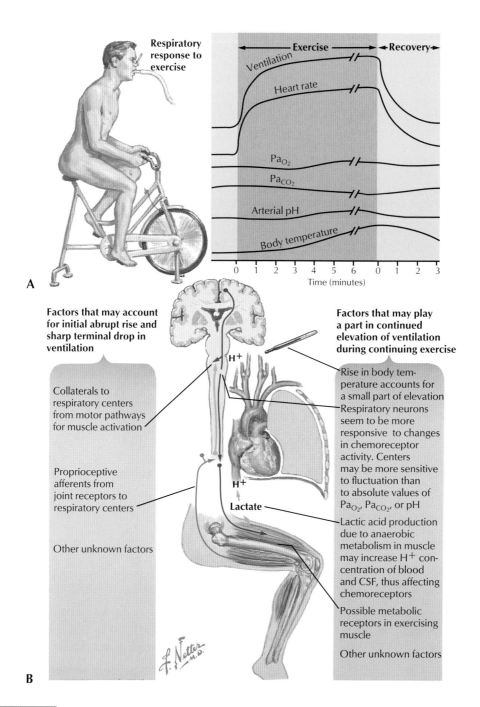

Respiratory response to exercise

Exercise — Recovery

Ventilation
Heart rate
Pa_{O_2}
Pa_{CO_2}
Arterial pH
Body temperature

Time (minutes)

A

Factors that may account for initial abrupt rise and sharp terminal drop in ventilation

Collaterals to respiratory centers from motor pathways for muscle activation

Proprioceptive afferents from joint receptors to respiratory centers

Other unknown factors

H^+

H^+

Lactate

Factors that may play a part in continued elevation of ventilation during continuing exercise

Rise in body temperature accounts for a small part of elevation

Respiratory neurons seem to be more responsive to changes in chemoreceptor activity. Centers may be more sensitive to fluctuation than to absolute values of Pa_{O_2}, Pa_{CO_2}, or pH

Lactic acid production due to anaerobic metabolism in muscle may increase H^+ concentration of blood and CSF, thus affecting chemoreceptors

Possible metabolic receptors in exercising muscle

Other unknown factors

B

Figure 16.6 **Respiratory Response to Exercise** Increased oxygen consumption and carbon dioxide production during exercise requires adjustments of cardiac output and respiration **(A)**. Factors accounting for the rapid adjustments in respiration at the onset and termination of exercise, as well as feedback mechanisms during continued exercise, are illustrated **(B)**.

The level of aerobic exercise being performed can be quantified by measuring the rate of oxygen consumption ($\dot{V}O_2$). $\dot{V}O_2$ max is the maximum level of aerobic exercise that can be achieved by an individual, with the limiting factor in such exercise being the ability to deliver sufficient oxygen to meet the metabolic needs of the exercising muscles. At $\dot{V}O_2$ max, oxygen consumption may rise from the resting level of 250 L/min to a maximum of 4 L/min (a 16-fold increase), whereas cardiac output in a young, fit individual might rise from 5 L/min at rest to 20 to 30 L/min (only a fourfold to sixfold increase). Thus increased cardiac output alone cannot account for the increased oxygen delivery. To fully account for the increased oxygen consumption, we must take into consideration the degree of oxygen extraction from blood. At rest, only 25% of oxygen in arterial blood is actually delivered to tissues, because oxygen content falls from approximately 20 mL O_2/dL in arterial blood to 15 mL O_2/dL in mixed venous blood. During maximum aerobic exercise, oxygen extraction from blood may increase by a factor of about 3, and mixed venous O_2 is greatly reduced. Thus, a fourfold to sixfold increase in cardiac output accompanied by a threefold increase in oxygen extraction accounts for the great increase in O_2 consumption. Note that in a healthy person, respiration can be adjusted during maximum aerobic exercise to maintain nearly 100% oxygen saturation of arterial blood.

Adaptation to High Altitude

The control of respiration is also important in **adaptation to high altitude**. At the lower atmospheric pressures associated with high altitudes, the PO_2 of inspired air is reduced, resulting in hypoxemia. The fall in Pa_{O_2} stimulates ventilation through peripheral chemoreceptors, but this effect is tempered by the resulting fall in Pa_{CO_2} and the accompanying alkalosis, which inhibit ventilation through central and peripheral chemoreceptor mechanisms. Over a period of time, renal compensatory mechanisms result in elevation of plasma HCO_3^-, and as pH returns to normal, ventilation again increases. In addition to increased ventilation, the following factors contribute to the adaptation to high altitude:

- Hypoxemia stimulates red blood cell production; the resulting **polycythemia** (and higher plasma hemoglobin) increases the oxygen-carrying capacity of blood.
- Elevation of **2,3-DPG** causes a rightward shift of the oxyhemoglobin dissociation curve, and thus oxygen more readily dissociates from hemoglobin at the tissue level.

Effects of High Altitude on Respiratory Mechanism

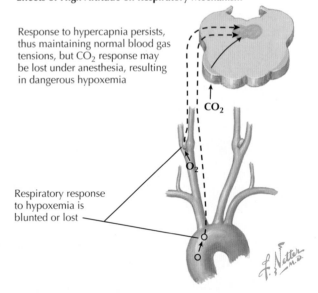

Response to hypercapnia persists, thus maintaining normal blood gas tensions, but CO_2 response may be lost under anesthesia, resulting in dangerous hypoxemia

CO_2

O_2

Respiratory response to hypoxemia is blunted or lost

Review Questions

CHAPTER 14: PULMONARY VENTILATION AND PERFUSION AND DIFFUSION OF GASES

1. The reduction in pulmonary vascular resistance that occurs when pulmonary artery pressure is increased is mainly a result of:

A. recruitment and distension of pulmonary capillaries.
B. autoregulation by myogenic mechanisms.
C. redistribution of pulmonary blood flow.
D. metabolic vasodilation.
E. active vasodilation of pulmonary arterioles.

2. Which of the following lung volumes or capacities **CANNOT** be measured by spirometry alone?

A. Tidal volume
B. Expiratory reserve volume
C. Inspiratory reserve volume
D. Total lung capacity
E. Vital capacity

3. With regard to the ventilation, the perfusion, or the ventilation/perfusion ratio in regions of the lung in a person in the standing position,

A. perfusion is highest at the top of the lung.
B. ventilation is lowest at the bottom of the lung.
C. the ventilation/perfusion ratio is greatest at the top of the lung.
D. the ventilation/perfusion ratio approaches infinity in areas of shunt.
E. the ventilation/perfusion ratio is zero in areas of dead space.

4. Diffusion of gas through a membrane:

A. is inversely related to the surface area for diffusion.
B. is directly related to the difference in partial pressure of the gas on each side of the membrane.
C. requires active transport.
D. is inversely related to the diffusion constant of the gas.
E. is directly related to the thickness of the membrane.

5. In zone 2 of the lung (in a standing person),

A. both pulmonary arterial and venous pressures exceed alveolar pressure.
B. both pulmonary arterial and venous pressures are below alveolar pressure.
C. alveolar pressure is higher than pulmonary venous pressure but less than arterial pressure.
D. the ventilation/perfusion ratio approaches zero.
E. the ventilation/perfusion ratio approaches infinity.

6. In dry air at an elevated altitude where atmospheric pressure is 700 mm Hg, atmospheric P_{O_2} will be approximately:

A. 130 mm Hg.
B. 140 mm Hg.
C. 147 mm Hg.
D. 157 mm Hg.
E. 200 mm Hg.

CHAPTER 15: THE MECHANICS OF BREATHING

7. The mechanical system that produces breathing is at rest (i.e., outward elastic recoil pressure of the chest wall is equal to and opposing inward elastic recoil pressure of the lungs) at:

A. residual volume.
B. functional residual capacity.
C. 60% of total lung capacity.
D. 70% of total lung capacity.
E. total lung capacity.

8. In the respiratory system as a whole, the greatest resistance to flow occurs in the:

A. respiratory bronchioles.
B. terminal bronchioles.
C. medium-sized airways.
D. bronchi.
E. trachea.

9. Dynamic compression of airways is responsible for:

A. effort independence of expiratory flow.
B. the vital capacity of the lung.
C. normal resting tidal volume.
D. the total lung capacity that can be achieved during inspiration.
E. peak flow during expiration.

10. In severe chronic obstructive pulmonary disease,

A. lung compliance is reduced.
B. elastic recoil of the lung is decreased.
C. total lung capacity is reduced.
D. functional residual capacity is reduced.
E. residual volume is reduced.

11. Which of the following is **NOT** a characteristic or function of surfactant?

A. Decreases pulmonary compliance
B. Reduces the work of breathing
C. Reduces surface tension of alveoli and small airways
D. Deficient in respiratory distress syndrome
E. Contains dipalmitoyl phosphatidyl choline

CHAPTER 16: OXYGEN AND CARBON DIOXIDE TRANSPORT AND CONTROL OF RESPIRATION

12. Which of the following changes will result in the greatest increase in oxygen content of arterial blood, assuming normal alveolar oxygen concentration?

A. An increase in hematocrit from 40 to 45
B. An increase in alveolar oxygen concentration from 100 to 150 mm Hg
C. A 10% increase in blood level of 2,3-DPG
D. A fall in blood pH from 7.4 to 7.35
E. A 10% increase in alveolar ventilation

13. Laboratory tests reveal that a patient has blood pH of 7.3, elevated arterial P_{CO_2}, and slightly elevated arterial plasma bicarbonate level. The acid-base status is:

A. metabolic acidosis.
B. metabolic alkalosis.
C. respiratory acidosis.
D. respiratory alkalosis.
E. impossible to determine from these values.

14. The major mechanism of long-term adaptation to high altitude is:

A. increased heart rate.
B. increased respiratory rate.
C. reduced 2,3-DPG in blood.
D. polycythemia.
E. reduced plasma bicarbonate.

15. In the control of respiration, central chemoreceptors respond mainly to changes in:

A. arterial pH.
B. arterial P_{CO_2}.
C. 2,3-DPG in blood.
D. arterial HCO_3^-.
E. arterial O_2.

16. The rapid adjustment of respiratory rate at the onset of exercise is mediated in part by:

A. a fall in arterial pH.
B. a rise in arterial P_{CO_2}.
C. 2,3-DPG in blood.
D. mechanoreceptors in joints.
E. a rise in body temperature.

17. Hemoglobin affinity for O_2 is

A. increased in the presence of 2,3-DPG.
B. increased when pH is reduced.
C. increased when temperature is elevated.
D. reduced when one of the binding sites on hemoglobin is already occupied by O_2.
E. higher for hemoglobin F compared with hemoglobin A.

Section 5

RENAL PHYSIOLOGY

The primary "job" of the kidneys is to maintain proper extracellular fluid volume and solute composition on an ongoing basis. This task is accomplished by intrarenal mechanisms, with input from the nervous and endocrine systems. The kidneys also excrete waste (i.e., excess fluid and electrolytes, as well as urea, bilirubin, drugs, and potential toxins), execute key endocrine functions, and play an important role in acid-base balance.

Chapter 17

Overview, Glomerular Filtration, and Renal Clearance

STRUCTURE AND OVERALL FUNCTION OF THE KIDNEYS

The kidneys perform a host of functions, including the following:

- Regulation of fluid and electrolyte balance: The kidneys regulate the volume of extracellular fluid (ECF) through reabsorption and excretion of NaCl and water. They are also the site of regulation of the plasma levels of other key substances (Na^+, K^+, Cl^-, HCO_3^-, H^+, Ca^{2+}, and phosphates). The following key renal processes are involved in the regulation of circulating substances:
  - *Filtration* of fluid and solutes from the plasma into the nephrons
  - *Reabsorption* of fluid and solutes from the renal tubules into the peritubular capillaries
  - *Secretion* of select substances from the peritubular capillaries into the tubular fluid, which facilitates their excretion; both endogenous (e.g., K^+, H^+, creatinine, norepinephrine, and dopamine) and exogenous (e.g., para-aminohippurate [PAH], salicylic acid, and penicillin) substances can be secreted into the tubular fluid and subsequently excreted in the urine
  - *Excretion* of excess fluid, electrolytes, and other substances (e.g., urea, bilirubin, and acid [H^+])
- Regulation of plasma osmolarity: "Opening" and "closing" of specific water channels (aquaporins) in the renal collecting ducts produces concentrated and dilute urine (respectively), allowing regulation of plasma osmolarity and ECF volume.
- Elimination of metabolic waste products: Urea (from protein metabolism), creatinine (from muscle metabolism), bilirubin (from breakdown of hemoglobin), uric acid (from breakdown of nucleic acids), metabolic acids, and foreign substances such as drugs are excreted in urine.
- Production/conversion of hormones: The kidney produces **erythropoietin** and **renin**. **Erythropoietin** stimulates red blood cell production in bone marrow. **Renin**, a proteolytic enzyme, is secreted into the blood and converts angiotensinogen to angiotensin I, which is then converted to angiotensin II by angiotensin-converting enzyme in lungs and other tissues. The renin-angiotensin system is critical for fluid-electrolyte homeostasis and long-term blood pressure regulation. The renal tubules also convert 25-hydroxyvitamin D to the active **1,25-dihydroxyvitamin D**, which can act on kidney, intestine, and bone to regulate calcium homeostasis.
- Metabolism: Renal production of ammonia through **ammoniagenesis** has an important role in acid-base homeostasis (discussed further in Chapter 21). The kidney, like the liver, has the ability to produce glucose through **gluconeogenesis.**

The kidneys are bilateral, retroperitoneal organs that receive their blood supply from the renal arteries (Fig. 17.1A). Each kidney is approximately the size of an adult fist and is surrounded by a fibrous capsule. The parenchyma is divided into the cortex and outer and inner medulla. The cortex contains renal corpuscles, which are **glomerular capillaries** surrounded by **Bowman's capsules**. The corpuscles are connected to **nephrons**, which are the tubules that are considered to be the functional units of the kidneys. The outer stripe of the outer medulla contains the thick ascending loops of Henle and collecting ducts, whereas the inner stripe contains the pars recta, thick and thin ascending loops of Henle, and collecting ducts (Fig. 17.2). Collecting ducts empty urine into the calyces and, ultimately, the ureter, which leads to the bladder. Thus a portion of the plasma fraction of blood entering the kidney is filtered through the **glomerular capillary membrane into Bowman's space**, flows into the nephrons, and becomes tubular fluid. After the tubular fluid is processed in the nephron, the remaining fluid (urine) flowing through the collecting ducts exits the renal pyramids into the minor calyces. The minor calyces combine to form the major calyces, which empty into the ureter (see Fig. 17.1B). The ureters lead to the bladder, where the urine is stored until excretion (micturition).

The Nephron

Each kidney contains more than 1 million nephrons. The kidneys have two populations of nephrons: **cortical** (superficial) and **juxtamedullary** (deep) nephrons. Most of the nephrons are cortical (~80%), with ~20% being juxtamedullary. The populations are similar in that they are composed of the same structures, but they differ in their location within the kidney and in the length of segments. The cortical nephrons originate from glomeruli in the upper and middle regions of

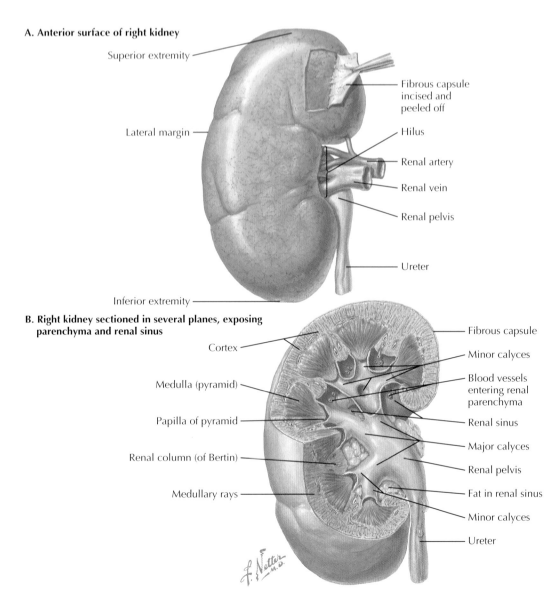

A. Anterior surface of right kidney

Superior extremity

Fibrous capsule incised and peeled off

Lateral margin

Hilus

Renal artery

Renal vein

Renal pelvis

Ureter

Inferior extremity

B. Right kidney sectioned in several planes, exposing parenchyma and renal sinus

Cortex

Fibrous capsule

Minor calyces

Medulla (pyramid)

Blood vessels entering renal parenchyma

Papilla of pyramid

Renal sinus

Renal column (of Bertin)

Major calyces

Renal pelvis

Medullary rays

Fat in renal sinus

Minor calyces

Ureter

Figure 17.1 **Anatomy of the Kidney**   The kidneys are bilateral organs with arterial blood supply from the abdominal aorta through the renal arteries **(A)**. The plasma is filtered at the glomeruli, which are located in the renal cortex. About 20% of the cardiac output enters the kidney (~1 L/min), and excess fluid and solutes are excreted as urine. The urine collects in the renal pelvis and exits the kidneys via the ureters **(B)**, which leads to the bladder where urine is stored until elimination.

the cortex, and their loops of Henle are short, extending only to the inner stripe of the outer medulla (see Fig. 17.2). The glomeruli of juxtamedullary nephrons are located deeper in the cortex (by the medullary junction) and are associated with long loops of Henle extending deep into the inner medulla.

Although all nephrons have the same basic structures, the location of the nephrons and the length of specific segments vary, with important consequences. The primary nephron segments are listed in sequential order in Table 17.1, along with functions and distinctive characteristics.

Blood Flow

Blood flow to the kidneys (**renal blood flow [RBF]**) is about 1 liter per minute (L/min), or ~20% of the cardiac output. The blood enters the kidneys via the renal arteries and follows the path shown:

- Interlobar arteries
- Arcuate arteries (at the corticomedullary junction)
- Interlobar/cortical radial arteries
- Afferent arteriole (site of regulation)
- **Glomerular capillaries**

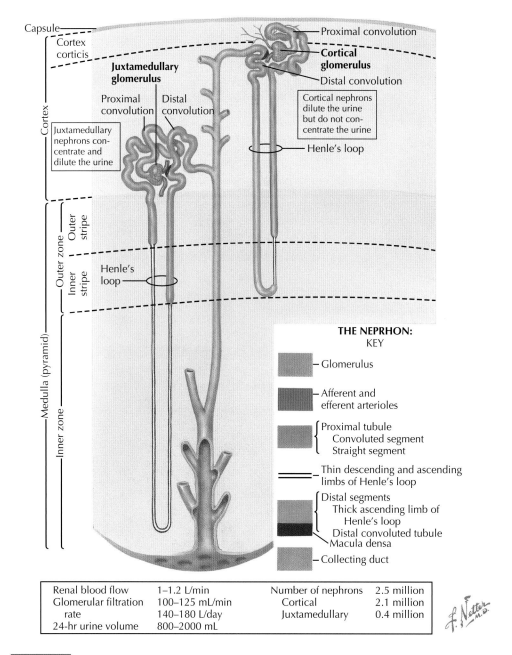

THE NEPRHON:
KEY

- Glomerulus
- Afferent and efferent arterioles
- Proximal tubule
  Convoluted segment
  Straight segment
- Thin descending and ascending limbs of Henle's loop
- Distal segments
  Thick ascending limb of Henle's loop
  Distal convoluted tubule
  Macula densa
- Collecting duct

Renal blood flow	1–1.2 L/min	Number of nephrons	2.5 million
Glomerular filtration	100–125 mL/min	Cortical	2.1 million
rate	140–180 L/day	Juxtamedullary	0.4 million
24-hr urine volume	800–2000 mL		

Figure 17.2 **Nephron Structure** The nephron is the functional unit of the kidney, and its structure differs depending on the location of the glomerulus. The glomeruli of cortical (superficial) nephrons are located in the upper cortical zone of the kidney and have loops of Henle that extend only to the outer zone of the medulla. The glomeruli of the juxtamedullary (deep) nephrons are located at the corticomedullary junction and have loops of Henle that extend deep into the inner medulla. There are about 5 times more cortical than juxtamedullary nephrons in the human kidney.

- Efferent arteriole
- Cortical peritubular capillaries (or vasa recta in deep nephrons)
- Venules
- Veins

The plasma fraction of the blood is filtered at the **glomerulus.** Blood enters the glomerular capillary from the *afferent arteriole* and exits the capillary by the *efferent arteriole.* Efferent arterioles associated with cortical nephrons lead to the **peritubular capillaries,** which collect material

reabsorbed from the nephrons; efferent arterioles of the juxtaglomerular nephrons lead to the **vasa recta** (straight vessels), which collect material reabsorbed from medullary tubules.

The Glomerulus

The glomerulus is a capillary system that filters blood to form the ultrafiltrate of plasma that flows into Bowman's space (Fig. 17.3). The glomerular capillary has a fenestrated endothelium and basement membrane, which allows this filtration but

Table 17.1 Nephron Segments: General Functions and Differences Between Segments in Cortical Versus Juxtamedullary Nephrons

Segments	Description and General Functions of Segment	Characteristics in Cortical Nephrons	Characteristics in Juxtamedullary Nephrons
Glomerulus	The capillary net that filters plasma, making ultrafiltrate; upon entering the proximal tubule, ultrafiltrate is called *tubular fluid*	Located superficially, in the outer and mid cortex; their efferent arterioles give rise to the peritubular capillaries	Located deep in the cortex, by the medullary junction; efferent arterioles give rise to the vasa recta, which are adjacent to deep nephrons and aid in concentration of urine
Proximal convoluted tubule	Has brush border villus membrane and is the main site of reabsorption of solutes and water	Shorter than proximal convoluted tubules in juxtamedullary nephrons	Longer than in cortical nephrons, allowing relatively more reabsorption of solutes
Proximal straight tubule	Additional reabsorption	Much longer than in deep nephrons	Shorter than in cortical nephrons
Thin descending loop of Henle	Impermeable to solutes but permeable to water; thus, it *concentrates* tubular fluid as water diffuses out	Much shorter than in deep nephrons	Very long, forming pyramids, crucial for concentrating tubular fluid
Thick ascending loop of Henle	Impermeable to water but has Na^+-K^+-$2Cl^-$ (NKCC-2) transporters that reabsorb more solutes and *dilute* the tubular fluid; sets up and maintains interstitial concentration gradient	Longer than deep nephrons, dilutes tubular fluid	Dilutes tubular fluid and is critical in producing the large concentration gradient in the inner medulla
Distal convoluted tubule	Electrolyte modifications; aldosterone acts on late distal tubule principal cells	Similar in cortical and deep nephrons	Similar in cortical and deep nephrons
Collecting ducts	Site of free water reabsorption through water channels (aquaporins) controlled by ADH; CDs are also important for acid-base balance: the α-intercalated cells allow H^+ secretion; β-intercalated cells have HCO_3^-/Cl^- exchangers, which allow HCO_3^- secretion when necessary	The CCDs reabsorb some Na^+ and Cl^- and secrete K^+ (from aldosterone-sensitive principal cells); less effect on urine concentration compared with deep nephrons because the ducts do not extend far into medulla; CCDs also have α- and β-intercalated cells for acid-base regulation	Because they extend deep into the medulla, the final concentration of urine occurs here; the IMCDs have principal cells (with aldosterone-sensitive Na^+ and K^+ channels), as well as intercalated cells (as seen in CCDs); medullary CDs are a key site of ADH-dependent urea reabsorption, which contributes to the high medullary interstitial fluid osmolarity

ADH, antidiuretic hormone; *CCD,* cortical collecting duct; *CD,* collecting duct; *IMCD,* inner medullary collecting duct.

keeps blood cells, proteins, and most macromolecules out of the glomerular ultrafiltrate. The glomerulus is surrounded by a single layer of epithelial cells (**podocytes**); this layer of podocytes contributes to the filtration barrier. Filtration by the glomerulus occurs according to size and charge. Because the basement membrane and podocytes are negatively charged, most proteins (which are also negatively charged) cannot be filtered. **Mesangial cells** are also present; they support the glomerulus but can also contract, decreasing the surface area for filtration.

The Juxtaglomerular Apparatus

Another important structural and functional aspect is the juxtaglomerular apparatus, which is the area where the distal convoluted tubule returns to its "parent" glomerulus. At this site, specialized **macula densa** cells in the distal convoluted tubule are in contact with the juxtaglomerular cells of the afferent arteriole, forming the **juxtaglomerular apparatus** (see Fig. 17.3). The macula densa cells of the juxtaglomerular apparatus are important in sensing tubular fluid flow and sodium delivery to the distal nephron, and because of their proximity to the afferent arteriole, macula densa cells can regulate renal plasma flow and **glomerular filtration rate (GFR)** (autoregulation). Macula densa cells also participate in the regulation of the release of the enzyme **renin** from juxtaglomerular cells adjacent to the afferent arterioles. The renin secretion aids in fluid and electrolyte homeostasis (see Chapter 20). Macula densa cells also receive input from adrenergic nerves through $β_1$-receptors.

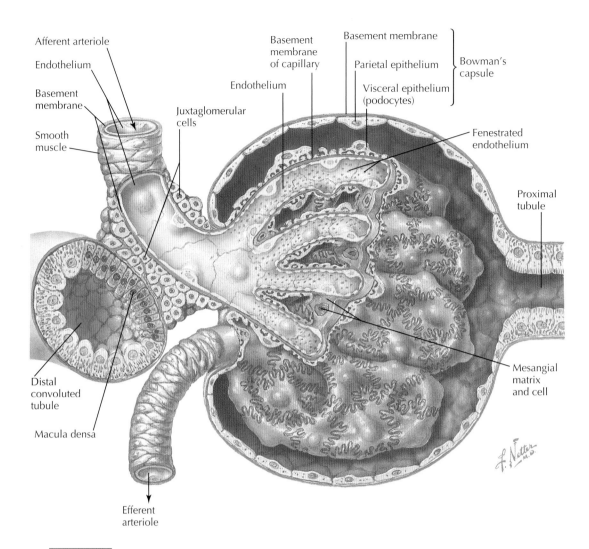

Figure 17.3 **Anatomy of the Glomerulus** Plasma is filtered at the glomerular capillaries into Bowman's space, and the ultrafiltrate then flows into the proximal tubule. The glomerular endothelial barrier prevents filtration of the cellular elements of the blood, so the ultrafiltrate does not contain blood cells or plasma proteins. The cells of the macula densa are in contact with the afferent arteriole through the juxtaglomerular cells, forming the juxtaglomerular apparatus. The macula densa monitors flow and NaCl delivery to the distal tubule and regulates renal plasma flow (autoregulation).

Renal Plasma Flow

Whereas whole blood enters the renal arteries, only *plasma* is filtered at the glomerular capillaries; thus, when discussing glomerular filtration, **renal plasma flow (RPF)** is an important factor. RPF can be determined by using the following equation:

$$RPF = RBF \times (1 - HCT)$$ **Eq. 17.1**

In the normal adult male, RBF = ~1 L/min, and hematocrit (HCT) is ~40% (0.4). Thus,

$$RPF = 1\,L/min \times 0.6 = 600\,mL/min$$ **Eq. 17.2**

To determine the **effective renal plasma flow (eRPF)**, which is the plasma flow entering the glomeruli and available for filtration, the plasma clearance of the organic acid **PAH** is

used. PAH is filtered at the glomeruli, and under normal circumstances, the remaining PAH in the peritubular capillaries is *secreted* into the proximal tubule, so that essentially no PAH enters the renal vein (Fig. 17.4; see Clinical Correlate 17.2).

GLOMERULAR FILTRATION: PHYSICAL FACTORS AND STARLING FORCES

Glomerular filtration is determined by the Starling forces and the permeability of the glomerular capillaries to the solutes in the plasma. With the exception of most proteins and protein-bound substances, plasma is freely filtered at the glomerular capillaries. Because the molecules must travel through several barriers to move from the capillary lumen to Bowman's space (fenestrated epithelium → basement membrane → between podocytes → filtration slit → Bowman's space), there are size

CLINICAL CORRELATE 17.1
Glomerulonephritis

The glomerulus is a key site for renal damage. Diseases and drugs that damage the glomerular basement membrane reduce the negative charge and allow large proteins (including albumin) to be filtered. Because there is no mechanism for reabsorbing large proteins in the nephron, the protein is excreted in the urine (proteinuria). In addition, diseases such as diabetes that increase mesangial matrix deposition increase rigidity and decrease the area of filtration of the glomerulus, thus reducing renal function.

Acute glomerulonephritis is usually caused by different factors in children and adults. In children, a common cause is streptococcal infection. In adults, acute glomerulonephritis can arise as a complication from drug reactions, pneumonia, immune disorders, and mumps. Acute glomerulonephritis can be asymptomatic (in about 50% of cases) or can be associated with edema, low urine volume, headaches, nausea, and joint pain. Treatment is aimed at reducing the inflammation, usually with steroids or immunosuppressive drugs, while determining and addressing the cause, when possible. In most cases, patients recover completely.

In contrast, **chronic glomerulonephritis** is associated with long-term inflammation of glomerular capillaries, resulting in thickened basement membranes, swollen epithelial cells, and narrowing of the capillary lumen. Major causes of chronic glomerulonephritis are diabetes, lupus nephritis, focal segmental glomerulosclerosis, and immunoglobulin A nephropathy. The rate of progression of kidney damage to chronic renal failure (i.e., a GFR less than 10 to 15 mL/min) is widely variable and can take as few as 5 years or more than 30 years, depending on the overall cause of the inflammatory process. Chronic glomerulonephritis can lead to other major systemic complications, including hypertension, heart failure, uremia, and anemia. Treatment is dependent on the cause of the damage, and in the case of diabetes-induced disease, angiotensin II receptor blockers or angiotensin-converting enzyme inhibitors are beneficial in slowing the renal damage. As the damage progresses toward end-stage renal failure, the GFR is insufficient to rid the body of waste, and uremia is one of the results. Patients usually start hemodialysis when their GFR is less than 20 mL/min. Patients can be maintained on dialysis for years, although many patients opt for renal transplantation, which is a common procedure.

Chronic Glomerulonephritis: Electron Microscopic Findings

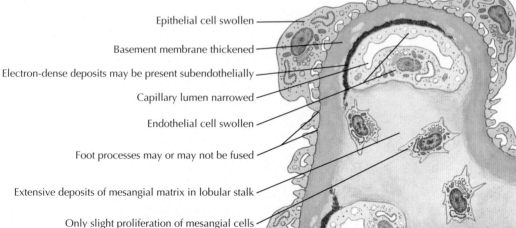

Epithelial cell swollen

Basement membrane thickened

Electron-dense deposits may be present subendothelially

Capillary lumen narrowed

Endothelial cell swollen

Foot processes may or may not be fused

Extensive deposits of mesangial matrix in lobular stalk

Only slight proliferation of mesangial cells

Late stage of chronic glomerulonephritis

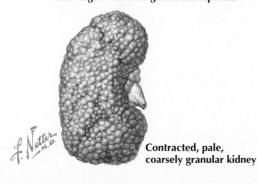

Contracted, pale, coarsely granular kidney

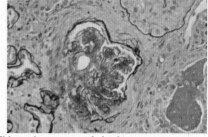

Glomeruli in various stages of obsolescence. Deposition of PAS-stained material, hyalinization, fibrous crescent formation, tubular atrophy, interstitial fibrosis

Chronic Glomerulonephritis The upper panel illustrates key features of chronic glomerular damage, including swollen epithelial cells, a grossly thickened basement membrane, fused foot processes, and increased matrix proteins. These abnormalities destroy the normal filtration barriers. The lower left panel depicts the effects of severe glomerulonephritis on the entire kidney, and the lower right panel provides a representative micrograph of damaged glomeruli. *PAS,* periodic acid–Schiff.

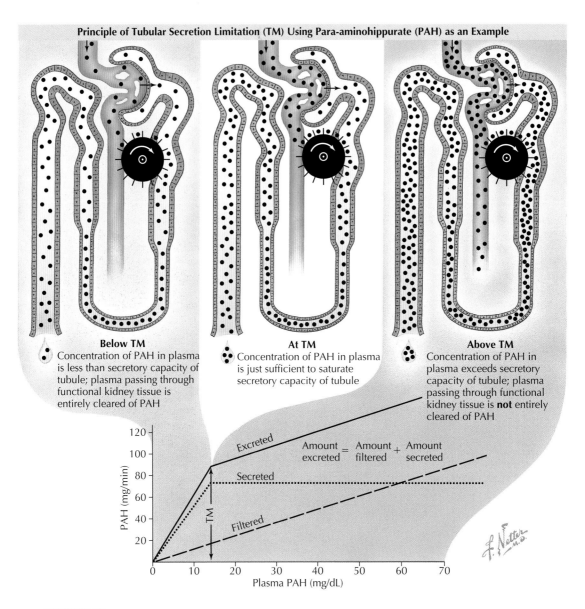

Principle of Tubular Secretion Limitation (TM) Using Para-aminohippurate (PAH) as an Example

Below TM
Concentration of PAH in plasma is less than secretory capacity of tubule; plasma passing through functional kidney tissue is entirely cleared of PAH

At TM
Concentration of PAH in plasma is just sufficient to saturate secretory capacity of tubule

Above TM
Concentration of PAH in plasma exceeds secretory capacity of tubule; plasma passing through functional kidney tissue is **not** entirely cleared of PAH

Excreted

$$Amount\ excreted = Amount\ filtered + Amount\ secreted$$

Secreted

Filtered

TM

PAH (mg/min)

Plasma PAH (mg/dL)

Figure 17.4 **Renal Handling of Para-aminohippurate** Para-aminohippurate (PAH) is filtered at the glomerulus and also secreted into the proximal tubule. When the plasma concentration of PAH is below the tubular transport maximum (TM), PAH is effectively cleared from the blood entering the kidney. However, if the plasma concentration exceeds the TM, PAH is not entirely removed and is found in the renal vein.

limitations, and ultimately the effective pore size is ~30 angstroms (Å). Small molecules such as water, glucose, sucrose, creatinine, and urea are freely filtered. As molecular size increases, or net negative charge of molecules increases (for example, among proteins), filtration becomes increasingly restricted.

Myoglobin, a small protein that is released from muscle after damage occurs, is only 20 Å, but its shape restricts free passage, and thus only about 75% is filtered. Most proteins are negatively charged or of high molecular weight and will not be filtered unless the glomerular barriers are damaged. Proteins entering the tubule cannot be reabsorbed and are excreted in urine (proteinuria).

Starling forces govern fluid movement into or out of the capillaries (see Chapter 1). The pressures that determine glomerular filtration are glomerular capillary hydrostatic pressure (HP_{GC}) forcing fluid out of the capillary, glomerular capillary oncotic pressure (π_{GC}) attracting fluid into the glomerular capillary, Bowman's space hydrostatic pressure (HP_{BS}) opposing capillary hydrostatic pressure, and Bowman's space oncotic pressure (π_{BS}) attracting fluid into Bowman's space (typically the amount of protein in Bowman's space is negligible, and thus π_{BS} is not significant). Thus, assuming π_{BS} is zero,

$$Net\ filtration\ pressure = (HP_{GC} - HP_{BS}) - \pi_{GC} \qquad \textbf{Eq. 17.3}$$

The glomerular capillaries are different from other capillaries (which have significantly reduced pressures at the distal end

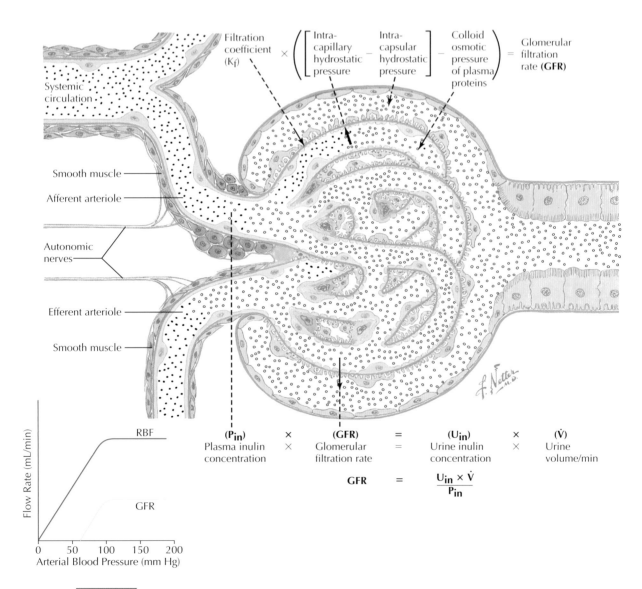

Figure 17.5 illustration labels:

Filtration coefficient (K_f) × [Intra-capillary hydrostatic pressure − Intra-capsular hydrostatic pressure] − (Colloid osmotic pressure of plasma proteins) = Glomerular filtration rate (GFR)

Systemic circulation

Smooth muscle

Afferent arteriole

Autonomic nerves

Efferent arteriole

Smooth muscle

(P$_{in}$) Plasma inulin concentration × (GFR) Glomerular filtration rate = (U$_{in}$) Urine inulin concentration × (V̇) Urine volume/min

$$GFR = \frac{U_{in} \times \dot{V}}{P_{in}}$$

RBF

GFR

Flow Rate (mL/min)

0   50   100   150   200
Arterial Blood Pressure (mm Hg)

Figure 17.5 **Glomerular Filtration**   Blood enters the glomerular capillaries from the afferent arterioles, and ~20% of the plasma is filtered into the nephrons; this is the filtration fraction. The glomerular filtration rate (GFR) can be described on the basis of the forces governing filtration *(top equation)*, or it can be calculated based on the clearance of inulin *(bottom equation)*. The graph illustrates that renal blood flow (RBF) and GFR remain fairly constant over a wide range of mean arterial blood pressures; this phenomenon occurs in part through autoregulation and tubuloglomerular feedback. (Black dots represent inulin.)

of the capillary) because the efferent arteriole (at the distal end of the glomerulus) can constrict and maintain pressure in the glomerular capillary. Thus there is very little reduction in HP$_{GC}$ through the capillary, and filtration can be maintained along its entire length. Afferent and efferent arteriolar resistance can be controlled by sympathetic nerves, circulating hormones (angiotensin II), myogenic regulation, and tubuloglomerular feedback signals, allowing control of glomerular filtration by both intrarenal and extrarenal mechanisms.

Glomerular Filtration Rate

GFR is considered the benchmark of renal function. GFR is the volume of plasma (without protein and cells) that is filtered across all of the glomeruli in the kidneys, per unit time. In a healthy adult, GFR is ~100 to 125 mL/min, with men

having a higher GFR than women. Many factors contribute to the regulation of GFR, which can be maintained at a fairly constant rate, over a range of mean arterial blood pressure (MAP) from 80 to 180 mm Hg (Fig. 17.5).

GFR is determined by the net filtration pressure and the permeability coefficient, **Kf** (mL/min × mm Hg; a function of water permeability of the glomerular capillary membrane and its total surface area, which reflects nephron number and size). The equation is:

$$GFR = Kf[(HP_{GC} - HP_{BS}) + \pi_{GC}] \qquad \textbf{Eq. 17.4}$$

Maintaining normal GFR is critical for eliminating excess fluid and electrolytes from the blood and regulation of overall homeostasis. Significant alteration of any of the parameters in Equation 17.4 can affect GFR. For example, a hemorrhage that reduces MAP below 80 mm Hg may decrease HP_{GC} enough to dramatically decrease or stop filtration. Filtration can also be reduced if the HP_{BS} is increased (for example, during distal blockage by kidney stones) or if Kf is reduced (for example, in persons with glomerulosclerosis).

In general, the nephrons are associated with filtration, reabsorption, secretion, and excretion. The following definitions and relationships hold true for freely filtered substances:

- The **filtered load** (FL_x) of a substance (i.e., the amount of a specific substance filtered per unit time) is equal to the plasma concentration of the substance (P_x) times GFR:

$$FL_x = P_x \times GFR \qquad \textbf{Eq. 17.5}$$

- The **urinary excretion** (E_x) of a substance is the urine concentration of the substance (U_x) times the volume of urine produced per unit time ($\dot{V}$):

$$E_x = U_x \times \dot{V} \qquad \textbf{Eq. 17.6}$$

- Most substances are reabsorbed (to some extent); reabsorption rate of a substance (R_x) is equal to the filtered load of the substance minus the urinary excretion of a substance:

$$R_x = FL_x - E_x \qquad \textbf{Eq. 17.7}$$

- Select substances are actively secreted (e.g., creatinine, PAH, H^+, and K^+). The secretion rate of a substance (S_x) is equivalent to the excretion rate minus the filtered load of the substance:

$$S_x = E_x - FL_x \qquad \textbf{Eq. 17.8}$$

Renal handling of key substances is discussed in Chapter 18.

RENAL CLEARANCE

Because GFR is a primary measure of the health of kidney function, GFR is routinely analyzed. This analysis can be performed in several ways. The physical factors and pressures can all be measured experimentally, but this measurement is not practical in patients. Instead, the principle of **renal clearance** is used. Renal clearance (C_x) is the *volume of plasma cleared of a substance per unit time*. The clearance equation incorporates the urine and plasma concentrations of the substance, along with the urine flow rate, and is usually reported in mL/min or L/day:

$$C_x = (U_x \times \dot{V})/P_x \qquad \textbf{Eq. 17.9}$$

This equation can be used to determine the GFR: the clearance of a substance is equated to the GFR if the substance is freely filtered but not reabsorbed or secreted. In this case, *the amount filtered will equal the amount excreted* ($FL_x = E_x$):

$$\text{because } FL_x = P_x \times GFR \text{ and } E_x = U_x \times \dot{V}$$
$$\text{when } FL_x = E_x \qquad \textbf{Eq. 17.10}$$

then,

$$P_x \times GFR = U_x \times \dot{V} \qquad \textbf{Eq. 17.11}$$

and, rearranging the equation,

$$GFR = (U_x \times \dot{V})/P_x \qquad \textbf{Eq. 17.12}$$

Thus, for such a substance, $GFR = C_x$.

The RPF feeds the glomerular capillaries, but not all of the plasma presented to the capillaries is filtered. The filtration fraction (FF) is the proportion of the RPF that becomes glomerular filtrate:

$$FF = GFR/RPF$$

In an average healthy adult, GFR = 100 to 125 mL/min and RPF = 600 mL/min; thus, FF is ~0.20 (i.e., ~20% of the plasma entering the kidneys is filtered). At the individual nephrons, the unfiltered plasma exits the efferent arteriole to the peritubular capillaries.

If the clearance of inulin (C_{in}) is 100 mL/min, this means that 100 mL of plasma is completely cleared of inulin each minute. Contrast the clearance of inulin to the clearance of glucose, which is 0 mL/min in a healthy person, indicating that no plasma is cleared of glucose (and therefore no glucose is found in the urine). The renal clearance of any filtered substance can be calculated, and comparison of the clearance with the GFR gives a general idea of whether net reabsorption or net secretion of the substance occurred, because the GFR is the total rate of filtration that is occurring at any given time.
- If the clearance of X is less than the GFR, net reabsorption has occurred.
- If the clearance of X is greater than the GFR, net secretion has occurred, because more was cleared from the plasma than can be accounted for by GFR alone.

Although no endogenous substance exactly meets these requirements (i.e., the substance is freely filtered but not reabsorbed or secreted, and therefore $FL_x = E_x$), the polyfructose molecule **inulin** does meet these criteria. Inulin is not broken down in the blood, is freely filtered, and is not reabsorbed or secreted by the kidney. To measure inulin clearance (and thereby determine GFR), inulin is infused intravenously, and when a stable plasma level is achieved, timed urine collections

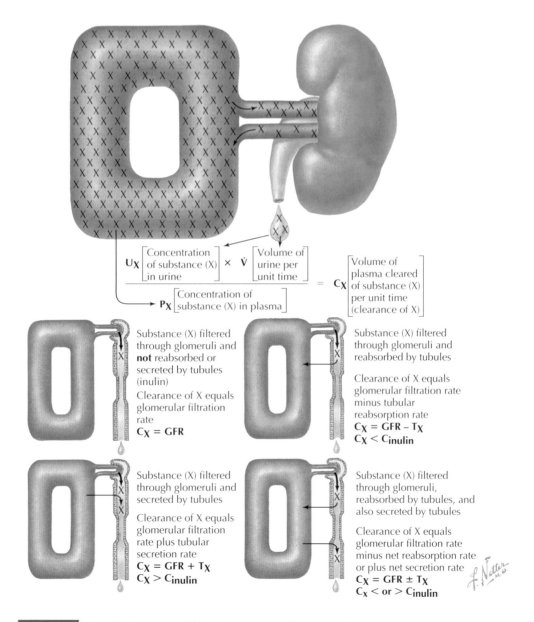

$$\dfrac{U_X \begin{bmatrix} \text{Concentration} \\ \text{of substance (X)} \\ \text{in urine} \end{bmatrix} \times \dot{V} \begin{bmatrix} \text{Volume of} \\ \text{urine per} \\ \text{unit time} \end{bmatrix}}{P_X \begin{bmatrix} \text{Concentration of} \\ \text{substance (X) in plasma} \end{bmatrix}} = C_X \begin{bmatrix} \text{Volume of} \\ \text{plasma cleared} \\ \text{of substance (X)} \\ \text{per unit time} \\ \text{(clearance of X)} \end{bmatrix}$$

Substance (X) filtered through glomeruli and **not** reabsorbed or secreted by tubules (inulin)

Clearance of X equals glomerular filtration rate

$C_X = GFR$

Substance (X) filtered through glomeruli and reabsorbed by tubules

Clearance of X equals glomerular filtration rate minus tubular reabsorption rate

$C_X = GFR - T_X$
$C_X < C_{inulin}$

Substance (X) filtered through glomeruli and secreted by tubules

Clearance of X equals glomerular filtration rate plus tubular secretion rate

$C_X = GFR + T_X$
$C_X > C_{inulin}$

Substance (X) filtered through glomeruli, reabsorbed by tubules, and also secreted by tubules

Clearance of X equals glomerular filtration rate minus net reabsorption rate or plus net secretion rate

$C_X = GFR \pm T_X$
$C_X <$ or $> C_{inulin}$

Figure 17.6 **Renal Clearance Principle** "Clearance" describes the volume of plasma that is cleared of a substance per unit time. The renal clearance (C) of a substance provides information on how the kidney handles that substance. Because inulin is freely filtered and not reabsorbed or secreted, all of the filtered inulin is excreted in the urine. Thus, C_{inulin} is equated with the glomerular filtration rate (GFR), and the net handling of other substances can be determined, depending on whether their clearance is greater than (indicating net secretion), less than (indicating net reabsorption), or equal to C_{inulin}.

are made. The calculated clearance of inulin (C_{inulin}) can be equated to the GFR (see Fig. 17.5):

$$C_{inulin} = GFR \qquad \textbf{Eq. 17.13}$$

Inulin infusion is not routinely performed to determine clearance because of the invasive nature of this procedure. Instead, the renal clearance of the *endogenous* substance **creatinine** is calculated to approximate GFR. Creatinine is a by-product of muscle metabolism and is freely filtered by the kidneys. It is not reabsorbed, but secretion of ~10% into the renal tubules from the peritubular capillaries occurs, and thus creatinine clearance overestimates GFR by ~10% (Fig. 17.6).

Plasma creatinine (P_{Cr}) is used clinically to estimate GFR. In most cases, the body produces creatinine at a constant rate, so the excretion rate is also constant. Because GFR is equated with the clearance of creatinine [GFR = ($U_{Cr} \times \dot{V}$) ÷ P_{Cr}], if creatinine excretion ($U_{Cr} \times \dot{V}$) is constant, the GFR is proportional to $1/P_{Cr}$. Thus, when the GFR decreases, less creatinine is filtered and excreted, and plasma creatinine builds up. As a clinical application, this allows a rapid approximation of the GFR by simply analyzing the P_{Cr}. P_{Cr} is normally ~1 mg%, so GFR is proportional to 1/1, or 100%. If P_{Cr} rises to 2, GFR is $\frac{1}{2}$, or 50%, and so on.

REGULATION OF RENAL HEMODYNAMICS

Intrinsic feedback systems, hormones, vasoactive substances, and renal sympathetic nerves regulate GFR through effects on renal hemodynamics (flow, resistance, and pressure).

Intrinsic systems include the **myogenic mechanism** and **tubuloglomerular feedback (TGF)**. Utilizing the myogenic mechanism, the renal arteries and arterioles respond directly to increases in systemic blood pressure by constricting, thereby maintaining constant filtration pressure in the glomerular capillaries. TGF is a regulatory mechanism that involves the macula densa of the juxtaglomerular apparatus. The kidney is unique in that the glomerular capillaries have arterioles (resistance vessels) at *either end* of the capillary network. Constriction of the afferent or efferent arterioles can produce immediate effects on the HP_{GC}, controlling GFR. Because the juxtaglomerular apparatus functionally couples the distal tubule with the afferent arteriole, the tubular flow past the macula densa can control afferent arteriolar resistance (see Fig. 17.3). Decreases in flow and tubular fluid sodium concentration in the distal tubule will decrease afferent arteriolar resistance and increase GFR in that nephron; conversely, if distal tubular flow or osmolarity is high, TGF will increase afferent arteriolar resistance, decreasing GFR. These systems allow minute-to-minute regulation of GFR over a wide range of systemic blood pressures (MAP 80 to 180 mm Hg).

Many substances (including nitric oxide and endothelin) regulate renal hemodynamics, but this section will focus on the **renin-angiotensin-aldosterone system (RAAS), atrial natriuretic peptide (ANP), sympathetic nerves/catecholamines,** and **intrarenal prostaglandins**. When **angiotensin II** is produced and the sympathetic nervous system is activated to preserve systemic blood pressure, the kidneys will respond to excessive constriction by intrarenal autoregulation, preserving blood flow to the glomeruli, except under extreme conditions. This balance between extrarenal and intrarenal control is necessary to maintain proper GFR.

Control of renal hemodynamics occurs through the following neural, humoral, and paracrine mechanisms:

- The **RAAS** is activated in response to low renal blood flow. Reduced stretch of renal vascular baroreceptors stimulates **renin** secretion by the juxtaglomerular cells at the ends of the afferent arterioles; similarly, reduced osmolarity or flow in the distal tubule is sensed by the macula densa cells and results in renin release by the juxtaglomerular cells (see Fig. 20.1). The renin will activate the RAAS, resulting in angiotensin II production and thus effects on GFR. Angiotensin II is produced both locally in the kidney and systemically.
- **Angiotensin II** exerts both direct and indirect effects on the GFR. It is a vasoconstrictor, and in the kidneys, it

acts directly on the renal arteries, and to a greater extent on the afferent and efferent *arterioles*, increasing resistance, reducing HP_{GC}, and decreasing GFR; angiotensin II actually has greater effect on the efferent arteriole than on the afferent arteriole. At the same time, it can constrict *glomerular* mesangial cells, reducing Kf, and further reducing GFR.

- **ANP** is released from the cardiac myocytes of the right atrium in response to stretch (at high blood volume). To regulate GFR, ANP dilates the afferent arteriole and constricts the efferent arteriole, increasing HP_{GC} and, thus, GFR. The enhanced GFR results in higher sodium and water excretion, reducing blood volume.
- **Sympathetic nerves** and **catecholamine secretion** (norepinephrine and epinephrine) are stimulated in response to reductions in systemic blood pressure and cause vasoconstriction of the renal arteries and arterioles. At tonic levels of sympathetic nerve activity, the intrarenal systems will counteract this constriction to ensure that the kidney vasculature remains dilated, preserving GFR. During high sympathetic nerve activity (hemorrhage and strenuous exercise), sympathetic nerve activity overrides the intrarenal regulatory mechanisms and reduces renal blood flow and GFR.
- **Intrarenal prostaglandins** (prostaglandin E_2 and prostacyclin) are vasodilators and serve to counteract primarily angiotensin II–mediated vasoconstriction, acting at the level of the arterioles and glomerular mesangial cells. Nonsteroidal anti-inflammatory drugs (NSAIDs) such as aspirin will block prostaglandin synthesis and restrict the compensatory renal vasodilation.

With blood loss from hemorrhage, the sympathetic nervous system and hormone systems (RAAS, antidiuretic hormone, and aldosterone) are activated to preserve systemic blood pressure and, ultimately, to restore blood volume. If the MAP falls below 80 mm Hg, the high level of vasoconstriction will overwhelm the intrarenal regulation of the GFR, and the GFR will drop. Although vasoconstriction is beneficial in terms of maintaining systemic blood pressure, this can result in acute renal failure (GFR < 25 mL/min) if blood volume is not restored quickly.

The fractional excretion (FE) of a substance (X) is the proportion of the filtered load of the substance ($P_X \times$ GFR) that is excreted in the urine ($U_x \times \dot{V}$):

$$FE_x = (U_x \times \dot{V})/(P_x \times GFR)$$

Because GFR is equivalent to inulin clearance and $C_{in} = (U_{in} \times \dot{V})/P_{in}$, the above equation can be simplified to

$$FE_x = [(U/P)_x/(U/P)_{in}] \times 100$$

CLINICAL CORRELATE 17.2
Analysis of Renal Function

This correlate will focus on the variety of calculations associated with renal function and provide examples of their solution.

A constant amount of inulin (in isotonic saline solution) was infused intravenously into a healthy 25-year-old man. After 3 hours, the man emptied his bladder completely, and then urine was collected after another 2 hours. A blood sample was obtained at the time of urine collection. Blood and urine were analyzed, with results shown below. Analyses of several parameters of renal function were performed.

	Urine	Plasma
Inulin concentration	1000 mg%	20 mg%
Creatinine concentration	55 mg%	1 mg%
PAH concentration	300 mg%	1 mg%
Sodium concentration	2.5 mEq/L	140 mEq/L

Urine volume (UV) = 240 mL

Urine collection time = 2 hours

HCT = 0.42

The following parameters can be calculated:

Urinary flow rate ($\dot{V}$), which is the rate at which urine is produced. Urine flow is dependent on general fluid homeostasis and fluid intake. Under normal circumstances, if fluid intake is increased, urine flow will increase. If a person ingests ~3 L of fluid in food and drink, the urinary losses will be slightly less, with the balance made up by insensible losses (e.g., breathing and sweating).

$$\dot{V} = \text{urine volume/time}$$
$$= 240 \text{ mL}/120 \text{ min}$$
$$= 2 \text{ mL/min}$$

Glomerular filtration rate (GFR), which is the volume of plasma filtered by the glomeruli per unit time. Normal GFR in an adult is ~100 mL/min, or ~144 L/day. The GFR in men is typically higher than in women.

GFR is determined by inulin clearance:

$$C_{in} = (U_{in} \times \dot{V})/P_{in}$$
$$= (1000 \text{ mg\%} \times 2 \text{ mL/min})/20 \text{ mg\%}$$
$$= 100 \text{ mL/min}$$

GFR can also be determined by creatinine clearance, which overestimates GFR by ~10% because of creatinine secretion:

$$C_{cr} = (U_{cr} \times \dot{V})/P_{cr}$$
$$= (55 \text{ mg\%} \times 2 \text{ mL/min})/1 \text{ mg\%}$$
$$= 110 \text{ mL/min}$$

Effective renal plasma flow (eRPF), which is the fraction of the renal plasma flow entering the glomeruli and available for filtration. eRPF is equated with the clearance of PAH (C_{PAH}):

$$eRPF = C_{PAH}$$
$$= (300 \text{ mg\%} \times 2 \text{ mL/min})/1 \text{ mg\%}$$
$$= 600 \text{ mL/min}$$

Effective renal blood flow (eRBF), which is the fraction of renal blood flow entering the glomeruli. It is usually ~20% of cardiac output.

$$eRBF = (eRPF)/(1 - HCT)$$
$$= (600 \text{ mL/min})/(1 - 0.42)$$
$$= 1034 \text{ mL/min, or } 1.034 \text{ L/min}$$

Filtration fraction (FF), which is the fraction of the renal plasma flow that is filtered per unit time.

$$FF = GFR/RPF$$
$$= (100 \text{ mL/min})/(600 \text{ mL/min})$$
$$= 0.17, \text{ or } 17\% \text{ of the RPF entering the kidney was filtered}$$

Filtered load of sodium (FL_{Na}), which is the amount of plasma sodium that is filtered per unit time.

$$FL_{Na} = \text{Plasma Na} \times GFR$$
$$= 140 \text{ mEq/L} \times 100 \text{ mL/min}$$
$$= 14 \text{ mEq/min}$$

Urinary excretion of sodium ($U_{Na}\dot{V}$ or E_{Na})

$$U_{Na}\dot{V} = \text{Urine concentration of Na} \times \dot{V}$$
$$= 2.5 \text{ mEq/L} \times 2 \text{ mL/min}$$
$$= 0.005 \text{ mEq/min}$$

Reabsorbed sodium (R_{Na})

$$R_{Na} = FL_{Na} - U_{Na}\dot{V}$$
$$= 14 \text{ mEq/min} - 0.005 \text{ mEq/min}$$
$$= 13.995 \text{ mEq/min}$$

Fractional excretion of sodium (FE_{Na}), which is the fraction of filtered sodium that is excreted. Usually ≥99% of filtered sodium is reabsorbed, so less than 1% of the amount filtered is excreted.

$$FE_{Na} = [(U/P)_{Na}/(U/P)_{in}] \times 100$$
$$= [2.5/140]/(1000/20)] \times 100$$
$$= 0.035\%$$

Fractional reabsorption of sodium (FR_{Na}), which is the fraction of filtered sodium that is reabsorbed back into the capillaries. Because the fraction reabsorbed plus the fraction excreted equals 1,

$$FR_{Na} = [1 - (U_{Na}\dot{V}/FL_{Na})] \times 100$$
$$= [1 - (0.005/14)] \times 100$$
$$= 99.97\%$$

Chapter 18

Renal Transport Processes

GENERAL OVERVIEW OF RENAL TRANSPORT

When the plasma filtered into Bowman's space enters the proximal tubule, the process of reabsorption begins. In general, nephrons reabsorb the majority of the fluid and solutes that pass though them, with the proximal tubule having the greatest reabsorptive capacity and the distal sites fine-tuning the process. In addition, the peritubular capillaries secrete select substances (creatinine, para-aminohippurate, and penicillin) into different segments of the renal tubule.

The proximal tubule (PT) is the site of bulk reabsorption of fluid and solutes. The PT is composed of three segments, S1, S2, and S3, which differ in the depth of the brush border and amount of mitochondria in the PT cells. These differences result in a high capacity for reabsorption through the segments. From S1 to S3 segments, the brush border becomes progressively deeper and the high concentration of cellular mitochondria observed in the S1 segment decreases. The high number of mitochondria in the S1 is consistent with a high rate of active transport in that segment. As the filtrate is reabsorbed and less is present in the tubule in subsequent segments, the deeper brush border increases the surface area, facilitating continued reabsorption.

SODIUM-DRIVEN SOLUTE TRANSPORT

Sodium, Chloride, and Water

The kidneys are integral in regulating plasma sodium concentration, thus allowing maintenance of fluid and electrolyte homeostasis (see Chapter 1). As seen with the intestinal absorption of essential nutrients (see Chapter 26), sodium is also a major driving force for the renal reabsorption of fluid, electrolytes, and a variety of other solutes. As sodium transporters reabsorb sodium (and other solutes), they generate the driving force for water reabsorption. When the water leaves the tubule, the concentration of electrolytes and solutes remaining in the tubular fluid increases, providing higher gradients for their diffusion into the cell.

In general, of the total filtrate coming into the nephrons, the **proximal tubule** reabsorbs:
- 65% to 70% of the Na^+ and H_2O
- 80% to 85% of the K^+
- 65% of the Cl^-
- 75% to 80% of the phosphate
- 100% of the glucose
- 100% of the amino acids

After this bulk reabsorption, "fine-tuning" of the reabsorption of the remaining tubular fluid occurs in subsequent segments of the nephron.

Approximately 65% to 70% of the water in tubular fluid is reabsorbed from the proximal tubule back into the peritubular capillaries, primarily driven by sodium reabsorption. The filtered load (FL) of sodium through the glomeruli is high (~25,000 milliequivalents [mEq]/day), and to maintain body fluid homeostasis, more than 99% of the FL_{Na} must be reabsorbed back into the blood. This reabsorption is accomplished by apical (luminal) secondary active transport of sodium down a concentration gradient established by the basolateral Na^+/K^+ ATPase pumps. Figure 18.1 illustrates the primary sites and transporters for sodium reabsorption along different segments of the nephron.

- *Proximal convoluted tubule (S1 and S2 segments):* Bulk flow occurs by secondary active sodium cotransport with several substances, including **glucose, amino acids, phosphate,** and **organic acids.** The proximal tubule also has Na^+/H^+ **antiporters** (or exchangers) that allow H^+ secretion into the proximal renal tubular fluid while sodium is reabsorbed.
- *Proximal straight tubule (S3 segment):* Na^+/H^+ **antiporters** continue to reabsorb sodium and secrete H^+ into the tubular fluid. The reabsorption of sodium and fluid also provides the electrochemical gradient that facilitates **chloride** reabsorption. Chloride concentration increases along the proximal tubule segments as water is reabsorbed. Chloride enters the cells in the S3 segment

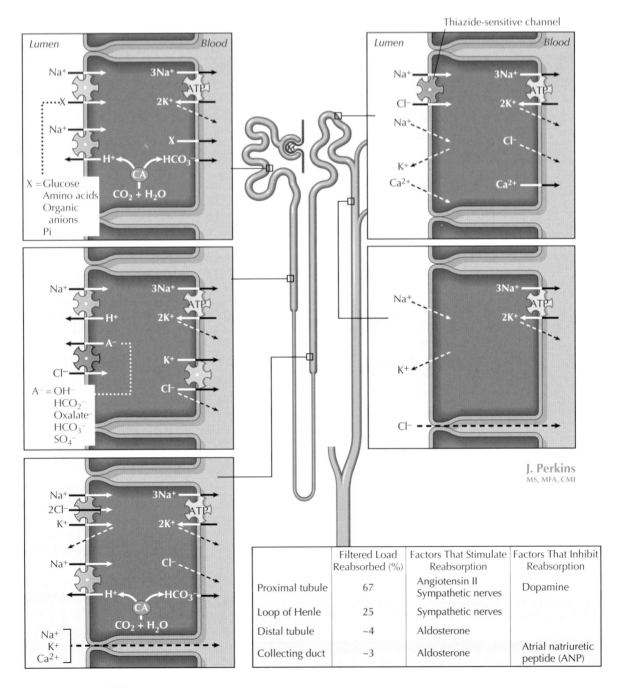

	Filtered Load Reabsorbed (%)	Factors That Stimulate Reabsorption	Factors That Inhibit Reabsorption
Proximal tubule	67	Angiotensin II Sympathetic nerves	Dopamine
Loop of Henle	25	Sympathetic nerves	
Distal tubule	~4	Aldosterone	
Collecting duct	~3	Aldosterone	Atrial natriuretic peptide (ANP)

J. Perkins
MS, MFA, CMI

Figure 18.1 **Nephron Sites of Sodium Reabsorption** Sodium reabsorption is critical for proper fluid and electrolyte homeostasis. More than 99% of the filtered load is reabsorbed through a variety of transport mechanisms. The gradient for sodium transport into the cells is maintained by basolateral Na⁺/K⁺ ATPase pumps. *CA*, carbonic anhydrase.

down its electrochemical gradient through antiporters, resulting in apical secretion of anions such as HCO_3^-, OH^-, SO_4^-, and oxalate. Cl^- reabsorption also occurs paracellularly, or between the cells. (The entire PT reabsorbs ~65% to 70% of FL_{Na}.)

■ *Thin descending limb of Henle:* This segment is *impermeable to sodium* and most other solutes but is permeable to water. The interstitial osmotic gradient drives water reabsorption through apical water channels

(aquaporin-1), which results in the tubular fluid being more *concentrated* as it enters the ascending limb of Henle (more information on this process is provided in Chapter 19).

■ *Thick ascending limb of Henle (TALH):* This segment is *impermeable to water*, but specialized apical **Na⁺-K⁺-2Cl⁻ cotransporters** (**NKCC-2**) facilitate reabsorption of electrolytes, thus *diluting* the tubular fluid entering the distal tubule. These transporters are the targets for loop diuretics such as furosemide and bumetanide. In addition, there is a backleak of K^+ out of the cells into the lumen, creating a lumen-positive transepithelial potential difference (compared with interstitial fluid). This mechanism allows paracellular movement of cations (Ca^{2+}, Mg^{2+}, Na^+, and K^+) out of the tubular lumen. In addition to NKCC-2, Na^+/H^+ antiporters are also present, which reabsorb Na^+ and secrete H^+ into the tubule. (The thick ascending limb reabsorbs ~20% to 25% of FL_{Na}.)

- Distal tubule (DT): The early DT has Na^+-Cl^- cotransporters that can be inhibited by thiazide diuretics. The principal cells of the late DT have apical Na^+ channels (epithelial sodium channels [ENaC]) and K^+ (big potassium [BK]) channels that are increased by the hormone aldosterone, resulting in greater Na^+ and water reabsorption and K^+ secretion. (The DT reabsorbs ~4% of FL_{Na}.)
- Collecting duct: Like the late DT, the principal cells of the collecting duct have ENaC and BK channels that are increased by aldosterone. Aldosterone also increases apical Na^+/H^+ exchangers in the α-intercalated cells. Through these mechanisms, the collecting duct (CD) reabsorbs ~3% of FL_{Na}.

Glucose Transport

Glucose is reabsorbed in the proximal tubules via insulin-independent sodium-glucose transporters (SGLT); reabsorption occurs via SGLT2 in S1 and S2 segments (proximal convoluted tubule) and via SGLT1 in the S3 segment (proximal straight tubule). Glucose exits the basolateral membranes via glucose transporter (GLUT-2)–facilitated transport.

Because of the large FL_{Na}, the reabsorption of sodium is not a rate-limiting step in the reabsorption of other solutes. For many solutes, the rate-limiting step is the number of specific transporters available for the solute. Glucose is a good example of this concept. Renal sodium-glucose carriers have a high transport maximum (TM), and under normal conditions, the FL of glucose is low enough that the transporters can carry all of the solute back into the blood, leaving none in the tubular fluid and urine (Fig. 18.2). Thus the renal clearance of glucose is normally zero.

However, if the plasma glucose level is high (such as in persons with diabetes), the FL of glucose will increase and the glucose in the tubular fluid may saturate the carriers. The renal threshold describes the point where the first nephrons exceed their TM; the unabsorbed glucose will remain in the tubular fluid and be excreted in the urine (glucosuria). When the plasma glucose concentration (and hence the FL of glucose) is under the renal threshold for reabsorption, all of the glucose in tubular fluid will be reabsorbed (see Fig. 18.2). However, when it exceeds the threshold, the transporters are saturated (i.e., TM is exceeded) and glucose appears in the urine.

The plasma concentration at which the renal threshold for glucose reabsorption is exceeded (and glucosuria is observed) is ~250 mg%. However, the calculated plasma threshold is 300 mg%. This difference between real and calculated values is explained by nephron heterogeneity (also called splay), whereby the cortical and juxtaglomerular nephron populations have lower and higher TMs for glucose, respectively. The average TM is the basis for the calculated threshold for plasma glucose levels (at which glucosuria occurs), despite the fact that some nephrons have a lower TM that will be exceeded when plasma glucose is greater than ~250 mg%.

This concept explains the glucosuria that can occur in persons with diabetes mellitus. When glucose cannot be efficiently transported into tissues, plasma glucose concentrations become elevated. The fasting plasma glucose level is much higher than normal in persons with diabetes (greater than 130 mg% compared with 80 to 90 mg%), resulting in an increased FL of glucose. Upon eating, the FL of glucose can easily exceed the TM of some nephrons, causing glucosuria. In addition, because glucose exerts an osmotic effect, the glucosuria will be associated with diuresis (i.e., loss of water through increased urine volume).

BICARBONATE HANDLING

Adequate plasma bicarbonate is necessary for acid-base homeostasis, and the kidneys reabsorb 100% of the filtered bicarbonate (HCO_3^-). However, this reabsorption occurs indirectly and involves the secretion of H^+ (through cation exchange and active H^+ pumps) in multiple nephron segments. In the tubular lumen, filtered HCO_3^- and secreted H^+ form CO_2 and H_2O (a reaction catalyzed by brush border carbonic anhydrase [CA]), which diffuse into the cell (Fig. 18.3). Once in the cell, the CO_2 and H_2O are converted back to carbonic acid (by intracellular CA); HCO_3^- is transported out of the cell via basolateral HCO_3^-/Cl^- exchangers or Na^+-HCO_3^- cotransporters, depending on the nephron segment. The H^+ generated from this process is secreted back into the tubular lumen and can be used to reabsorb more HCO_3^-; in the CDs the H^+ can be buffered and excreted (see Chapter 21). This mechanism is present in three segments of the nephron, facilitating reabsorption of filtered bicarbonate in the PT (80% of filtered load), TALH (15%), and CD (5%).

Under normal conditions, the renal clearance of HCO_3^- is 0, meaning that there is no HCO_3^- in the urine. The regulation of bicarbonate handling is an integral part of acid-base homeostasis and will be discussed in Chapter 21.

POTASSIUM HANDLING

As with all of the major electrolytes, potassium balance is important to overall homeostasis, and dietary intake must be matched by urinary and fecal excretion. Plasma K^+

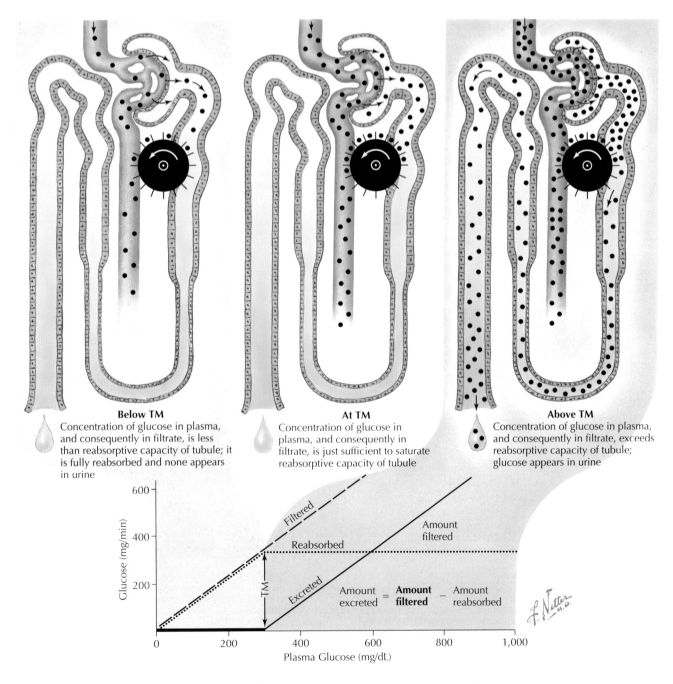

Below TM
Concentration of glucose in plasma, and consequently in filtrate, is less than reabsorptive capacity of tubule; it is fully reabsorbed and none appears in urine

At TM
Concentration of glucose in plasma, and consequently in filtrate, is just sufficient to saturate reabsorptive capacity of tubule

Above TM
Concentration of glucose in plasma, and consequently in filtrate, exceeds reabsorptive capacity of tubule; glucose appears in urine

Figure 18.2 **Renal Handling of Glucose** Glucose is freely filtered at the glomerulus and is 100% reabsorbed in the proximal tubules by sodium-glucose cotransporters (insulin-independent). However, if blood glucose levels become elevated, as in persons with diabetes, the maximal tubular reabsorption rate (TM) is exceeded, and glucose appears in the urine *(far right panel).*

concentration must be maintained at relatively low levels (3 to 5 mEq/L), and the kidneys can regulate K^+ excretion (and reabsorption). Figure 18.4 illustrates potassium handling through the nephron and the effects of dietary K^+ intake.

Potassium handling varies along the nephron:

■ *Proximal tubule:* Potassium reabsorption occurs by **paracellular movement** (between cells), not by entry into the cells. Reabsorption initially occurs via **solute drag**, which is initiated by water reabsorption. In the S2 and S3 segments, the positive potential in the tubular lumen allows additional paracellular potassium reabsorption by **diffusion** down the electrochemical gradient (this mechanism accounts for ~70% reabsorption of filtered potassium).

■ *Thick ascending limb of Henle:* NKCC-2 transporters in the TALH use the sodium and chloride gradients to

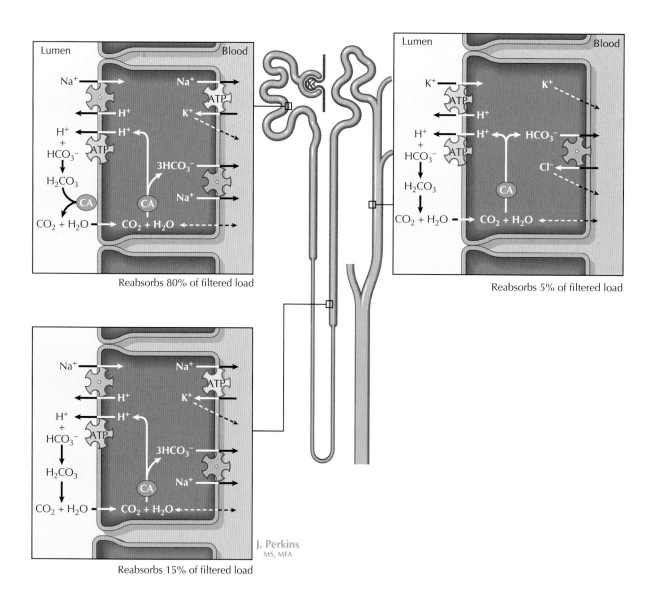

Figure 18.3 **Renal HCO₃⁻ Reabsorption** Bicarbonate is freely filtered at the glomerulus and is reabsorbed along the nephron through a process involving secretion of H⁺. Under normal conditions, 100% of the filtered bicarbonate is reabsorbed. The upper right panel represents an α-intercalated cell of the collecting duct (also found in the late distal tubules). *CA*, carbonic anhydrase.

facilitate transport of K⁺ (accounting for ~20% of filtered potassium).

- *Late distal tubules:* In response to an elevation in plasma potassium concentration, aldosterone is secreted from the adrenal cortex and acts at principal cells of the late DT to increase basolateral Na⁺/K⁺ ATPases, as well as apical potassium (and sodium) channels. The additional Na⁺/K⁺ ATPases increase intracellular potassium, and the potassium is secreted (through BK channels) into the tubular fluid, following its concentration gradient.
- *Collecting ducts:* As seen in the DT, potassium is secreted from the principal cells into the CDs through aldosterone-sensitive apical K⁺ channels. In addition, in the α-intercalated cells of the CD, an apical H⁺/K⁺ ATPase facilitates K⁺ reabsorption in exchange for H⁺ secretion

into the tubular fluid. Under normal conditions there is a net secretion of K⁺ and excretion in the urine. Net reabsorption can occur during dietary K⁺ depletion.

Renal potassium handling is influenced by the following factors:

- *Dietary potassium intake:* When high potassium intake results in elevation of plasma K⁺, aldosterone is secreted. As previously described, aldosterone stimulates K⁺ secretion from the principal cells of the late DT and CDs into the tubular fluid. Conversely, when dietary K⁺ intake is low, secretion of K⁺ from the principal cells is reduced. In this situation, K⁺ *reabsorption* from the α-intercalated cells of the CDs predominates.

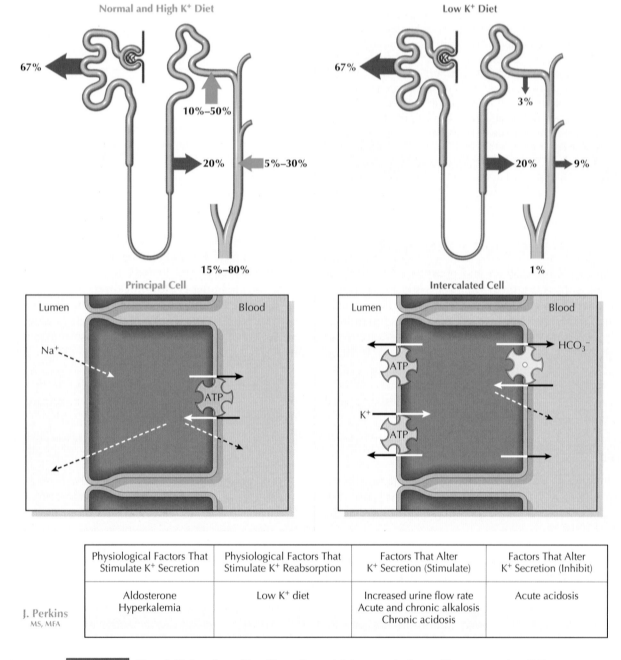

Normal and High K⁺ Diet

67%

10%–50%

20%   5%–30%

15%–80%

Principal Cell

Lumen   Blood

Na⁺

ATP

Low K⁺ Diet

67%

3%

20%   9%

1%

Intercalated Cell

Lumen   Blood

HCO₃⁻

ATP

K⁺

ATP

Physiological Factors That Stimulate K⁺ Secretion	Physiological Factors That Stimulate K⁺ Reabsorption	Factors That Alter K⁺ Secretion (Stimulate)	Factors That Alter K⁺ Secretion (Inhibit)
Aldosterone Hyperkalemia	Low K⁺ diet	Increased urine flow rate Acute and chronic alkalosis Chronic acidosis	Acute acidosis

J. Perkins
MS, MFA

Figure 18.4 Renal Potassium Handling To maintain normal plasma K⁺ concentration (3.5 to 5 mEq/L), the kidney must control K⁺ excretion, and the amount of K⁺ excreted changes with dietary intake. Diets low in K⁺ stimulate avid K⁺ reabsorption throughout the nephron, whereas diets high in K⁺ stimulate distal K⁺ secretion (in *green*).

■ *Plasma volume:* In addition to responding to increased plasma K⁺ concentrations, aldosterone is also released in response to decreased plasma volume through the renin-angiotensin-aldosterone system. As previously noted, the increased aldosterone facilitates K⁺ secretion from the principal cells of late DT and CD (see Chapter 20).

■ *Acid-base status:* To maintain normal systemic acid-base balance, excess acid (principally from dietary intake) must be buffered and excreted in the urine (see Chapter

21). To facilitate acid excretion, the α-intercalated cells of the CDs have apical H⁺/K⁺ ATPases that secrete H⁺ into the tubular fluid (where it is buffered and excreted) and pump K⁺ into the cells; the K⁺ exits the cells via basolateral K⁺-Cl⁻ symports.

■ *Tubular fluid flow rate:* When tubular fluid flow is high (e.g., in response to volume expansion), the concentration gradient for K⁺ (from collecting duct cell to the lumen) is elevated, and K⁺ secretion increases.

Targeting specific renal sodium transporters can be effective for controlling hypertension, because blocking reabsorption increases sodium and water excretion, reducing blood volume. Loop diuretics such as furosemide and bumetanide inhibit the NKCC-2 transporters in the TALH, whereas thiazide diuretics act on the Na^+/Cl^- antiporters in the distal tubule. Extended use of loop diuretics and thiazides can also result in urinary K^+ loss, and thus plasma K^+ must be monitored when these drugs are used. Potassium-sparing diuretics, such as amiloride, target the (aldosterone-sensitive) ENaCs in the principal cells of the cortical collecting tube. Both thiazides and potassium-sparing diuretics also limit urinary calcium loss.

CALCIUM AND PHOSPHATE TRANSPORT

Regulation of plasma calcium and phosphate is critical for proper bone and tissue growth in the fetus and growing child and continues to be important in the adult for bone health. The kidneys control plasma levels of calcium and phosphate by altering their rate of reabsorption. Most of the calcium and phosphate in the body (99% and 85%, respectively) is found in bone matrix. The continual remodeling of the bone is facilitated by active vitamin D3 and parathyroid hormone (PTH); renal phosphate and calcium reabsorption are both regulated by PTH (see Chapter 31).

Calcium Handling

About 40% of plasma Ca^{2+} is bound to proteins, leaving 60% free for filtration at the glomeruli. The kidneys reabsorb ~99% of the filtered Ca^{2+} at the following sites throughout the nephron (Fig. 18.5A):

- *Proximal tubule:* Ca^{2+} reabsorption is paracellular via solvent drag initiated by bulk reabsorption of Na^+ and water. This mechanism accounts for ~70% of Ca^{2+} reabsorption.
- *Thick ascending limb of Henle:* Reabsorption is paracellular, again in parallel with Na^+ and water reabsorption. In addition, the lumen-positive transepithelial potential favors paracellular reabsorption of divalent cations in this segment (~20% of reabsorption). Because Ca^{2+} follows sodium reabsorption, changes in sodium reabsorption (such as with loop diuretics) will usually reduce Ca^{2+} reabsorption.
- *Distal tubule:* Although the DT accounts for only ~8% to 9% of Ca^{2+} reabsorption, this is the site of hormonal control by PTH. In response to decreases in plasma Ca^{2+},

PTH (released from the parathyroid glands) increases apical Ca^{2+} channels (TRPV5) and basolateral Ca^{2+} ATPase pumps and Na^+/Ca^{2+} exchangers (which transport Ca^{2+} out of the cell into the interstitium; see Fig. 18.5A and Chapter 31).

Phosphate Handling

Phosphates (Pi) are required for bone matrix formation, as well as for intracellular high-energy mechanisms (e.g., adenosine triphosphate formation and utilization). The majority of plasma Pi (>90%) is available for filtration, and Pi reabsorption and excretion are highly dependent on diet and age. As with glucose, Pi has a TM that can be saturated. Under normal dietary conditions, transporters are present only in the proximal tubules, and ~75% of the filtered phosphate is reabsorbed by apical Na^+-Pi cotransporters (see Fig. 18.5B). The remaining 25% of the Pi load is excreted; part of the Pi can be used to buffer H^+, forming titratable acids (see Chapter 21). In growing children and in persons with diets low in Pi, Na^+-Pi cotransporters are also present in the proximal straight tubules and DTs, facilitating reabsorption of up to 90% of the filtered Pi load.

Renal Pi reabsorption is primarily controlled by diet and PTH, both of which affect the number of Na^+-Pi cotransporters in the apical membranes:

- *Diet:* High dietary Pi causes a reduction in the number of Na^+-Pi cotransporters in the apical membrane, increasing Pi excretion. Conversely, with low dietary Pi, transporters are increased on the proximal tubule brush border, as well as in sites distal to the proximal tubule, allowing avid Pi uptake and reduced urinary Pi excretion.
- *PTH:* PTH is secreted from the parathyroid glands in response to low plasma Ca^{2+}, as well as indirectly, in response to high plasma Pi concentrations. PTH decreases apical Na^+-Pi cotransporters, reducing reabsorption and increasing urinary excretion of Pi.

Plasma calcium and phosphate regulation are intertwined because of the constant bone resorption and deposition. In response to low plasma Ca^{2+}, **vitamin D** increases intestinal calcium and phosphate absorption, whereas PTH induces bone resorption; both actions increase Ca^{2+} *and* phosphate in extracellular fluid (ECF). At the kidneys, PTH increases calcium reabsorption but compensates for the additional ECF phosphate by decreasing phosphate transporters, thereby increasing urinary phosphate excretion.

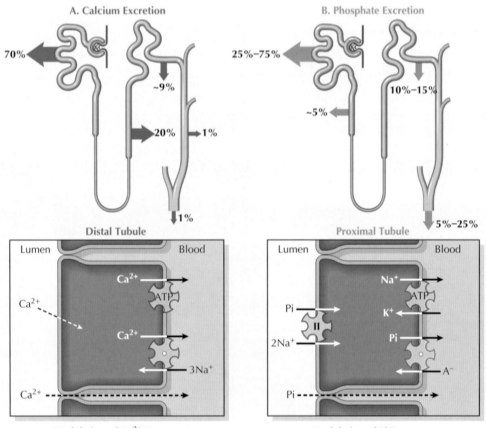

A. Calcium Excretion

70%

~9%

20% ┤1%

↓1%

Distal Tubule

Lumen   Blood

Ca²⁺ → | ATP
Ca²⁺
Ca²⁺ → | ⊙
        ← 3Na⁺
Ca²⁺ - - - - - - - - - - →

**Modulation of Ca²⁺ Transport
(Decreased Excretion)**

Factor	Nephron Site	Mechanism
↑ PTH	DCT	Activate Ca²⁺ channels
↓ ECF	Proximal tubule	Solvent drag
↑ Pi intake	DCT	↑ PTH secretion

B. Phosphate Excretion

25%–75%

10%–15%

~5%

5%–25%

Proximal Tubule

Lumen   Blood

Na⁺ → | ATP
Pi →
2Na⁺ → | II   K⁺ ←
                    Pi →
                    | ⊙
                    ← A⁻
Pi - - - - - - - - - - →

**Modulation of Pi Transport
(Increased Excretion)**

Factor	Nephron Site	Mechanism
↑ PTH	Proximal tubule	↓ Apical symporter
↓ ECF	Proximal tubule	↑ Solvent drag/symporter
↑ Pi intake	Proximal tubule	↓ Apical symporter

J. Perkins
MS, MFA, CMI

Figure 18.5 **Renal Calcium and Phosphate Handling** **A,** Calcium is reabsorbed along much of the nephron, and very little is excreted. Regulation of distal calcium reabsorption is by parathyroid hormone (PTH), which increases apical calcium channels (TRPV5), as well as basolateral Ca²⁺ ATPase pumps and Na⁺/Ca²⁺ exchangers. **B,** Under normal conditions, ~75% of the filtered load of phosphate is reabsorbed, with all of the reabsorption occurring in the proximal tubule via Na⁺-Pi cotransporters. This occurrence is highly dependent on the dietary intake of phosphate, as well as PTH levels. In response to PTH, proximal tubular reabsorption of phosphate is inhibited, and phosphate excretion increases. This effect also occurs with diets high in phosphate. Low-phosphate diets significantly increase Pi reabsorption, recruiting transporters in sites distal to the proximal convoluted tubule (in *green*), which can reduce phosphate excretion to 5% or 10%. *DCT,* distal convoluted tubule; *ECF,* extracellular fluid.

CLINICAL CORRELATE 18.1
Kidney Stones (Renal Calculi)

Kidney stones are solid aggregates of minerals that form in the kidney (nephrolithiasis). Stones can also form in the ureters (ureterolithiasis). The size of stones is variable, and many small stones pass through the ureters and urethra without causing any problems. However, if stones grow large enough (2 to 3 mm), they can block the ureter and cause intense pain and vomiting. The most common stones are calcium oxalate, and it is the presence of oxalate (not the calcium) that drives mineral precipitation. Treatment depends on size of the stone and duration of the blockage. Typically, unless a person has severe symptoms, small stones will be left to pass without intervention; however, long-term blockage of more than 30 days can result in renal failure, and in these cases, intervention with stent placement and laser or ultrasound may be performed.

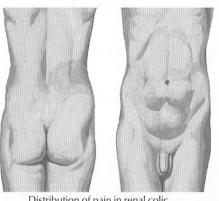

Distribution of pain in renal colic

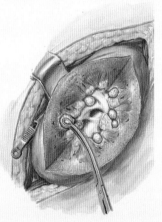

Kidney split and widely laid open for removal of multiple stones

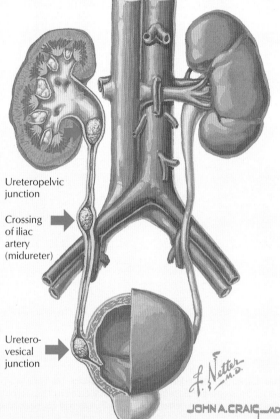

Ureteropelvic junction

Crossing of iliac artery (midureter)

Uretero-vesical junction

Common sites of obstruction

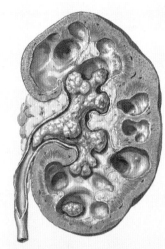

Staghorn calculus plus smaller stone

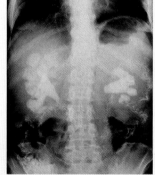

Bilateral staghorn calculi

Renal Calculi

CLINICAL CORRELATE 18.2
Hyponatremia

Hyponatremia is defined as the state of low plasma sodium (<135 mEq/L). Hyponatremia can be caused by several mechanisms that result in low sodium concentrations and reduced plasma osmolarity. During hyponatremia, fluid shifts *into* cells, reestablishing normal ECF osmolarity but causing cellular swelling. This action can have important effects, especially on brain tissues, which are confined to a bony space and are unable to tolerate swelling:

- Rapid shift of fluid into cells of the brain can result in acute cerebral swelling, which can lead to disoriented mental status, seizures, coma, and death. In these cases, therapy is aimed at reducing ECF volume, which will draw fluid out of cells. Water restriction and/or antidiuretic hormone (ADH) V2 antagonists (which increase urinary free water excretion) can help correct the disturbance.
- If the hyponatremia is established over time (for example, in persons with Addison's disease), brain tissues compensate for fluid shifts by decreasing intracellular content of osmolytes (organic solutes such as inositol and glutamine). This action reduces the osmotic force that would draw fluid into the cells and allows the cells to maintain normal volume. Because of this effect, treatment of hyponatremia should involve slow restoration of salt and fluid balance to normal levels. Otherwise, the brain cells will shrink, inducing an acute, potentially critical, intracellular imbalance. Gradual correction of this type of hyponatremia will allow the osmolytes to increase in brain cells.

Exercise-associated hyponatremia (EAH) can occur as a result of fluid and electrolyte losses through sweat during long-term exercise (such as marathons and triathlons). Although most people do not experience a serious drop in ECF Na^+ concentration, critical cases of EAH are most likely to occur as a result of a combination of the following factors:

- An initial imbalance of fluid and electrolyte losses due to overhydration during the exercise.
- Acute syndrome of inappropriate ADH secretion that can occur because of large fluid losses. Recall that both increased osmolarity and fluid losses can stimulate ADH but that the system is more sensitive to changes in ECF osmolarity than to changes in fluid volume. However, if the volume loss continues and becomes severe, when a dehydrated athlete drinks too much hypotonic fluid, the increase in ADH will cause excessive free water reabsorption by the CDs, rapidly decreasing the ECF sodium concentration. In this scenario, the need to compensate for the volume will override regulation of the sodium levels; ADH secretion will continue, and plasma Na^+ can fall to critically low levels (<125 mEq/L).

Early symptoms include bloating, nausea, vomiting, and headaches, which can progress to disorientation, seizures, and death if the person is not immediately treated. Hyponatremia can be prevented by restricting water intake (drinking only when thirsty). Although overhydration during the exercise is a direct cause of hyponatremia, risk factors for developing EAH include low body weight, female sex, and inexperience with marathons. ADH V2 receptor antagonists are used to treat severe hyponatremia.

Postsurgical acute hyponatremia frequently occurs in elderly patients. The stress of surgery can cause an acute syndrome of inappropriate ADH secretion, rapidly increasing free water reabsorption and reducing ECF Na^+ concentration. As stated earlier, treatment should begin immediately, with water restriction (to limit further fluid retention) and a V2 antagonist. Correcting the hyponatremia restores normal function.

Chapter 19

Urine Concentration and Dilution Mechanisms

THE LOOP OF HENLE AND COLLECTING DUCT CELLS

Extracellular osmolarity and volume homeostasis require that the kidneys be able to concentrate and dilute urine. The loops of Henle, collecting duct (CD) cells, and vasa recta capillaries are integral to these functions.

Differences in solute and water permeability along the loop of Henle allow for the concentration and dilution of the tubular fluid and result in establishment of a gradient in interstitial fluid osmolarity from the cortex through the medulla. This gradient facilitates solute-free water reabsorption in the thin descending limb of Henle and CDs, and its maintenance is facilitated by the slow blood flow through the vasa recta capillaries that surround the medullary tubules and CDs.

The descending and ascending limbs of the loops of Henle have the following specific permeability characteristics:

- *Thin descending limbs of Henle's loop are concentrating segments.* These segments are permeable to water but impermeable to reabsorption of solutes (urea can be secreted into the tubule, further concentrating the tubular fluid).
- *Thick ascending limbs of Henle's loop are diluting segments.* These segments are impermeable to water, but Na$^+$-K$^+$-2Cl$^-$ (NKCC-2) and Na$^+$-Cl$^-$ cotransporters reabsorb electrolytes, thus diluting the fluid before it enters the distal tubule.

With this mechanism in place, the tubular fluid entering the distal tubule has an osmolarity of ~100 milliosmoles per liter (mosm/L). In response to elevated extracellular fluid (ECF) osmolarity or reduced ECF volume, **antidiuretic hormone (ADH)** is secreted from the posterior pituitary gland into the blood and stimulates the insertion of **aquaporin (AQP)**-2 water channels into the apical membranes of the CD cells. This action allows solute-free water reabsorption and thus concentrated urine. However, this concentration of tubular fluid and urine can occur only if an interstitial osmotic gradient exists to draw water from the tubular lumen to the interstitial space.

URINE-CONCENTRATING MECHANISM

The Medullary Interstitium

The ability to reabsorb solute-free water in both the descending limb of Henle and the CDs is possible because of the osmolar concentration gradient within the medullary interstitial fluid and the presence of specific water channels (AQPs; see Chapter 2) in the apical membranes of both nephron segments. Water can move only when an osmotic gradient exists (Fig. 19.1). In this generalized interstitial gradient, the osmolarity is ~300 mosm/L at the corticomedullary border and rises to ~1200 mosm/L in the deepest part of the medullary interstitium (see the "Medullary Countercurrent Multiplier" section). With this gradient in place, if water channels are present in the apical membranes, water from the tubules readily diffuses into the interstitium (with its higher osmolar concentration) and then into the vasa recta network.

> AQPs found in the renal tubules allow for water reabsorption. **AQP-1** channels are always found in apical membranes of the proximal convoluted tubule, proximal straight tubule, and thin descending limb of Henle. In the solute-impermeable thin descending limb of Henle, these channels serve to concentrate the tubular fluid flowing through this segment. In the CDs, insertion of **AQP-2** channels into apical membranes is stimulated by ADH and facilitates solute-free water absorption and urine concentration. AQP-1 is stored in intracellular vesicles before insertion.

Medullary Countercurrent Multiplier

The "countercurrent multiplier" mechanism enables the establishment of an interstitial osmolar gradient from the cortex to the inner medulla and depends on the coordinated effects of the ascending and descending limbs of the loop of Henle and their selective permeability to solutes and water (Video 19.1). Figure 19.2 illustrates the repeated cycle in which solutes are transported out of the thick ascending limb of Henle (which raises the osmolar concentration of the interstitial fluid) and then water is reabsorbed from the thin descending limb (which further concentrates the tubular fluid

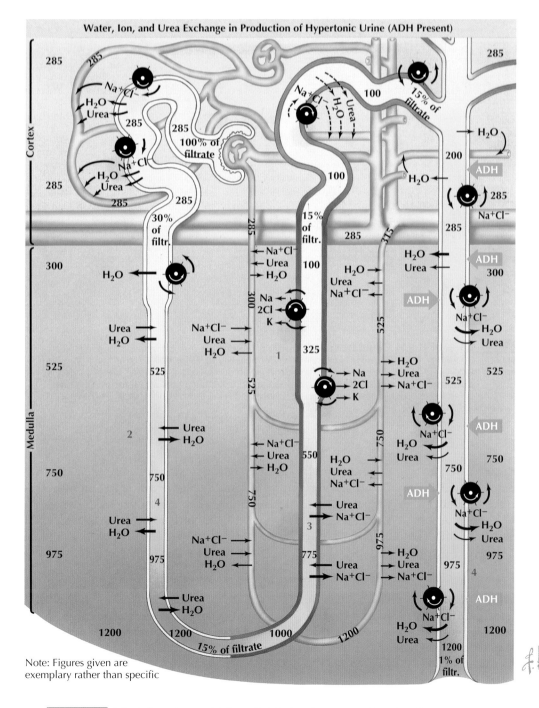

Water, Ion, and Urea Exchange in Production of Hypertonic Urine (ADH Present)

Note: Figures given are exemplary rather than specific

Figure 19.1 **Medullary Interstitial and Tubular Concentration Gradients** The concentration gradient is established by the following steps: *(1)* solutes, but not water, are transported out of the thick ascending limb of Henle via the NKCC-2 transporters (diluting limb); *(2)* free water is reabsorbed from the thin descending limb of Henle, which increases the tubular fluid osmolarity in the descending limb (concentrating limb); *(3)* as the more concentrated tubular fluid flows from the descending limb to the ascending limb, further transport of solutes into the interstitium occurs; and *(4)* urea recycling contributes to the gradient because it remains in the tubular fluid in the thin descending limb of Henle, contributing to the tubular fluid osmolarity while water is reabsorbed, and when antidiuretic hormone (ADH) is present, both water *and* urea reabsorption increase in the medullary collecting ducts and the urea is recycled to the inner medulla, contributing to the interstitial concentration gradient. *filtr.,* filtrate.

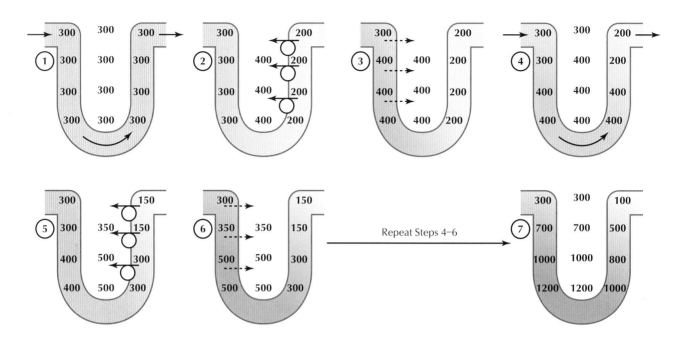

Figure 19.2 **Loop of Henle Countercurrent Multiplier System** A hyperosmotic renal medullary interstitium is created by transport of ions and selective permeability to water and electrolytes in specific segments of the loops of Henle. The steps (1-7) involved in the countercurrent multiplier are explained in the text, and values are in milliosmoles per liter. The hyperosmotic interstitium allows for the concentration of urine in the collecting ducts. (Modified from Hall JE: *Guyton and Hall Textbook of Medical Physiology,* ed 12, Philadelphia, 2011, Elsevier.)

entering the thick ascending limb). This cycle is repeated until the full interstitial gradient (from 300 mosm/L in the cortex to 1200 mosm/L in the deep medulla) is established.

Creation of the gradient begins with solute transport in the thick ascending loop of Henle and occurs through the following process:

■ The NKCC-2 *transporters in the thick ascending limb of Henle* transport solutes into the interstitium, increasing interstitial fluid osmolarity and decreasing tubular fluid osmolarity. This process can produce a gradient of 200 mosm/L between tubular fluid and interstitium (Fig. 19.2, step 2). Note that the thick ascending limb is impermeable to water.
■ *This increased interstitial osmolarity promotes free water reabsorption out of the descending limb* until osmolar equilibrium is reached between the tubular fluid of the descending limb and that area of the interstitium (Fig 19.2, step 3). Note that the descending limb is impermeable to solutes. The reabsorption of water continually increases osmolarity of the tubular fluid as it moves through the descending limb. The osmolarity of the interstitial fluid and the 200 mosm/L gradient between it and the fluid in the ascending tubule are maintained by the continued transport of solutes out of the ascending limb.
■ As *tubular fluid flows from the descending to the ascending limb, more concentrated tubular fluid now flows into the*

ascending limb as a result of these processes (Fig 19.2, step 4). With the higher concentration of solutes entering the ascending limb, more solutes can be transported into the interstitium, further increasing interstitial fluid osmolarity (Fig. 19.2, step 5). This increase in interstitial osmolarity facilitates more free water reabsorption from the descending limb (Fig. 19.2, step 6), further concentrating the tubular fluid entering the thick ascending limb. This cycle (movement of concentrated tubular fluid into the thick ascending limb, transport of solutes into the interstitium, and water movement out of the descending limb of Henle) is repeated until the full interstitial gradient is established. The osmolar concentration in the interstitium deep in the medulla is dependent on the length of the loops of Henle (the longer the loops, the higher the concentrating ability). In humans, the highest interstitial concentration attained (deep in the inner medulla) is ~1200 mosm/L, which allows the concentration of tubular fluid to reach ~1200 mosm/L at the bottom of the loops of Henle.

■ Finally, *urea recycling* contributes to developing and maintaining the interstitial osmolar gradient, because:
   ■ urea remains in the tubular fluid while water is reabsorbed from the descending limb, contributing to the tubular fluid osmolarity (Fig. 19.1, red number 3), and
   ■ ADH increases both water *and* urea reabsorption in the medullary (but not cortical) CDs; the urea is recycled into the inner medulla, contributing to the interstitial concentration gradient (Fig. 19.1, red number 4).

Ⓠ The ability to concentrate urine differs between species. Animals that live in desert environments (such as desert rodents and camels) have exceptionally long loops of Henle and thus have the ability to concentrate their urine to more than 2000 mosm/L, allowing tremendous fluid retention. The medullary interstitium of these animals appears to have additional osmotic agents (~20% extra "osmolytes" such as sorbitol and myo-inositol) to help accomplish this action.

Concentration of the Urine

As discussed in Chapter 1, to maintain plasma (and thus cellular) osmolarity, fluid volume must be controlled. Either a small increase in plasma osmolarity (~1%) or a significant decrease (a greater than 10% loss) in plasma volume (from, for example, hemorrhage) will elicit release of ADH from the posterior pituitary gland. This hormone binds to V2 receptors on principal cells of the renal CDs (Fig. 19.3). In the principal cells of the CDs, ADH increases apical AQP-2 water channels, effectively concentrating solutes in the tubular fluid, which at this point is considered urine.

The urine-concentrating mechanism depends on the plasma level of ADH and the osmolar concentration of the interstitial fluid surrounding the CDs. Plasma ADH is tightly regulated, and water channels are continually being inserted and removed from the apical membranes of the CD cells to maintain ECF balance through the regulated retention and excretion of water in the urine. ADH-dependent urea recycling plays an important role in the ability to concentrate urine, because the additional urea added to the interstitium when ADH is elevated helps maintain the interstitial gradient for water reabsorption. Therefore, dehydration → higher plasma osmolarity → higher ADH levels → increased AQP and water reabsorption → more concentrated urine. Under extreme conditions, the urine can be concentrated maximally to 1200 mosm/L, with minimal urinary water loss.

DILUTION OF URINE

An excess of ECF increases renal blood flow and glomerular filtration rate (GFR) and decreases plasma osmolarity (inhibiting pituitary ADH release). The elevation in renal blood flow increases vasa recta capillary blood flow, which washes solutes out of the interstitium. This disruption of the gradient is compounded by the increase in GFR, which increases tubular flow rate. This action reduces solute reabsorption (especially in the thick ascending limb), limiting

concentration of the interstitial fluid (Figure 19.4). The decrease in ADH results in loss of apical AQP-2 in the CDs and thus less medullary water and urea reabsorption. The reduced urea reabsorption further limits the interstitial osmolarity. The result is the production of an increased amount of urine that is hypotonic (diuresis). When the excess hypotonic fluid is excreted, ECF volume and osmolarity will return to normal, and the interstitial concentration gradient will be reestablished over several hours.

FREE WATER CLEARANCE

Increased urine output in response to expansion of plasma volume entails both sodium excretion (natriuresis) and water excretion (diuresis). The ability to excrete hypotonic urine is as important to fluid homeostasis as the ability to excrete concentrated urine, and impairment of this mechanism can be life-threatening (see Chapter 18, Clinical Correlate on hyponatremia). The concept of **free water clearance** is useful in quantifying water excretion during diuresis; it is defined as water excretion in excess of the water required for iso-osmotic excretion of the solutes present in the urine. Free water clearance is determined by subtracting the osmolar clearance from the urine flow rate,

$$C_{H_2O} = \dot{V} - (Uosm/Posm) \times \dot{V} \qquad \textbf{Eq. 19.1}$$

Thus, in dilute urine (e.g., urine osmolarity/plasma osmolarity [Uosm/Posm] is less than 1) the C_{H_2O} is a positive number, implying that water was excreted. In contrast, if the urine is concentrated (e.g., Uosm/Posm is greater than 1), the C_{H_2O} is negative, implying that water was retained. If urine osmolarity equals plasma osmolarity, the C_{H_2O} equals zero.

Ⓠ Micturition is the process of emptying the bladder (urination). Micturition is under voluntary control, because the external sphincter of the bladder is skeletal muscle. However, the micturition reflex system is under both sympathetic and parasympathetic control. While the bladder is filling, the sympathetic nerves relax the smooth muscle of the bladder wall, accommodating the urine, and contract the internal urethral sphincter smooth muscle. When the bladder becomes "full," mechanoreceptors signal a spinal reflex arc that stimulates parasympathetic contraction of the bladder (detrusor muscle) and relaxation of the internal sphincter. The external urethral sphincter is skeletal muscle and is voluntarily relaxed, allowing urination.

Mechanism of Antidiuretic Hormone in Regulating Urine Volume and Concentration

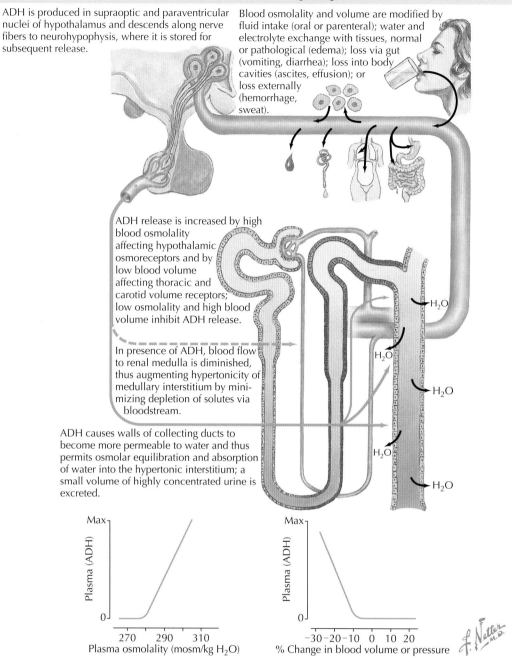

ADH is produced in supraoptic and paraventricular nuclei of hypothalamus and descends along nerve fibers to neurohypophysis, where it is stored for subsequent release.

Blood osmolality and volume are modified by fluid intake (oral or parenteral); water and electrolyte exchange with tissues, normal or pathological (edema); loss via gut (vomiting, diarrhea); loss into body cavities (ascites, effusion); or loss externally (hemorrhage, sweat).

ADH release is increased by high blood osmolality affecting hypothalamic osmoreceptors and by low blood volume affecting thoracic and carotid volume receptors; low osmolality and high blood volume inhibit ADH release.

In presence of ADH, blood flow to renal medulla is diminished, thus augmenting hypertonicity of medullary interstitium by mini-mizing depletion of solutes via bloodstream.

ADH causes walls of collecting ducts to become more permeable to water and thus permits osmolar equilibration and absorption of water into the hypertonic interstitium; a small volume of highly concentrated urine is excreted.

H_2O

H_2O

H_2O

H_2O

H_2O

Max

Plasma (ADH)

0

270   290   310
Plasma osmolality (mosm/kg H_2O)

Max

Plasma (ADH)

0

−30 −20 −10   0   10 20
% Change in blood volume or pressure

Figure 19.3 **The Renal Response to Antidiuretic Hormone Secretion**   In response to dehydra-tion, ADH is secreted from the posterior pituitary gland into the circulation. It acts on the kidneys to increase water channels in the collecting ducts, allowing solute-free water absorption, and also increases urea reab-sorption in the medullary collecting ducts. The additional urea is added to the inner medullary interstitium and contributes to the high interstitial osmolar concentration.

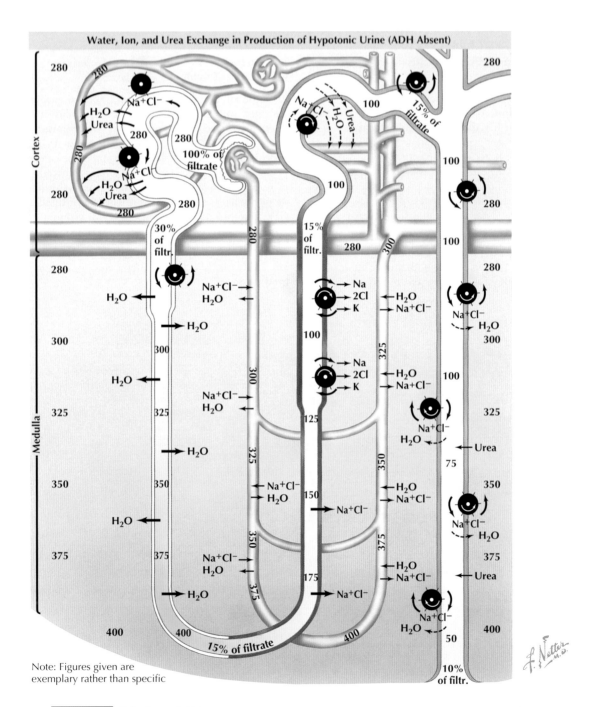

Note: Figures given are exemplary rather than specific

Figure 19.4   **Dilution of Urine**   Excess extracellular fluid will decrease antidiuretic hormone (ADH) secretion and thus reduce water channels in the collecting ducts. Volume expansion increases blood flow through the vasa recta, which contributes to disruption of the interstitial concentration gradient. The disruption of the gradient and the lack of water channels produce diuresis. *filtr.*, filtrate.

CLINICAL CORRELATE 19.1
Chronic Pyelonephritis

Pyelonephritis is inflammation of the renal pelvis caused by bacterial infection. Although acute kidney infections are usually caused by urinary tract infections, they can recur, and with each occurrence they can further damage the kidney. Urinary tract infections typically arise from contamination from bowel microorganisms, although with recurring infections that reach the kidney, potential underlying causes such as kidney stones or other anatomic abnormalities should be considered. Increased risk of pyelonephritis is associated with diabetes, pregnancy, prostate enlargement, compromised immunity, and sexual behavior and spermicide use.

The complications arising from **chronic pyelonephritis** relate to the area that is infected. The smooth muscle lining of the renal pelvis exhibits peristaltic activity that helps direct the newly collected urine toward the ureters for transit to the urinary bladder. The infection causes abscesses and necrosis of the pelvic tissue, which can lead to fibrosis and scarring.

As more of the medulla (tubules and parenchyma) becomes damaged with repeated infections, the ability to maintain the interstitial concentration gradient is compromised. Because high interstitial osmolarity provides the gradient that allows ADH-dependent free water reabsorption and urine concentration, loss of the deep gradient restricts the ability to excrete concentrated urine and thus can cause polyuria (excessive urine production), despite dietary water restriction. The loss of tubules also reduces GFR, and thus renal function as a whole is diminished.

Pyelonephritis is treated with antibiotics over a period of several weeks. Increased fluid intake is recommended to flush any lower urinary tract bacteria out though increased urine production.

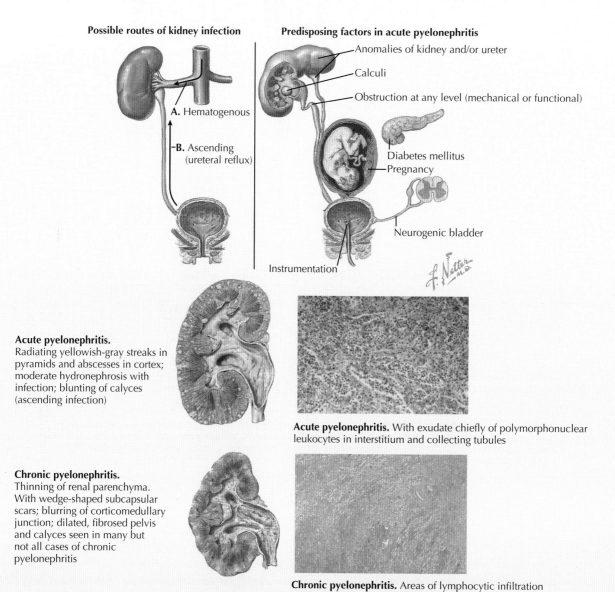

Possible routes of kidney infection

A. Hematogenous

B. Ascending (ureteral reflux)

Predisposing factors in acute pyelonephritis

Anomalies of kidney and/or ureter

Calculi

Obstruction at any level (mechanical or functional)

Diabetes mellitus

Pregnancy

Neurogenic bladder

Instrumentation

Acute pyelonephritis.
Radiating yellowish-gray streaks in pyramids and abscesses in cortex; moderate hydronephrosis with infection; blunting of calyces (ascending infection)

Acute pyelonephritis. With exudate chiefly of polymorphonuclear leukocytes in interstitium and collecting tubules

Chronic pyelonephritis.
Thinning of renal parenchyma. With wedge-shaped subcapsular scars; blurring of corticomedullary junction; dilated, fibrosed pelvis and calyces seen in many but not all cases of chronic pyelonephritis

Chronic pyelonephritis. Areas of lymphocytic infiltration alternating with areas of relatively normal parenchyma

Chronic and Acute Pyelonephritis

Chapter 20

Regulation of Extracellular Fluid Volume and Osmolarity

INTRARENAL REGULATION OF SODIUM AND FLUID REABSORPTION

Because of its importance in extracellular fluid (ECF) homeostasis, renal sodium handling is closely regulated. A number of intrarenal factors can alter sodium (and, thus, fluid) reabsorption in response to changes in ECF:

- **Glomerular filtration rate (GFR):** Increases in GFR will increase the filtered load (FL) of sodium, and because the percentage of sodium reabsorbed in the proximal tubule (65% to 70%) does not change, the absolute amount of sodium entering the loop of Henle increases. Assuming that fractional reabsorption of sodium in the distal segments is unchanged, more sodium will be excreted. Conversely, if GFR decreases, the absolute amount of sodium entering the loop of Henle will decrease, as will sodium excretion (if fractional reabsorption in later segments is unchanged).

- **Tubular fluid flow rate:** Mechanisms within the kidney serve to keep tubular fluid flow rate within a normal range. This is necessary because low flow rates will result in decreased delivery of sodium to the loop of Henle, a lower osmolar gradient in the interstitium, and a reduced ability to reabsorb water in the collecting ducts (CDs). On the other hand, rapid tubular flow rates will disrupt the osmolar gradient in the medullary interstitium and lead to enhanced sodium and water excretion. For these reasons, **tubuloglomerular feedback** mechanisms are important in controlling the flow rate. When tubular flow is rapid, the macula densa cells of the juxtaglomerular apparatus secrete ATP into the interstitial fluid adjacent to the afferent arteriole. ATP is converted to adenosine, which stimulates afferent arteriolar constriction, reducing glomerular capillary hydrostatic pressure and thus decreasing the pressure for filtration. Thus GFR and tubular flow are reduced.

- **Baroreceptors** located in the walls of the afferent arterioles respond to changes in blood pressure. When blood pressure decreases, the reduced stretch of the baroreceptors stimulates juxtaglomerular cells to release renin into the afferent arteriole (this mechanism is intrarenal and does not involve the central nervous system vasomotor center). The effects of renin are discussed in the next section.

- **Medullary blood flow:** If blood flow in the vasa recta increases, the medullary interstitial concentration gradient will be reduced, resulting in decreased solute reabsorption in the thick ascending limb of Henle and decreased water reabsorption in the thin descending limb of Henle. The reduction in sodium reabsorption in the loop of Henle increases the delivery of sodium and water to the distal tubules (DTs) and CDs. Because the interstitial concentration gradient is reduced, the ability to concentrate urine in the CDs is decreased, resulting in natriuresis/diuresis.

NEUROHUMORAL CONTROL OF RENAL SODIUM REABSORPTION

In addition to the intrarenal controls noted earlier, several important neural and humoral mechanisms are involved in the regulation of renal sodium reabsorption:

- **Sympathetic nerves** increase sodium reabsorption through several mechanisms. Sympathetic nerves innervate afferent and efferent arterioles (via α-adrenergic receptors). During sympathetic stimulation, the arterioles constrict, decreasing GFR, and thus decreasing sodium excretion; this effect will preserve ECF volume. Direct sympathetic innervation of the proximal tubule also occurs, which increases basolateral Na^+/K^+ ATPase activity, thus facilitating sodium reabsorption.

- The **renin-angiotensin-aldosterone system** has an important role in regulation of the renal retention of sodium and water and thus in the regulation of ECF volume and solute composition. In response to low tubular sodium concentration and low tubular fluid flow rate, juxtaglomerular cells produce the proteolytic enzyme **renin** and secrete it into the afferent arterioles (Fig. 20.1B). As illustrated in Figure 20.1A, renin hydrolyzes angiotensinogen (a plasma protein secreted by the liver) to angiotensin I, which is converted to **angiotensin**

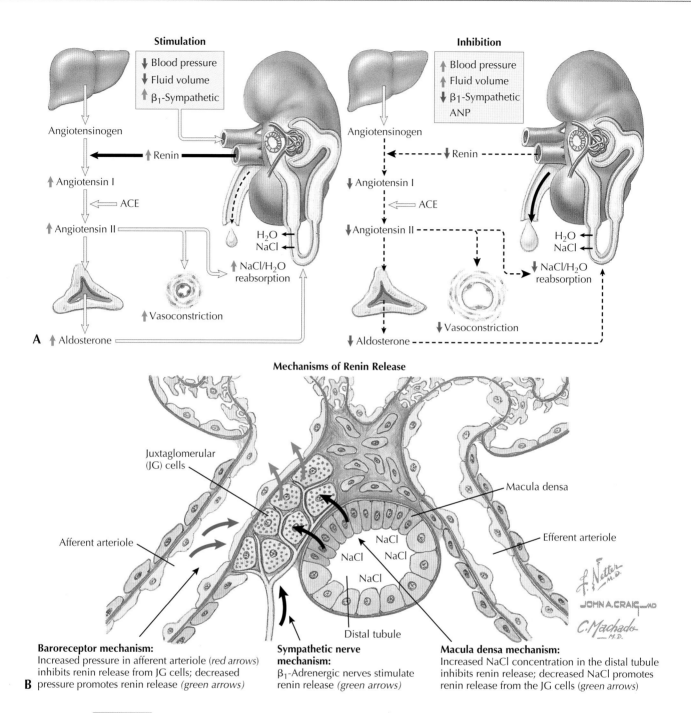

Stimulation

- ↓ Blood pressure
- ↓ Fluid volume
- ↑ β₁-Sympathetic

Angiotensinogen

↑ Angiotensin I

⇐ ACE

↑ Angiotensin II

↑ Aldosterone

↑ Vasoconstriction

↑ NaCl/H₂O reabsorption

H₂O NaCl

A

Inhibition

- ↑ Blood pressure
- ↑ Fluid volume
- ↓ β₁-Sympathetic
- ANP

Angiotensinogen

↓ Angiotensin I

⇐ ACE

↓ Angiotensin II

↓ Aldosterone

↓ Vasoconstriction

↓ NaCl/H₂O reabsorption

H₂O NaCl

Mechanisms of Renin Release

Juxtaglomerular (JG) cells

Afferent arteriole

Macula densa

Efferent arteriole

NaCl NaCl NaCl NaCl

Distal tubule

Baroreceptor mechanism:
Increased pressure in afferent arteriole (*red arrows*) inhibits renin release from JG cells; decreased pressure promotes renin release (*green arrows*)

B

Sympathetic nerve mechanism:
β₁-Adrenergic nerves stimulate renin release (*green arrows*)

Macula densa mechanism:
Increased NaCl concentration in the distal tubule inhibits renin release; decreased NaCl promotes renin release from the JG cells (*green arrows*)

Figure 20.1 **Mechanism of Renin Secretion and Factors Regulating the Renin-Angiotensin-Aldosterone System** Renin is secreted from the juxtaglomerular cells in response to reduced sodium concentration and flow in the distal tubule **(B)**. The cascade of events initiated to promote renin secretion and sodium and water reabsorption is illustrated in part **A** *(left)*. Inhibition of the **renin-angiotensin-aldosterone system** also is depicted in part **A** *(right)*. ACE, angiotensin-converting enzyme.

II by angiotensin-converting enzyme in the lung (and other tissues); angiotensin II stimulates release of the mineralocorticoid **aldosterone** from the adrenal cortex. In the kidney, angiotensin II has dual effects: it directly stimulates apical Na⁺/H⁺ antiporters in the proximal tubule (increasing sodium and water reabsorption) and has vasoconstrictor effects on afferent and efferent arterioles, which result in a lower GFR and, thus, sodium and water retention.

- **Aldosterone** acts at the principal cells of the late DTs and CDs, stimulating Na⁺ and water reabsorption (and K⁺ secretion). In the principal cells, aldosterone increases the synthesis of specific transport proteins: Na⁺/K⁺ ATPases are inserted in the basolateral membranes, and epithelial sodium channels (ENaC) are inserted in the apical membranes, resulting in an increase in sodium and water reabsorption from the tubular fluid. In addition, aldosterone also increases synthesis and insertion of apical potassium channels (big K [BK]) in the principal cells.

- **Atrial natriuretic peptide (ANP)** is primarily produced by myocytes in the right atrium of the heart and is released into the blood in response to atrial stretch (as occurs with increased blood volume). ANP opposes the actions of angiotensin II by increasing GFR (through dilation of the afferent arteriole and constriction of the efferent arteriole) and reducing Na^+ reabsorption in the DT (by inhibiting Na^+-Cl^+ cotransport) and CD (by inhibiting ENaC). Thus ANP causes natriuresis and diuresis, reducing ECF volume.
- **Urodilatin** is an intrarenal natriuretic peptide (closely related to ANP) that is produced in medullary DT and

CD cells in response to an elevation in blood volume or pressure. It has effects similar to those of ANP but is not found outside the kidney.

RENAL RESPONSE TO CHANGES IN PLASMA VOLUME AND OSMOLARITY

As discussed earlier, control of ECF is a continual process, with changes in plasma osmolarity and volume signaling multiple neural and humoral systems to regulate the renal concentration and dilution of the urine. The integration of these systems in the response to ECF volume contraction and expansion is illustrated in Figures 20.2 and 20.3. When plasma volume is

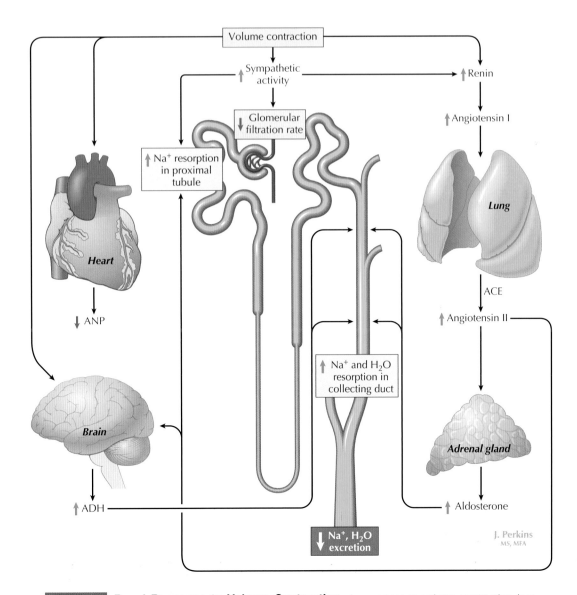

Figure 20.2 **Renal Response to Volume Contraction** In response to volume contraction (e.g., dehydration), the renin-angiotensin-aldosterone system is activated, stimulating renal sodium and fluid retention; antidiuretic hormone (ADH) secretion from the posterior pituitary is stimulated to increase water reabsorption in the renal collecting ducts, and the sympathetic nerves are stimulated to decrease glomerular filtration rate and increase renal sodium reabsorption. *ACE,* angiotensin-converting enzyme; *ANP,* atrial natriuretic peptide.

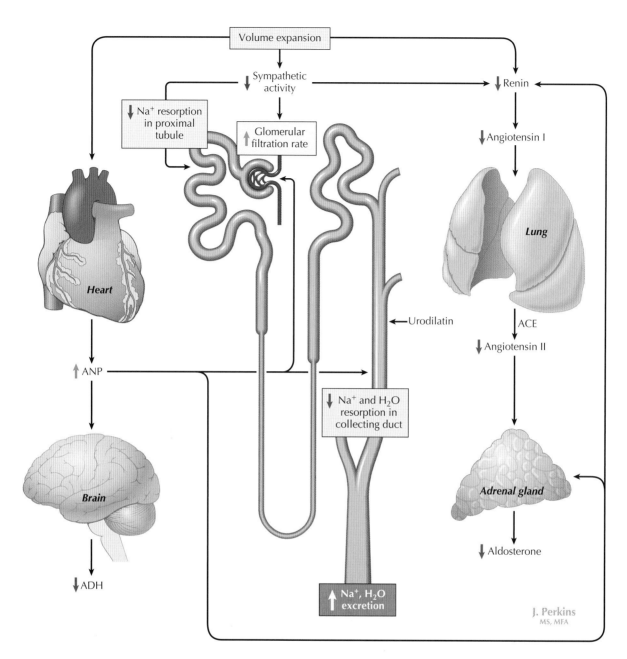

Figure 20.3 **Renal Response to Volume Expansion** In response to volume expansion, the sodium- and fluid-retaining mechanisms are decreased (i.e., the **renin-angiotensin-aldosterone system** and antidiuretic hormone [ADH]), and the increased stretch on the cardiac right atrium releases atrial natriuretic peptide (ANP), which acts at the kidneys to decrease sodium and water retention, resulting in diuresis and natriuresis and elimination of the excess fluid. *ACE,* angiotensin-converting enzyme.

contracted (see Fig. 20.2), fluid and sodium conservation systems are activated. During volume contraction, the kidneys respond to the following mechanisms:

- An increase in sympathetic nervous system activity, which increases renal vascular resistance and decreases GFR; proximal tubular sodium (and water) reabsorption increases.
- Activation of the **renin-angiotensin-aldosterone system**, which increases angiotensin II and aldosterone, enhancing sodium (and water) reabsorption in the proximal tubules (via angiotensin II) and DTs and CDs (via aldosterone).
- An increase in antidiuretic hormone (ADH), which inserts aquaporin-2 water channels in the apical membranes of the principal cells of the CDs, enhancing solute-free water absorption.

These systems limit further volume contraction by decreasing the loss of fluid in urine. When plasma volume is expanded, these mechanisms are reversed, allowing elimination of fluid and reduction of plasma volume and ECF (Fig. 20.3). As noted in the previous section, the increase in **ANP** from the right cardiac atrium has a key role in producing natriuresis and diuresis by:

- Inhibiting aldosterone secretion.
- Decreasing ADH.

- Increasing GFR by relaxing mesangial cells (and also through renovascular effects).
- Decreasing sodium (and water) reabsorption in the CDs by reducing aldosterone-sensitive Na^+ channels (ENaC).

Because of the potent vasoconstrictor and sodium-retaining effects of angiotensin II on the kidney, inhibition of angiotensin II is a major therapeutic target for treatment of hypertension. **Angiotensin-converting enzyme inhibitors** (e.g., captopril and enalapril) prevent conversion of angiotensin I to angiotensin II, whereas **angiotensin receptor blockers** (e.g., losartan and candesartan) block angiotensin II AT_1 receptors. Both pharmacologic interventions increase urinary sodium and water excretion, facilitating blood pressure control.

To maintain homeostasis, almost all of the filtered sodium must be reabsorbed, because loss of even a small percentage of the filtered sodium can result in severe sodium deficiency. Although the fine-tuning by aldosterone in the late DTs and CDs increases sodium reabsorption only about 2% to 3%, in persons with aldosterone insufficiency (Addison's disease), these losses can lead to severe ECF sodium depletion, ECF volume contraction, and circulatory collapse.

CLINICAL CORRELATE 20.1
Diabetes Insipidus

In response to hypovolemia, ADH is secreted from the posterior pituitary gland and stimulates solute-free water reabsorption in the CDs, allowing the kidneys to concentrate urine and retain water. Insufficiency of ADH secretion results in **diabetes insipidus (DI)**, a disease in which large volumes of hypotonic urine are excreted. **Central DI** is usually caused by trauma, disease, or surgery affecting the posterior pituitary gland or hypothalamus. In rare cases, DI may be masked by concomitant ACTH deficiency; administration of corticosteroids can unmask DI, as seen below. **Nephrogenic DI** is rare and involves a reduction in ADH V2 receptors or reduction in the aquaporin-2 water channels in the CDs of the kidney, reducing sensitivity of the cells to ADH.

Fluid intake must increase to compensate for the urinary losses, which can range from 3 to 18 liters per day. Mortality is rare, although children and elderly people are at greater risk of death resulting from severe dehydration, cardiovascular collapse, and hypernatremia. ADH analogs such as desmopressin (DDAVP) are used to treat central DI. The analogs act like endogenous ADH and increase water channels in the CDs. Nephrogenic DI can respond to indomethacin as well as to dihydrochlorothiazide, which is a diuretic that has the paradoxical effect of increasing water reabsorption in persons with DI.

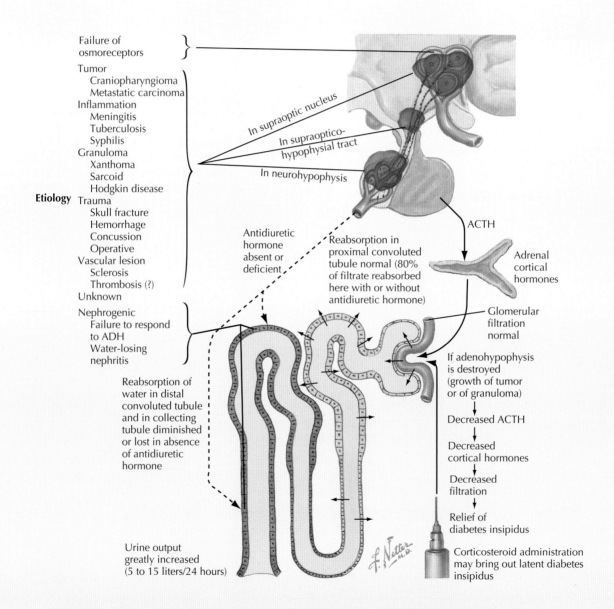

Central Diabetes Insipidus The causes of central diabetes insipidus, with the effects on the kidneys, are depicted. *ACTH*, adrenocorticotropic hormone; *ADH*, antidiuretic hormone.

Chapter 21

Regulation of Acid-Base Balance by the Kidneys

CONTROL OF EXTRACELLULAR FLUID PH

Why is systemic acid-base equilibrium important, and what is the role of the kidneys in maintaining this equilibrium? As noted in Chapter 16, the blood (and extracellular fluid [ECF]) pH must be maintained within a narrow range (7.35 to 7.45) to allow for normal cellular functions. This physiological pH range corresponds to a narrow range in H$^+$ concentration (45 to 35 nanomoles per liter [nmol/L]) and signals the necessity for tight control of pH in the ECF.

Because acid balance is critical for controlling ECF pH, the daily entry of H$^+$ into the ECF must equal the losses. A net gain of H$^+$ can occur as a result of ingestion of acid contained in foods (e.g., acidic drinks and proteins), cell and protein metabolism, hypoventilation, and diarrhea (loss of HCO$_3^-$ results in a gain in H$^+$, as described later). Net losses of acid can result from hyperventilation, vomiting, and, of course, urinary acid excretion. Under normal conditions, the daily gain of acid from normal metabolism (such as sulfuric acid, phosphoric acid, and keto acids) and diet (proteins) will be equaled by acid excreted in the urine. CO$_2$ is a volatile acid and can be excreted by the lungs, but when it is in the blood it is dissolved and contributes to the overall acid pool. A general scheme is illustrated in Figure 21.1.

In general, we ingest and produce ~40 to 80 *millimoles* of acid each day. This amount is huge compared with the ~40 *nanomolar* level (at pH 7.4) that is maintained in the ECF. This excess acid must be:

- **buffered in the ECF,** to prevent a fall in pH to below physiological levels (e.g., below 7.35), and
- **secreted** into the renal tubules, buffered, and **excreted** in the urine.

Buffering of Acid

The body handles excess acid using intracellular and extracellular buffers in a continuous regulatory "dance," escorting acid in and out of cells and through the blood to the kidneys for excretion.

Extracellular Buffering

Bicarbonate (HCO$_3^-$) is the major ECF buffer; it is available for consuming free H$^+$ through the following reaction:

$$HCO_3^- + H^+ \overset{\leftharpoonup}{\rightarrow} H_2CO_3 \overset{CA}{\rightleftharpoons} CO_2 + H_2O \qquad \textbf{Eq. 21.1}$$

The carbonic acid can be converted to CO$_2$ and H$_2$O in the presence of carbonic anhydrase (CA). This action occurs in the ECF and tissues, and the CO$_2$ and H$_2$O produced can diffuse into and out of tissues as part of the process of bicarbonate reabsorption and H$^+$ excretion in the renal tubules. Dissolved CO$_2$ in the blood can be blown off in the lungs. CO$_2$ produced by the bicarbonate buffer system does not contribute to the *net gain* in acid (because H$^+$ is consumed in this process). However, CO$_2$ does contribute to the net acid gain during **hypoventilation.**

As described in Chapter 16, the **Henderson-Hasselbalch** equation describes the relation between acid-base status and pH as:

$$pH = 6.1 + \log[\text{base/acid}] \qquad \textbf{Eq. 21.2}$$

where the base is plasma bicarbonate (normally ~24 millimoles per liter [mmol/L] in ECF) and acid is the P$_{CO_2}$ × 0.03 (solubility constant; normally 40 mm Hg × 0.03 mmol/L/mm Hg = 1.2 mmol/L in ECF). Thus, under normal conditions,

$$pH = 6.1 + \log[24/1.2]$$
$$pH = 7.4 \qquad \textbf{Eq. 21.3}$$

The kidneys control the amount of base (free bicarbonate) present in the ECF. This task is accomplished by generating new bicarbonate or by excreting excess bicarbonate (e.g., in alkalosis). Although the kidneys also control the amount of acid present in the ECF, if the ECF becomes acidotic or alkalotic, respiration can also regulate ECF acid status. ECF phosphates and proteins also contribute to the buffering, but only

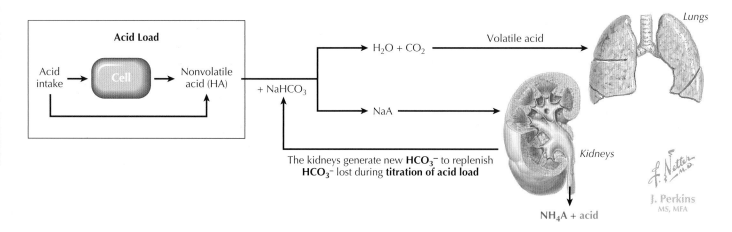

Figure 21.1 **General Scheme for Eliminating Excess Acid** New acid is added daily from diet and metabolism and must be buffered (primarily by extracellular fluid bicarbonate) and then excreted in the urine as titratable acids or ammonium.

to a very small extent, allowing the calculation of ECF pH based on bicarbonate concentration and P_{CO_2}.

What makes a good buffer? Good buffers have a pK (the negative log of the ionization constant of an acid at which there are equal concentrations of the acidic and basic forms) that is close to the physiological pH of 7.4. In the ECF, the best buffer would be phosphate (HPO_4^{2-}), which has a pK of 6.8 (which, at a pH of 7.4, results in a base/acid ratio of 4:1 [HPO_4^{2-}: $H_2PO_4^-$]). However, there is relatively little phosphate in the ECF (~1 mM), so it is not an effective ECF buffer. Instead, **bicarbonate** (pK = 6.1 with a base to acid ratio of 20:1 [HCO_3^-: H_2CO_3] at pH of 7.4) is the main ECF buffer because of its high ECF concentration (~24 mEq/L). The large amount of free bicarbonate allows ready buffering of additional acid load.

Intracellular Buffering

Although phosphates offer minor buffering in the ECF, they provide major buffering capacity *within* the cells because of their high intracellular concentration. Proteins also contribute to the intracellular buffering. Movement of H^+ into and out of cells occurs through cation exchange (H^+/K^+ and Na^+/H^+ antiporters). The intra- and extracellular buffering process minimizes the effects of generated and ingested acid on pH of the ECF and allows shuttling of acid to the kidneys, where it can be excreted.

HCO_3^- AND H^+ HANDLING THROUGH THE RENAL TUBULE

HCO_3^-

As previously described, the process of bicarbonate reabsorption occurs in the proximal tubule, thick ascending limb of Henle (TALH), and collecting duct (CD) and is dependent on the secretion of H^+ into the tubular lumen (see Fig. 18.3). The

bicarbonate reformed within the cells is transported into the ECF via basolateral HCO_3^-/Cl^- exchangers, and this mechanism effectively reclaims 100% of the filtered bicarbonate back into the ECF; under normal conditions there is no urinary excretion of bicarbonate.

The exception to this situation occurs in alkalosis, when the acid-base balance depends on the removal of HCO_3^- from the ECF. This task is accomplished in the CDs, where the **β-intercalated cells** have the ability to secrete HCO_3^- into the tubule for excretion via HCO_3^-/Cl^- exchangers. This transporter is active only during alkalosis and results in HCO_3^- loss and H^+ accumulation, thus reducing ECF pH.

H^+

H^+ is secreted into the proximal tubule and TALH in exchange for Na^+, which facilitates the reabsorption of bicarbonate in those segments (see Chapter 18). It is not until the distal segments of the nephron that H^+ is secreted in *excess* of filtered bicarbonate, with significant secretion occurring from the **α-intercalated cells** of the collecting ducts (see Fig. 18.3, *upper right box*). These cells have apical H^+ ATPase and H^+/K^+ ATPase pumps and actively secrete H^+ into the tubular fluid. When H^+ is secreted by these cells, bicarbonate is reabsorbed via basolateral HCO_3^-/Cl^- exchangers. Acid secreted into the CDs must be buffered by ammonia or phosphate (to minimize acidification of the urine and allow continued secretion of H^+). Factors that can regulate H^+ secretion in the nephron are provided in Table 21.1.

RENAL MECHANISMS CONTRIBUTING TO NET ACID EXCRETION

The ingested/generated acid load is always buffered in the plasma and is then dissociated from the buffers in the kidneys in the process of bicarbonate metabolism. Free H^+ is secreted into the lumen of the proximal tubules, TALH loop, and CDs.

Table 21.1	Factors Influencing H⁺ Secretion by the Nephron
Factor	**Principal Site of Action**
INCREASED H⁺ SECRETION—PRIMARY	
↓HCO₃⁻ concentration (↓pH)	Entire nephron
↑Arterial PCO_2	Entire nephron
INCREASED H⁺ SECRETION—SECONDARY	
↑Filtered load of HCO₃⁻	Proximal tubule
↓Extracellular fluid volume	Proximal tubule
↑Angiotensin II	Proximal tubule
↑Aldosterone	Collecting duct
Hypokalemia	Proximal tubule
DECREASED H⁺ SECRETION—PRIMARY	
↑HCO₃⁻ concentration (↑pH)	Entire nephron
↓Arterial PCO_2	Entire nephron
DECREASED H⁺ SECRETION—SECONDARY	
↓Filtered load of HCO₃⁻	Proximal tubule
↑Extracellular fluid volume	Proximal tubule
↓Aldosterone	Collecting duct
Hyperkalemia	Proximal tubule

From Hansen J: *Netter's Atlas of Human Physiology,* Philadelphia, 2002, Elsevier.

When secreted, it binds to **ammonia** to become **ammonium;** it can also be incorporated into phosphates to become a **titratable acid (TA)**. This production of ammonium and TA buffers the H⁺ in the CDs, controlling urine pH.

Production of Titratable Acids

The primary form of TA is phosphoric acid ($H_2PO_4^-$). Phosphate is a strong buffer (pK 6.8) but is not readily available in the ECF because of its low concentration (~1 mM). However, in the tubular fluid, the amount of filtered phosphate (FLPi) is significant (FLPi = 1 mmol/L × ~140 L/day [glomerular filtration rate] = ~140 mmol Pi/day), and part of this FLPi can be used for buffering and excreting H⁺ (bicarbonate cannot be used in the tubule because it is completely reabsorbed). The amount of phosphate available to form TA depends on (1) the amount of basic phosphate (HPO_4^{2-}) available to bind with H⁺ and (2) the renal handling of phosphate.

- **Phosphate buffering:** According to the Henderson-Hasselbalch equation, at a blood pH of 7.4, there will be four times the amount of base (HPO_4^{2-}) to acid ($H_2PO_4^-$), and this base is available for buffering excess H⁺. Thus the ~140 mmol/day of FLPi includes about 112 mmol/day of HPO_4^{2-} that could be used as buffer; however, not all of the HPO_4^{2-} is available because of the tubular handling of phosphate.

- **Renal handling of phosphate:** In an average healthy adult, ~75% of the FLPi is reabsorbed and therefore unavailable for generating TA. Thus only 25% of the filtered HPO_4^{2-} can be used, or ~28 mmol/day (112 mmol/day × 0.25 = 28 mmol/day).

TAs can form in the proximal tubule, but when the blood becomes acidic, additional H⁺ will be actively secreted by the α-intercalated cells and can bind to the HPO_4^{2-}, creating more TA (**$H_2PO_4^-$**), which is then excreted. TAs are a main source of acid excretion, and although TA production can increase (to buffer an additional acid load), their maximal rate of excretion is fixed because TA formation depends on the amount of phosphate available; even at maximal rate of excretion, there is not enough HPO_4^{2-} to eliminate the daily acid load. Thus, under normal conditions, acid is also buffered by NH_3, and when the acid load increases, ammoniagenesis will increase to buffer the load.

Ammoniagenesis

The proximal tubule cells are capable of producing ammonia from glutamine, which is extracted from the tubular fluid and peritubular capillary blood. The ammonia can buffer H⁺ by forming ammonium (NH_4^+) that can ultimately be excreted in the urine. As with TA production, a key aspect of this reaction is that the process of excreting NH_4^+ produces new HCO_3^- that is reabsorbed into the plasma.

In the proximal tubular cells, the glutamine is hydrolyzed to produce glutamate and one molecule of ammonia (NH_3). The glutamate is further metabolized to α-ketoglutarate, producing another NH_3 molecule. The two NH_3 molecules are immediately combined with two H⁺ ions, forming two ammonium (NH_4^+) ions. Additional α-ketoglutarate metabolism yields two HCO_3^- ions. Thus, a single glutamine molecule generates two HCO_3^- ions, which are reabsorbed as *new HCO_3^-*, and two NH_4^+ ions, which are secreted into the tubular fluid (see Fig. 21.2, *upper right*).

The NH_4^+ produced in the proximal tubule is not directly excreted. Instead, it is reabsorbed in exchange for K⁺ by the NKCC-2 transporters in the TALH. The NH_3 stays in the interstitial fluid, increasing the medullary interstitial concentration of NH_3. The dissociated H⁺ is secreted back into the TALH in exchange for Na⁺. The interstitial NH_3 gradient promotes secretion of NH_3 into the tubular lumen of the CDs; NH_3 immediately binds free H⁺ in the CDs, re-forming NH_4^+ that is excreted in the urine (see Fig. 21.2, *lower right*).

A key concept is that a new bicarbonate ion is generated for every H⁺ ion that is *excreted* as a TA or ammonium. This

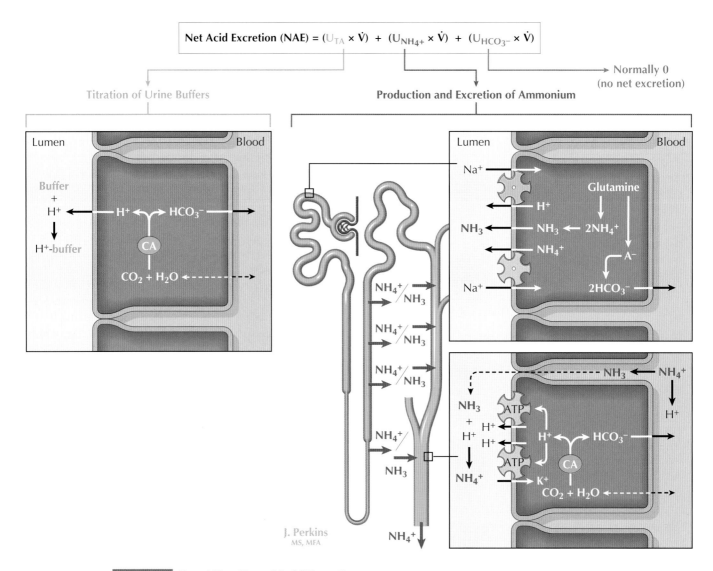

$$\text{Net Acid Excretion (NAE)} = (U_{TA} \times \dot{V}) + (U_{NH_{4+}} \times \dot{V}) + (U_{HCO_{3-}} \times \dot{V})$$

Titration of Urine Buffers

Production and Excretion of Ammonium

Normally 0
(no net excretion)

J. Perkins
MS, MFA

Figure 21.2 Renal Handling of Acid Excretion H^+ is excreted as titratable acids (mainly phosphoric acids) and ammonium (NH_4^+). With either mechanism, excretion of an H^+ results in generation of a new HCO_3^- that enters into the blood. Panels on the right illustrate H^+ secretion by the proximal tubule (*top right*) and in the α-intercalated cells of the collecting ducts (*lower right*). The titratable acids and ammonium produced in this segment are excreted in the urine. *CA,* carbonic anhydrase.

phenomenon occurs because the secretion of H^+ and excretion as ammonium (NH_4^+) or as a TA results in HCO_3^- reabsorption (Fig. 21.2).

Net Acid Excretion

Acid balance is determined by the difference between acid intake and urinary excretion of acid. Under normal conditions, intake equals excretion. **Net acid excretion (NAE)** describes the total amount of acid that is excreted in the urine,

$$\text{NAE} = \text{TA} + NH_4^+ - HCO_3^- \qquad \textbf{Eq. 21.4}$$

where TA is titratable acids. Under most conditions, urinary HCO_3^- is zero (all HCO_3^- is normally reabsorbed). However, when HCO_3^- appears in the urine, it implies that H^+ was added to the ECF (recall the 1:1 relationship between

bicarbonate reabsorbed or excreted and H^+ transported in the opposite direction). In Equation 21.4, any HCO_3^- in the urine is subtracted from the TA and NH_4^+ to account for the newly accumulated acid. HCO_3^- excretion is indicative of alkalosis or renal tubular acidosis, and in both conditions, there is an equimolar gain of acid for the HCO_3^- excreted.

Urine pH

Normal urine pH varies between 4.4 and 8 according to the acid-base status, with average values around 5.5 to 6.5. The minimal attainable urine pH is 4.4, which represents a 1000-fold greater concentration of H^+ than the blood pH of 7.4 and is the greatest concentration difference against which the H^+ pumps of the α-intercalated cells can effectively secrete H^+. This maximal level of urine acidity is only attained in the state of severe metabolic acidosis.

Table 21.2   **Sample Values for Uncompensated Acid-Base Imbalances**

	pH	P_{CO_2} (mm Hg)	HCO_3^- (mM or mEq/L)
Primary respiratory acidosis	7.32	50	24
Primary metabolic acidosis	7.32	40	18
Primary respiratory alkalosis	7.54	28	24
Primary metabolic alkalosis	7.54	40	34

ACIDOSIS AND ALKALOSIS

When the pH of the ECF falls outside the normal physiological range, acidosis (i.e., pH <7.35) or alkalosis (i.e., pH >7.45) results. A disturbance is designated as:

- **respiratory** (acidosis or alkalosis) if it is caused by abnormal CO_2, or
- **metabolic** (acidosis or alkalosis) if the pH change is consistent with the alteration in HCO_3^-.

The acid-base status is assessed by examining the plasma values of pH, P_{CO_2}, and HCO_3^-, which are the key components of the Henderson-Hasselbalch equation. Under normal conditions in arterial blood, these values will be:

- pH =7.4
- P_{CO_2} = 40 mm Hg
- HCO_3^- = 24 mM

When pH is altered, the primary disturbance can be identified by determining which component (P_{CO_2} or HCO_3^-) is altered in the direction consistent with the change in pH. Increased P_{CO_2} or decreased HCO_3^- will produce respiratory and metabolic acidosis, respectively, whereas decreased P_{CO_2} or increased HCO_3^- will produce respiratory and metabolic alkalosis, respectively (Table 21.2 and Fig. 21.3).

In general, compensation for metabolic disturbances will include an altered respiratory rate—that is, hyperventilation to blow off excess CO_2 in acidosis or hypoventilation to retain CO_2 in alkalosis (discussed in Chapter 16). Compensation for respiratory disturbances will include altered renal bicarbonate and acid handling. **Metabolic compensation** occurs over several hours through the renal excretion of acid (in acidosis) or bicarbonate (in alkalosis) to return the pH toward normal. The examples in Table 21.2 represent *uncompensated* disturbances. The following paragraphs focus mainly on metabolic disturbances.

Acidosis

Acidosis can result from either gain of acid or loss of bicarbonate. Net acid gain can arise from either decreased respiration (increasing CO_2) or from the accumulation of acids from the following metabolic sources (**metabolic acidosis**):

- Keto acids, which are generated by β-oxidation of fatty acids (a phenomenon that occurs during starvation and poorly controlled diabetes)
- Phosphoric acid, which is generated during renal failure as a result of the inability to excrete the acids because of a low glomerular filtration rate
- Lactic acid, which is released from damaged tissues during hypoxia or heart failure
- Ingestion of substances such as antifreeze (ethylene glycol) and Sterno (containing methanol)

Bicarbonate losses are always caused by metabolic disturbances and mainly occur as a result of:

- increased fecal elimination associated with prolonged diarrhea, because HCO_3^- is secreted into the intestinal lumen in exchange for Cl^- and is lost with other constituents of the chyme.
- increased urinary excretion because of insufficient HCO_3^- reabsorption in the proximal tubule.
- proximal (type 2) renal tubular acidosis, in which failure of the Na^+/H^+ exchanger in the proximal tubule will decrease the secretion of H^+, thus reducing proximal tubular HCO_3^- reabsorption (leading to bicarbonate loss in the urine) and acidosis. The α-intercalated cells of the CDs are still functional, so some of the acid load can be excreted.

Whether it is because of the addition of acid or the loss of bicarbonate, the resulting acid load must be buffered systemically and then excreted by the kidneys. To compensate for acid gain, the plasma level of free bicarbonate falls as it forms carbonic acid (H_2CO_3) to buffer the free acid. When pH is less than 7.35, respiration will increase to "blow off" CO_2 and reduce acid load. On a long-term basis, the main compensation will occur in the kidneys—increased acid secreted into the renal tubules will combine with phosphates and ammonia to form TAs and ammonium for excretion (see Fig. 21.2).

Although an immediate increase in TAs can occur because some excess HPO_4^{2-} is present, the ability to excrete TAs is limited by the amount of phosphate reabsorbed. The main urinary buffering of excess H^+ is by ammonia (and formation of NH_4^+). Ammoniagenesis increases during acidosis over a period of hours to days, allowing a great increase in ammonium excretion. The increase in NAE will raise pH toward normal values, although in severe cases of diabetic ketoacidosis, the kidneys may not be able to attain normal pH values until blood glucose is restored.

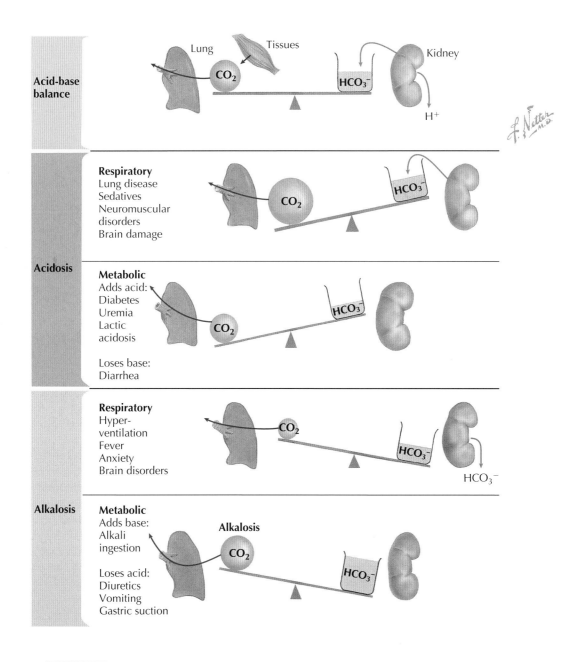

Figure 21.3 **Role of the Lungs and Kidneys in the Regulation of Acid-Base Balance** Under normal conditions *(top panel)*, the acid-base status is in balance, with kidneys absorbing all filtered HCO_3^- and the daily gain of acid excreted by the kidneys. In both respiratory and metabolic imbalances, acid gain (acidosis) or loss (alkalosis) is compensated for by the renal excretion of acid or HCO_3^-, respectively. When blood pH is outside the physiological window (<7.35 or >7.45), respiratory compensation also occurs by hyperventilation or hypoventilation.

ANION GAP

The anion gap is used to differentiate between acidosis resulting from acid gain and acidosis caused by bicarbonate loss. This diagnostic tool utilizes the fact that the only way to lose bicarbonate is through HCO_3^-/Cl^- exchangers located in the β-intercalated cells of the CDs and in the gastrointestinal tract. Thus if HCO_3^- is lost, Cl^- is gained, which is reflected in the plasma concentrations of these anions. The anion gap is the difference in concentration between the major plasma cation, Na^+, and the major plasma anions, Cl^- and HCO_3^-. When Cl^- and HCO_3^- concentrations are subtracted from the Na^+ concentration, the "anion gap" is normally ~8 to 12 mEq/L. The anion gap represents the sum of the concentrations of about 10 anions present in the plasma and includes proteins, lactate, citrate, phosphates, sulfates, and so on:

$$AG = Na^+ - (Cl^- + HCO_3^-) \qquad \textbf{Eq. 21.5}$$

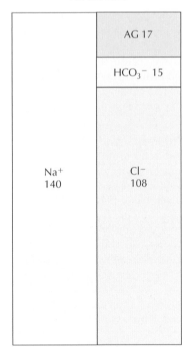

Normal

Acidosis From Acid Load
AG increases

Acidosis From Base Loss
AG normal

Figure 21.4 **Plasma Anion Gap** In cases of metabolic acidosis, measurement of the anion gap (AG) will reveal whether the disruption in pH was from acid loading or base loss. (Values [in mEq/L] are approximate.)

As illustrated in Figure 21.4, *middle panel*, when acidosis results from acid loading, the anion gap increases as a result of the lower plasma bicarbonate (used to buffer the acid). In contrast, when HCO_3^- is lost, it is replaced by Cl^-, and the anion gap is normal (see Fig. 21.4, *right panel*). Thus the anion gap can help determine the overall cause of the metabolic acidosis, with the specific cause of the acid load or base loss remaining to be determined.

Alkalosis

Alkalosis results from the loss of acid or gain of HCO_3^-. Chronic hyperventilation (for example, at high altitude) will cause the loss of CO_2 and result in **respiratory alkalosis.** **Metabolic alkalosis** can result from:

- vomiting, which results in the loss of significant amounts of H^+ in the form of HCl, and volume contraction.
- sodium bicarbonate overdose.
- chronic diuretic usage, which can cause hypokalemia, volume contraction, aldosterone excess, and chloride depletion, all of which are associated with alkalosis.

During alkalosis, HCO_3^-/Cl^- exchange by transporters in β-intercalated cells of the renal CDs is activated, allowing secretion and excretion of bicarbonate until normal pH is attained.

Why don't the renal HCO_3^-/Cl^- exchangers always work during alkalosis? It is important to recall that under normal conditions the HCO_3^-/Cl^- exchangers in the β-intercalated cells are not active because HCO_3^- is being retained, not excreted. However, these exchangers are needed during alkalosis and depend on the lumenal concentration of Cl^- to effectively transport the HCO_3^- into the tubular fluid. The importance of distal lumenal Cl^- is illustrated in contraction alkalosis, as seen with chronic vomiting and volume depletion.

In response to the alkalosis caused by loss of H^+ during vomiting, the renal HCO_3^-/Cl^- exchangers are expressed but are unable to function because of the low level of Cl^- in the CD fluid. The low Cl^- is a result of fluid- and sodium-retaining mechanisms (e.g., sympathetic nervous system, renin-angiotensin system, aldosterone, and antidiuretic hormone) stimulated in response to volume depletion. Chloride is avidly reabsorbed with the sodium and water, leaving little entering the distal areas of the nephron. The problem can be corrected by isotonic fluid replacement, which will decrease sympathetic nervous system and renin-angiotensin system activation and aldosterone and antidiuretic hormone release and increase Na^+ and Cl^- delivery to the CDs. The HCO_3^-/Cl^- exchanger will then be able to function, and excess HCO_3^- will be excreted, correcting the alkalosis.

CLINICAL CORRELATE 21.1
Metabolic Ketoacidosis

A common cause of ketoacidosis is poorly controlled diabetes. When glucose is not able to efficiently enter cells, there are two immediate consequences: elevated blood glucose and increased cellular β-oxidation of fatty acids to provide an alternate energy source in the absence of intracellular glucose. If the rate of oxidation is high, metabolites including ketoacids (or ketone bodies) enter the blood, decreasing the pH. Several problems ensue:

■ **Acidosis:** The ketoacids (acetoacetic acid and β-hydroxybutyric acid) are filtered at the glomeruli and eliminated in the urine, but the production overwhelms excretion. Plasma bicarbonate may not be sufficient to buffer the acid load, and in severe cases the free bicarbonate will be heavily consumed and blood pH will be dangerously reduced (i.e., below 7.2). Contributing to this problem, ketone bodies interfere with the production of ammonia from glutamine in the proximal tubular cells and thus decrease new bicarbonate generation (further decreasing plasma bicarbonate levels).

■ **Glucosuria:** Glucose is excreted in the urine because the elevation in filtered glucose load overwhelms the renal sodium-glucose transporters. The glucose pulls water with it, creating polyuria (high urine volume).

■ **Volume expansion and cellular contraction:** When glucose increases in the blood, it exerts additional osmotic force, drawing water into the vascular space. As a result the ECF expands and the intercellular fluid volume decreases. Blood pressure may increase, and the concentration of other plasma constituents, including electrolytes and proteins, is diluted. Ultimately, diabetic ketoacidosis may actually result in dehydration as a result of osmotic diuresis and vomiting.

■ **Toxicity:** The ketoacids in the blood are toxic and may cause coma and death if blood glucose level is not corrected. The most severe consequences are seen primarily in type 1 (insulin-dependent) diabetes, and death from ketoacidotic coma is rarely seen in people with type 2 (insulin-resistant) diabetes. With insulin treatment, β-oxidation of fatty acids and therefore ketoacid production are reduced and excess acid is excreted in the urine.

Review Questions

CHAPTER 17: OVERVIEW, GLOMERULAR FILTRATION, AND RENAL CLEARANCE

1. If the clearance of a freely filtered substance is **less than** the clearance of inulin, then the substance:

A. is filtered and completely reabsorbed.
B. is filtered and completely secreted.
C. underwent net reabsorption.
D. underwent net secretion.
E. was neither secreted nor reabsorbed.

2. Diabetic nephropathy is associated with thickened glomerular capillary basement membranes. The decrease in glomerular filtration rate results from:

A. a reduction in renal blood flow.
B. a reduction in glomerular capillary hydrostatic pressure.
C. a reduction in the permeability of the glomerular filtration barrier.
D. an increase in the glomerular capillary hydrostatic pressure.
E. an increase in the permeability of the glomerular filtration barrier.

3. The kidneys perform all of the following functions **EXCEPT:**

A. production of aldosterone.
B. excretion of excess acid.
C. regulation of fluid and electrolyte homeostasis.
D. activation of vitamin D.
E. ammoniagenesis.

4. Renal blood flow in a young man was determined to be 1 L/min. If his inulin clearance is 125 mL/min and his hematocrit is 0.4, what is his renal filtration fraction?

A. 15%
B. 20%
C. 25%
D. 30%
E. 35%

5. Select the **TRUE** statement regarding renal architecture.

A. The vasa recta are capillaries that surround the cortical nephrons.
B. There are equal numbers of cortical and medullary (juxtaglomerular) nephrons in the kidneys.
C. The collecting ducts contain both principal cells and intercalated cells.
D. All of the glomeruli are located in the renal medulla.
E. The juxtaglomerular apparatus links the distal tubule of one nephron with the glomerulus of an adjacent nephron.

CHAPTER 18: RENAL TRANSPORT PROCESSES

6. Reabsorption of glucose occurs in the:

A. proximal tubule.
B. thin descending limb of Henle.
C. thick ascending limb of Henle.
D. distal tubule.
E. collecting duct.

7. Potassium handling by the kidneys is affected by all of the following **EXCEPT:**

A. dietary potassium.
B. plasma potassium concentration.
C. plasma aldosterone.
D. plasma antidiuretic hormone.
E. acidosis.

8. "Loop" diuretics such as furosemide or bumetanide produce a diuresis by targeting which of the following transporters?

A. Na^+-glucose cotransporters
B. Na^+/H^+ antiporters
C. Na^+/K^+ ATPase
D. NKCC-2 (Na^+-K^+-$2Cl^-$) cotransporters
E. Epithelial sodium channel (ENaC)

9. Select the **TRUE** statement about renal sodium handling.

A. The thick ascending limb of Henle is impermeable to sodium.
B. Approximately 99.9% of filtered sodium is reabsorbed.
C. The entire proximal tubule (S1-S3) reabsorbs approximately 40% of the filtered sodium.
D. Aldosterone acts at the proximal tubule to increase sodium reabsorption.
E. Antidiuretic hormone (ADH) acts at the collecting ducts to increase sodium reabsorption.

10. Select the **TRUE** statement about renal bicarbonate handling.

A. All (100%) of the filtered bicarbonate load is reabsorbed in the proximal tubule via apical HCO_3^-/Cl^- exchangers.
B. Intrarenal bicarbonate buffers the H^+ in tubular fluid until the acid can be excreted.
C. Carbonic anhydrase in the tubular fluid and renal tubular cells is critical for bicarbonate reabsorption.
D. Under normal conditions, some bicarbonate will be found in urine.
E. Bicarbonate can be secreted into tubular fluid only in the principal cells of the collecting ducts.

CHAPTER 19: URINE CONCENTRATION AND DILUTION MECHANISMS

11. ADH-sensitive water channels (aquaporins) are located in the:

A. proximal tubule.
B. thin descending limb of Henle.
C. thick ascending limb of Henle.
D. distal tubule.
E. collecting duct.

12. All of the following factors contribute to establishing or maintaining the medullary interstitial concentration gradient **EXCEPT:**

A. distal tubule sodium reabsorption.
B. NKCC-2 cotransporters on the thick ascending limb of Henle.
C. solute-free water reabsorption in the descending limb of Henle.
D. the counter-current multiplier effect.
E. urea recycling.

13. A patient with chronic pyelonephritis has the following laboratory values:

Urine flow = 3 mL/min
P_{osm} = 300 mosm/L
U_{osm} = 200 mosm/L

The free water clearance in this person is:

A. −3 mL/min.
B. −1 mL/min.
C. 0 mL/min.
D. +1 mL/min.
E. +3 mL/min.

14. All of the following factors can disrupt the interstitial osmotic gradient **EXCEPT:**

A. increased glomerular filtration rate.
B. increased renal blood flow.
C. increased vasa recta blood flow.
D. dehydration.
E. loop diuretics.

CHAPTER 20: REGULATION OF EXTRACELLULAR FLUID VOLUME AND OSMOLARITY

15. Select the **TRUE** statement regarding the renin-angiotensin-aldosterone system.

A. Renin secretion is stimulated by high tubular fluid sodium concentration.
B. Angiotensin I is converted to angiotensin II exclusively in the pulmonary circulation.
C. Angiotensin II directly stimulates proximal tubular sodium reabsorption.
D. High tubular fluid flow rate in the distal tubule will stimulate renin secretion.
E. Aldosterone stimulates proximal tubule sodium reabsorption.

16. A severely dehydrated patient is admitted to the emergency department. Which of the following would **NOT** be elevated in this patient?

A. Plasma atrial natriuretic peptide
B. Plasma renin
C. Plasma ADH
D. Sympathetic nervous system activity
E. Plasma aldosterone

17. Diabetes insipidus:

A. usually involves a reduction in renal V2 receptors for ADH.
B. is associated with increased water channels (aquaporins) in the collecting ducts.
C. results in excretion of large quantities of hypertonic urine.
D. occurs mainly in children.
E. results in excretion of large quantities of hypotonic urine.

18. A decrease in sodium concentration in the distal tubule:

A. inhibits renin secretion from the macula densa cells.
B. increases renin secretion from the macula densa cells.
C. promotes ATP release from the juxtaglomerular cells.
D. promotes renin release from the juxtaglomerular cells.
E. decreases constriction of the efferent arteriole.

CHAPTER 21: REGULATION OF ACID-BASE BALANCE BY THE KIDNEYS

19. Renal tubular hydrogen secretion *increases:*

A. when extracellular fluid volume is increased.
B. when arterial P_{CO_2} is decreased.
C. when plasma bicarbonate concentration is low as a result of metabolic acidosis.
D. when plasma aldosterone is low.
E. when plasma ADH is low.

20. Net acid excretion is dependent on all of the following factors **EXCEPT:**

A. excretion of ammonium.
B. excretion of bicarbonate.
C. excretion of sodium.
D. excretion of titratable acids.
E. distal H^+ secretion.

21. Use the following blood values to determine the acid-base disorder.

pH	P_{CO_2}	HCO_3^-	Na^+	K^+	Cl^-
7.28	26 mm Hg	14 mEq/L	136 mEq/L	5.0 mEq/L	100 mEq/L

A. Metabolic alkalosis
B. Metabolic acidosis from acid gain
C. Metabolic acidosis from diarrhea
D. Respiratory acidosis
E. Respiratory alkalosis

22. When compared with normal values, in *acute uncompensated respiratory alkalosis,* plasma:

A. pH and P_{CO_2} are high and HCO_3^- is normal.
B. pH and P_{CO_2} are low and HCO_3^- is high.
C. pH is high and P_{CO_2} and HCO_3^- are low.
D. pH is high, P_{CO_2} is low, and HCO_3^- is normal.
E. pH is high, P_{CO_2} is low, and HCO_3^- is high.

23. The major titratable acids excreted in the urine are in the form of:

A. $H_2PO_4^-$.
B. lactic acid.
C. K_2PO_4.
D. H_2CO_3.
E. ketoacids.

24. Renal ammoniagenesis:

A. occurs in the distal tubular cells.
B. uses glutamine as the substrate.
C. is stimulated by ADH.
D. is elevated during alkalosis.
E. is always at maximal activity.

Section 6

GASTROINTESTINAL PHYSIOLOGY

What makes the gastrointestinal (GI) tract interesting? We are acutely aware of this area of our bodies several times a day: we have hunger pangs, feel "full," become thirsty, hear intestinal sounds, expel gases, and feel the urge to defecate. Furthermore, GI-related distress and disease is high on the list of personal and medical complaints. Homeostasis in the entire organism relies on the input of nutrients (gas exchange, food, and water), and thus the GI tract is a critical component of the supply side.

Chapter 22

Overview of the Gastrointestinal Tract

STRUCTURE AND OVERALL FUNCTION OF THE GASTROINTESTINAL TRACT

The gastrointestinal (GI) tract can be thought of as one long tube with an input (mouth) and output (anus), with specialized areas and direct input from associated organs (e.g., the liver, pancreas, and gallbladder) (Fig. 22.1). Sphincters along the tract separate major sections of the tract, allowing the regulation of flow of food into the stomach and **chyme** (food and digestive juices) out of the stomach and through the intestines.

Here is a brief tour down the tract:

- **Mouth:** Mechanical digestion occurs through **mastication** (chewing), and chemical digestion begins on the predominant dietary carbohydrate, **starch,** and to a small degree on **lipids.** The salivary glands secrete saliva, which helps buffer and digest the food and lubricate it so it can be swallowed.
- **Esophagus:** During swallowing, relaxation of the upper esophageal sphincter allows the bolus of food to enter the esophagus. The esophagus is a tube through which the bolus of food passes through the chest and into the stomach. At the end of the esophagus the **lower esophageal sphincter (LES)** relaxes to allow the bolus into the stomach.
- **Stomach:** The stomach is a pouch in the tract that is specialized for the storage of food, secretion of digestive enzymes and hydrochloric acid (HCl), and mixing of the food and digestive juices to make **chyme.** The **pyloric sphincter**, which is located at the distal end of the stomach, regulates the entry of chyme into the small intestine.
- **Small intestine:** The small intestine is composed of three sections, the **duodenum** (~1 foot long), the **jejunum** (~8 feet long), and the **ileum** (~10 feet long). Secretions from the liver, gallbladder, and pancreas enter the small intestine via the common bile duct and through the sphincter of Oddi, into the duodenum. A key aspect of the small intestine is the presence of finger-like projections in the intestinal mucosa (the **villus brush border**), which increases the surface area for absorption. Through the tract, there is only one layer of intestinal epithelial cells, or **enterocytes**, between the lumen of the tract and the systemic blood supply. Although this layer of enterocytes effectively keeps the lumenal contents in the small intestine, when absorption occurs, movement of nutrients into the blood occurs rapidly and efficiently. However, the enterocytes are vulnerable to the harsh environment of the gut lumen and need protection, which is conferred by intestinal buffers and mucus. Despite this protection, intestinal cells are sloughed off and replaced by new cells every 3 to 4 days.

  The majority of digestion occurs in the jejunum by pancreatic enzymatic action. When nutrients are digested to their constitutive elements (e.g., monosaccharides, monoglycerides, and amino acids), absorption can occur all along the small intestine, although most nutrients are absorbed by the mid jejunum. The **terminal part of the ileum** is also the site of vitamin B12 absorption and recycling of bile. Chyme that remains in the tract leaves the small intestine through the **ileocecal sphincter** and moves into the large intestine (colon).

- **Large intestine:** The large intestine does not have a brush border. The primary functions of the large intestine are dehydration of chyme to produce feces and storage of the feces until defecation. The large intestine consists of the cecum, appendix, colon, rectum, and anal canal. At the end of the rectum, the **internal** and **external anal sphincters** regulate expulsion of feces.

Another important aspect of the tract is that its musculature is smooth muscle, except for the mouth, upper esophagus, and external anal sphincter, which have skeletal muscle. Skeletal muscle allows voluntary control of both input (chewing and swallowing) and output (defecation). The remainder of the tract has **longitudinal** and **circular** bands of smooth muscle that allow for propulsion and mixing of the chyme. Figure 22.2 illustrates these muscle layers, as well as the **oblique** layer that is unique to the stomach. The oblique layer aids in the important mixing function of stomach motility.

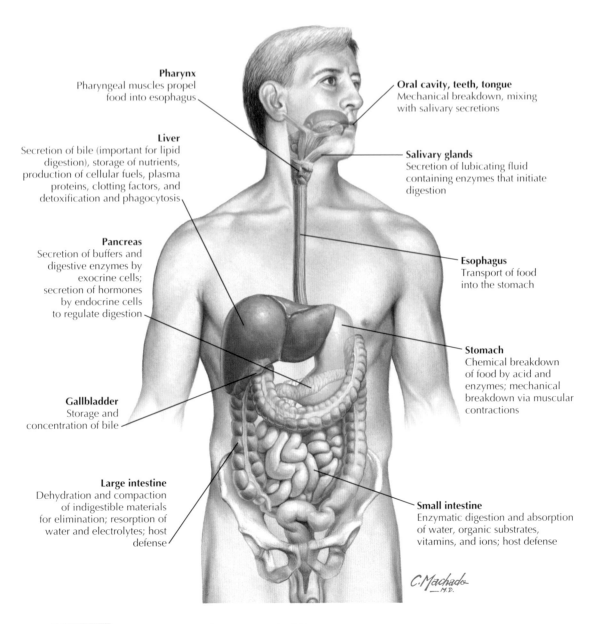

Pharynx
Pharyngeal muscles propel food into esophagus

Liver
Secretion of bile (important for lipid digestion), storage of nutrients, production of cellular fuels, plasma proteins, clotting factors, and detoxification and phagocytosis

Pancreas
Secretion of buffers and digestive enzymes by exocrine cells; secretion of hormones by endocrine cells to regulate digestion

Gallbladder
Storage and concentration of bile

Large intestine
Dehydration and compaction of indigestible materials for elimination; resorption of water and electrolytes; host defense

Oral cavity, teeth, tongue
Mechanical breakdown, mixing with salivary secretions

Salivary glands
Secretion of lubicating fluid containing enzymes that initiate digestion

Esophagus
Transport of food into the stomach

Stomach
Chemical breakdown of food by acid and enzymes; mechanical breakdown via muscular contractions

Small intestine
Enzymatic digestion and absorption of water, organic substrates, vitamins, and ions; host defense

C.Machado
_M.D.

Figure 22.1 **Overview of the Gastrointestinal Tract**   The gastrointestinal tract begins at the mouth and is made up of discrete areas that aid the digestion and absorption of nutrients. Each area of the tract contributes to the efficient processing of nutrients. The gastrointestinal-associated organs (e.g., the liver, pancreas, and gallbladder) provide important secretions including bile, enzymes, and buffers.

Other distinct aspects of the GI tract include large blood and lymphatic networks, copious secretion of substances from a variety of glands, and a nervous system that is unique to the GI tract (Fig. 22.3).

Blood Supply

The GI tract has a large blood supply to support digestion, absorption, and propulsion. Oxygen, nutrients, and hormones are supplied by blood to aid in the many secretory, absorptive, and propulsive functions. In addition, efficient absorption of

nutrients through the intestinal cells requires high blood flow, which ensures there is always a gradient for nutrient entry into the bloodstream.

During active digestion, blood flow to the GI tract increases to as much as 3 times normal to facilitate metabolic and digestive needs. Cardiac output is increased, and its distribution to various tissues is altered, favoring the GI tract. The proportions of cardiac output perfusing various tissues are described in Section 3 (Chapter 9) for the normal resting state.

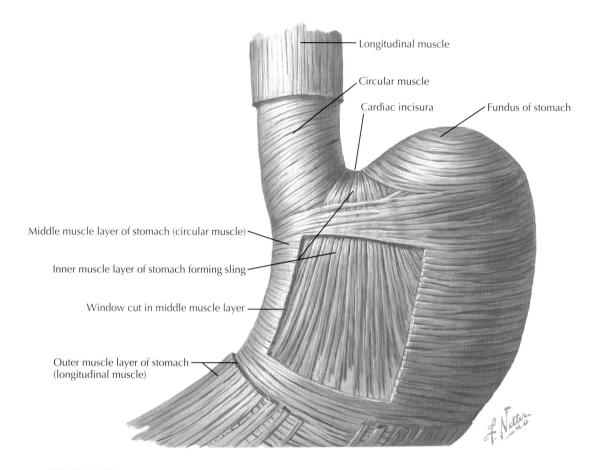

Longitudinal muscle

Circular muscle

Cardiac incisura

Fundus of stomach

Middle muscle layer of stomach (circular muscle)

Inner muscle layer of stomach forming sling

Window cut in middle muscle layer

Outer muscle layer of stomach (longitudinal muscle)

Figure 22.2 **Musculature of the Stomach** The stomach has longitudinal, circular, and oblique muscle layers. The oblique layer helps mix and grind the chyme.

Blood supply to the GI tract is from the splanchnic bed (superior and inferior mesenteric arteries), with the capillary system extending into all of the villi. Blood flows out of the tract through the **portal veins** directly to the liver (Fig. 22.4). This is the **"first pass"** effect, whereby absorbed substances are routed directly to the liver (without entering into the general circulation), where much processing occurs.

Glandular Secretions

To efficiently move and digest the chyme, tremendous amounts of substances are secreted into the lumen of the tract. Each day, glands in the mouth, pancreas, and small intestine produce liters of saliva, mucus, buffers, and enzymes to facilitate the digestive process.

Lymphatics

Lymph lacteals (i.e., lymphatic capillaries) are found throughout the intestines and extend into the villi of the small intestine, allowing absorption of lipids. The lymphatic system is also extensive in the liver, where it absorbs fluids and proteins and transports them to the systemic venous blood via the thoracic ducts into the subclavian veins. Lymphatic transport of proteins contributes substantially to plasma oncotic

pressure and provides binding proteins for circulating substances, including hormones, calcium, and iron.

General Functions of the GI Tract

- **Digestion:** Both mechanical and chemical digestion occur in the GI tract. **Mechanical** digestion refers to the physical breakdown of the food by initial chewing and grinding in the stomach. **Chemical** digestion occurs through the action of specific enzymes and gastric acid (HCl).
- **Endocrine:** A great number of hormones are produced by GI-associated cells in the stomach, intestines, pancreas, and liver. Some hormones are secreted into the general circulation and affect other parts of the GI tract, as well as other organs. Hormones also are produced that act on adjacent GI tissues in a paracrine manner.
- **Expulsion:** After absorption of nutrients through the tract, the fecal waste matter (fiber, other undigested matter, bacteria, and fat from sloughed cells) is expelled through defecation.
- **Protection:** The GI tract protects us from the bacteria and antigens that we ingest by degrading these substances by means of gastric HCl, as well as through

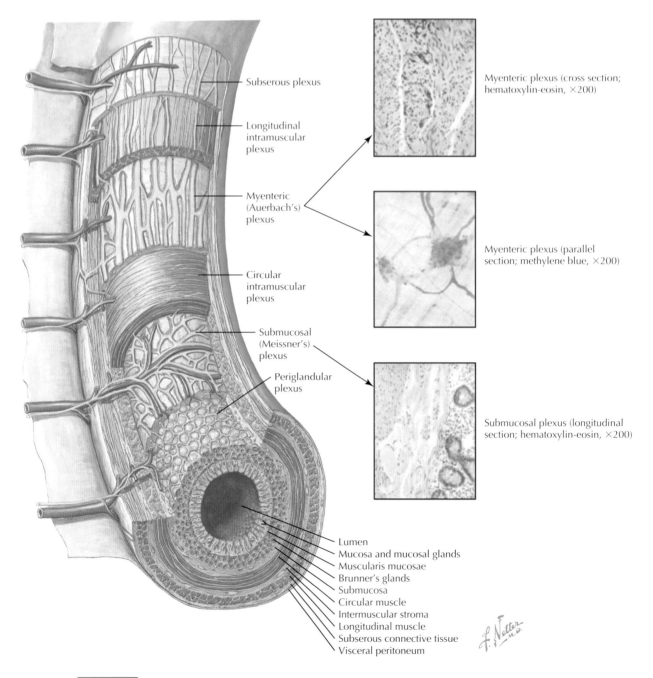

Subserous plexus

Longitudinal intramuscular plexus

Myenteric (Auerbach's) plexus

Circular intramuscular plexus

Submucosal (Meissner's) plexus

Periglandular plexus

Myenteric plexus (cross section; hematoxylin-eosin, ×200)

Myenteric plexus (parallel section; methylene blue, ×200)

Submucosal plexus (longitudinal section; hematoxylin-eosin, ×200)

Lumen
Mucosa and mucosal glands
Muscularis mucosae
Brunner's glands
Submucosa
Circular muscle
Intermuscular stroma
Longitudinal muscle
Subserous connective tissue
Visceral peritoneum

Figure 22.3 **The Enteric Nervous System** The myenteric plexus (nerve net) is found between the longitudinal and circular muscle layers of the gastrointestinal tract, and stimulation of these nerves affects muscle contraction. The submucosal plexus is located between the circular muscle and submucosa, and stimulation can affect gastrointestinal secretions. The enteric nerves can function without extrinsic nerves through local mechanisms.

immune activity (immunoglobulin A and Peyer's patches).

■ **Motility:** Several different forms of propulsion move chyme through the tract, allowing digestion, absorption, and expulsion to occur. If an area loses the ability to move chyme, serious disease can develop.

■ **Absorption:** The ultimate goal for the elaborate GI system is getting nutrients into the body. The intestines absorb most of the nutrients that are digested, and most

absorption occurs in the first half of the small intestine (through the jejunum). This absorption is made possible by the large surface area provided by the villi of the enterocytes of the small intestine.

■ **Secretion:** The secretion of mucus, buffers, hormones, and enzymes facilitates lubrication, digestion, motility, and absorption through the tract.

■ **Storage:** Both the stomach and large intestine act as storage sites. The stomach accommodates food as it is

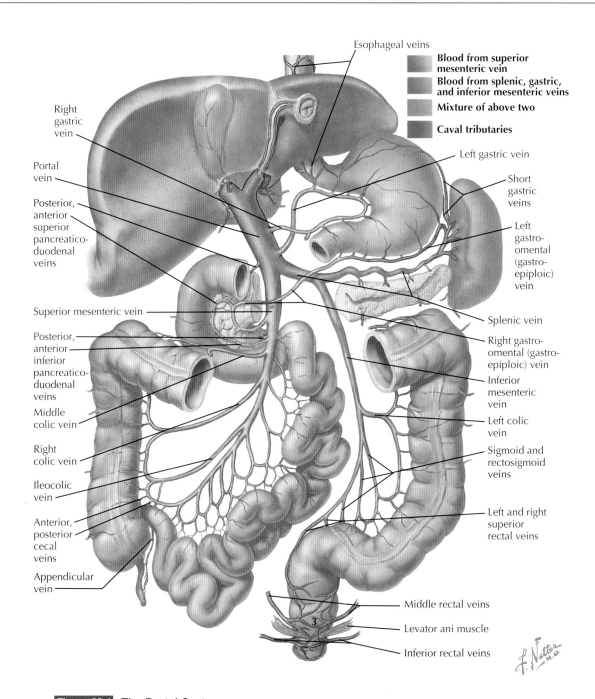

Esophageal veins

Blood from superior
mesenteric vein
Blood from splenic, gastric,
and inferior mesenteric veins
Mixture of above two
Caval tributaries

Right
gastric
vein

Left gastric vein

Portal
vein

Short
gastric
veins

Posterior,
anterior
superior
pancreatico-
duodenal
veins

Left
gastro-
omental
(gastro-
epiploic)
vein

Superior mesenteric vein

Splenic vein

Posterior,
anterior
inferior
pancreatico-
duodenal
veins

Right gastro-
omental (gastro-
epiploic) vein

Inferior
mesenteric
vein

Middle
colic vein

Left colic
vein

Right
colic vein

Sigmoid and
rectosigmoid
veins

Ileocolic
vein

Anterior,
posterior
cecal
veins

Left and right
superior
rectal veins

Appendicular
vein

Middle rectal veins

Levator ani muscle

Inferior rectal veins

Figure 22.4 **The Portal System** The portal vessels carry venous blood directly from the intestines to the liver through the portal vein.

mixed with the gastric acid and enzymes, promoting breakdown of food into chyme that is small enough to pass through the pyloric sphincter into the duodenum. This storage is facilitated by **receptive relaxation**, which is mediated by the vagus nerve (see Chapter 23). If receptive relaxation did not occur, the quantity of food that could be ingested during a meal and enter the stomach would be dramatically limited. On the opposite end, the large intestine dehydrates the chyme to produce feces, and the type of motility there allows retention of the feces until it is expelled.

The GI tract produces many hormones that regulate a host of actions, including GI secretion and motility, feeding behavior, and insulin secretion. In fact, the first substance described by Starling and Bayliss (in 1902) as a "circulating hormone" was **secretin,** which is secreted from duodenal endocrine cells in response to acidic chyme. In addition to important effects on the GI tract, secretin has recently been shown to act at the hypothalamus, pituitary, and kidney to aid in osmoregulation.

THIRST AND HUNGER: BEHAVIORAL RESPONSES TO MAINTAIN THE MILIEU INTERIEUR

Thirst

To maintain a homeostatic internal environment, the behavioral responses to thirst and hunger ensure adequate ingestion of fluid and nutrients. **Thirst** is a response to the elevated plasma osmolarity associated with cell and vascular dehydration. The loss of fluid can be a result of urine and fecal fluid loss, as well as **insensible loss** of water from sweating and breathing (Fig. 22.5). The feeling of thirst can be stimulated when hypothalamic osmoreceptors sense elevated plasma osmolarity or arterial baroreceptors sense low blood pressure (often associated with reduced extracellular fluid [ECF] volume). Both the baroreceptors and osmoreceptors regulate the thirst response (mediated through the hypothalamus) and the release of **antidiuretic hormone** (**ADH**, also known as **vasopressin**). ADH has effects on renal retention of water (discussed in Section 5) and vascular smooth muscle contraction (discussed in Section 3). Thirst is controlled at both central and local levels. Dry mouth can be relieved temporarily by rinsing the mouth with fluid; however, if the overall cause is dehydration, the thirst response will return rapidly.

The control of plasma osmolarity is so important that a small (2%) change (e.g., from 300 to 306 mOsm/L) can induce thirst and ADH secretion. In contrast, a 12% to 15% reduction in ECF volume is needed to stimulate the same ADH response if osmolarity is unchanged. Thus, when thirst is stimulated by dehydration, sufficient fluid has been lost to raise plasma osmolarity and stimulate ADH.

Appetite and Hunger

Although the mechanisms involved in hunger are not as clearly understood as those involved in thirst, hunger is a powerful behavioral stimulus to ensure nutrient intake (Fig. 22.6). The main site of control, as with thirst, resides in the hypothalamus, which has both the hunger and satiety centers. The basic control of the centers is through essential nutrients such as glucose and fatty acids. Stimulation of appetite can occur through low blood glucose, as well as through sensory input (e.g., smelling or seeing food). The GI tract also plays a

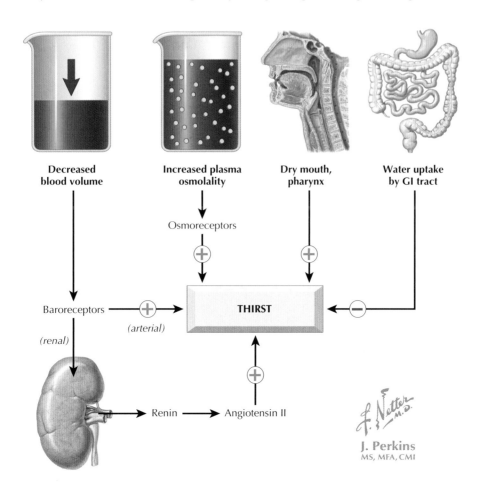

Decreased blood volume · **Increased plasma osmolality** · **Dry mouth, pharynx** · **Water uptake by GI tract**

Osmoreceptors

Baroreceptors (arterial) — ⊕ → **THIRST** ← ⊖

(renal)

Renin → Angiotensin II → ⊕

J. Perkins
MS, MFA, CMI

Figure 22.5 **Thirst** The thirst response is initiated by either a small increase in plasma osmolarity (1% to 2%) or a significant decrease in blood volume (12% to 15%). The mechanism is depicted in this illustration. (Modified from Widmaier E, Raff H, Strang K: *Vander, Sherman, and Luciano's Human Physiology,* ed 9, New York, 2003, McGraw Hill.)

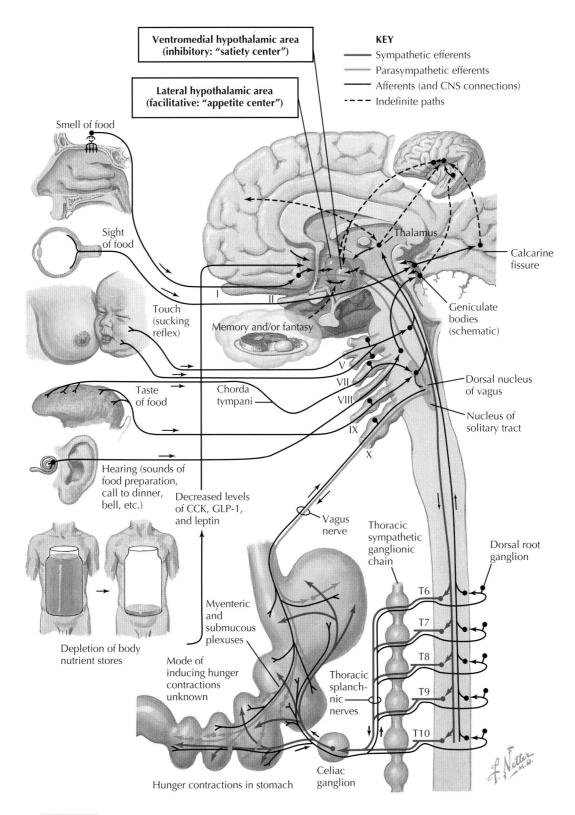

KEY
— Sympathetic efferents
— Parasympathetic efferents
— Afferents (and CNS connections)
--- Indefinite paths

Ventromedial hypothalamic area
(inhibitory: "satiety center")

Lateral hypothalamic area
(facilitative: "appetite center")

Smell of food

Sight
of food

Thalamus

Calcarine
fissure

Touch
(sucking
reflex)

I

II

Memory and/or fantasy

Geniculate
bodies
(schematic)

Taste
of food

Chorda
tympani

V

VII

VIII

IX

X

Dorsal nucleus
of vagus

Nucleus of
solitary tract

Hearing (sounds of
food preparation,
call to dinner,
bell, etc.)

Decreased levels
of CCK, GLP-1,
and leptin

Vagus
nerve

Thoracic
sympathetic
ganglionic
chain

Dorsal root
ganglion

T6

T7

T8

T9

T10

Depletion of body
nutrient stores

Myenteric
and
submucous
plexuses

Mode of
inducing hunger
contractions
unknown

Thoracic
splanch-
nic
nerves

Hunger contractions in stomach

Celiac
ganglion

Figure 22.6 **Appetite and Hunger** The feelings of hunger and satiety are not completely understood; however, general pathways are illustrated. The hypothalamus (see Chapter 27) plays important roles in both hunger (lateral hypothalamic area) and satiety (ventromedial hypothalamic area) and responds to endocrine and nerve input. Multiple hormonal systems facilitate the feeling of satiety: cholecystokinin (CCK), peptide YY, and glucagon-like peptide-1 (GLP-1) are released in response to ingested food, decreasing hunger. In addition, the hormone leptin is released in response to insulin and acts at the satiety center of the hypothalamus. In contrast, hunger "pains" are elicited through vagal stimulation of the enteric nerves, and the release of other gastrointestinal hormones, including ghrelin and orexin, stimulates the appetite center of the hypothalamus. *CNS,* central nervous system.

role by releasing hormones that stimulate hunger (e.g., ghrelin, galanin, and orexin) and decrease hunger (e.g., glucagon-like peptide-1 [GLP-1], cholecystokinin, and peptide YY) at the level of the hypothalamus. Another factor in satiety is leptin, a hormone that can be released from adipose tissues in response to elevated levels of insulin and glucose in the blood. Leptin receptors are present on the hypothalamus (as well as other tissues), and leptin stimulates the satiety center. In the hypothalamus, neuropeptide Y is implicated in hunger and is elevated by orexigenic peptides (e.g., ghrelin and orexin) and suppressed by anorexic hormones and peptides (e.g., GLP-1 and leptin). In addition to such specific factors, the hunger response is affected by body mass and functions to provide nutrients for energy expenditure.

Sweat is hypo-osomotic. Its concentration of salts (mainly sodium and chloride) is less than that found in plasma. Thus, sweating depletes the ECF of more fluid volume than salt, which increases the plasma osmolarity. Increased osmolarity stimulates the thirst response and reduction of fluid losses in urine and feces, helping to maintain adequate blood volume and pressure.

FLUID SHIFTS AND pH THROUGH THE TRACT

Although only ~2 liters (L) of fluid is ingested daily, the fluid flux in and out of the GI tract is actually about 9 L. Secretion of fluids is necessary for digestion of the chyme, and the additional volume is reabsorbed into the bloodstream with the added nutrients (Fig. 22.7). With the addition of secretions, the pH of the chyme changes. In the stomach, gastric acid secretion lowers the pH to ~2, allowing for indiscriminant breakdown of ingested food. To protect the small intestine from the corrosive effects of the acid, buffers (pH ~8) are secreted into the duodenum from the liver, pancreas, and gallbladder, as well as from intestinal crypt cells. This action raises the pH of the chyme in the first few centimeters of the duodenum to ~5. The addition of the buffers through the tract brings the pH of the chyme up to ~7.4 in the proximal jejunum. This increase in pH is necessary for optimal action of pancreatic enzymes.

Whereas we often are conscious of the salivary secretions that occur in anticipation of and during feeding, the other fluid fluxes tend to go unnoticed, although at times rumbling intestinal sounds called **borborygmi** may be heard. These sounds are the result of fluid and gas fluxes in the lumen.

THE ENTERIC NERVOUS SYSTEM

The GI tract is unique in having an intrinsic nervous system that is made up of the **myenteric** and **submucosal plexuses** (see Fig. 22.3). This **enteric nervous system (ENS)** is able to function independently, based on input from mechanoreceptors, chemoreceptors, and osmoreceptors located in the

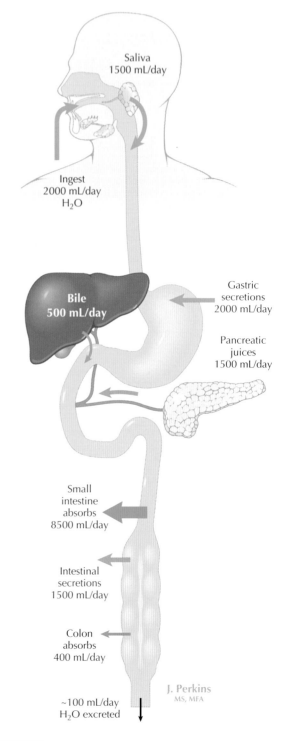

Figure 22.7 **Fluid Shifts Through the Gastrointestinal Tract** Humans ingest ~2 L of fluid in food and drink each day, and the gastrointestinal tract adds ~8 L of secretions to facilitate digestion and absorption of the nutrients. These secretions, including buffers, acid, and enzymes, are absorbed back into the blood (primarily by absorption in the small intestine) so that only ~200 mL of fluid is excreted in feces each day.

luminal epithelium of the tract. The ENS also receives input from the central nervous system (CNS), autonomic nervous system, and hormones, which help fine-tune and regulate the ENS. Without autonomic innervation, the ENS would still function, but in a less coordinated manner.

Located between the circular and longitudinal muscle layers, the **myenteric (Auerbach) plexus** regulates contraction and relaxation of the musculature, producing motility and mixing of the luminal contents. The **submucosal (Meissner) plexus** is located between the circular muscle and submucosa and regulates local fluid secretions.

THE GI TRACT AS AN ENDOCRINE ORGAN

Endocrine cells in the gastric and intestinal epithelia synthesize and release a variety of hormones into the bloodstream. The hormones act on other areas of the GI system, such as the liver and pancreas, in addition to the stomach and intestines. These hormones regulate GI function, as well as hunger, satiety, and insulin secretion. In addition, some hormones act on adjacent cells in a paracrine manner. An example is the action of somatostatin (produced in gastric pit cells) on adjacent parietal cells. GI hormones will be covered in Chapter 24.

GUT MICROBIOTA

The presence of bacterial microflora is a normal feature of the lower GI tract. More than a thousand species of bacteria (mostly gram-positive) are found in the intestines; many have been studied for their importance in nutrition and intestinal health. In the "healthy" gut, commensal bacteria have several functions that benefit the host. Bacteria can metabolize some undigested components of chyme, including some fibers and carbohydrates. They can produce short-chain fatty acids, such as acetic, butyric, and proprionic acid, and can synthesize biotin and vitamin K2. These products can be absorbed by the epithelial cells of the small and large intestines and can be used within the cells or enter the portal blood. Butyric acid is a source of energy for the colonic epithelial cells, and the major short-chain fatty acids also have important trophic effects on epithelial cell growth and differentiation. Bacteria also provide a protective function by stimulating the development and maintenance of immune competence in the gut. The immune function is important because the gut represents a broad interface between the internal and external environments. Changes in the composition of the bacterial flora (e.g., resulting from antibiotic treatment, chronic stress, and disease) can affect gut immunity, which may lead to acute or chronic inflammation of the intestinal mucosa and diarrhea.

Probiotics are dietary supplements (or dairy products such as yogurt) that contain species of live bacteria that are considered beneficial to gut health. Although the idea of adding useful bacteria to the GI tract has been explored for more than a century, it has been difficult to substantiate the effectiveness of probiotic supplementation. Many ongoing studies are evaluating the efficacy of probiotics and methods of delivery.

CLINICAL CORRELATE 22.1
Fecal Microbiota Transplants

Treatment of severe diarrhea using fecal matter was first described in traditional Chinese medicine by Ge Hong in the fourth century C.E. In the late 1950s, in response to the difficulty in treating cases of fulminant pseudomembranous colitis (caused by *Clostridium difficile* infection), physicians tested the effects of transplanting feces from healthy donors into affected colons. They postulated that repopulating the colon with healthy bacteria would resolve the colitis. The treatment was successful but had limited use until around 2000, when the incidence of highly virulent antibiotic-resistant *C. difficile* infections dramatically increased. Since then, fecal microbiota transplants (FMTs) have been used when *C. difficile* infections cannot be resolved or when they recur after antibiotic treatment. FMT has been shown to resolve *C. difficile* infection with a success rate of 90% and also has been shown to eliminate *C. difficile* spores, preventing reinfection.

The transplants of fecal bacteria (or fecal extracts) are infused by enema, colonoscope, or nasogastric tube; there does not appear to be a difference in outcomes with the various methods. *C. difficile* infections are usually resolved with one infusion, whereas treatment for ulcerative colitis requires multiple infusions to achieve remission.

As the link between imbalances in gut microbiota and GI disorders (including irritable bowel syndrome and ulcerative colitis) has become more evident, the use of FMT as a therapeutic option in critical cases has increased, with positive results. Clinical trials are being conducted to examine the therapeutic potential of FMT in various GI diseases, as well as in systemic disorders that might be associated with imbalances in gut microflora.

INTEGRATED REGULATION OF GI FUNCTION

Proper GI function results from input and integration at multiple levels:

- **Enteric nervous system (ENS):** The myenteric and submucosal nerve plexuses are unique to the GI tract. These nerve nets receive input from the parasympathetic and sympathetic nervous systems, as well as hormones, peptides, and lumenal receptors that sense the chemical composition of the chyme.
- **Central nervous system (CNS):** Sensory input to the CNS provides initial stimulus for salivary and gastric acid secretion and is integral to many reflexes. Simply smelling or seeing food initiates a central response. The CNS acts through the sympathetic and parasympathetic nervous systems.
- **Parasympathetic nervous system (PNS):** In general, the PNS promotes secretion and motility in the GI tract, and many of the actions of the PNS are through the vagus nerve.
- **Sympathetic nervous system (SNS):** The SNS slows secretion and motility in the GI tract. Consider what happens

in "fight or flight" reactions, which require more blood flow to skeletal muscle and less perfusion of the intestines.

- **Lumenal receptors:** The GI tract has many different receptors that react to the chyme in the lumen of the tract and act locally on the ENS to regulate motility and secretions through the following general scheme:

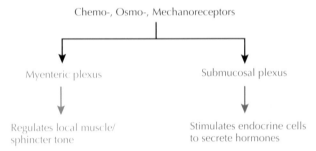

- **Mechanoreceptors** sense stretch of the smooth muscle, and the generated signal is transduced through the myenteric plexus, stimulating contractions.
- **Chemoreceptors** sense the chemical composition of the chyme and regulate motility and buffer secretion to control luminal pH during the influx of acidic chyme into the duodenum.
- **Osmoreceptors** sense the osmolarity of the chyme in the small intestine. This function is important because there is only a one-cell barrier between the chyme in the lumen of the small intestine and the capillaries, and hypertonic chyme can exert an osmotic force, drawing fluid out of the cells (and ultimately the plasma). Thus the osmoreceptors control the amount of chyme entering the small intestine, as well as the amount of secretions produced to buffer the chyme.
- **Hormones:** As previously stated, the GI tract produces a variety of endocrine, paracrine, and autocrine factors that increase the efficiency of the tract (both for motility and secretions).

Lastly, although all of these areas are important in the coordinated function of the GI tract, it bears noting that the **parasympathetic nerves**, primarily via the **vagus**, are responsible for multiple effects early in the digestive process. Some of the actions they stimulate or participate in include the following:

- Initiation of salivation upon seeing, smelling, or tasting food during the cephalic phase of salivation (through facial and glossopharyngeal nerves)
- Initiation of acid production when food is in the mouth during the cephalic and gastric phases of gastric acid secretion (vagus)
- Stimulation of pancreatic enzyme secretion during the cephalic and gastric phases of pancreatic secretion (vagus)
- Stimulation of pancreatic enzyme and buffer secretion during the intestinal phase of pancreatic secretion (vagus)
- Stimulation of primary peristalsis (through stimulation of vagal efferent nerves by the swallowing center of medulla) and possibly secondary esophageal peristalsis (which is mainly a local action through the myenteric nerves; the vagus may help but is not necessary)
- Initially relaxing the sphincter of Oddi during cephalic and gastric phases (vagus)
- Causing receptive relaxation of stomach and duodenum (vagus) to accommodate entry of food/chyme
- Stimulation of the synthesis of bile by the liver (vagus)
- Stimulation of intestinal motility (ileal motility and colonic mass movements) (vagus to upper colon, pelvic nerves on lower colon; with gastrin and cholecystokinin)

Thus, if parasympathetic (mainly vagal) innervation of these areas is impaired, the digestive and propulsive processes will be dysregulated.

Chapter 23

Motility Through the Gastrointestinal Tract

ELECTRICAL POTENTIALS

Motility in the gastrointestinal (GI) tract involves both mixing and propulsive movements and is determined by the mechanical activity of the intestinal smooth muscle. The movements serve to mix, transport, and eliminate matter. To accomplish the muscle contraction, electrical potentials must be generated.

The electrical activity in the GI tract is unique. Unlike in other tissues, there are undulations in the **resting membrane potential** known as **slow waves,** which are generated by the **interstitial cells of Cajal (ICC).** The ICC are located throughout the GI tract between the longitudinal and circular muscle layers, and although the undulations in their resting potential were long thought to be caused by the small changes in membrane charge resulting from Na^+/K^+ ATPase activity, this is not the case. Generation of the slow waves is more complex and involves the coupling of the smooth muscle to ICC and a third cell type, forming the SIP (smooth muscle, ICC, platelet-derived growth factor receptor–α^+ cell) syncytium. These SIP cells are regulated by ion channels (notably K^+) and receptor-mediated processes to produce the slow waves. The ICC, as part of the SIP syncytium, act as pacemakers, determining the number of waves that occur per minute along different segments of the GI tract. Because of gap junctions between cells, the slow waves (also called the **basic electrical rhythm**) can be propagated over relatively long segments of the tract.

Under resting conditions, the slow waves undulate between −70 and −80 millivolts (mV) and do not cause contractions. However, if the slow waves are depolarized (i.e., made less negative) by nerve activity or circulating hormones, the amplitude of the waves may increase, and when peaks of the slow wave cross the threshold of −40 mV, the cells will generate one or more **action (or spike) potentials**.

Neurotransmitters such as acetylcholine and substance P (tachykinin) released from parasympathetic nerves terminating on the myenteric plexus depolarize the slow waves, generating action potentials and causing contractions (Figs. 23.1 and 23.2). With greater neurotransmitter release, the amplitude of the waves is higher, producing greater depolarization.

Greater depolarization results in more action potentials, increasing the strength of the contraction. Some GI hormones (cholecystokinin [CCK] and gastrin) can also depolarize the slow waves, causing contractions. In addition, local mechanoreceptors that sense stretch or chemoreceptors that sense composition of chyme signal the myenteric plexus to fire excitatory motor neurons, depolarizing the slow waves and causing contraction (Fig. 23.3). The characteristic frequency of the slow waves for any section of the tract determines the maximal rate at which contractions could occur, and thus the basic electrical rhythm sets the maximal rate of propulsion possible (Table 23.1).

Serotonin (5-hydroxytryptamine) is an important neurotransmitter released from both enterochromaffin cells and interneurons in the GI tract, with about 90% of the total serotonin in the body coming from GI enterochromaffin cells. Release of serotonin from the enterochromaffin cells (and into blood) depolarizes the slow waves in smooth muscle, evoking peristaltic-type contractions. Although this function is normal, if the GI tract is irritated, more serotonin is released, causing additional slow wave depolarization and increasing contractions. This process can cause both vomiting and diarrhea, depending on the area of the tract affected.

The **action potentials** generated by depolarization of the slow waves are caused by the entry of **calcium** into the smooth muscle via voltage-gated channels. The calcium binds to calmodulin, initiating events that result in smooth muscle contraction.

Inhibitory motor neurons also exist that are stimulated by sympathetic nerves and release **vasoactive intestinal peptide (VIP)** and nitric oxide, which hyperpolarize the slow waves (i.e., make them more negative), relaxing the smooth muscle (see Fig. 23.3). The interplay between excitatory and inhibitory motor neurons results in a variety of propulsive and mixing movements through the tract (see "Site-Specific Propulsion").

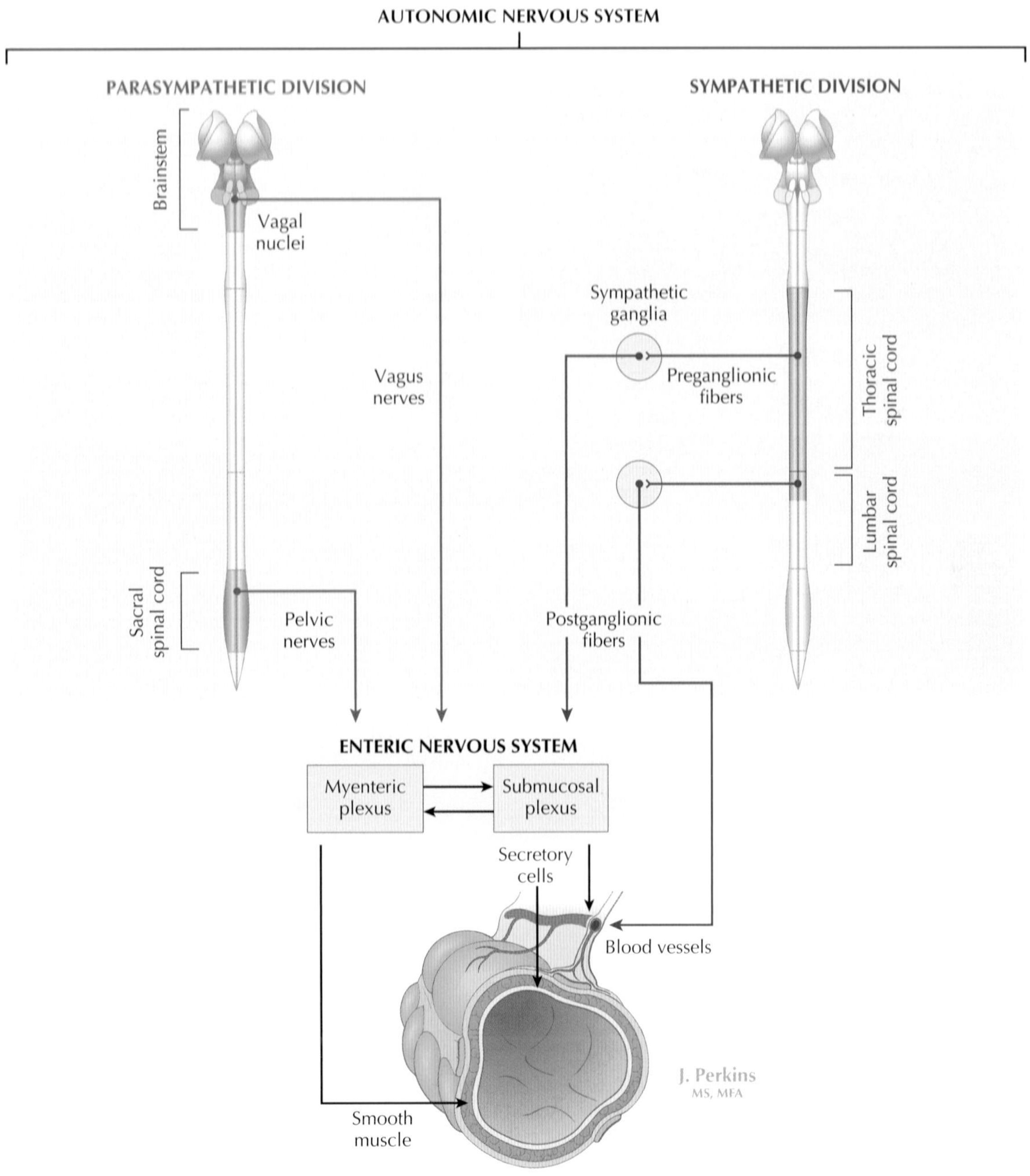

AUTONOMIC NERVOUS SYSTEM

PARASYMPATHETIC DIVISION

SYMPATHETIC DIVISION

Brainstem

Vagal nuclei

Vagus nerves

Sympathetic ganglia

Preganglionic fibers

Thoracic spinal cord

Lumbar spinal cord

Sacral spinal cord

Pelvic nerves

Postganglionic fibers

ENTERIC NERVOUS SYSTEM

| Myenteric plexus | → ← | Submucosal plexus |

Secretory cells

Blood vessels

J. Perkins
MS, MFA

Smooth muscle

Figure 23.1 **Regulation of the Enteric Nervous System by the Autonomic Nervous System**
This diagram illustrates basic connections between the enteric nervous system (myenteric and submucosal plexuses) and the autonomic nervous system. In general, stimulation from the parasympathetic nerves enhances motility and secretions through the enteric nervous system, whereas sympathetic stimulation reduces motility and secretions. The sympathetic nerves can also act directly on intestinal blood vessels and stimulation will constrict vessels, reducing flow to the affected area.

INTERDIGESTIVE HOUSEKEEPING: THE MIGRATING MYOELECTRIC COMPLEX

Fasting, or the interdigestive state, is characterized by long periods of quiescence, with short periods of waves of contractions. The contractions, called the **migrating myoelectric complex (MMC)**, originate in the mid stomach and continue to the terminal ileum during each cycle. They serve to sweep the undigested material and bacteria out of the stomach and small intestine and into the colon, thus protecting the delicate upper GI tract (small intestine) from damage and sequestering most of the bacteria in the colon, which is a drier environment that is less prone to bacterial overgrowth.

The MCC has four phases that occur in 75- to 120-minute cycles during fasting, based on contractile activity. Phases I, II,

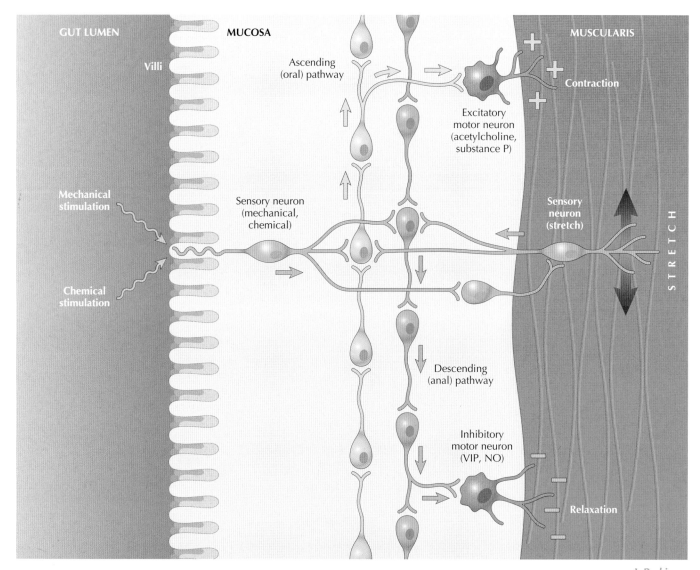

Figure 23.2 **Local Control of Motility** In response to the presence of chyme in the lumen of the small intestine, mechanoreceptors and chemoreceptors transduce signals to ascending and descending neurons. The ascending *(green)* pathway leads to excitatory motor neurons behind the bolus of chyme, stimulating depolarization of the slow waves, generation of an action potential, and contraction. At the same time, the descending *(pink)* pathway ends in inhibitory motor neurons, which hyperpolarize the slow waves in the muscle, causing relaxation in front of the bolus. The overall result is a peristaltic contraction, moving the bolus toward the anus. *NO,* nitric oxide; *VIP,* vasoactive intestinal peptide.

and IV have little activity. Phase III is the most important of the four, when the hormone **motilin** is released from Mo cells of the small intestine into the circulation. Motilin stimulates strong sequential contractions, sweeping bacteria and undigested matter down the tract from mid stomach into the colon. Phase III lasts only 6 to 10 minutes in each cycle. Motilin acts through the enteric and autonomic nerves to stimulate contractions, which proceed for a few feet aborally and then start up again slightly further down the intestine. This pattern is repeated until the complex reaches the terminal ileum, and then a new complex is initiated in the stomach.

These cycles begin about 3 hours after the last meal and repeat until ingestion of food. With ingestion, the normal propulsive patterns resume.

SITE-SPECIFIC PROPULSION

Mouth and Esophagus

Food is first moved by chewing and then **swallowing**, which has three stages. In the voluntary **oral** stage, the bolus of food

J. Perkins
MS, MFA

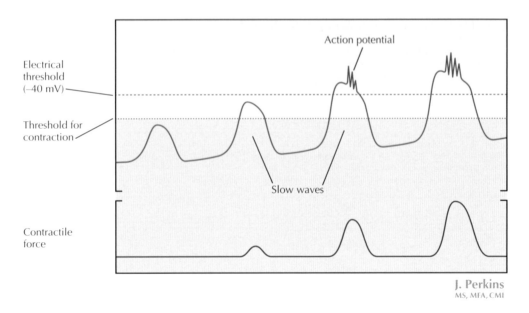

Figure 23.3 **Slow Waves**   Slow waves are present from mid stomach through the rectum and are the resting membrane potential of the gastrointestinal smooth muscle. Depolarization of the slow waves above −40 mV stimulates action potentials, which cause contraction of the smooth muscle.

Table 23.1   **Slow Waves**

Area	Frequency of Slow Waves	General Actions When Depolarized Above Threshold
Stomach	3/minute	Mixing
Duodenum	12/minute	Propulsion
Ileum	10/minute	Propulsion
Proximal colon	3/hour 11/minute	Haustra formation and storage Mass movements and propulsion
Distal colon	10/hour 17/minute	Haustra formation and storage Mass movements and propulsion

is moved to the back of the mouth into the pharyngeal region, stimulating touch receptors and initiating the swallowing reflex (Table 23.2). In the **pharyngeal** stage, the bolus of food moves to the back of the pharynx, the larynx moves toward the epiglottis to prevent food entering the trachea, and the upper esophageal sphincter relaxes. The pharyngeal muscles contract, propelling the bolus through the upper esophageal sphincter into the esophagus. During the pharyngeal phase, the swallowing reflex prevents respiration. The last stage is the **esophageal** stage. When the bolus enters the esophagus, voluntary control of the movement is lost. A wave of **primary esophageal peristalsis** is generated by the swallowing center in the medulla (efferent vagal nerves terminate on the myenteric plexus → excitatory signal → depolarized slow waves → contraction). The alternating contraction and relaxation helps carry the bolus of food through the esophagus to the stomach.

During swallowing, if the bolus is dry and does not move quickly through the esophagus, **secondary esophageal peristalsis** is initiated by local mechanoreceptors sensing stretch. The **lower esophageal sphincter** has a high resting muscle tone to prevent reflux of gastric acid. This basal tone is maintained by enteric nerves, hormones, and vagal cholinergic fibers and can be increased by sympathetic stimulation. As the bolus nears the lower esophageal sphincter, noncholinergic vagal fibers acting on inhibitory interneurons release VIP and nitric oxide, causing relaxation of the sphincter (Fig. 23.4). The bolus then passes into the stomach. If the lower esophageal sphincter is unable to relax, disease can occur (see Clinical Correlate 23.4). In addition, the esophagus, like the remainder of the tract, is highly vascularized, and damage to the esophageal capillaries (e.g., by acid reflux or portal hypertension) can result in esophageal bleeding, which can be life-threatening (see Clinical Correlate 25.2 in Chapter 25).

Stomach

As the bolus enters the stomach, **receptive relaxation** increases the stomach size, allowing accommodation of the meal. Receptive relaxation is vagally mediated through the release of VIP. As the stomach fills, the stretch and chemical contents stimulate contractions, which help mix the food with the gastric secretions, forming chyme. As the chyme mixes, the contents separate, with carbohydrates and readily digested substances in the lower part (antrum), large chunks in the body of the stomach, and most fats floating at the top. The **transit time** through the stomach depends on the amount and type of food ingested. Small, easily digested meals (e.g., meals that are high in carbohydrates such as pasta and sugar) move through the stomach quickly (i.e., in 30 to 60 minutes). Meals with more solid foods (such as meats) and high fat content

Table 23.2 Main Reflexes Through the Gastrointestinal Tract

Reflex	Action	Mediated By
Swallowing	Contraction of the pharynx and upper esophagus, inhibition of respiration	Touch receptors in the pharynx to the swallowing center of the medulla; back to the pharynx and upper esophagus through cranial nerves and to the remainder of the esophagus through vagal nerves
Reflexive relaxation of the stomach and duodenum	Relaxes the fundus and body of the stomach when food, water, or gas is present; also occurs in the duodenum when a bolus enters	Vagal fiber release of VIP
Vomiting	Expulsion of upper intestinal and gastric contents by reverse peristalsis	Irritation of the pharynx, esophagus, stomach, or intestines via vagal and sympathetic afferents to the vomiting center; or stimulation of the chemoreceptor trigger zone (in the medulla near area postrema) to the vomiting center
Peristalsis	Contraction behind a bolus, relaxation ahead of a bolus	Mechanoreceptor and hormone action on enteric nerves
Gastrocolic	Mass movements in the colon after a meal	PNS and hormones (CCK, gastrin)
Gastroileal	Increased segmentation in the ileum in response to gastric emptying	PNS and hormones (CCK and gastrin)
Ileogastric	Chyme in the ileum decreases gastric emptying	Enteric and autonomic nerves
Enteroenteric (also known as intestinointestinal)	If one area of SI is overdistended (e.g., by bacterial infection), the remainder of the SI will relax, causing cessation of motility	Enteric and autonomic nerves
Colonocolonic	Distension of one area of the colon will relax other areas	Enteric and sympathetic nerves
Rectosphincteric (also known as the defecation reflex)	Feces entering the rectum will cause peristalsis and relax the internal anal sphincter	Both local (enteric nerves) and PNS

These reflexes do not act in an all-or-nothing manner; many act at the same time to create efficient movement of the chyme.
CCK, cholecystokinin; PNS, parasympathetic nervous system; SI, small intestine; VIP, vasoactive intestinal peptide.

(such as fried foods) take much longer (e.g., 3 to 4 hours) to clear the stomach.

Figure 23.5 illustrates the waves of contractions that begin in the mid stomach and move the chyme toward the antrum and pylorus (which acts as a sphincter). Regulation of the tone of the pyloric sphincter occurs through both neural and hormonal pathways (Table 23.3).

For the most part, pyloric tone is high, so as the stomach is filling and contractions are starting, most of the acidic chyme moving into the antrum undergoes retropulsion away from the sphincter (see Fig. 23.5B, 1 and 2). As the waves of contraction progress into the antrum, the **antral cycle** occurs, with relaxation of the sphincter and ejection of some chyme into the duodenal bulb (see Fig. 23.5B, 3). This cycle (ejection into duodenum) is under tight control, and as chyme enters the antrum and duodenum, the release of hormones decreases

Table 23.3 Effects of Nerves and Hormones on Gastric Emptying

Effector	Action on Pyloric Sphincter
Sympathetic nerves	Constriction

Parasympathetic vagal nerves
   Excitatory via ACh motor neuron
   Inhibitory via VIP motor neuron
	
	Constriction
	Relaxation

Hormones—GIP, CCK, secretin	Constriction

ACh, acetylcholine; CCK, cholecystokinin; GIP, glucose insulinotropic peptide; VIP, vasoactive intestinal peptide.

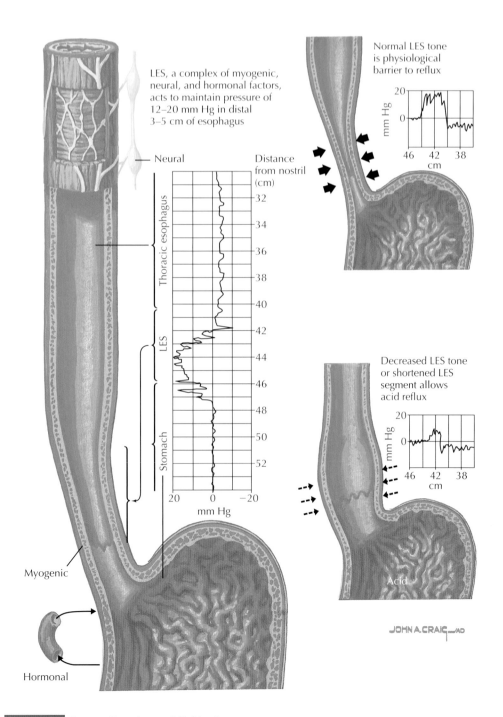

LES, a complex of myogenic, neural, and hormonal factors, acts to maintain pressure of 12–20 mm Hg in distal 3–5 cm of esophagus

Normal LES tone is physiological barrier to reflux

Decreased LES tone or shortened LES segment allows acid reflux

Neural

Distance from nostril (cm)

Thoracic esophagus

LES

Stomach

Myogenic

Hormonal

JOHN A. CRAIG—AD

Figure 23.4 **Lower Esophageal Sphincter**   The resting tone of the lower esophageal sphincter (LES) is usually very high, preventing reflux of stomach contents. When food is swallowed, esophageal peristalsis is initiated by the vagus nerve and is propagated by the enteric nerves. As the bolus of food reaches the LES, local nitric oxide and vasoactive intestinal peptide are released, and the sphincter relaxes, allowing the bolus to enter the stomach. If LES tone is decreased when at rest, acid reflux can occur.

CLINICAL CORRELATE 23.1
Bariatric Surgery

Receptive relaxation is crucial for the storage function of the stomach, which allows the contents to be well mixed with gastric acid and enzymes and some digestion to occur. If vagal innervation to the fundus is lost, the accommodation will not occur

and stomach pressures will quickly rise, imparting a feeling of "fullness" and the inability to physically get more into the stomach. Reducing the storage ability of the stomach is the rationale behind gastric bypass and stomach stapling in severely obese persons.

■ **Adjustable gastric banding** restricts the size of the pylorus so that only small amounts of food can comfortably enter the

Continued

CLINICAL CORRELATE 23.1
Bariatric Surgery—cont'd

stomach. This procedure restricts food intake because of discomfort but does allow food to enter the intact stomach.

- **Vertical banded gastroplasty** was once a common procedure involving both stomach stapling and banding to create a small pouch while leaving the remainder of the tract intact. It is now considered dangerous and is performed less frequently today.
- **Gastric bypass**, a common surgery, involves keeping only a small part of the fundus and attaching it to the early jejunum. This procedure leaves little space for food (about a tablespoonful). In this procedure, the main body of the stomach is "stapled" off and the duodenum is reattached to the jejunum, allowing the digestive juices (primarily from the pancreas and liver) to mix with the food. The overall effect is a reduction in food intake and weight loss.

Bariatric surgery is not without risk and can be associated with vomiting, diarrhea, reflux, leaking from the surgical sites, and

infection. Although it is usually recommended only in morbidly obese patients, it has been determined that bariatric surgery not only results in weight loss but that a significant number of patients (>50%) with type 2 diabetes (T2D) experience normalization of glucose levels and lowering of triglycerides and cholesterol in the blood. This remission of T2D is associated with an increase in circulating glucagon-like peptide–1 and peptide YY with feeding, which enhances satiety. The glucagon-like peptide–1 is also associated with an increase in insulin secretion from the pancreatic β-cells (insulin sensitivity and sometimes insulin secretion are reduced in T2D). Because of these positive outcomes, bariatric surgery is being explored as a possible therapy for T2D in obese patients; however, noninvasive therapy is still preferred.

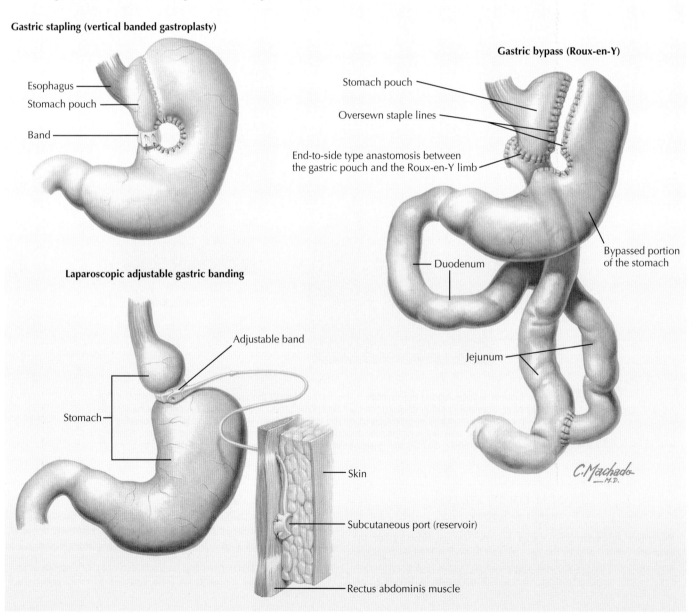

Gastric stapling (vertical banded gastroplasty)

Esophagus
Stomach pouch
Band

Laparoscopic adjustable gastric banding

Adjustable band
Stomach
Skin
Subcutaneous port (reservoir)
Rectus abdominis muscle

Gastric bypass (Roux-en-Y)

Stomach pouch
Oversewn staple lines
End-to-side type anastomosis between the gastric pouch and the Roux-en-Y limb
Duodenum
Bypassed portion of the stomach
Jejunum

C.Machado
M.D.

A. Factors affecting gastric emptying

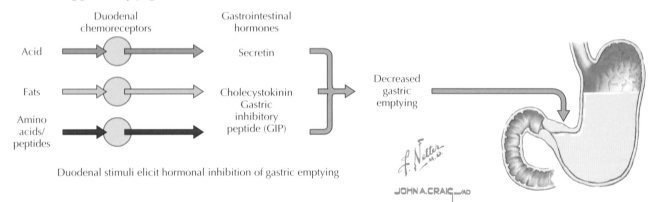

Duodenal stimuli elicit hormonal inhibition of gastric emptying

B. Sequence of gastric motility

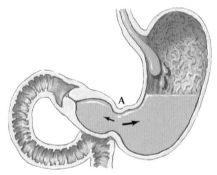

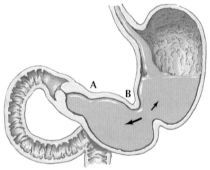

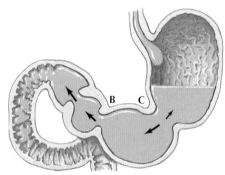

1. Stomach is filling. A mild peristaltic wave (A) has started in antrum and is passing toward pylorus. Gastric contents are churned and largely pushed back into body of stomach.

2. Wave (A) fading out as pylorus fails to open. A stronger wave (B) is originating at incisure and is again squeezing gastric contents in both directions.

3. Pylorus opens as wave (B) approaches it. Duodenal bulb is filled, and some contents pass into second portion of duodenum. Wave (C) starting just above incisure.

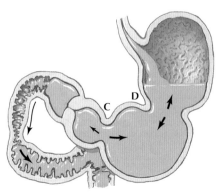

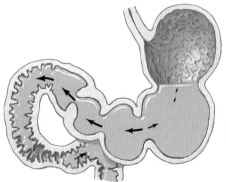

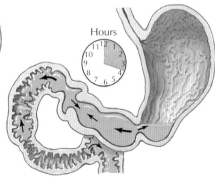

4. Pylorus again closed. Wave (C) fails to evacuate contents. Wave (D) starts higher on body of stomach. Duodenal bulb may contract or may remain filled as peristaltic wave originating just beyond it empties second portion.

5. Peristaltic waves are now originating higher on body of stomach. Gastric contents are evacuated intermittently. Contents of duodenal bulb area pushed passively into second portion as more gastric contents emerge.

6. 3 to 4 hours later, stomach is almost empty. Small peristaltic wave empties duodenal bulb with some reflux into stomach. Reverse and antegrade peristalsis present in duodenum.

Figure 23.5 **Gastric Motility**   The presence of food in the stomach stimulates gastric motility through release of vasoactive intestinal peptide stimulated by vagal nerves. The stomach contractions force the gastric contents toward the pylorus, where most of the contents are pushed back into the body of the stomach *(1)*. As spurts of chyme enter the duodenum *(3)*, duodenal hormones are released into the blood, circulate back to the stomach, and decrease gastric emptying (**A**). Thus there is a coordinated effect of local and autonomic nerves and hormones controlling gastric motility. The entire sequence is depicted in **B**, images *1-6*.

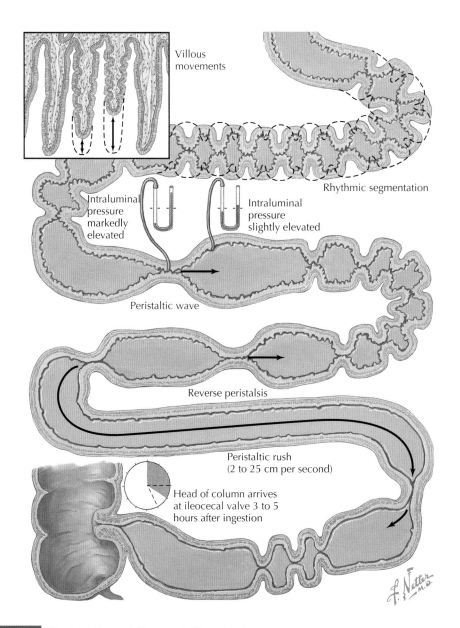

Villous movements

Rhythmic segmentation

Intraluminal pressure markedly elevated

Intraluminal pressure slightly elevated

Peristaltic wave

Reverse peristalsis

Peristaltic rush (2 to 25 cm per second)

Head of column arrives at ileocecal valve 3 to 5 hours after ingestion

Figure 23.6 **Peristalsis and Segmentation** Motility in the small intestine is primarily under local control of the myenteric plexus and consists of both peristalsis and segmentation. Segmentation forms pockets of chyme and serves to mix and propel the chyme, although normally peristalsis creates aboral movement (away from mouth). Peristaltic rushes can occur when the intestines are irritated, moving chyme rapidly through the intestines.

gastric emptying by constricting the sphincter (see Table 23.3). Thus relaxation via vagal inhibitory signals allows the antral cycle to occur, while the other factors prevent too much acidic chyme from entering the duodenum at once.

Small Intestine

Two types of propulsion occur in the small intestine: peristalsis and segmentation. **Peristalsis** follows the "law of the intestines," according to which contraction takes place behind the bolus of chyme and relaxation takes place ahead of the chyme (Fig. 23.6 and Video 23.1). This process is accomplished by the simultaneous stimulation of excitatory and inhibitory motor neurons (see Fig. 23.2). In this manner, the

chyme is propelled aborally (away from the mouth). In the small intestines, **peristaltic rushes** also can occur, which quickly propel the chyme over long segments. Peristaltic rushes occur when there is irritation or bacteria in the segments; the movement facilitates removal of the irritant. Because peristaltic rushes quickly sweep the chyme through the intestines, much less absorption occurs, which can result in **diarrhea** (see Fig. 23.6). **Reverse peristalsis** (seen in vomiting) also can occur, when intestinal and stomach contents are moved rapidly toward the mouth.

Segmentation is the formation of moving pockets of chyme by relaxation of the longitudinal muscle and close constriction of circular muscle (see Fig. 23.6). Unlike peristalsis, which

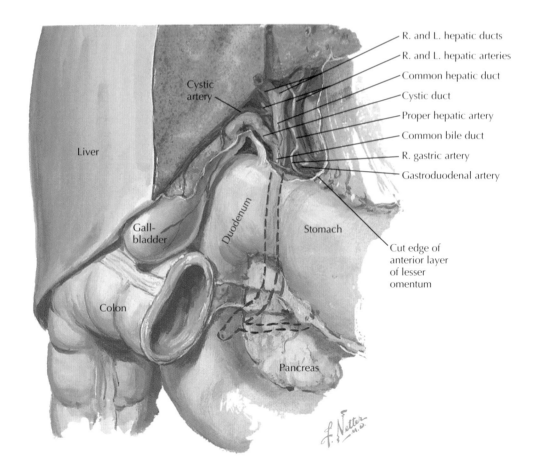

R. and L. hepatic ducts
R. and L. hepatic arteries
Common hepatic duct
Cystic duct
Proper hepatic artery
Common bile duct
R. gastric artery
Gastroduodenal artery

Cystic artery

Liver

Gall-bladder

Duodenum

Stomach

Cut edge of anterior layer of lesser omentum

Colon

Pancreas

Figure 23.7 **Structure of the Gallbladder** The gallbladder is the storage site for bile and is located adjacent to the liver. It can store between 25 and 50 mL of bile, which is released by vagal stimulation and cholecystokinin. When food and chyme are in the stomach, the vagus stimulates initial contraction. Subsequently, when chyme enters the duodenum, cholecystokinin is released, causing strong contractions and emptying of the gallbladder. Bile is concentrated (i.e., salts and water are removed) in the gallbladder, allowing more storage. *L.,* left; *R.,* right.

contracts behind the bolus, segmentation contracts in the middle of the bolus, spreading the chyme proximally and distally. These constrictions occur rhythmically and are the most frequent type of contractions in the small intestine. Because the simultaneous constriction moves in waves, the "pockets" slowly move aborally, mixing contents backward and forward. Although much mixing occurs, there is also net aboral movement. Segmentation and peristalsis occur in adjacent segments of the small intestine throughout the digestion and absorption of the chyme. This pattern is orchestrated by the myenteric nerves (with fine-tuning by autonomic nerves and hormones).

Gallbladder

The gallbladder is a specialized structure for concentration and storage of bile (approximately 30-60 milliliters [mL]) during the interdigestive state (Fig. 23.7). When chyme first enters the duodenum, CCK is secreted into the blood in response to the presence of fats and glucose in the chyme. CCK, in combination with vagal stimulation, causes rhythmic

contractions of the gallbladder, forcing bile out through the cystic duct and into the common bile duct. CCK also relaxes the **sphincter of Oddi**, allowing the contents of the common bile duct to enter the duodenum. The vagal nerves can also relax the sphincter of Oddi indirectly by stimulating CCK.

Large Intestine (Colon)

The colon also has specialized forms of propulsion: segmental propulsion and mass movements. The muscle structure of the large intestine is different from that of the small intestine, in that three bands of longitudinal muscle, called the **taenia coli**, run the length of the organ. When they contract, they form sacs called **haustra**; this occurrence is **segmental propulsion** (Fig. 23.8). The haustra stay in formation for long periods, which helps "store" the chyme as it is dehydrated and made into feces. The ileocecal sphincter regulates the amount of chyme entering the colon. The sphincter is usually tonically closed, but movement of chyme in the terminal ileum relaxes the sphincter, allowing small amounts of chyme to enter the colon. The **ileocecal sphincter** is controlled by

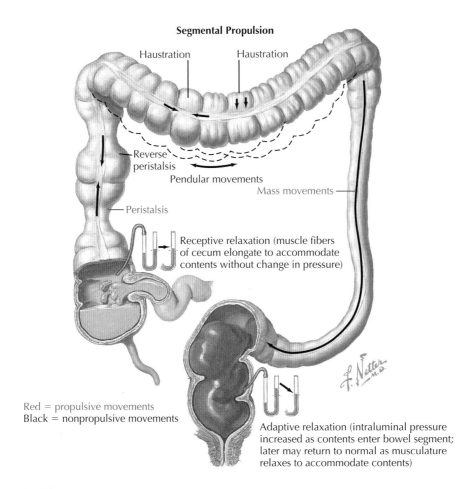

Segmental Propulsion

Haustration   Haustration

Reverse peristalsis

Pendular movements

Mass movements

Peristalsis

Receptive relaxation (muscle fibers of cecum elongate to accommodate contents without change in pressure)

Red = propulsive movements
Black = nonpropulsive movements

Adaptive relaxation (intraluminal pressure increased as contents enter bowel segment; later may return to normal as musculature relaxes to accommodate contents)

Figure 23.8 **Motility in the Colon** The colon is specialized for storage, and when bands of longitudinal muscle (taenia coli) contract, they form haustra. This process is called segmental propulsion, which is very slow, and allows time for additional sodium and fluid absorption (colonic salvage). The slow movement dries out the chyme, producing feces. Several times a day mass movements will occur, creating peristaltic movements that force the chyme/feces toward the rectum. Factors that increase intestinal motility can cause diarrhea by minimizing formation of haustrae and increasing mass movements.

the enteric nerves, as well as hormones (CCK and gastrin) and autonomic nerves.

Although segmental propulsion is very slow, **mass movements** periodically occur (usually one to three times daily). Mass movements are peristaltic in nature, but contractions extend over a longer period (seen in Fig. 23.8, descending colon). These strong contractions occur when the distal colon is relaxed, and they force the feces to travel quickly through the descending colon into the rectum. The mass movements are stimulated by the parasympathetic nerves (vagal nerves in the proximal colon and pelvic nerves in the distal colon and rectum), as well as by the hormones CCK and gastrin. These factors are all active when there is chyme in the upper GI tract (during active digestion), and their stimulation of mass movements serves to clear the lower GI tract of chyme and feces in preparation for new waste coming down the tract. Conversely, stimulation of the sympathetic nerves (from the superior mesenteric in the proximal colon and from the inferior mesenteric and superior hypogastric nerves in the distal colon) inhibits colonic movements, consistent with the overall actions of the sympathetic nervous system on the GI system (Fig. 23.9).

DEFECATION AND THE RECTOSPHINCTERIC (DEFECATION) REFLEX

When feces are moved into the rectum, the rectal distension is sensed by local mechanoreceptors that signal the myenteric nerves to relax the **internal anal sphincter** and initiate peristalsis, further pushing feces into the rectum. This local effect is reinforced by the parasympathetic nervous system (PNS), which initiates stronger contractions via pelvic nerves. At the same time, an afferent signal is sent centrally, stimulating the **urge to defecate**, which is the **rectosphincteric**, or **defecation, reflex** (Fig. 23.10). This reflex is rapid, and in response to the urge to defecate, we voluntarily constrict the **external anal sphincter** to prevent defecation. If defecation cannot occur, some relaxation of the rectum occurs, the internal anal sphincter constricts again, and the external anal sphincter can be relaxed, until another movement pushes feces into the rectum and the reflex recurs. At the appropriate time for defecation, the external anal sphincter voluntarily relaxes, intraabdominal pressure increases (the **Valsalva maneuver**), and defecation occurs. This reflex takes time to fully develop in children and is usually intact by 2 to 4 years of age.

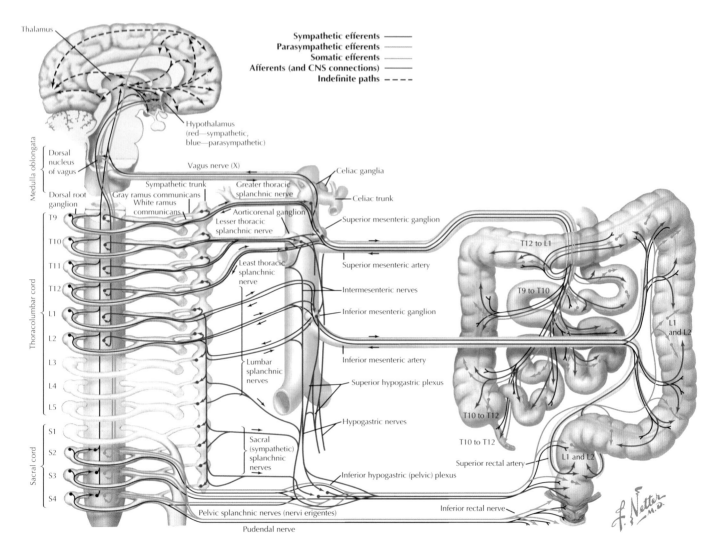

Figure 23.9 **Autonomic Innervation of the Colon** The small intestines and colon are innervated by both sympathetic and parasympathetic nerves. Sympathetic fibers from the spinal cord lead to the celiac, superior mesenteric ganglia, and inferior mesenteric ganglia. Parasympathetic fibers arise from the vagus nerves (which innervate the transverse colon) and pelvic nerves (which innervate the descending colon, sigmoid, and rectum). Sympathetic stimulation reduces motility and secretions, whereas parasympathetic stimulation increases motility, secretions, and relaxation of the internal anal sphincter. *CNS*, central nervous system.

REGULATION OF MOTILITY

Motility is regulated by the enteric nerves, autonomic nerves, and hormones. The enteric nerves respond to luminal receptors that sense the presence and composition of the chyme, as well as signals from autonomic nerves and hormones. As previously stated, the PNS primarily promotes motility, whereas the sympathetic nervous system slows or stops motility. A general rule is that most of the GI hormones inhibit gastric motility and/or emptying, thus allowing the duodenum to process the chyme that is released (Fig. 23.11). However, gastrin is an exception; it tends to stimulate gastric motility and activity of the antral cycle, thus increasing gastric emptying. These hormones work together to regulate the exit of chyme from the stomach. The relaxation of the pylorus by gastrin and the PNS is moderated by the effects of

gastric inhibitory peptide, CCK, and secretin, which increase pyloric tone.

In addition, CCK and gastrin have stimulatory effects on motility in the lower small intestine and colon. They initiate mass movements in the colon, facilitating removal of the undigested waste matter. Motilin has the specific function of stimulating contractions during fasting (the MMC) (see Fig. 23.11).

VOMITING AND DIARRHEA

Vomiting (emesis) is a reflex action that is under central regulation by the vomiting center in the medulla oblongata. Stimuli for vomiting also initiate salivation, and retching usually

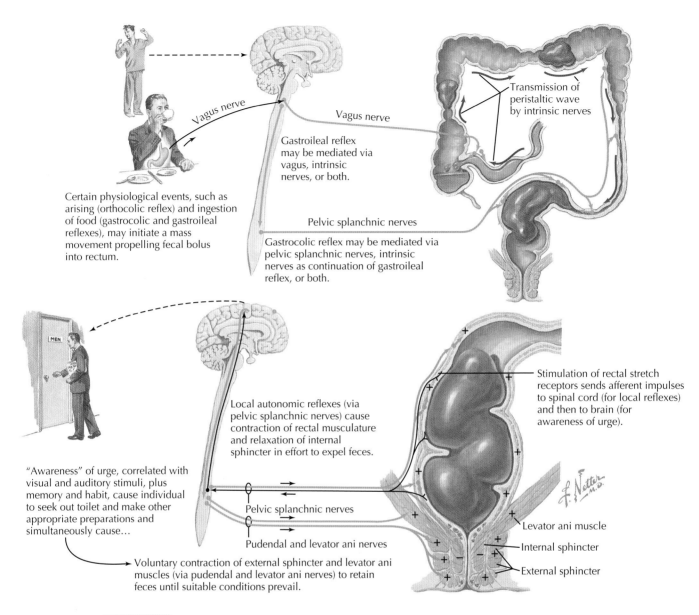

Certain physiological events, such as arising (orthocolic reflex) and ingestion of food (gastrocolic and gastroileal reflexes), may initiate a mass movement propelling fecal bolus into rectum.

Vagus nerve

Gastroileal reflex may be mediated via vagus, intrinsic nerves, or both.

Pelvic splanchnic nerves

Gastrocolic reflex may be mediated via pelvic splanchnic nerves, intrinsic nerves as continuation of gastroileal reflex, or both.

Transmission of peristaltic wave by intrinsic nerves

Stimulation of rectal stretch receptors sends afferent impulses to spinal cord (for local reflexes) and then to brain (for awareness of urge).

Local autonomic reflexes (via pelvic splanchnic nerves) cause contraction of rectal musculature and relaxation of internal sphincter in effort to expel feces.

"Awareness" of urge, correlated with visual and auditory stimuli, plus memory and habit, cause individual to seek out toilet and make other appropriate preparations and simultaneously cause...

Pelvic splanchnic nerves

Pudendal and levator ani nerves

Voluntary contraction of external sphincter and levator ani muscles (via pudendal and levator ani nerves) to retain feces until suitable conditions prevail.

Levator ani muscle

Internal sphincter

External sphincter

Figure 23.10 **Defecation** The sequence of events leading to defecation is detailed in this illustration. The defecation (or rectosphincteric) reflex is triggered by distension of the rectum by feces. This distension is sensed by the mechanoreceptors that activate the myenteric nerves, and impulses are transduced to the spinal cord (and back) to relax the internal anal sphincter. The spinal cord conveys signals to the brain to stimulate the urge to defecate. These reflex actions cause the voluntary contraction of the external anal sphincter until defecation occurs. During defecation, the external anal sphincter relaxes, the person increases intraabdominal pressure, and feces are eliminated from the rectum.

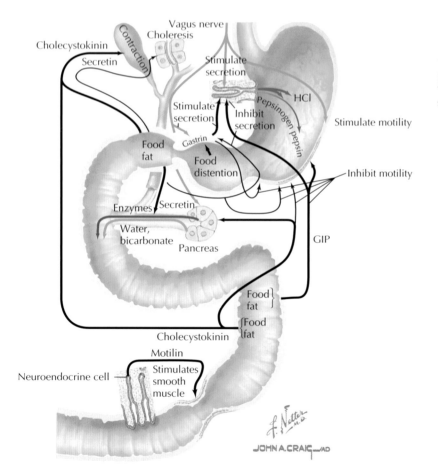

Figure 23.11 **Effect of Major Gastrointestinal Hormones on Gastrointestinal Motility** The main effects of gastrointestinal hormones on gastrointestinal motility are depicted in this illustration. Although many hormones have effects on gastric motility, a few stimulate motility in the small and large intestines, including motilin (small intestine only) and gastrin (in both the small and large intestines, not shown). *GIP,* gastric inhibitory peptide; *HCl,* hydrochloric acid.

precedes vomiting. During retching, intestinal and gastric contents rise into the esophagus but not into the mouth because of a closed upper esophageal sphincter. The reflex can be stimulated by factors within the GI tract, as well as centrally through the chemoreceptor trigger zone (Box 23.1).

The events induced by all stimuli are consistent. **Reverse peristalsis** occurs (with aboral constriction and relaxation proximal to the bolus) from the middle of the small intestine to the pylorus, which relaxes to allow chyme to enter the stomach. The stomach fills, and strong abdominal muscle contractions force gastric contents into the esophagus, causing retching. With further stimulus, the upper esophageal sphincter relaxes, allowing contents to rise into the mouth. Because the vomitus includes intestinal and stomach contents, acid, bile, and electrolytes are excreted. Chronic vomiting can result in esophageal damage and metabolic disorders. Because serotonin released from enterochromaffin cells induces motility associated with nausea and vomiting, ondansetron, a 5-hydroxytryptamine antagonist, is an effective antiemetic.

Box 23-1 **Examples of Stimuli Initiating the Vomiting Reflex**

- An irritant in the stomach or small intestine, an enteric virus, or bacteria (local effect through enteric chemoreceptors)
- A systemic irritant sensed by the chemoreceptor trigger zone on the floor of the fourth ventricle of the brain (near area postrema)
- A head injury (concussion; central effect)
- Abnormal stimulation of vestibular organs (central effect via the autonomic nervous system to the chemoreceptor trigger zone, then to the vomiting center)

CLINICAL CORRELATE 23.2
Inflammatory Bowel Disease

The causes of irritable bowel syndrome (IBS) and inflammatory bowel disease (IBD) are not clear. Although both IBS and IBD result in diarrhea and abdominal pain, it is important to distinguish between these disorders affecting the GI tract, because they result in distinct pathologic presentations and require different therapeutic management. In the United States, the incidence of IBD is between 28 to 199 per 100,000 persons.

The primary types of IBD are **Crohn's disease** and **ulcerative colitis (UC)**. Although both are autoimmune disorders involving chronic inflammation of the GI tract, an important distinction is that whereas Crohn's disease can occur anywhere along the tract (from the mouth to the anus, although the rectum is usually spared) and involves transmural inflammation, UC occurs *only* in the colon (hence its name) and involves only the mucosa and submucosa. Both IBDs are characterized by acute flare-ups alternating with periods of remission.

Crohn's disease may present in any area of the GI tract, although it often affects ileal, ileocolic, and colic regions (part **C**). Marked inflammation infiltrates the bowel tissue, giving it a cobblestone appearance (part **C**), with intraabdominal abscesses and bowel-skin or bowel-viscus fistulae (part **A**). In contrast, the granular, superficial ulcer formation found in UC begins in the rectum and moves proximally toward the cecum; the disease rarely, if ever, infiltrates the ileum. In addition to diarrhea, symptoms of IBD include tenesmus (i.e., feeling the urge to defecate without being able to do so), urgency, pain, and sometimes mucus, pus, or blood in diarrhea (although blood in stool is seen more often in UC than in Crohn's disease). Systemically, patients may present with fever, fatigue, and weight loss; with long-term disease, other autoimmune-related symptoms such as arthritis, uveitis, and pyoderma gangrenosum can be present. Because Crohn's disease and UC are autoimmune diseases, the primary therapy includes immunosuppression with oral steroids and antibiotics and progresses to intravenous steroids and biologic agents (such as tumor necrosis factor–α inhibitors). Unlike Crohn's disease, which has no cure, in persons with severe UC, removal of the colon and rectum will cure the patient. Follow-up surgery is performed to make a new "rectum" via ileal pouch anal anastomosis, which is then attached at the end to the anus (part **D**).

Environmental factors appear to play a large role in the diseases. Because the gut microbiota modulates GI and systemic immunity, factors that create an imbalance in the GI flora (such as diet, stress, and use of nonsteroidal antiinflammatory drugs and antibiotics) are risk factors for the development of IBD. IBD also has a genetic component, with persons of Jewish descent (especially Ashkenazi Jews) exhibiting a higher incidence of IBD. Interestingly, although smoking is a high risk factor in Crohn's disease, it decreases the risk of UC.

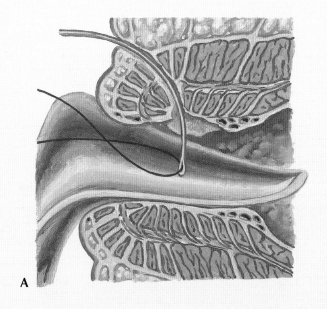

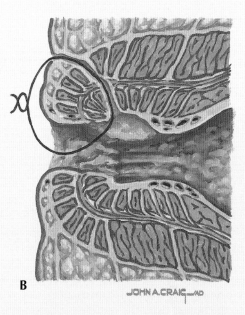

JOHN A. CRAIG—MD

Inflammatory Bowel Disease Crohn's disease can affect any part of the GI tract, from mouth to anus, causing transmural infiltration and inflammation and producing abscesses and fistulae (parts **A, B,** and **C**). Part **D** illustrates the formation of a new rectum by ileal pouch anal anastomosis; this procedure cures ulcerative colitis.

CLINICAL CORRELATE 23.2
Inflammatory Bowel Disease—cont'd

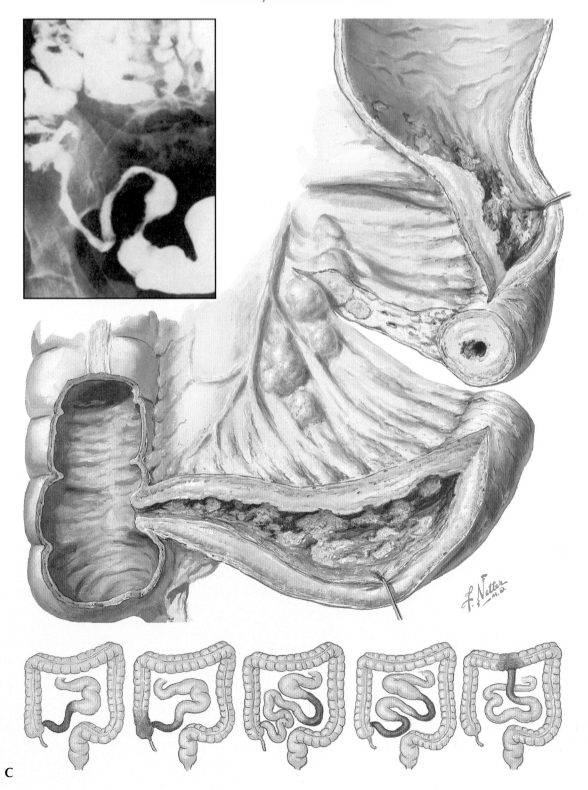

C

Continued

CLINICAL CORRELATE 23.2
Inflammatory Bowel Disease—cont'd

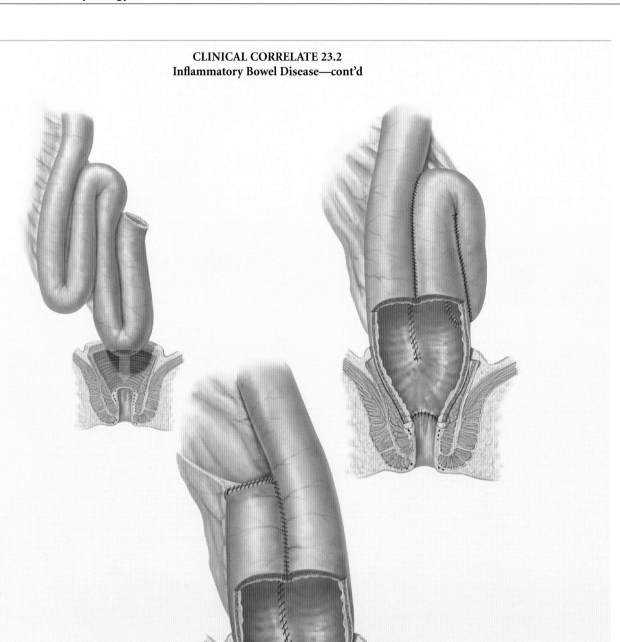

D

K. Carter

Think about homeostasis. With chronic vomiting or diarrhea, significant loss of fluids and electrolytes occurs. Vomiting causes a disproportionate loss of H^+ and K^+ (from stomach acid). If vomiting continues over an extended time, in addition to dehydration, the loss of H^+ and K^+ can lead to **metabolic alkalosis** and **hypokalemia** (low plasma K^+).

In contrast, chronic diarrhea results in dehydration and a disproportionate loss of HCO_3^-, which can result in **metabolic acidosis** (see Chapter 21).

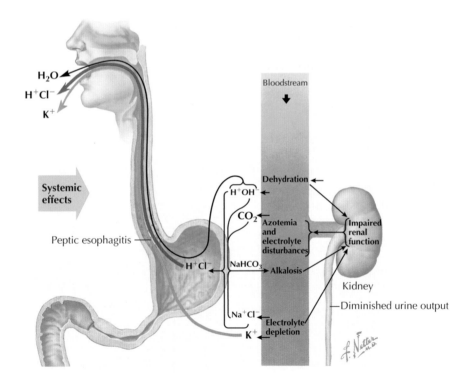

Vomiting Vomiting leads to a disproportionate loss of H^+, Cl^-, and K^+, as well as loss of fluid from the extracellular space. Prolonged vomiting results in dehydration, alkalosis, and hypokalemia (low plasma K^+).

Diarrhea occurs when chyme and fecal matter move too quickly through the colon, which can result from irritation of the lower GI tract (from bacteria or virus), inflammatory diseases (such as IBS, UC, and Crohn's disease [see Clinical Correlate 23.2]), or nervous stress. Movement in the colon is usually very slow (segmental propulsion), allowing absorption of most of the remaining water and salt from the chyme, producing feces. If abnormal motility occurs, the fecal matter will move through the colon with minimal absorption, producing diarrhea. When diarrhea occurs in response to bacterial irritation, rapid motility serves to wash the irritant out of the tract, and the diarrhea abates with the loss of the bacteria or virus.

In general, **osmotic diarrhea** is the most common type of diarrhea. It is caused by hypermotility, which forces the chyme/feces through the lower GI tract too rapidly for proper absorption of fluid and electrolytes. In contrast, **secretory diarrhea** is specifically caused by increased secretion of solutes into the lumen of the gut. The hallmark example of secretory diarrhea occurs with **cholera toxin** (see Clinical Correlate 23.3).

REFLEXES

All of the reflexes ultimately serve to promote the efficient digestion and elimination of ingested material. One of the most important reflexes, the rectosphincteric reflex, was discussed earlier in relation to defecation. A variety of GI-related reflexes are described in Table 23.2.

CLINICAL CORRELATE 23.3
Cholera

Cholera is disease caused by the bacterium *Vibrio cholerae*, which is contracted in a fecal-oral manner, usually through ingestion of contaminated water. In the small intestine, the bacteria produce cholera toxin, which has devastating effects on the intestinal epithelium. The toxin modifies the G protein G_s, and as a result, GTP activity is inhibited, causing prolonged activation of adenyl cyclase. The resulting high level of cAMP activates the cystic fibrosis transmembrane regulator gene (see Clinical Correlate 2.1 in Chapter 2), which causes active Cl^- secretion into the lumen of the small intestine. The Cl^- efflux is accompanied by Na^+ efflux (preserving electroneutrality), and water follows the electrolytes. This *isotonic* secretion of electrolytes and water depletes the vascular space while maintaining plasma osmolarity. In this manner, the vascular space is rapidly contracted, causing severe dehydration and shock. Antibiotic therapy and intravenous hydration is the most desirable therapy; otherwise, oral rehydration therapy (containing water, electrolytes, and glucose) from an uncontaminated source can be administered. The glucose in the oral rehydration therapy promotes sodium and electrolyte absorption and actually draws more of the secreted Na^+ and Cl^- back into the vascular space, along with water, which is effective in reducing the diarrhea and maintaining hydration. Without antibiotics, and with availability of uncontaminated water, the bacteria will be cleared from the GI tract in 7 to 10 days, as the enterocytes slough off and are replaced by new cells.

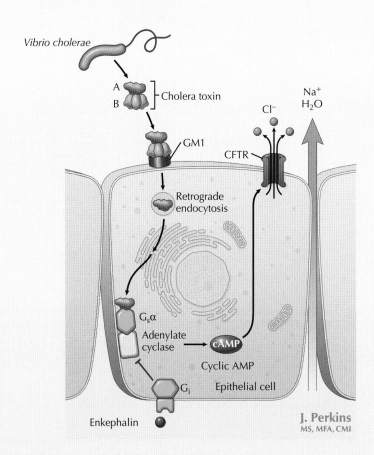

Actions of Cholera Toxin in Intestinal Cells *CFTR,* cystic fibrosis transmembrane regulator.

CLINICAL CORRELATE 23.4
Smooth Muscle Disorders:
Achalasia and Hirschsprung's Disease

The enteric nervous system (ENS) can be thought of as the central processing center that responds to a variety of signals and makes the appropriate connections. Thus, if the ENS is absent from an area in the tract, disease ensues. Smooth muscle diseases such as achalasia and Hirschsprung's disease involve aganglionic areas of the digestive tract (i.e., loss of nerves in some areas).

Achalasia is a loss of the ENS in the lower part of the esophagus and always involves the **lower esophageal sphincter**. The aganglionic region is not able to stimulate contractions and cannot transduce the signal to relax the lower esophageal sphincter. As a result, the sphincter tone is very high. Because the lower esophageal sphincter is not able to relax, food and fluid can take hours, rather than seconds, to leave the esophagus. Over time (months to years) this condition causes dilation of the esophagus, which is very painful and can eventually lead to anorectic behavior. Also, because the ingested food sits in the esophagus for long periods, partial digestion occurs, and the contents can further damage the esophageal mucosa. The cause of achalasia is unknown, and treatment usually only partially alleviates the problem. Pharmacologic intervention with nitrites (e.g., isosorbide dinitrate) or calcium channel blockers (e.g., nifedipine or verapamil) is used but has limited efficacy; the primary treatment is by balloon dilation or surgically cutting the sphincter (**esophagomyotomy**). This intervention allows food and fluids to pass through the lower esophageal sphincter into the stomach but usually produces gastric reflux; as a result, patients must be treated with oral medication for **gastroesophageal reflux disease**.

Hirschsprung's disease (or **megacolon**) is a congenital disease that involves the loss of the ENS in the distal part of the colon and always involves the internal anal sphincter. The normal defecation reflex mediated by the myenteric plexus (feces in rectum → stretch → relaxation of the internal anal sphincter) does not occur, so very little defecation occurs and feces build up in the rectum and colon. This buildup of feces dilates the colon. Symptoms include little or no bowel movements and vomiting. Drugs are not effective; surgical removal of the aganglionic region typically restores the ability to defecate.

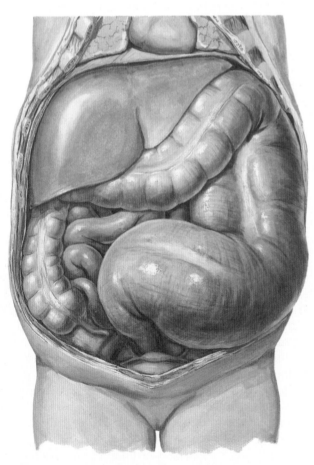

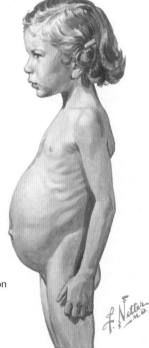

Tremendous distention and hypertrophy of sigmoid and descending colon; moderate involvement of transverse colon; distal constricted segment

Typical abdominal distention

Hirschsprung's Disease (Megacolon)

Chapter 24

Gastrointestinal Secretions

When food is placed in the mouth, in anticipation of nutrients entering the gastrointestinal (GI) tract, the **parasympathetic nervous system (PNS)** will initiate secretions including saliva, gastric juices, and bile from the liver. If nutrients enter the tract, secretions will increase as local controls are stimulated. If nutrients do not enter the stomach (i.e., if food is spit out), PNS activity will return to normal and the GI secretions will stop.

SALIVARY SECRETION AND REGULATION

Salivary Gland Secretion

Salivary secretions make foods easier to ingest (and digest) by lubricating, cooling, and adding digestive enzymes to the food. Three sets of salivary glands produce the secretions: the parotid, submandibular, and sublingual glands. Secretion from these glands is under the control of cranial nerves VII and IX (Figs. 24.1 and 24.2 and Table 24.1). A few unique aspects of salivary gland secretion bear noting:

- A large volume of fluid is produced by relatively small glands.
- Saliva is always hypotonic to plasma.
- Compared with plasma, saliva is relatively rich in K^+.
- Salivary secretion is primarily regulated by cranial parasympathetic nerves, which control blood flow to the glands.

Approximately 1.5 L of saliva are secreted daily, with most of this secretion occurring when food enters the mouth. The functions of saliva are to:

- Lubricate the food (water and mucins)
- Start digestion
- Protect the mouth and GI tract by cooling hot foods
- Protect the tract through the antimicrobial actions of white blood cells, immunoglobulin A, and opsonins
- Aid in hemostasis through the presence of platelet-activating factor

- Provide oral hygiene by buffering and cleansing the oral cavity (swallowing and spitting allow us to get rid of bacteria, whereas low flow while we sleep allows bacteria to accumulate, causing halitosis [bad breath])

Digestion of starches and lipids begins in the mouth. The serous secretions of the parotid and submandibular glands secrete **salivary α-amylase** (or **ptyalin**), which begins starch digestion by breaking α1-4 linkages of glucose polymers (oligosaccharides). Salivary α-amylase has >97% homology with pancreatic α-amylase. In addition, **Von Ebner's glands** of the tongue produce a serous secretion that includes the enzyme **lingual lipase**, which is active at low pH and can hydrolyze dietary triglycerides in the stomach, yielding free fatty acids and diglycerides. Lingual lipase is an acid lipase and has ~90% homology with gastric lipase; these acid lipases differ from pancreatic lipase (which hydrolyzes triglycerides to two fatty acids and one monoglyceride).

Production of Saliva

Salivary glands are exocrine glands, with secretions draining through ducts. The primary secretion is produced at the blind ends of the ducts, in the acini (Fig. 24.3). The glands are highly vascularized, and the parasympathetic nerves stimulate active secretion by increasing blood flow to the acini. Isotonic **ultra-filtrate** from the plasma diffuses through the acinar cells into the lumen, where this **"primary secretion"** is mixed with the additional products of the serous and mucous cells (α-amylase and mucins, respectively). The secretion drains down the striated ducts, and the electrolyte content is modified depending on the rate of flow through the ducts. Thus, as the rate of flow increases, there is less time for Na^+ and Cl^- reabsorption, and concentration of these electrolytes in saliva is higher (see Fig. 24.3). Under conditions of reduced vascular volume (such as dehydration and hemorrhage), Na^+ and water reabsorption by the ductal cells is increased by **antidiuretic hormone (ADH)** and the mineralocorticoid **aldosterone** (see Chapter 20). ADH increases water reabsorption (Na^+ follows), whereas aldosterone increases Na^+ reabsorption (water follows). Thus, when plasma volume is low, saliva production is decreased to

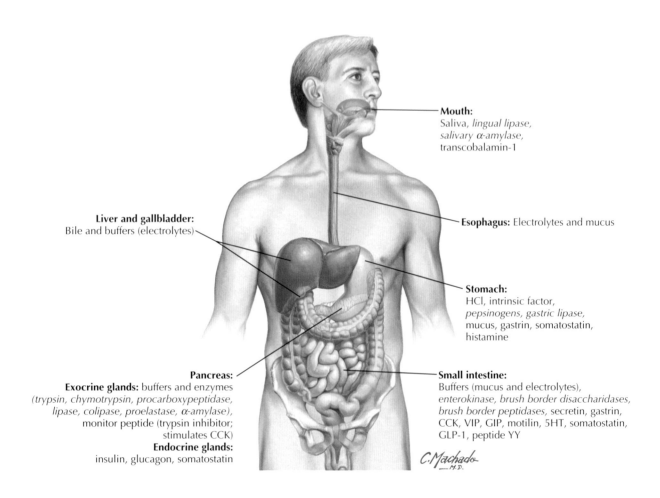

Mouth:
Saliva, *lingual lipase, salivary α-amylase,* transcobalamin-1

Esophagus: Electrolytes and mucus

Liver and gallbladder:
Bile and buffers (electrolytes)

Stomach:
HCl, intrinsic factor, *pepsinogens, gastric lipase,* mucus, gastrin, somatostatin, histamine

Pancreas:
Exocrine glands: buffers and enzymes *(trypsin, chymotrypsin, procarboxypeptidase, lipase, colipase, proelastase, α-amylase),* monitor peptide (trypsin inhibitor; stimulates CCK)
Endocrine glands:
insulin, glucagon, somatostatin

Small intestine:
Buffers (mucus and electrolytes), *enterokinase, brush border disaccharidases, brush border peptidases,* secretin, gastrin, CCK, VIP, GIP, motilin, 5HT, somatostatin, GLP-1, peptide YY

C. Machado
—M.D.

Figure 24.1 **Major Secretions into the Gastrointestinal Tract**   To enable digestion and absorption of ingested nutrients, liters of secretions are added daily to the gastrointestinal tract from associated glands, organs, and tissues. The general sources of digestive enzymes, acid, buffers, and bile are shown. *CCK,* cholecystokinin; *HCl,* hydrochloric acid; *5HT,* serotonin; *GIP,* gastric inhibitory peptide; *GLP-1,* glucagon-like peptide–1; *VIP,* vasoactive intestinal peptide.

conserve fluid. As indicated, regardless of flow rate, the saliva is always *hypotonic* to plasma.

When the saliva enters the mouth, it mixes with secretions from Von Ebner's glands (and other minor glands), adding lipase to the fluid. In addition, **transcobalamin-1** (TC-1, also known as R-binder) found in the saliva binds to the essential vitamin B12 (cobalamin) in the stomach, protecting it from degradation by pepsins. In the stomach, **intrinsic factor** (IF, secreted by the gastric parietal cells) also binds to the B12, and the TC-1–B12–IF complex enters the duodenum. In the duodenum, pancreatic proteases cleave the TC-1 and the remaining B12–IF complex forms dimers. This dimer is protected from pancreatic proteases until it is absorbed in the terminal ileum. If any part of the pathway (especially IF) is lost, B12 cannot be absorbed.

Neural Control of Saliva

Salivary flow is dependent on glossopharyngeal and facial parasympathetic nerves. These nerves innervate the blood

CLINICAL CORRELATE 24.1
Pernicious Anemia

Pernicious anemia (PA) is a form of megaloblastic anemia resulting from the lack of the essential vitamin B12. B12 is a critical cofactor in red blood cell (RBC) maturation. B12 deficiency causes an accumulation of 5-methyltetrahydrofolate, ultimately limiting DNA synthesis in erythroblastic cells, which produces immature, fragile erythrocytes called megaloblasts. This disease presents with wide-ranging symptoms, including anemia, fatigue, depression, diarrhea, and neuropathic pain. Unlike anemia resulting from loss of mature RBCs, untreated PA can lead to neurologic impairment (probably from accumulation of methylmalonyl coenzyme A) and death. When diagnosed early, PA is treated with B12 injections to replenish liver stores, and the prognosis is good. If diagnosis is delayed, the disease may progress despite B12 injections and death can ensue.

The primary cause of PA is autoimmune destruction of parietal cells, resulting in lack of IF and, therefore, vitamin B12 deficiency. PA can also occur after gastric bypass surgery.

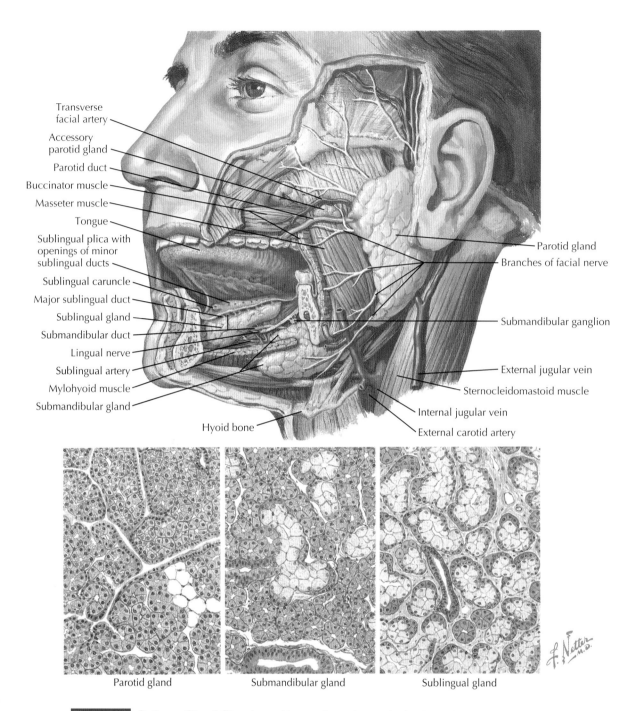

Transverse facial artery
Accessory parotid gland
Parotid duct
Buccinator muscle
Masseter muscle
Tongue
Sublingual plica with openings of minor sublingual ducts
Sublingual caruncle
Major sublingual duct
Sublingual gland
Submandibular duct
Lingual nerve
Sublingual artery
Mylohyoid muscle
Submandibular gland
Hyoid bone

Parotid gland
Branches of facial nerve
Submandibular ganglion
External jugular vein
Sternocleidomastoid muscle
Internal jugular vein
External carotid artery

Parotid gland   Submandibular gland   Sublingual gland

Figure 24.2 Salivary Gland Structure Humans have three paired salivary glands: the parotid, sub-mandibular, and sublingual glands. These glands are specialized to secrete mainly serous (parotid), mucous (sublingual), or mixed (submandibular) saliva (see representative histology). About 1.5 L of saliva are secreted daily, with the greatest amount of secretion occurring while eating.

Table 24.1 Salivary Gland Secretions

Gland	Secretion	Cranial Nerve
Parotid	Watery fluid, without mucins (serous), ~25% of total	Glossopharyngeal (IX)
Submandibular	Mixed serous/mucous fluid, ~70% of total	Facial (VII)
Sublingual	Mucous fluid, 5% of total	Facial (VII)

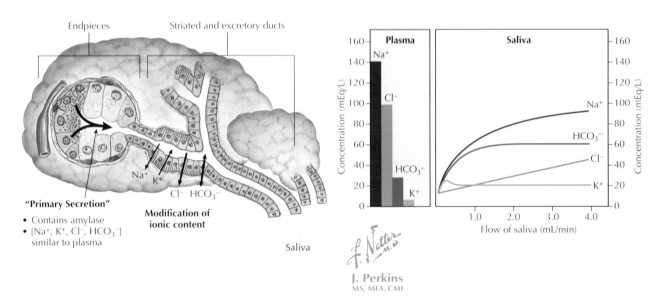

Figure 24.3 **Salivary Acinar Cells**   During salivary secretion, blood flow to the acini is increased by parasympathetic stimulation, and ultrafiltrate from plasma (mostly serous fluid) enters the acini. Filtrate from the cells enters the lumen of the acinar cells, mixing with secreted mucus and α-amylase, creating the primary secretion. This secretion is modified as it passes through the ducts into the mouth. Lingual lipase (secreted from the Von Ebner's glands of the tongue) is added to the saliva in the mouth. The graph illustrates the effect of increasing salivary secretion on composition of saliva. As flow increases, less modification occurs and the saliva more closely resembles the primary (original) secretion.

Table 24.2   **Regulation of Salivary Flow**

Increase Salivary Flow	Reduce Salivary Flow
Parasympathetics (CN VII and IX), ACh, VIP	Sympathetics, NE
CNS (in cephalic, "sensory" phase)	Hormones (ADH, aldosterone) conserve water and salt when volume depleted
Nausea	Sleep
Esophageal distension	Dehydration (activates hormones)
Chewy, flavorful foods	Drugs, chemotherapy
Dry, acidic foods	Aging (decreases ANS tone; glands can atrophy)
Meats, sweets, and bitter foods	

ACh, acetylcholine; *ADH,* antidiuretic hormone; *ANS,* autonomic nervous system; *CN,* cranial nerve; *CNS,* central nervous system; *NE,* norepinephrine; *VIP,* vasoactive intestinal peptide.

vessels of salivary glands and, when stimulated, increase blood flow, thus increasing ultrafiltrate secretion into the acini. If the parasympathetic nerves are severed, salivation will significantly decrease and the glands will atrophy. The sympathetic nerves also innervate blood vessels around the glands, as well as the myoepithelial cells in the acini. Sympathetic stimulation (e.g., during stress) reduces blood flow but contracts the myoepithelial cells, releasing preformed mucous saliva. After that, flow will decrease (resulting in dry mouth). The salivatory nucleus in the brainstem transduces a sensory, central nervous system effect, for example, when visual or olfactory input stimulates salivation (Table 24.2).

Esophagus

Glands in the esophageal mucosa produce serous and mucous secretions, which lubricate the bolus of food and help propel it to the stomach. This process is especially important if food gets stuck and secondary peristalsis occurs. The increased salivation and esophageal secretion aid movement of the food into the stomach.

SECRETIONS OF THE GASTRIC GLANDS AND REGULATION OF HYDROCHLORIC ACID SECRETION

The Gastric Glands

Gastric secretions facilitate and regulate digestion, lubricate the bolus of food, and protect the gastric mucosa. The mixing of gastric juices with food produces **chyme**. The gastric glands are made of a variety of cells (Fig. 24.4):

- Stem/regenerative cells differentiate to replace other cells in the glands.
- Mucous neck cells protect the gastric mucosa by secreting mucus.
- Parietal cells secrete hydrochloric acid (HCl) and intrinsic factor.
- Chief cells secrete pepsinogens and gastric lipase.
- Endocrine cells secrete somatostatin, and mast cells secrete histamine.

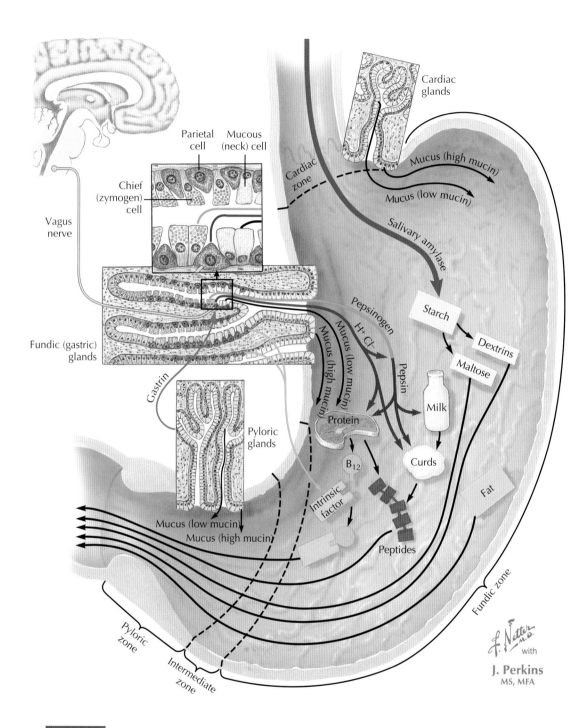

Figure 24.4 **Digestion in the Stomach** The stomach secretes concentrated hydrochloric acid (HCl), which breaks ingested food into smaller fragments and activates the enzymes pepsin and lipase, which digest proteins and fats, respectively. The parietal cells, which secrete HCl, also secrete the protein intrinsic factor, which binds to vitamin B12 to prevent the vitamin's digestion by pancreatic proteases in the small intestine. In general, about 25% of nutrient digestion occurs before the chyme reaches the duodenum.

■ The stem cells can also migrate up through the gastric pits to the lumen of the stomach, differentiating to form **superficial epithelial cells.** These cells secrete mucus and bicarbonate (HCO_3^-), which protect the gastric epithelium from the corrosive acid in the lumen. The superficial epithelial cells are replaced every few days.

Because of the harsh physical environment, cells lining the stomach (and intestines) are continually replaced. This process requires substantial energy expenditure, as well as a constant supply of proteins and lipids to make new cells.

Secretions from the Cells of the Gastric Glands

The gastric secretions are initiated by parasympathetic (vagal) nerves when food enters the mouth. The presence of food/chyme in the stomach stimulates additional secretion. These secretions primarily serve to digest the food; however, they

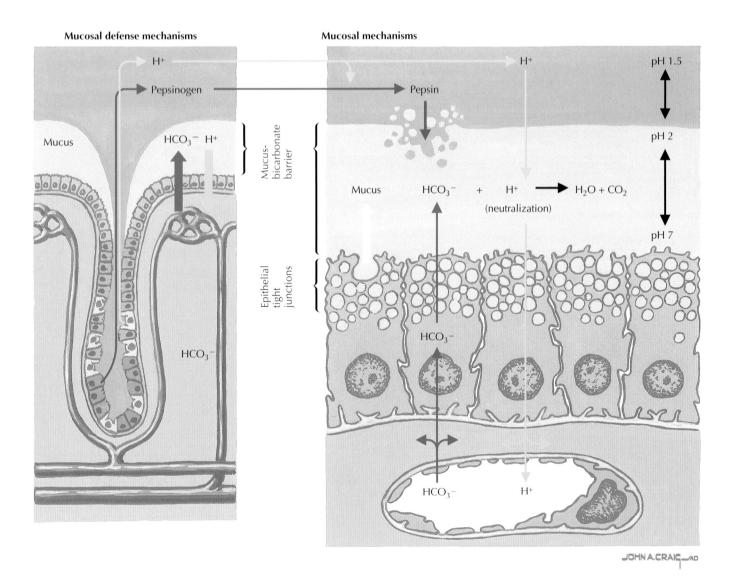

Mucosal defense mechanisms

Mucosal mechanisms

Figure 24.5 **Protection of Gastric Mucosa** Thick mucus from the surface epithelial cells traps the secreted bicarbonate on the surface of the gastric mucosa. The bicarbonate-containing mucus layer protects the cells from the caustic acid (~pH 2.0), keeping the pH at the cell surface ~7.0.

also contribute to the regulation of acid secretion and intestinal motility and protect vitamin B12 from digestion. The gastric secretions consist of the following substances:

■ HCl: Digests food, kills ingested bacteria, and converts inactive pepsinogens to pepsins
■ IF: An essential gastric secretion required for B12 absorption in the terminal ileum
■ Pepsinogens: The inactive form of pepsins, which are proteases that are activated in the acid environment; thus, protein digestion begins in the stomach
■ Gastrin: A gut hormone secreted from G cells located in the antrum of the stomach and the duodenum; gastrin stimulates HCl secretion and gastric motility (mixing), as well as motility of the lower GI tract (including mass movements)
■ Lipase (gastric lipase): Continues the process of lipid digestion

■ Mucus: Thick mucus is secreted with HCO_3^-; the HCO_3^- remains trapped in the mucus layer at the epithelial cell surface, effectively buffering the cells from the acid environment in the lumen (Fig. 24.5)
■ Other factors: Somatostatin inhibits HCl secretion, and histamine stimulates HCl secretion

In general, factors that stimulate gastric activity (e.g., vagus and gastrin) elicit secretion from chief, parietal, and mucus cells.

Production and Regulation of Gastric HCl

Acid secretion is important for the indiscriminate digestion of ingested matter. However, because of its corrosive nature, multiple factors regulate its secretion from **parietal cells**. One factor that keeps HCl secretion under control is parietal cell structure. During stimulation, parietal cells are well organized,

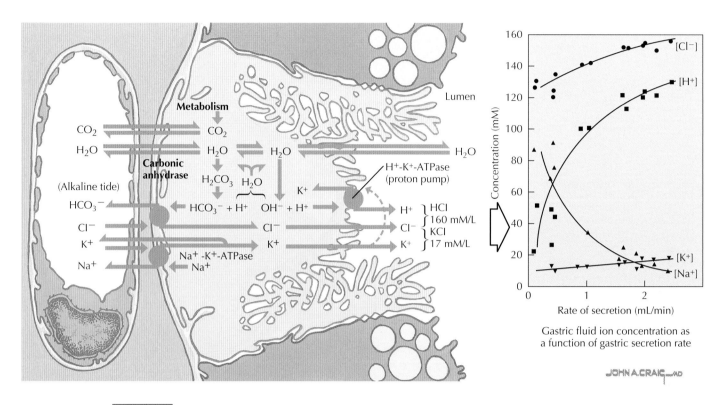

Figure 24.6 **Parietal Cell Hydrochloric Acid Production**   The apical H$^+$/K$^+$ ATPase (proton pump) causes the active secretion of H$^+$ into the lumen of the gastric glands. Meanwhile, the HCO$_3^-$/Cl$^-$ exchanger on the basolateral membrane brings Cl$^-$ into the parietal cell; Cl$^-$ then diffuses through apical channels into the lumen along its electrochemical gradient. Thus, concentrated hydrochloric acid (HCl) is formed. HCl production is stimulated by vagal efferent fibers, gastrin, and histamine, which are active when there is food in the stomach. The H$^+$/K$^+$ ATPase is the target for proton pump inhibitors to decrease gastric acid release in patients with gastroesophageal reflux disease or ulcers.

with tubulovesicles fusing with the plasma membrane to form intricate canaliculi that increase surface area for secretion of H$^+$ (Fig. 24.6). In contrast, unstimulated cells have less surface area exposed to the lumen of the gastric glands, keeping potential H$^+$ secretion minimal.

Parietal cells secrete H$^+$ against a million-fold concentration gradient in the lumen. The gradient is maintained by **tight junctions**, which prevent H$^+$ leak back into the blood. The transport of H$^+$ against the gradient is accomplished by the apical **H$^+$/K$^+$ ATPase**, or **proton, pump**. This active antiport pump is the **rate-limiting step** in H$^+$ transport and is the site of regulation of acid secretion. The following aspects of acid production and secretion are important:

- The production of H$^+$ from intracellular conversion of CO$_2$ to carbonic acid in the reaction (CO$_2$ + H$_2$O → H$_2$CO$_3$ → H$^+$ + HCO$_3^-$) (carbonic anhydrase is required for this conversion); H$^+$ is then available for secretion by the proton pump
- The entry of chloride into the cells via the HCO$_3^-$/Cl$^-$ exchanger at the basolateral membrane, and diffusion

through apical chloride channels into the lumen of the gastric gland (down its concentration gradient)
- The exit of HCO$_3^-$ from the cell into the blood via basolateral HCO$_3^-$/Cl$^-$ exchangers; in the actively secreting parietal cell, the large amount of HCO$_3^-$ entering the blood is described as an **alkaline tide**
- The movement of potassium back into the lumen with Cl$^-$

This process primarily produces HCl in the lumen but also produces some potassium chloride. Regulation of acid production takes place through stimulation and inhibition of the **proton pump**, which is the target for most therapeutic interventions to reduce acid secretion (for gastroesophageal reflux disease, ulcers, and the like). Proton pump inhibitors (PPIs) such as omeprazole and histamine H2 blockers (e.g., ranitidine and cimetidine) are used for this purpose.

Constituents of the gastric juices vary with the rate of secretion (see Fig. 24.6). Quiescent, nonstimulated cells produce fluid at a low rate; the fluid contains higher levels of Na$^+$ and Cl$^-$ (which are more bufferlike) and low concentrations

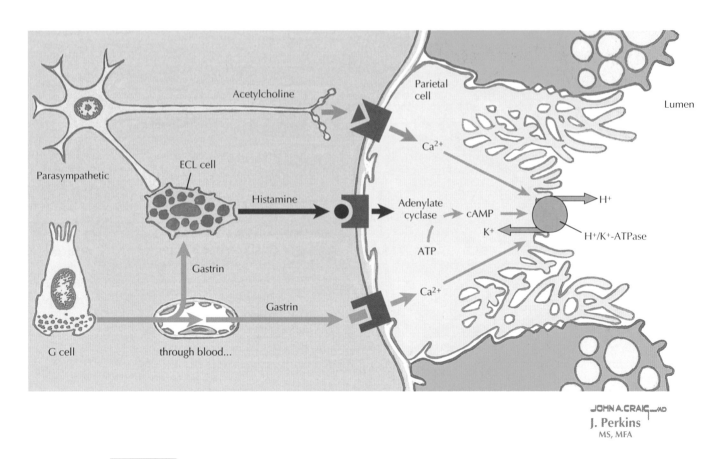

Figure 24.7 **Signal Transduction Mechanisms Regulating HCl Secretion** Stimulation of the proton pump can be achieved by both calcium and cAMP-mediated messenger systems. *ECL*, enterochromaffin-like.

of H^+. However, when cells are stimulated, Na^+ concentrations drop and H^+ concentrations increase in the juice, producing more HCl.

Regulation of Gastric Acid Secretion

Gastric acid secretion has three phases:

- **Cephalic phase:** Food in the mouth initiates vagal release of acetylcholine at the parietal cell, stimulating acid secretion. Vagal fibers also terminate on enterochromaffin-like (ECL) cells and G cells, stimulating histamine and gastrin secretion, respectively; histamine and gastrin promote acid secretion.
- **Gastric phase:** Stretch, initiated by food in the stomach, releases gastrin from antral G cells. The gastrin in the bloodstream circulates back to the stomach, stimulating the proton pump of the parietal cells. At the same time, the vagus also stimulates the proton pump directly, as well as by increasing gastrin-releasing peptide and histamine, which also act on the pump.
- **Intestinal phase:** When chyme enters the duodenum and jejunum, hormones are released that feed back to

regulate acid secretion (see the following factors that stimulate and inhibit acid secretion).

Several factors **stimulate** acid secretion:

- The *vagus* acts directly at parietal cells and indirectly through stimulation of gastrin and histamine release and inhibition of local somatostatin release (Fig. 24.7).
- *Histamine*, from ECL cells, diffuses through the mucosa to act on adjacent parietal cells and is a major stimulus for acid secretion.
- *Gastrin*, which is carried through the blood, acts at cholecystokinin (CCK) type B receptors on ECL cells to release histamine and also acts directly on parietal cells.
- *Insulin*, which is carried through the blood, acts on G cells and promotes gastrin secretion.
- *Caffeine* (a phosphodiesterase inhibitor) increases cAMP in the parietal cells, increasing proton pump activity.
- *Stress*, although not well understood, appears to increase acid secretion in certain people and may be a cofactor in ulcer formation (with *Helicobacter pylori*).

Several factors **inhibit** acid secretion:

- *Somatostatin*, which is released from endocrine cells in the gastric glands, acts in a paracrine manner on the parietal cells, as well as on G cells, to inhibit gastrin.
- *Glucose-dependent insulinotropic peptide* or *gastric inhibitory peptide* (GIP), which is released from the duodenum and jejunum, acts at the G cells to suppress gastrin.
- *Secretin*, which is released from the duodenum and jejunum, acts at the G cells to suppress gastrin.
- *Peptide YY*, which is released from various areas of the GI tract (mainly ileum and colon) in response to fats, helps shut off acid and pancreatic secretions when chyme is leaving the upper GI tract. The effects on gastric acid may be through suppression of acetylcholine release from cholinergic fibers and/or stimulation of paracrine somatostatin.

Although these factors regulate acid secretion, it is important to understand that acid secretion is not an all-or-nothing event; for the most part, these factors work in concert to modulate the amount of acid being secreted during the digestive period. Thus, for example, while the vagus, gastrin, and histamine are stimulating acid secretion, secretin works to moderate gastrin release, and somatostatin and gastric inhibitory peptide act on the parietal cells, keeping acid secretion at a reasonable rate. The dynamic nature of the system is critical to proper function. One reason that potentiation of acid secretion can occur is that various factors use different second messenger systems (see Fig. 24.7).

SECRETIONS FROM THE SMALL INTESTINE: BUFFERS, ENZYMES, AND HORMONES

The **crypts of Lieberkühn**, located at the bottom of the villi in the small intestine, have a variety of cell types that produce and secrete buffers, enzymes, and hormones that facilitate digestion and absorption of the nutrients in the chyme (Fig. 24.8). **Brunner's glands** are located in the duodenum and secrete thick mucus, which helps to protect the early part of the small intestine from the acidic chyme leaving the stomach. These glands are stimulated by secretin and activity of the vagus. The **Paneth cells** are deep in the crypts and are stimulated by secretin to secrete ions and water, which buffer the chyme. They also secrete **lysozyme**, which has antimicrobial

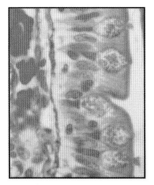

Goblet cells and striated border of human jejunal villus (azan stain, ×650)

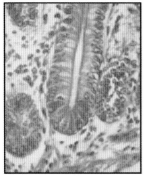

Floor of crypt of Lieberkühn with granulated, oxyphilic cells of Paneth (hematoxylin-eosin, ×325)

A

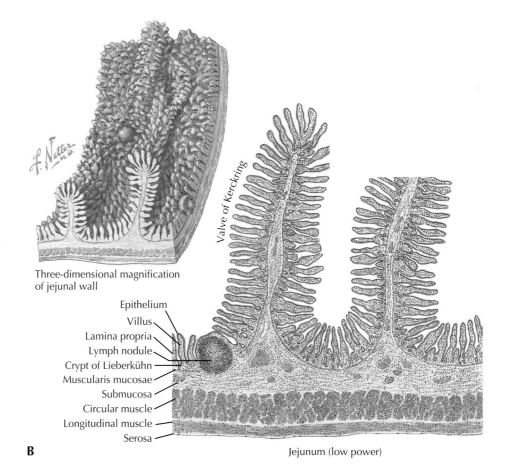

Three-dimensional magnification of jejunal wall

Valve of Kerckring

Epithelium
Villus
Lamina propria
Lymph nodule
Crypt of Lieberkühn
Muscularis mucosae
Submucosa
Circular muscle
Longitudinal muscle
Serosa

Jejunum (low power)

B

Figure 24.8 **Crypts of Lieberkühn and Cell Types** **A,** Micrographs of goblet and Paneth cells in small intestinal mucosa. **B,** Brush border lining of the small intestine, illustrating crypts of Lieberkühn within villus projections.

Table 24.3   **Major Gastrointestinal Hormones**

GI Hormones	Site of Secretion	Primary Stimuli	General Actions
Gastrin	G cells in antrum of stomach and duodenum	Stretch, peptides and amino acids, vagus (through GRP)	↑**Gastric H⁺** ↑Gastric mixing ↑Lower GI tract motility
Secretin	S cells of the duodenum	Acidic chyme	↑**Pancreatic buffer (HCO₃⁻) secretion** ↑**Biliary and small intestine buffer secretion** ↓Gastric H⁺ (by ↓gastrin) ↓Gastric emptying
Cholecystokinin	I cells of the duodenum and jejunum	Small peptides and amino acids, fats	↑**Pancreatic enzyme secretion** Contracts gallbladder and relaxes sphincter of Oddi ↑Pancreatic and biliary buffer secretion ↓Gastric emptying ↑Lower GI tract motility ↑Satiety
Gastric inhibitory peptide, aka glucose-dependent insulinotropic peptide	Duodenum and jejunum	Fatty acids, glucose, amino acids	↓**Gastric H⁺ secretion** ↑**Pancreatic insulin secretion** ↓Gastric emptying
Motilin	Mo cells of the duodenum	Fasting	↑**Phase III contractions of the MMC**
Glucagon-like peptide–1	L cells of the jejunum and ileum	Chyme	↑**Satiety** ↑**Insulin release** ↓Glucagon release

GI, gastrointestinal; *GRP*, gastrin-releasing peptide; *MMC*, migrating myoelectric complex.
Modified with permission from Hansen J: *Netter's Atlas of Human Physiology*, Philadelphia, 2002, Elsevier.

actions. **Goblet cells** are located throughout the small intestine and secrete mucus. Cells in the duodenum also release the enzyme **enterokinase**, which cleaves the inactive zymogen trypsinogen, forming the active enzyme trypsin. This activity occurs in the duodenum, activating the powerful protease in the presence of chyme.

The small intestine produces and secretes a variety of endocrine hormones that regulate digestion and motility (Table 24.3; the major actions of the hormones are in bold). The effects on digestion are detailed in Chapter 27.

PANCREATIC BUFFER AND ENZYME SECRETIONS

The pancreas synthesizes and releases both endocrine secretions (into blood) and exocrine secretions (through ducts into the duodenum). The exocrine secretions help buffer the acid chyme and contain enzymes; buffering of the chyme is essential for the action of the pancreatic enzymes. Approximately 70% to 75% of digestion occurs in the proximal small intestine by the pancreatic and brush border enzymes. The exocrine secretions enter the common bile duct from the pancreatic ducts, mixing with liver and gallbladder secretions. The bile duct enters the duodenum through the **sphincter of Oddi**, at the **papilla of Vater** (Fig. 24.9*A*).

Secretin, which is released into the blood from duodenal S cells in response to acidic chyme, binds to its receptors on the pancreatic centroacinar cells and stimulates **electrolyte and fluid secretion** (for buffering; see Fig. 24.9*C*). As with the salivary glands, as the secretion goes through the ducts, modification of electrolytes occurs. As the rate of secretion increases, the composition of the fluid changes, becoming rich in HCO_3^- (see Fig. 24.9*B*).

CCK is released into the blood from the duodenal I cells and acts on the pancreatic acinar cells. Along with the vagus, CCK stimulates **pancreatic enzyme secretions**, which include the following substances:

- **Pancreatic proteases:** Trypsin, chymotrypsin, and elastase, the major proteases, are stored and released as inactive zymogens (trypsinogen, chymotrypsinogen, and proelastase). Storage of the proteases as zymogens protects the pancreas and ducts from digestion. When the zymogens enter the duodenum, **enterokinase** (in the duodenal brush border) cleaves trypsinogen to trypsin, and then trypsin can activate additional trypsinogen (autoactivation), as well as the other proteases.

- **Pancreatic α-amylase:** Starch digestion begins in the mouth with salivary α-amylase. Digestion continues in the small intestine by pancreatic α-amylase, which

CLINICAL CORRELATE 24.2
Ulcers

Gastric and duodenal ulcerations were thought to result mainly from ingestion of **nonsteroidal anti-inflammatory drugs** (**NSAIDs;** e.g., aspirin) and stress until about 25 years ago, when Australian investigators discovered *Helicobacter pylori* residing in the gastric and duodenal mucosa of patients with ulcers. It is now known that *H. pylori* accounts for almost all non–NSAID-related gastric ulcers and nearly all duodenal ulcers. How can the bacteria live in the acidic (~pH 2) environment of the stomach? *H. pylori* is resistant to HCl and can burrow into the mucosal layer because it secretes **urease,** which breaks urea into CO_2 and ammonia (NH_3). This action disrupts the bicarbonate-mucus layer protecting the gastric mucosa and allows acid and pepsins to infiltrate the mucosa. In addition, the ammonia stimulates proinflammatory cytokines in the mucosal tissue. This mechanism, together with the corrosive effects of the acid and pepsin, initiates ulceration.

Interestingly, although many people harbor the bacteria (as a result of fecal-oral contamination from food or water), a critical number of bacteria must be present to initiate enough urease production to disrupt the mucosa and expose the epithelium to ulceration. The condition is treated with long-term use of antibiotics and PPIs.

NSAIDS, including aspirin, ibuprofen, and similar drugs, can also cause gastric ulcers. Because prostaglandins are cytoprotective (by decreasing parietal cell acid production and increasing mucus production), the suppression of prostaglandins (especially prostaglandin E_2) by NSAIDs can lead to ulcerations. The ulcerative effect is enhanced when the drugs are taken on an empty stomach or in the presence of caffeine, which stimulates acid secretion. Stress, coffee, and NSAIDs (particularly when taken on an empty stomach) may have additive effects in aggravating existing ulcers. Treatment is with PPIs and behavior modification.

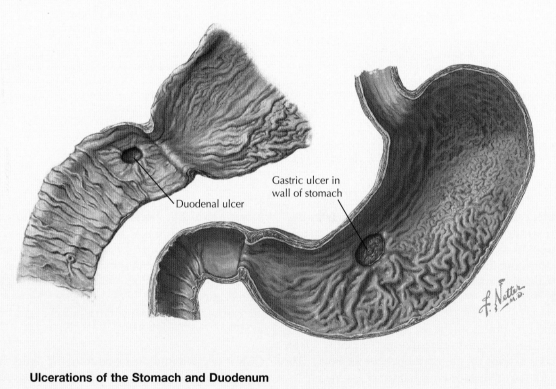

Duodenal ulcer

Gastric ulcer in wall of stomach

Ulcerations of the Stomach and Duodenum

hydrolyzes the starch molecules to maltose, maltotriose, and isomaltose. The amylase is activated by Cl^-.

■ **Pancreatic lipase** and **colipase:** Pancreatic lipase hydrolyzes triglycerides to monoglycerides and free fatty acids; colipase is a cofactor for this process. Other lipases convert cholesterol esters to cholesterol and fatty acid and phospholipids to lysophospholipids and fatty acid. However, pancreatic lipase cannot readily access the lipids in the small intestine because of the presence of bile, which surrounds these lipids. Colipase is released as precolipase and is activated in the duodenum by trypsin. The colipase displaces the bile, allowing the pancreatic lipase to access the lipids (see Chapter 26).

GALLBLADDER FUNCTION

The function of the gallbladder is to store and release bile. Early in feeding, when food enters the stomach, vagal nerves

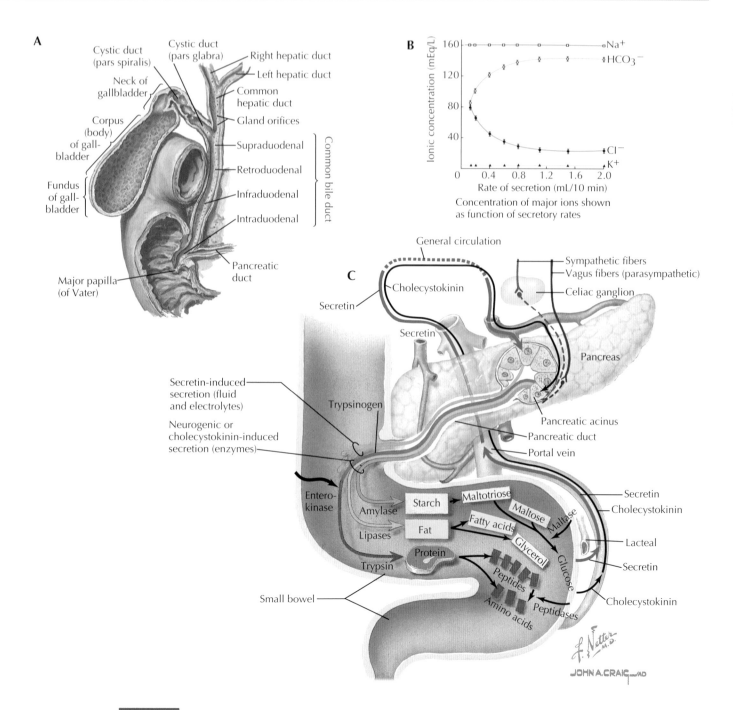

A Pancreatic, liver, and gallbladder secretions flow through the common bile duct and into the duodenum through the sphincter of Oddi at the papilla of Vater. **B,** As the rate of pancreatic secretions increases, the concentration of electrolytes changes, producing high sodium bicarbonate buffer. **C,** In response to food/chyme in the stomach and duodenum, vagal efferents and hormones stimulate pancreatic buffer and enzyme secretions.

Figure 24.9 **Structure and Secretion of the Pancreas** **A,** Pancreatic, liver, and gallbladder secretions flow through the common bile duct and into the duodenum through the sphincter of Oddi at the papilla of Vater. **B,** As the rate of pancreatic secretions increases, the concentration of electrolytes changes, producing high sodium bicarbonate buffer. **C,** In response to food/chyme in the stomach and duodenum, vagal efferents and hormones stimulate pancreatic buffer and enzyme secretions.

are stimulated to relax the sphincter of Oddi. This relaxation releases the contents of the bile duct (bile and electrolytes) into the duodenum. Later, when chyme enters the duodenum, CCK is released, which fully relaxes the sphincter and causes gallbladder contractions. Stored bile is released into the duodenum and begins lipid emulsification (see Chapter 26).

COLONIC SECRETIONS

In the colon, secretion of mucus, HCO_3^-, H^+, and K^+ occurs. The mucus is required to lubricate chyme as it is dehydrated and formed into feces. With colonic reabsorption of sodium (through Na^+/H^+ exchangers), H^+ is secreted into the lumen. Cl^- is also absorbed, in part through lumenal Cl^-/HCO_3^-

exchangers. Thus, both H^+ and HCO_3^- are secreted into the lumen of the colon. Potassium secretion through channels occurs as a secondary effect of aldosterone (e.g., during volume contraction), as aldosterone increases ENaC on the lumenal membrane, reclaiming additional sodium. Because of the relatively high HCO_3^- content of feces, chronic diarrhea can cause metabolic acidosis (see Chapter 21).

Gases are formed throughout the GI tract, diffusing in and out of cells, and for the most part we are unaware of the fluxes. CO_2 is constantly produced, but in addition, bacterial action can produce O_2, H_2, N_2, and methane, for example. Typically, a person produces and expels (as flatus) about 200 mL of colonic gas each day. This amount can change with diets high in carbohydrates and meat protein, which increases gas production.

CLINICAL CORRELATE 24.3
Zollinger-Ellison Syndrome

Gastrinomas are tumors (usually pancreatic) that secrete gastrin but do not have any of the normal feedback systems that control gastrin release. These tumors cause **Zollinger-Ellison syndrome**, in which the constant secretion of gastrin causes uncontrolled high levels of gastric HCl release and ulcerations of the stomach mucosa. In addition, the high acid secretion acidifies the duodenum and early jejunum, impeding the actions of digestive enzymes and decreasing the ability of bile to make micelles. This process especially affects lipid digestion and absorption, because the low duodenal pH denatures pancreatic lipase, significantly reducing lipid digestion. As the pH increases in the jejunum, digestion can occur, but the overall effect is maldigestion and malabsorption, resulting in steatorrhea and excess bile salts in the feces. Treatment consists of removal of the tumor and the use of PPIs to further reduce acid secretion, allowing ulcerations to heal.

Chapter 25

Hepatobiliary Function

OVERVIEW OF LIVER FUNCTIONS

The liver has numerous important functions that involve its metabolic and secretory activity and its unique vascular structure. **Hepatocytes** make up the parenchyma of the liver, and the basic functions of the organ include:

- **Carbohydrate, lipid, and protein metabolism:** The liver receives newly absorbed nutrients through the portal vein, along with the contents of systemic blood, and processes them according to need. The liver produces albumin, fibrinogen, immunoglobulins, binding proteins, cholesterol, lipoproteins, bile, and other important molecules.
- **Cholesterol production and excretion:** The body requires cholesterol, and although this substance can be synthesized by many cells in the body, the liver can produce it at a high rate when necessary (with hydroxymethylglutaryl–coenzyme A [HMG-CoA] reductase being the first enzyme used in this process). Cholesterol is also used to synthesize bile, and thus when bile is excreted in the feces, cholesterol is removed from the body.
- **β-Oxidation of fatty acids:** Although many tissues use β-oxidation of fatty acids as an alternate energy source when glucose is not present, the liver has a high capacity for β-oxidation during the interdigestive period.
- **Bile acid production and secretion:** Bile is necessary for efficient lipid absorption, because lipids alone cannot efficiently pass through the water that bathes the enterocytes. Bile is polar, allowing it to incorporate lipids into **micelles**; the micelles can move the lipids through the unstirred water layer adjacent to the enterocytes (see Chapter 26). Without bile, the bulk of the hydrophobic lipids would not be able to get near the brush border.
- **Endocrine functions:** The hepatocytes produce and secrete hormones into blood, including insulin-like growth factor–1 (IGF-1), hepatocyte growth factor, angiotensinogen, and cytokines. It converts thyroxine to active triiodothyronine (see Chapter 28) and participates in activation of vitamin D (see Chapter 31).

- **Detoxification:** The liver contains reticuloendothelial cells, known as **Kupffer cells**, that are fixed macrophages in the endothelial lining of hepatic sinusoids. As blood passes through the liver, old and damaged erythrocytes undergo phagocytosis by Kupffer cells. Hormones, drugs, and other chemicals are metabolized by the hepatocytes.
- **Vitamin and iron storage:** The liver stores several elements crucial to normal body functions (e.g., vitamin B12, folic acid, and iron). Vitamins can be stored for weeks to months, and in the case of essential vitamins like B12 (cobalamin), storage by the liver provides a source of the vitamin if the dietary supply is depleted. The liver is also the site of large iron stores, bound to the protein **ferritin** (the liver contains 25% of the body's iron). When needed, the iron (bound to transferrin) is released into the blood and enters the bone marrow for the production of hemoglobin.

Because it performs this great variety of functions, the liver is a crucial organ for maintaining proper blood glucose levels, excreting waste products, processing proteins, contributing to immune function, and even indirectly regulating blood pressure and fluid homeostasis (through synthesis of plasma proteins; Fig. 25.1). When part of the liver is damaged, it has the ability to regenerate functional hepatocytes, and this compensatory action allows it to maintain adequate metabolic function. In addition, the ability of the liver to achieve a high level of metabolic activity is dependent on the blood flow to and from the liver.

The regenerative capacity of the liver is so great that one donor liver can be used for two liver transplant patients, with the liver sections regaining almost 100% of the original liver size. Because the demand for donor livers is far greater than the supply, since 1989, living donor liver transplantation (using one lobe) has been performed widely. The remnant liver in the donor recovers its original size and function in 4 to 6 weeks, as does the donated lobe in the patient.

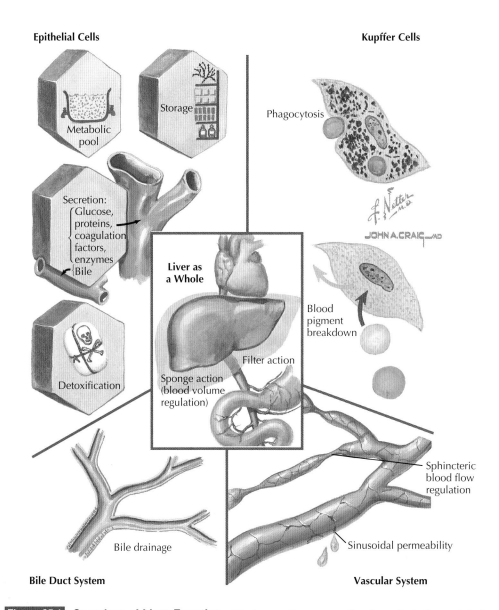

Epithelial Cells

Metabolic pool

Storage

Secretion:
Glucose,
proteins,
coagulation
factors,
enzymes
Bile

Detoxification

Liver as a Whole

Filter action

Sponge action
(blood volume
regulation)

Kupffer Cells

Phagocytosis

Blood
pigment
breakdown

Sphincteric
blood flow
regulation

Sinusoidal permeability

Bile drainage

Bile Duct System

Vascular System

Figure 25.1 **Overview of Liver Function** The liver performs many functions, including metabolism of carbohydrates, proteins, and lipids; phagocytosis and removal of waste and bacteria; detoxification of blood; synthesis of bile; blood volume control; and vitamin storage.

LIVER STRUCTURE AND BLOOD FLOW

The liver is like a sponge, with the hepatocytes surrounded by sinusoidal capillaries and lymph vessels (Fig. 25.2). This structure permits the free flow of blood to the cells, and the cellular products secreted into the perisinusoidal **space of Disse** can be released back into blood or into the lymphatic vessels. **Kupffer cells** are affixed to the sinusoidal membranes and are present throughout the liver. These cells phagocytose bacteria and damaged red blood cells. The hepatocytes metabolize hormones and drugs and clear the waste, "detoxifying" the blood. The hepatocytes also produce bile and electrolyte solutions that drain into the **bile canaliculi**, which lead to the bile duct.

Under normal conditions the liver contains ~400 mL of blood, which is ~8% of the total blood volume, illustrating the "blood

storage" aspect of the liver. The system works properly because hepatic vein pressure is very low, near 1 mm Hg, and thus there is no obstruction of blood flow through the liver sinusoids. This allows the free flow of blood through and out of the liver into the vena cava. If obstruction of flow through the liver occurs because of damage (e.g., cirrhosis or hepatitis) or increased hepatic vein pressure (e.g., congestive heart failure), blood will back up and pressure will rise, causing **portal hypertension** (see Clinical Correlate 25.2). The volume of blood associated with the liver can also increase to 1 L, causing **hepatomegaly** (an enlarged liver).

The blood supply to the liver comes from the hepatic artery and portal vein. The systemic arterial blood from the hepatic artery enters the liver at a rate of ~450 mL per minute. The

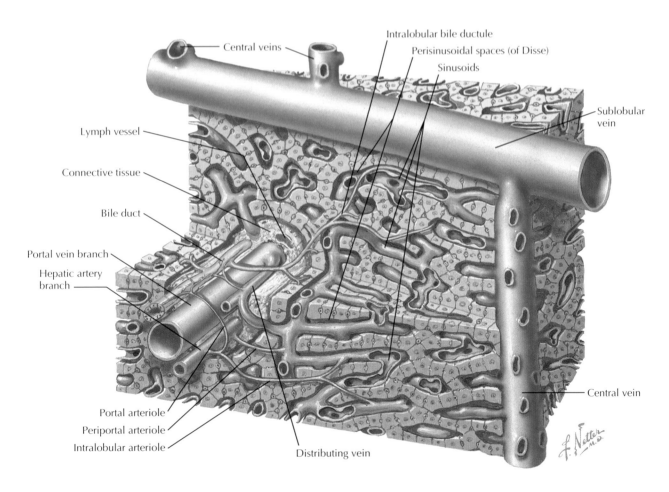

Central veins — Intralobular bile ductule — Perisinusoidal spaces (of Disse) — Sinusoids — Sublobular vein — Lymph vessel — Connective tissue — Bile duct — Portal vein branch — Hepatic artery branch — Central vein — Portal arteriole — Periportal arteriole — Intralobular arteriole — Distributing vein

Figure 25.2 **Structure of the Liver**   The liver is highly vascularized to accomplish its primary function of filtering portal and systemic blood. The capillary sinusoids surround the liver cells (hepatocytes), allowing efficient access of blood to the cells and transfer of products back into the blood. The sinusoids also contain the Kupffer cells, which are fixed macrophages that phagocytose damaged or aged red blood cells.

portal vein carries venous blood from the intestines (~1 L per minute), and the arterial and venous blood intermingles in the sinusoidal capillaries. Total blood flow through the liver represents approximately 30% of the cardiac output. As the blood from the portal vein enters the liver, nutrients, bacteria, and foreign bodies are processed, which is the "first pass" effect, allowing the absorbed materials to be "cleared" by the liver before the blood leaves through the hepatic vein into the systemic circulation. Although not all such substances will be cleared or metabolized in one pass, the bulk of the substances will be handled.

Ascites is fluid in the peritoneal cavity of the abdomen. Ascites can be caused by increased vascular pressure within the liver. Because the liver capillaries have a high permeability, when the pressure increases, fluid is forced out of the space of Disse and lymphatics into the peritoneal space. Increased venous pressure, as in congestive heart failure, can cause ascites, as well as peripheral **edema**. Edema results from elevated capillary hydrostatic pressure, which makes the net balance of Starling forces favor filtration of fluid out of the capillaries into the interstitial space (in the lower extremities).

BASIC METABOLISM OF CARBOHYDRATES, LIPIDS, AND PROTEINS

Carbohydrates

The liver acts as a blood glucose reservoir, storing glucose as glycogen and releasing it when blood levels are low. Carbohydrates are absorbed in the intestine as monosaccharides and are carried in the portal blood to the liver. Most of the glucose passes through the liver rapidly and is released into the systemic blood, where elevated insulin will facilitate its entry into tissues. In the liver, excess monosaccharides are handled by the following processes:

- **Conversion of other monosaccharides to glucose:** Fructose and galactose can be converted into glucose.
- **Glycogen synthesis and storage:** Excess glucose is polymerized and stored as glycogen. Stored hepatic glycogen can provide glucose for 12 to 17 hours during fasting. When blood glucose levels are low, glucagon and other hyperglycemic hormones such as epinephrine and growth hormone stimulate glycolysis to break down glycogen and release glucose into the blood. If the glycogen

CLINICAL CORRELATE 25.1
Bilirubin and Jaundice

Red blood cells (RBCs) are viable for about 120 days, and new cells are continually being produced to replace damaged old cells. Mononuclear phagocytic system (MPS) cells throughout the body (especially Kupffer cells in the liver) remove and break down the old RBCs so that the iron can be recycled and waste can be eliminated. **Bilirubin** is a by-product of RBC (hemoglobin) degradation and is eliminated from the body by incorporation into bile (and subsequent loss in feces) and by urinary excretion as urobilinogen. The addition of bilirubin to bile, and its excretion in feces (as stercobilin), contributes to the pigmentation of the feces. The bile pathway of excretion is important for ridding the body of excess bilirubin.

Jaundice (synonymous with icterus) occurs when the plasma bilirubin level is elevated. In patients, jaundice is observed as a yellowing tinge in the whites of the eyes and in skin tone and nail beds; stool may be pale. **Primary jaundice** arises from hepatic dysfunction, as seen with obstructive liver disease (cirrhosis), blockage of bile ducts (by tumor or gallstones), or inflammation (hepatitis C). **Secondary jaundice** occurs from extrahepatic causes, such as abnormal lysis of RBCs (hemolytic disease). In most cases, when the cause of the jaundice is treated, the jaundice will abate as the excess bilirubin eventually clears (through fecal and urinary excretion). An important distinction is that with secondary jaundice, liver function is normal.

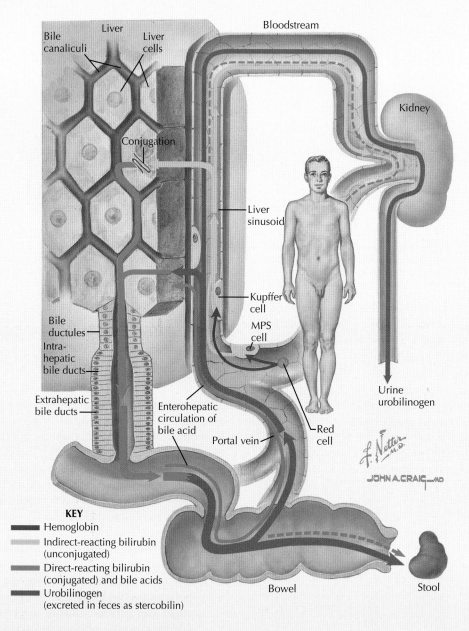

KEY
- ━━ Hemoglobin
- ▬▬ Indirect-reacting bilirubin (unconjugated)
- ━━ Direct-reacting bilirubin (conjugated) and bile acids
- ━━ Urobilinogen (excreted in feces as stercobilin)

Bilirubin Production and Excretion

stores are not used, excess glucose (which is not released into the blood) will eventually be converted to triglycerides (TGs) and transported to adipose tissue for storage (see the "Lipids" section).

- **Gluconeogenesis:** The liver (and to a lesser extent, the kidney) has the ability to make glucose from substrates such as glycerol, pyruvate, and the amino acids glutamine and alanine. Gluconeogenesis provides an alternate energy source and occurs primarily during fasting and starvation.
- **Formation of chemical compounds:** Excess glucose can also be converted into other chemical compounds (e.g., pyruvic acid, lactic acid, and acetyl CoA) that can be used in metabolic pathways such as the citric acid cycle.

The ingestion of carbohydrates raises blood glucose levels, but the effects can vary according to the type of carbohydrate and its quantity in the food. The **glycemic index** of a specific food refers to the degree to which it raises blood glucose in comparison with pure glucose, which has a glycemic index of 100. **Glycemic load** is a parameter that takes into account the glycemic index of a food and puts it into the context the quantity of food consumed. Thus, glycemic load is the glycemic index of a food times the grams of carbohydrate per serving. Diets that produce high glycemic loads are believed to be associated with development of type II diabetes. Although fructose has a relatively low glycemic index, the addition of high fructose corn syrup, a common ingredient in prepared foods, can greatly increase the glycemic load of food and beverages, as is the case with the addition of other sugars.

Lipids

Most lipids are packaged into **chylomicrons** in the enterocytes (see Chapter 26). The chylomicrons enter the lymph lacteals in the small intestine, ultimately entering the systemic blood in the large vessels in the thoracic cavity. Thus the first entry of absorbed lipid into the liver is from the systemic, not portal, circulation. Liver lipid metabolism includes the following:

- **β-Oxidation of fatty acids:** Although many tissues use β-oxidation as a source of energy when the need arises, the rate of use of β-oxidation is very high in the liver.
- **Formation of most lipoproteins:** Very low density lipoprotein (VLDL), low-density lipoprotein (LDL), and high-density lipoprotein (HDL) are formed in the liver. VLDL and LDL transport TGs and cholesterol to tissues. LDL is implicated in development of cardiovascular disease because it is incorporated into atherosclerotic plaques. HDL transports lipids from tissues to the liver and is considered beneficial in terms of cardiovascular health.
- **Synthesis of cholesterol and phospholipids:** Cholesterol and phospholipids are necessary for making membranes, and cholesterol is also the precursor for steroid hormones and bile. Because of these important functions, the liver ensures a supply of these substrates by forming them from other lipids. A liver enzyme, HMG-CoA reductase, catalyzes the rate-limiting step in cholesterol synthesis, and pharmacologic intervention by **statins** (cholesterol-lowering drugs) inhibits this enzyme.
- **Conversion of unused glycogen to TGs:** When liver glycogen is not used, it is converted to TGs, which are transported to adipose tissues in VLDLs.

Proteins

Protein metabolism in the liver is essential for survival; the liver processes dietary amino acids for systemic use and participates in the processing of nitrogen wastes for excretion. The major functions of the liver are:

- **Deamination of amino acids:** Deamination is the first step in removal of excess amino acids; aminotransferases remove the amino group from the amino acids, creating ammonia (NH_3).
- **Production of urea:** The NH_3 combines with CO_2 to form urea, thereby buffering NH_3, a toxin, and allowing urinary excretion.
- **Synthesis of plasma proteins:** About 90% of plasma proteins are made in the liver. These proteins include:
  - albumin, which contributes to oncotic pressure of plasma.
  - immunoglobulins, which contribute to immune functions.
  - fibrinogen, which is necessary for blood clotting.
- **Interconversion of amino acids:** Needed amino acids are synthesized from other available amino acids.

PRODUCTION AND SECRETION OF BILE

Bile is critical for transport of lipids through the **unstirred water layer** to the enterocytes. The unstirred water layer is the area in the gut lumen closest to the villous lining. Little flow occurs near the cells, and the water and mucus create a barrier that impedes access of the hydrophobic lipids to the enterocytes (carbohydrates and proteins have no trouble passing through this area). This challenge of transporting hydrophobic lipids is met by bile, which is amphipathic (i.e., it has both hydrophilic and hydrophobic regions) and thus is able to move lipids across the unstirred water. Bile and lipids form **micelles**, which act like taxis to shuttle the lipids through the unstirred water layer to the enterocytes (discussed further in Chapter 26). **Bile solids** are composed of bile salts (50%), phospholipids (40%), and smaller amounts of cholesterol (~4%), bilirubin (~2%), and water and electrolytes. Secreted bile is composed of the bile solids, water, and electrolytes.

In the liver, primary **bile acids** (e.g., cholic and chenodeoxycholic acids) are synthesized in the hepatocytes from sterol rings (from cholesterol). Bile synthesis and secretion has several important aspects:

■ In the liver, one side of the primary bile acid is conjugated with an amino acid (either taurine or glycine), forming a **bile salt**. This conjugation increases the water solubility of bile in the lower pH found in the duodenum. The primary bile salts are secreted into the bile canaliculi, or ductules, and into the common bile duct (see Fig. 25.2).

■ Bile salts are osmotic, and their secretion will draw water and then solutes (e.g., sodium chloride and HCO_3^-) from the cells; this process is called **solvent drag** and contributes to the buffering capacity of the bile when it enters the duodenum.

■ After micelles are formed and the lipids are dropped off at the enterocytes, the majority of the bile remains in the lumen of the small intestine until the terminal ileum, where Na^+-dependent transporters recycle the primary bile into the portal vein back to the liver. This bile recycling will occur three to five times for each meal and permits efficient absorption of the lipids without synthesis of large amounts of bile, because it is reused.

■ However, with each cycle about 10% of the bile is not absorbed but is lost in the feces, which is the major pathway by which cholesterol is removed from the system. Other substances, including bilirubin, are also excreted. The synthesis of bile acids is under feedback control by bile salts that enter the liver from the portal circulation. A reduced level of bile salts will increase **cholesterol 7α-hydroxylase**, the rate-limiting enzyme in bile synthesis.

■ In the intestines, some of the primary bile acids undergo dehydroxylation by bacteria, forming **secondary bile acids** (deoxycholic and lithocholic acids), which are less efficient at crossing the unstirred water layer and hence are less readily absorbed.

■ When the stomach is emptied and chyme is no longer present in the duodenum, the sphincter of Oddi closes. As the bile is recycled through the portal system back through the liver, it will stop at the sphincter of Oddi and back up into the relaxed gallbladder, where it will be stored until the next meal.

BASIC ENDOCRINE FUNCTIONS

The liver produces or modifies several endocrine and paracrine substances, including:

■ **IGF-1**, which is released by the liver into the circulation in response to growth hormone. IGF-1 mediates many of the somatic effects of growth hormone.

■ **Angiotensinogen**, which is the precursor to angiotensin I and II (angiotensin II plays an important role in fluid and electrolyte homeostasis).

■ **Thrombopoietin**, which stimulates stem cells in bone marrow to differentiate to megakaryocytes, which give rise to platelets. Platelets participate in blood clotting.

■ **Hepatocyte growth factor**, which acts locally to stimulate regeneration of liver cells and is especially important when the organ is damaged.

■ **Vitamin D metabolism:** The liver hydroxylates cholecalciferol (from diet or synthesized in skin) to form 25-hydroxycholecalciferol. This substance is still inactive and must be further hydroxylated in the kidney to become the active form of vitamin D, 1,25-dihydroxycholecaliferol. Vitamin D is a key regulator of intestinal calcium absorption.

CLINICAL CORRELATE 25.2
Portal Hypertension and Esophageal Varices in Obstructive Liver Disease

Increased blood pressure in the portal vein can result from suprahepatic pathology (e.g., an increase in systemic venous pressures, as with congestive heart failure) or hepatic disease (e.g., obstructive liver disease). Obstructive liver disease is most commonly a result of cirrhosis or fibrous scarring of the liver, which severely decreases the flow of blood through the organ. In the United States, alcoholism and hepatitis C infection are the most common causes of cirrhosis. Interestingly, alcoholism causes cirrhosis of the liver or pancreatitis, but not both diseases, in individual patients.

In the cirrhotic liver, the obstruction of blood flow through the liver increases pressure in the portal system. This **portal hypertension** results in increased back pressure in vessels coming from the stomach and esophagus, causing enlargement and thinning of these vessel walls, thus forming **varices**. The thin walls, high pressure, and increased radius in the varices make them susceptible to rupture, and because of the superficial nature of the vessels serving the esophagus, rupture may cause severe bleeding into the esophageal lumen, requiring immediate medical attention. Varices are treated either by sclerotherapy (i.e., injecting a solution to block the vessels) or by rubber-band ligation of the varices, a procedure in which a band is wound around the varices, cutting off blood flow. The atrophied area sloughs off, leaving a healed scar. Portal hypertension can also result in **ascites** and **hemorrhoids**, which also are caused by the backup in pressure in the portal venous system.

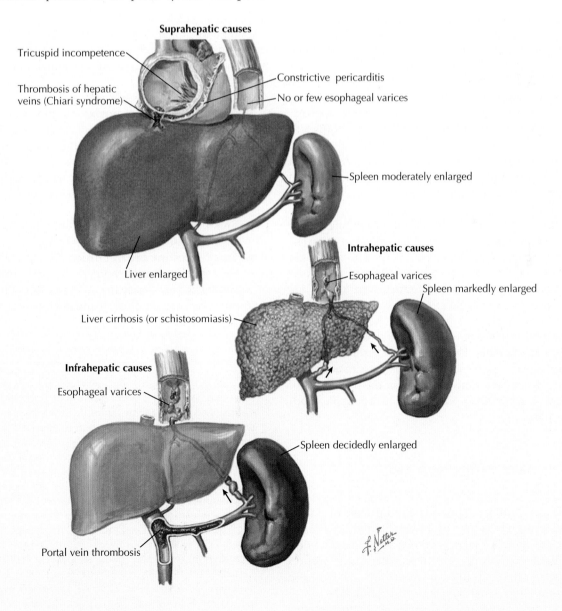

Suprahepatic causes

Tricuspid incompetence

Thrombosis of hepatic veins (Chiari syndrome)

Constrictive pericarditis

No or few esophageal varices

Spleen moderately enlarged

Liver enlarged

Liver cirrhosis (or schistosomiasis)

Intrahepatic causes

Esophageal varices

Spleen markedly enlarged

Infrahepatic causes

Esophageal varices

Spleen decidedly enlarged

Portal vein thrombosis

Esophageal Varices

Chapter 26

Digestion and Absorption

ANATOMY AND NUTRIENT ABSORPTION

Absorption of nutrients in the small intestine depends on an intact villous lining, which facilitates efficient nutrient uptake into the body. The lumen of the small intestine is composed of circular folds, villi, and microvilli (Fig. 26.1). This structure increases the surface area of the small intestine to approximately 250 square meters and dramatically increases the efficiency of the system so that most of the nutrient absorption will occur by the mid to late part of the jejunum. As a rule of thumb, ~25% to 30% of digestion occurs preduodenally, and ~70% to 75% occurs in the small intestine.

The upper half of the villi is the site of membrane-bound enzymes (brush border enzymes) for final digestion of carbohydrates and proteins. The upper half of the villi is also the site of absorption. The bottom part of the villi forms the crypts of Lieberkühn, from which buffers and mucus are secreted, as discussed in Chapter 24. Finally, because laminar flow occurs through the intestinal lumen, the slowest movement of chyme is near the enterocytes, which secrete copious mucus to protect the cells. This process creates an **unstirred water layer** through which molecules must pass to access the enterocytes. This passage is no problem for the movement of most nutrients but presents a problem for hydrophobic, lipid-based molecules.

CARBOHYDRATE DIGESTION AND ABSORPTION

Much of the carbohydrate in our diet is in the form of **starch, sucrose** (table sugar), **lactose** (milk sugar), and, to a lesser extent, fructose. **Starch** is a large, branched, long-chain polysaccharide synthesized by plants; it represents a large portion of the carbohydrates consumed in most diets. The glucose moieties within the starch molecule are bound by α-1,4 glycosidic linkages, with α-1,6 linkages at branch points. **Sucrose** and **lactose** are **disaccharides**. Most people also ingest large quantities of the monosaccharide **fructose**, which is present in fruits and foods containing added high-fructose corn syrup.

Although we ingest oligosaccharides and disaccharides, we can absorb only monosaccharides. Carbohydrate digestion begins in the mouth and continues through to the small intestine (Fig. 26.2):

- **Mouth:** Salivary α-amylase begins starch digestion, breaking α-1,4 linkages and creating smaller oligosaccharides (maltose and isomaltose, both of which are disaccharides consisting of two linked glucose molecules, as well as larger oligosaccharides and polysaccharides). The amylase is inactivated as the chyme becomes more acidic in the stomach.
- **Small intestine:** Pancreatic α-amylase continues the digestion, forming more maltose and isomaltose.
- **Small intestine brush border:** As the chyme contacts the villous brush border, specific **brush border saccharidases** digest the maltose, isomaltose, sucrose, and lactose to their constitutive monosaccharides (Table 26.1).

Starch, glycogen, and cellulose are all polymers of glucose. Starch is a storage form of glucose in plants and is a major source of energy in our diet. Although animals store glucose in their muscles in the form of glycogen, the meat we consume contains little or no glycogen, because it is broken down after animals are slaughtered. Cellulose is the structural component of the cell walls of green plants and has no caloric value to humans. Cellulose is plant fiber and is composed of glucose molecules bound by β-1,4 linkages in large, straight chains. Humans do not have an enzyme that can digest β-linkages, so this "fiber" remains in the gastrointestinal lumen. Fiber is an osmotic agent and draws water through the large intestine, loosening stools and adding bulk. High-fiber diets have other health-related advantages, including a blood cholesterol–lowering effect and enhanced glucose tolerance (i.e., an antidiabetogenic effect). In contrast to other mammals, ruminants such as cows are able to use cellulose as a food source because of the presence of symbiotic bacteria in their GI tracts that are able to digest the cellulose to its constituent glucose molecules.

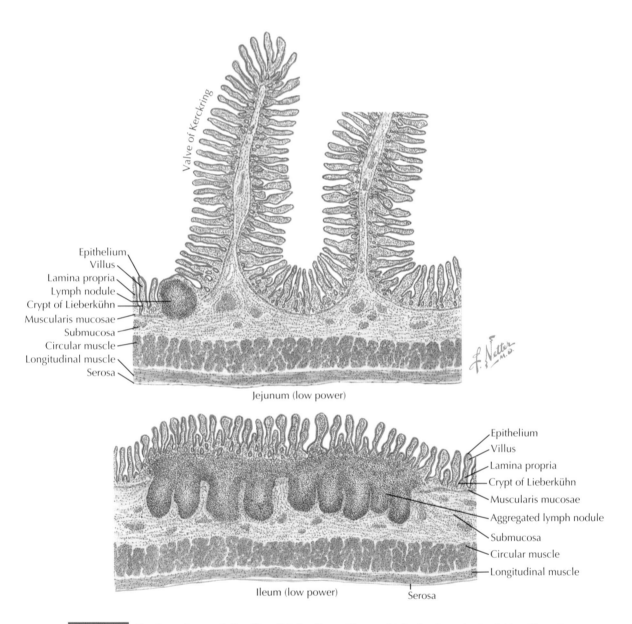

Valve of Kerckring

Epithelium
Villus
Lamina propria
Lymph nodule
Crypt of Lieberkühn
Muscularis mucosae
Submucosa
Circular muscle
Longitudinal muscle
Serosa

Jejunum (low power)

Epithelium
Villus
Lamina propria
Crypt of Lieberkühn
Muscularis mucosae
Aggregated lymph nodule
Submucosa
Circular muscle
Longitudinal muscle

Ileum (low power)   Serosa

Figure 26.1 **Surface Area of the Small Intestine** The small intestine has circular folds, villi, and microvilli (i.e., hairlike projections on villi) that increase the surface area for absorption to about 600 times the area of a straight tube, thus allowing efficient absorption of nutrients. The upper half of the villi is the site of the brush border saccharidases and proteases (for final digestion of carbohydrates and proteins), and the upper half is also where absorption occurs.

In the small intestine, glucose and galactose are transported into the enterocytes with sodium via sodium-glucose linked transporter (SGLT)-1 transporters (secondary active transport); fructose has its own transport protein, glucose transporter (GLUT)-5 (facilitated transport). The monosaccharides leave the enterocytes through the basolateral membranes by facilitated transport through GLUT-2, diffuse through the interstitial space into the capillary, and are transported through the portal vein to the liver for processing and release into the systemic circulation. It is important to recognize that glucose transport in the GI tract (including the liver) is independent of insulin. This is also true of glucose transport in the brain, pancreatic β cells, and renal tubules. Carbohydrates are efficiently digested and quickly absorbed.

Because glucose is absorbed into the enterocytes by SGLT-1 transporters, the rapid absorption of glucose facilitates sodium absorption and hence chloride and water absorption. This principle is the basis of **oral rehydration therapy** for dehydration, or fluid loss in enteric diseases such as cholera.

PROTEIN DIGESTION AND ABSORPTION

Dietary protein digestion begins in the stomach and continues in the small intestine in the following manner:

- **Stomach: Pepsinogens** are secreted from the gastric chief cells and are activated to pepsins by stomach acid.

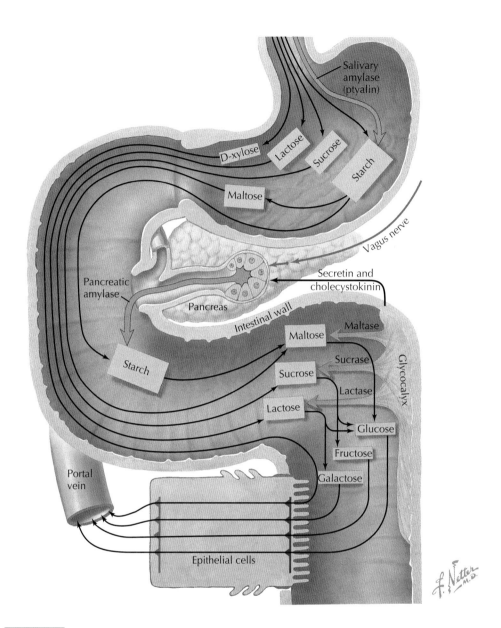

Figure 26.2 **Carbohydrate Digestion and Absorption** Starch digestion begins in the mouth with salivary α-amylase, which produces smaller malto-oligosaccharides. In the small intestine lumen, pancreatic α-amylase continues to digest the starches. At the same time, brush border saccharidases perform the final digestion of maltose and isomaltose (by maltase and isomaltase, producing glucose molecules), and of the disaccharides sucrose (by sucrase, producing glucose and fructose) and lactose (by lactase, producing glucose and galactose). The monosaccharides then are transported into the enterocytes by secondary active transport with sodium (glucose and galactose) and facilitated transport (fructose; not shown). All of the glucose transport in the small intestine and liver is independent of insulin.

Table 26.1 **Brush Border Saccharides and Products**

Brush Border Enzyme	Substrate	Product(s)
Maltase	Maltose	Glucose
Isomaltase	Isomaltose	Glucose
Sucrase	Sucrose	Glucose + fructose
Lactase	Lactose	Glucose + galactose

Gastric pepsins are endopeptidases (like the pancreatic enzymes trypsin and chymotrypsin) and hydrolyze inner peptide bonds, creating smaller oligopeptides. Pepsins are inactivated in the higher pH of the duodenum.

- **Small intestine: Pancreatic proteases** are responsible for digesting the oligopeptides so they become smaller peptides. As stated in Chapter 24, the pancreatic proteases are released as zymogens into the duodenum primarily in response to cholecystokinin. In the duodenum, **enterokinase** (secreted into the lumen from cells in the

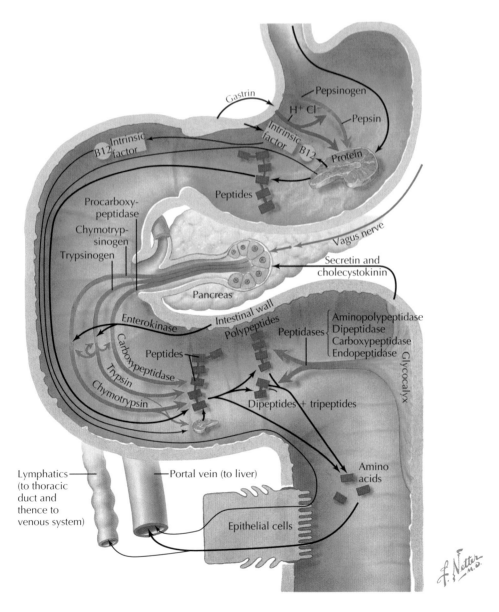

Figure 26.3 **Protein Digestion and Absorption** Protein digestion begins in the stomach with HCl and pepsins, which break protein down to smaller polypeptides. In the lumen of the small intestine, pancreatic proteases (e.g., trypsin, chymotrypsin, and carboxypeptidase) are activated and continue the digestion to create smaller peptide chains. Final digestion occurs by the brush border proteases, which produce dipeptides, tripeptides, and single amino acids. These molecules are transported into the cells via secondary active transport with sodium (amino acids) or are coupled with H+ (dipeptides and tripeptides). In the cells, the dipeptides and tripeptides are digested by cytoplasmic peptidases to form amino acids.

crypts of Lieberkühn) converts trypsinogen to the active protease **trypsin**, and the trypsin activates the other endopeptidases (chymotrypsin and elastase) and the exopeptidases (carboxypeptidase A and carboxypeptidase B). Once activated, trypsin can also activate trypsinogen. The pancreatic proteases continue to perform hydrolysis of peptide bonds, making smaller oligopeptides (Fig. 26.3).

- **Small intestine brush border:** A variety of **brush border peptidases** hydrolyze the peptides to amino

acids and dipeptides and tripeptides, which can then be absorbed.

- **Enterocytes: Cytoplasmic peptidases** digest the dipeptides and tripeptides to form amino acids, which can exit the cells and enter the capillaries.

In the small intestine, most of the proteins are absorbed into the enterocytes in dipeptide and tripeptide form via H+ symporters that are specific for the peptides. Amino acids have different Na+-dependent transporters for basic, acidic, neutral, and imino acids. Once inside the intestinal cells, the dipeptides and tripeptides are hydrolyzed by cytoplasmic peptidases to amino acids, which leave the cells by facilitated transport into the capillaries. A small amount of the dipeptides and tripeptides can be transported through the cell and into the blood, but the mechanism for this transport is unclear.

As previously stated, storage and secretion of the pancreatic proteolytic enzymes as zymogens helps prevent digestion of the pancreatic tissue. The pancreas also produces **trypsin inhibitor,** which prevents the small amount of trypsin that might become activated in the pancreas and its ducts from damaging the tissues. Similarly, in the small intestine, where a large amount of active trypsin can be found, damage of the intestinal mucosa is prevented by the secretion of a trypsin inhibitor (specifically, pancreatic secretory trypsin inhibitor) by the intestinal epithelium. Pancreatic secretory trypsin inhibitors have also been shown to stimulate growth of intestinal epithelial cells.

LIPID DIGESTION AND ABSORPTION

Almost all (~98%) of dietary lipids are triglycerides (TGs), with the remainder being cholesterol esters and phospholipids. Lipids are easily hydrolyzed to molecules that can be absorbed; however, their hydrophobicity does not allow easy access to the absorptive cells of the small intestine brush border. As a result, a complex mechanism exists for efficiently moving lipids through the unstirred water layer to the enterocytes. It should be noted that although not a lot of lipid digestion occurs in the upper GI tract in adults, lingual and gastric lipases do have a role in hydrolysis of lipids in the neonate. Lipid digestion occurs in the following locations:

- **Mouth: Lingual lipase** is secreted from Von Ebner's glands of the tongue into the saliva and begins hydrolysis of TGs to diglycerides and free fatty acids (FFAs). The enzyme remains active in the stomach.
- **Stomach: Gastric lipase** is secreted from the chief cells of the gastric pits and also hydrolyzes TGs to diglycerides and FFAs. Again, in adults, gastric lipase appears to be a minor factor in lipid digestion.
- **Small intestine:** Various **lipases** are secreted in active form from the pancreas in response to the action of cholecystokinin on the pancreatic acinar cells. **Pancreatic lipase** hydrolyzes TGs to monoglycerides and FFAs; **procolipase** is also secreted and is activated in the lumen of the duodenum by trypsin, forming colipase, which facilitates the enzymatic action of pancreatic lipase; **cholesterol ester hydrolase** hydrolyzes cholesterol esters to cholesterol and FFA; and **phospholipase A2** hydrolyzes phospholipids to lysophospholipids and FFAs. These cleavage products are all able to diffuse into the enterocytes; however, they must first be incorporated into micelles to traverse the unstirred water layer.

Micelle Formation

Most lipids are found in the last portion of chyme entering the small intestine after a meal. In the duodenum, the lipid droplets are surrounded by bile and emulsified, but they need to be digested by the pancreatic lipases before absorption. Colipase is critical for lipase action because it displaces the bile from the lipids, providing the lipase access to hydrolyze the lipids. Micelles are aggregates of bile salts and the products of this lipid digestion. The polar (hydrophilic) ends of the bile are oriented outward, and the hydrophobic core allows incorporation of small lipids. The function of micelles is to solubilize lipids and allow them to move through the unstirred water layer to the apical surface of enterocytes, where the lipids are absorbed.

Micelle formation and lipid absorption entail the following steps (Fig. 26.4):

1. The detergent action of bile emulsifies the fat globules, forming smaller fat droplets, which increases the surface area for digestion.
2. Colipase displaces bile from the lipid, allowing pancreatic lipase to hydrolyzes the TGs to monoglycerides and FFAs. Cholesterol ester hydrolase and phospholipase A2 hydrolyze cholesterol esters and phospholipids (these enzymes are not affected by bile).
3. When a critical amount of bile and digested lipids is available, these substances form a **micelle**, with the hydrophilic ends of bile salts on the outside and the lipophilic ends inside surrounding the lipid products.
4. After the micelles diffuse through the unstirred water layer to the enterocytes, the lipids leave the micelle and diffuse through the enterocyte membrane, leaving the bile salts in the lumen.
5. The bile salts continue down the small intestine to the terminal ileum, where most primary bile salts enter the intestinal cells through Na^+-dependent transporters. The bile diffuses out of the cell into the portal circulation back to the liver.

Intracellular Lipid Processing

Once the lipids diffuse into the enterocytes, triglycerides, cholesterol esters, and phospholipids are re-formed in the smooth endoplasmic reticulum (see Fig. 26.4). The lipids are assembled into minute lipid droplets called **chylomicrons** using apolipoproteins (lipid-binding proteins) that are synthesized in the enterocytes (especially apolipoprotein B). The chylomicrons are exported from the enterocytes by exocytosis and enter the lymph lacteals; they are too large to enter capillaries. The chylomicrons enter the systemic blood via the thoracic ducts into the left subclavian vein and travel through the systemic circulation to the liver for processing. A small number of short-chain fatty acids are not incorporated into chylomicrons and are able to diffuse into the portal blood to the liver, but this pathway is minor.

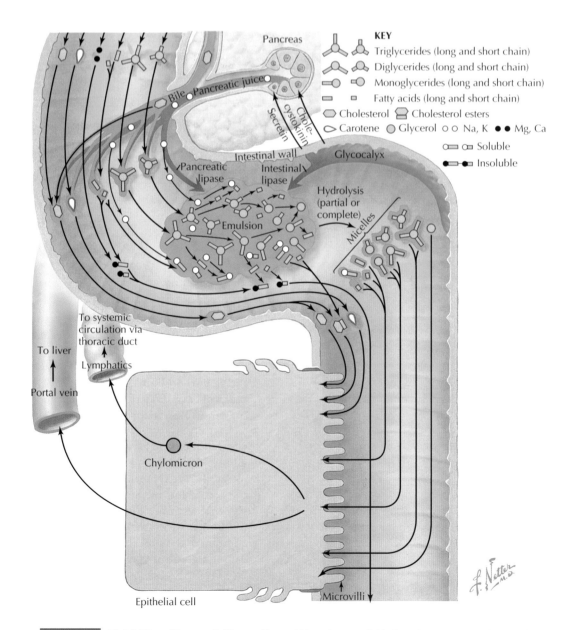

Figure 26.4 **Lipid Digestion and Absorption** Although some lipid digestion occurs preduodenally by lingual and gastric lipases, most digestion occurs in the lumen of the small intestine by pancreatic lipase. The overall process of digestion and absorption is complex, because the lipids must first be incorporated with bile into micelles and shuttled through the unstirred water layer to the enterocytes. The lipids diffuse into the enterocytes and are re-esterified with free fatty acids in the smooth endoplasmic reticulum and packaged into chylomicrons for transport into the lymph, eventually passing through the thoracic duct into venous circulation. Some soluble fats (glycerol and short-chain and medium-chain fatty acids) can access the enterocytes without micelles and are small enough to be taken up into the portal circulation.

Apolipoprotein B-48 is critical for **chylomicron** formation and export from enterocytes. When it is absent, lipids cannot exit the intestinal cells and thus fat builds up in the enterocytes. As the villous cells are replaced and sloughed off into the lumen, the lipid is excreted into the feces. This condition of **abetalipoproteinemia** results in the inability to absorb dietary lipids along with fat-soluble vitamins.

ELECTROLYTE AND WATER ABSORPTION

In addition to the ~2 L of fluid ingested each day, approximately 7 or more additional liters are added to the GI tract in various segments to facilitate digestion and absorption of the nutrients. Absorption of the 9 L of fluid occurs as nutrients and electrolytes are absorbed. Within the small intestine, several mechanisms of absorption occur in various segments (Fig. 26.5):

A. Jejunum

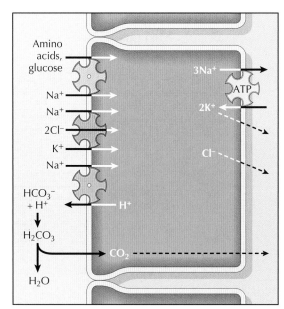

B. Ileum

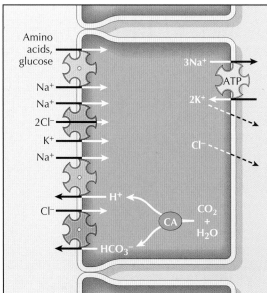

C. Colon

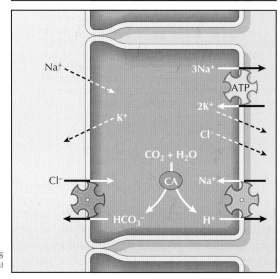

J. Perkins
MS, MFA, CMI

Figure 26.5 **Electrolyte and Fluid Transport** **A,** Sodium absorption in the jejunum and ileum occurs though a variety of mechanisms, with sodium movement facilitating entry of other molecules such as glucose, galactose, amino acids, and chloride. This mechanism also creates an osmotic gradient for water absorption. **B,** The ileum secretes HCO_3^- into the gut lumen in exchange for Cl^-, and the Cl^- exits through the basolateral membrane by facilitated transport. **C,** The colon is sensitive to the effects of aldosterone, which enhances sodium absorption (through epithelial sodium channels [ENaC]) and potassium secretion (through big potassium [BK] channels). The water follows the osmotic gradient, dehydrating the chyme and producing feces. *CA,* carbonic anhydrase.

- **Jejunum:** The large surface area in the small intestine makes digestion and absorption extremely efficient. The bulk of the nutrients, fluid, and electrolytes is absorbed by the mid to late portion of the jejunum, although significant electrolyte and fluid absorption occurs in the ileum. **Sodium** absorption is a driving force for nutrient and water absorption, and the basolateral Na^+/K^+ ATPase maintains the low intracellular concentration necessary for the lumenal Na^+ to enter the cells (via secondary active transport with nutrients, NKCC-2, and Na^+/H^+ exchangers). **Bicarbonate** is formed by the intracellular CO_2 metabolism and exits the cell through the basolateral membrane by facilitated transport (not shown on figure); the H^+ formed in the process enters the lumen in exchange for Na^+. This process permits additional sodium absorption. The **water** follows the osmotic gradient created by the electrolyte movement (see Fig. 26.5*A*).
- **Ileum:** As noted, the Na^+-dependent transporters for sugars and amino acids are also present in the ileum, and this redundancy helps ensure nutrient absorption if absorption is reduced in the jejunum. **Sodium** is absorbed in the same manner as in the jejunum; however, in the ileum, the bicarbonate formed in the cells is *secreted* into the lumen via HCO_3^-/Cl^- **antiporters,** facilitating **chloride** absorption; the **chloride** leaves the enterocytes at the basolateral side via facilitated transport (see Fig. 26.5*B*).
- **Colon:** Aldosterone stimulates lumenal **sodium** absorption via epithelial sodium channels (ENaC) and potassium *secretion* via **BK channels.** As water follows the sodium, the chyme is dehydrated, producing feces. The colon usually absorbs ~400 to 500 mL of water per day, which follows sodium absorption, and water absorption in the colon can increase to ~1 L per day when aldosterone is elevated. Also, as observed in the ileum, the bicarbonate formed in the cells is secreted into the lumen via HCO_3^-/Cl^- antiporters, allowing additional chloride absorption. **Chloride** leaves enterocytes at the basolateral side via facilitated transport (see Fig. 26.5*C*).

DIVALENT CATION ABSORPTION

Calcium and iron are absorbed early in the small intestine, from the duodenum through the early to mid portions of the jejunum (i.e., the first half of the jejunum). The low pH in the early part of the small intestine favors the reduced (Fe^{2+}) form of iron and keeps the cations from forming insoluble salts.

Calcium

Calcium absorption is regulated by active **vitamin D** (1,25-dihydroxycholecalciferol), which increases the calcium channels (TRPV-6) at the lumenal membrane, as well as the cytosolic binding protein **calbindin**. Calbindin binds Ca^{2+}, keeping the intracellular free Ca^{2+} levels very low (10^{-7} M), thus maintaining a gradient for Ca^{2+} entry into the cell and allowing Ca^{2+}-dependent messenger systems to function properly. The Ca^{2+} exits the basolateral side through active Ca^{2+} ATPase pumps and Na^+/Ca^{2+} antiporters (Fig. 26.6A). The activity of the Ca^{2+} ATPase is also increased by vitamin D.

Iron

To be absorbed, iron is either in the ferrous Fe^{2+} form or in the organic heme form. Once heme is absorbed, intracellular heme oxygenase frees the iron, which is then bound to ferritin in the cell or is transported out of the cell into the blood, where it is bound to transferrin. In the case of Fe^{2+} absorption, lumenal membranes in the jejunum have divalent metal transport proteins that transport free Fe^{2+} into the cell, where it is bound to ferritin or transported out of the cell into the blood, where it is bound to transferrin (see Fig. 26.6B).

In the blood, the transferrin-bound iron is transported to the bone marrow for hemoglobin and red blood cell synthesis or to the liver and spleen for storage. Typically, ~1 mg of iron is absorbed daily, which balances the ~1 mg loss of iron (from intracellular ferritin stores) that occurs when enterocytes slough off each day. If iron utilization increases (such as in growing children and menstruating women), iron intake must increase or anemia will result.

VITAMIN ABSORPTION

Fat-soluble vitamins A, D, E, and K are transported in micelles to the enterocytes, although vitamins A and D are less hydrophobic and are also able to access the enterocytes independently. Table 26.2 details the sites and mechanism of vitamin absorption in the small intestine.

Vitamin B12 Absorption

Vitamin B12 (cobalamin) is an essential vitamin and is protected from digestion until it is absorbed in the terminal ileum. The process of protecting B12 begins in the mouth:

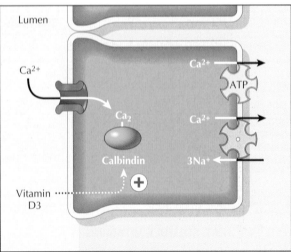

A. Ca^{2+}

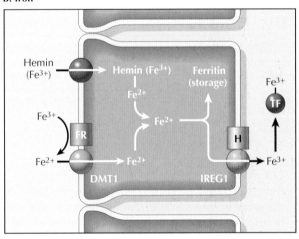

B. Iron

J. Perkins
MS, MFA

Figure 26.6 Calcium and Iron Absorption A, Calcium entering the enterocytes through the TRPV-6 Ca^{2+} channel is bound by cytosolic calbindin. Transport out of the cell is via Ca^{2+} ATPase and Na^+/Ca^{2+} exchangers. Active vitamin D enhances Ca^{2+} absorption by increasing calbindin production and Ca^{2+} pump activity. **B,** Ingested iron is in both organic (heme) and inorganic forms. Heme can enter enterocytes, and intracellular heme oxygenase releases the iron (Fe^{2+}), which is then bound in ferritin stores or shuttled out of the cell through the transmembrane protein hephaestin (H) and bound to transferrin (TF) in the blood. It is transported to the liver for storage or to the bone marrow for hemoglobin and red blood cell production. Inorganic iron in the ferric form (Fe^{3+}) can be reduced to Fe^{2+} by ferrireductase (FR) in the lumenal membrane, and then the Fe^{2+} can enter the enterocytes through the divalent metal transporter 1 (DMT1) and be handled as previously described. *IREG1,* iron-regulated transporter-1.

- **Transcobalamin-1** (TC-1, also known as R-binder), which is present in the saliva, binds vitamin B12 in the stomach, protecting it from digestion by pepsins.
- **Intrinsic factor (IF)** is secreted by gastric parietal cells. IF binds to B12–TC-1 complexes in the stomach. When the complex enters the duodenum, the TC-1 is cleaved from it by trypsin.

Table 26.2   **Vitamin Absorption**

Vitamin	Site of Absorption	Mechanism
WATER-SOLUBLE VITAMINS		
Vitamin C	Ileum	Na⁺-coupled/secondary active
Thiamin (B1)	Jejunum	Na⁺-coupled/secondary active
Riboflavin (B2)	Jejunum	Na⁺-coupled/secondary active
Biotin	Jejunum	Na⁺-coupled/secondary active
Vitamin B12	Ileum	Facilitated diffusion
Pyridoxine (B6)	Jejunum and ileum	Passive diffusion
FAT-SOLUBLE VITAMINS		
Vitamin A	Jejunum and ileum	Passive diffusion
Vitamin D	Jejunum and ileum	Passive diffusion
Vitamin E	Jejunum and ileum	Passive diffusion
Vitamin K	Jejunum and ileum	Passive diffusion

From Hansen J: *Netter's Atlas of Human Physiology*, Philadelphia, 2002, Elsevier.

■ The **B12-IF complex dimerizes,** protecting the B12 from pancreatic proteases. The dimer continues down the small intestine to the terminal ileum, where it binds to a transport protein and enters the enterocyte. The B12 binds to **transcobalamin II** in the cytosol, exits the cell, and is transported to the bone marrow (where it promotes red blood cell maturation) or to the liver for storage (Fig. 26.7).

💡 Surgical removal of the **terminal ileum** causes multiple problems, including loss of both bile recycling and vitamin B12 absorption. In this circumstance, bile production by the liver is upregulated, but production is not sufficient for absorption of fat contained in a normal meal, and steatorrhea may result. To treat vitamin B12 deficiency, injections are administered every few months (infrequent injections are sufficient because excess B12 is stored in the liver).

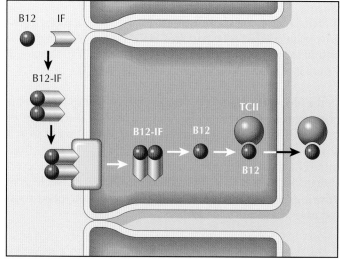

Vitamin B12

J. Perkins
MS, MFA

Figure 26.7   **Vitamin B12 Absorption**   Vitamin B12 (cobalamin) is protected from digestion in the small intestine by binding to intrinsic factor (IF) released from the gastric parietal cells. B12-IF dimers are recognized by receptors on the enterocytes in the terminal ileum, and the complex is internalized. The B12 is bound to transcobalamin II (TCII) for transport through the blood to the liver for storage or to bone marrow for red blood cell maturation.

CLINICAL CORRELATE 26.1
Celiac Disease (Gluten Enteropathy)

Celiac disease (also called **sprue**) is an autoimmune condition in which exposure to **gluten proteins** in wheat results in the plasma cell production of antibodies (immunoglobulin A and G). The antibodies cross-react with intestinal tissue, producing inflammation, which results in flattening of the villous lining and crypt hyperplasia. The surface area for absorption is diminished, and brush border enzyme activity is impaired. The loss of brush border enzymes with villous atrophy reduces the final digestion of carbohydrates and protein and causes general malabsorption of micronutrients and macronutrients. The presence of undigested nutrients lower in the GI tract facilitates bacterial production of gases, causing cramps and odiferous gas and feces. Diarrhea, weight loss, anemia, and vitamin deficiency can occur.

Although it is suggested that 1 in 250 persons have some degree of gluten enteropathy, most cases are considered "silent" sprue, with few or no GI symptoms. In addition to GI complications, the immune reaction to wheat can also cause respiratory problems (e.g., rhinitis and asthma) and skin problems (e.g., eczema, dermatitis herpetiformis, and/or hives). Primary treatment involves complete avoidance of gluten in the diet.

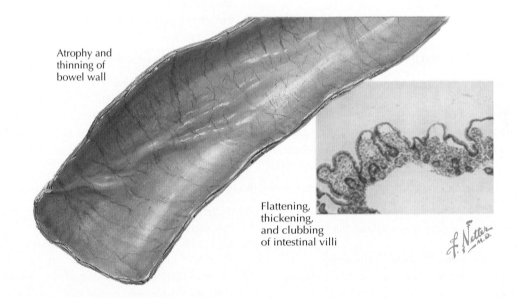

Atrophy and thinning of bowel wall

Flattening, thickening, and clubbing of intestinal villi

Celiac Disease (Gluten Enteropathy)

Review Questions

CHAPTER 22: OVERVIEW OF THE GASTROINTESTINAL TRACT

1. All of the following are functions of the GI tract **EXCEPT:**

A. secretion of endocrine hormones.
B. secretion of digestive enzymes.
C. absorption of nutrients from chyme.
D. regulation of systemic blood flow.
E. storage of chyme.

2. Motility and secretion through the GI tract is regulated by all of the following factors **EXCEPT:**

A. duodenal hormones.
B. chemoreceptors.
C. autonomic nerves.
D. enteric nerves.
E. growth hormone.

3. In the absence of *extrinsic* innervation to the gastrointestinal tract,

A. propulsive movements would not occur in the small intestine.
B. propulsive movements would occur in the small intestine, but secretion into the small intestine would be abolished.
C. secretion would be abolished in the entire GI tract.
D. motility and secretion throughout the tract would be stimulated by local mechanoreceptors, chemoreceptors, and osmoreceptors as well as hormones.
E. motility and secretion would be affected only in the large intestine.

4. Input from vagal efferent nerves to the gastrointestinal tract stimulates all of the following **EXCEPT:**

A. lower intestinal motility (ileum, proximal colon).
B. salivation.
C. the cephalic phase of gastric acid secretion.
D. receptive relaxation in the stomach.
E. hepatic bile production.

5. Bacteria in the GI tract:

A. are found only in the colon.
B. can modulate immune function in the GI tract.
C. metabolize compounds to long-chain fatty acids.
D. are all probiotic, unless antibiotics are used.
E. have no trophic actions in the gut.

CHAPTER 23: MOTILITY THROUGH THE GASTROINTESTINAL TRACT

6. Action potentials in the GI smooth muscle:

A. occur only in the upper GI tract.
B. are stimulated when the slow waves are depolarized above −40 mV.
C. occur only in the lower GI tract.
D. result from influx of sodium ions into the smooth muscle cells.
E. are stimulated only by extrinsic nerves.

7. Slow waves constituting the basal electrical rhythm:

A. are hyperpolarized by stretch, acetylcholine, and gastrin.
B. occur at a consistent rate throughout the GI tract.
C. are undulations in the resting membrane potential.
D. are depolarized by sympathetic nerve stimulation.
E. are absent in the colon.

8. In the small intestine, peristalsis:

A. occurs primarily in response to the chemical composition of the chyme.
B. proceeds in both forward and backward directions from the site of contraction.
C. involves receptive relaxation of the muscle proximal (toward the mouth) to the bolus of chyme.
D. involves contraction of the muscle proximal to the bolus of chyme.
E. is dependent on parasympathetic innervation.

9. Select the **FALSE** answer about the migrating myoelectric complex (MMC).

A. The MMC is a series of contractions that "sweep" undigested fibers and bacteria into the colon.
B. The parasympathetic nervous system controls the MMC contractions.
C. In each cycle, the main contractions of the MMC occur during phase III.
D. The hormone motilin stimulates phase III contractions.
E. The MMC occurs during fasting and stops upon feeding.

10. A 25-year-old woman presents with recurrent episodes of abdominal pain, abdominal distension, and occasional vomiting that occur 2 hours after a meal. She usually has to miss work, and her symptoms resolve after a day or two, during which she has no bowel movements. A CT scan of the abdomen shows patchy segments of the colon that are thickened, as well as a stricture in the mid jejunum. What is the most likely diagnosis?

A. Small bowel obstruction
B. Crohn's disease
C. Ulcerative colitis
D. Celiac disease
E. Irritable bowel syndrome

CHAPTER 24: GASTROINTESTINAL SECRETIONS

11. Secretions entering the lumen of the stomach include all of the following substances **EXCEPT:**

A. hydrochloric acid.
B. lipase.
C. mucus.
D. gastric inhibitory peptide.
E. intrinsic factor.

12. Parietal cell gastric acid secretion is regulated:

A. directly by secretin.
B. directly by peptide YY.
C. directly and indirectly by the vagus nerve.
D. indirectly by somatostatin.
E. indirectly by gastrin.

13. Select the **TRUE** statement about gastric acid secretion.

A. Histamine suppresses acid secretion.
B. Blocking the H^+/K^+ ATPase pump increases alkaline tide.
C. Basolateral HCO_3^-/Cl^- exchangers decrease intracellular Cl^- in parietal cells.
D. Parasympathetic stimulation of acid secretion is limited to the cephalic phase of acid secretion.
E. Proton pump (H^+/K^+ ATPase) activity is the rate-limiting step in acid secretion.

14. Which of the following processes is **NOT** a mechanism by which sodium ions enter the intestinal epithelial cells from the lumenal membrane?

A. Diffusion down the sodium concentration gradient
B. Active transport via Na^+/K^+ ATPase
C. Co-transport with certain amino acids
D. Co-transport with monosaccharides
E. Exchange for H^+

15. Select the **TRUE** statement about pancreatic secretions.

A. The release of pancreatic enzymes into the duodenum is primarily stimulated by gastrin.
B. The release of pancreatic enzymes into the duodenum is primarily stimulated by the sympathetic nerves.
C. Secretin is released in response to low duodenal pH and acts at the pancreas to stimulate the release of electrolyte buffer solution.
D. Pancreatic proteases are released in active form.
E. Pancreatic lipase digests all lipids.

16. Vagal stimulation early in feeding will stimulate all of the following secretions **EXCEPT:**

A. pepsins.
B. gastrin.
C. intrinsic factor.
D. glucagon-like peptide–1 (GLP-1).
E. hydrochloric acid.

CHAPTER 25: HEPATOBILIARY FUNCTION

17. The liver performs all of the following functions **EXCEPT:**

A. synthesis of cholesterol.
B. vitamin production.
C. β-oxidation of fatty acids.
D. bile acid production.
E. metabolism of proteins.

18. Obstruction of blood flow through the liver (from cirrhosis or hepatitis) will:

A. have no effect on gastrointestinal digestion or absorption.
B. have no effect on bile secretion.
C. increase bile secretion.
D. increase portal vein pressure.
E. reduce hepatic blood content.

19. Metabolism of lipids within the hepatocytes includes:

A. formation of most lipoproteins, including very low density and low-density lipoproteins.
B. net breakdown of phospholipids.
C. net breakdown of cholesterol.
D. formation of steroid hormones.
E. none of the above.

20. Bile salt(s):

A. hydrolyze dietary fats.
B. are essential for creating micelles.
C. are efficiently absorbed into the portal blood all along the small intestine.
D. do not contribute to intestinal buffering of acidic chyme.
E. production is not affected by loss of the terminal ileum.

21. What effects would intestinal villous atrophy (from celiac disease) have on liver function?

A. Conversion of fructose to glucose in the glycolytic pathway would be increased.
B. β-Oxidation of fatty acids would be decreased.
C. Plasma protein production would be decreased.
D. Urea production would be increased.
E. Deamination of amino acids would not occur.

CHAPTER 26: DIGESTION AND ABSORPTION

22. Which of the following mechanisms would delay or diminish the absorption of proteins?

A. Inhibition of gastric somatostatin release
B. An increase in Na^+/K^+ ATPase activity in the basolateral membranes of the enterocytes
C. Stimulation of gastric pepsinogen release
D. A pH of 3 in the duodenum and jejunum
E. Secretion of cholecystokinin

23. Select the **TRUE** statement about carbohydrate digestion.

A. Digestion of carbohydrates begins in the intestines.
B. Loss of the intestinal brush border has no effect on carbohydrate digestion.
C. Disaccharidases are secreted into the duodenum from the crypts of Lieberkühn.
D. Starch digestion is initiated by α-amylase.
E. The majority of carbohydrate digestion occurs before the duodenum.

24. When chyme is present in the stomach and duodenum, all of the following actions occur **EXCEPT:**

A. stimulation of intestinal buffers by gastrin.
B. stimulation of hepatic bile production by vagal nerves.
C. stimulation of pancreatic enzyme secretion by cholecystokinin.
D. stimulation of intestinal buffer secretion by secretin.
E. conversion of trypsinogen to trypsin by enterokinase.

25. In a patient with celiac disease:

A. gastric ulcerations are common.
B. fibrosis decreases intestinal absorption and reduces portal blood flow.
C. flattening of intestinal villi decreases enzymatic digestion and reduces the area for absorption.
D. hypersecretion of mucus obstructs absorption.
E. hyposecretion of mucus allows acidification of the intestinal lumen, decreasing enzymatic digestion.

26. The absorption of vitamin B12 (cobalamin) is dependent on all of the following **EXCEPT:**

A. the integrity of the terminal ileum.
B. the concentration of B12 in the enterocytes.
C. the presence of TC-1 in saliva.
D. the presence of intrinsic factor in gastric juice.
E. the presence of vitamin B12–intrinsic factor binding sites in the ileum.

27. A 42-year-old woman undergoes resection and removal of the last third of her ileum because of ileal obstruction. What would be a consequence of this maneuver?

A. Liver bile production would be downregulated.
B. Micelles would not be formed.
C. Vitamin B12 would not be absorbed.
D. Bowel movements will be decreased.
E. Water absorption would increase.

Section 7

ENDOCRINE PHYSIOLOGY

The endocrine system, along with the nervous system, is responsible for controlling the internal environment and has a central role in growth, development, and propagation of the species. It consists of a number of endocrine glands, as well as regions of the brain and other organs not typically thought of as endocrine glands, all of which secrete hormones into the bloodstream as a mechanism for regulation of function in target tissues. An understanding of normal physiological function requires knowledge of this complex system, and appreciation of the consequences of dysfunction of the endocrine system is essential to the practice of medicine. Nearly all physiological processes are affected by hormones, and many of the diseases or dysfunctions of the endocrine system are commonly encountered by medical practitioners; examples of the more common endocrine disorders include diabetes, thyroid diseases, and some forms of infertility.

Chapter 27

General Principles of Endocrinology and Pituitary and Hypothalamic Hormones

HORMONE SYNTHESIS

Hormones are substances that are secreted by a gland or tissue into the blood and bind to receptors in other tissues, where they affect specific physiological processes. The organization of the endocrine system is illustrated in Figure 27.1. Chemically, hormones are **peptides** (e.g., insulin and growth hormone), **steroids** (e.g., estrogen, testosterone, and cortisol), or **amines** or their derivatives (e.g., epinephrine and thyroxine). Neurohormones are a subclass of hormones secreted by neurons (e.g., vasopressin and oxytocin).

Peptide Synthesis

Genes that encode peptide hormones are transcribed to produce specific mRNAs that are subsequently translated on ribosomes to form a preprohormone. The preprohormone is a protein hormone precursor that includes a "signal peptide" that directs its transport to the endoplasmic reticulum, where the signal peptide is cleaved, leaving a prohormone. The prohormone is transported to the Golgi apparatus, where it is sequestered into secretory vesicles. Within these vesicles, the prohormone is further cleaved to create the final active form of the hormone. Secretion of peptide hormones occurs when the contents of these vesicles are released.

Steroid Synthesis

Steroid hormones are synthesized from the precursor cholesterol, which is derived from dietary sources or synthesized de novo from acetyl-CoA. The steroidogenic pathways involved in synthesis of steroid hormones from cholesterol are considered in Chapters 29 and 32. The following major steroid hormones are derived from those pathways:

- Cortisol
- Aldosterone
- Testosterone and other androgens
- Estradiol and other estrogens
- Progesterone

Cortisol and aldosterone are synthesized in the adrenal cortex (Chapter 29), which also produces adrenal androgens.

Testosterone is produced by the testes, and estradiol and progesterone are synthesized in the ovaries (Chapter 32). The active form of vitamin D is another steroid hormone; it is formed from the inactive precursor, cholecalciferol (which is absorbed from the diet or synthesized in the skin), and processed by the liver and kidneys to form active vitamin D (Chapter 31).

MECHANISMS OF HORMONE ACTION

Hormones are secreted into the blood by endocrine glands or tissues. When they reach their target tissues, they bind to membrane or nuclear receptors, initiating a chain of events that ultimately results in the physiological effects of the hormone.

Steroid hormones (e.g., testosterone, estradiol, and progesterone), thyroid hormone, and active vitamin D are lipophilic and therefore readily enter into the target cell, where they bind to nuclear receptors and initiate gene transcription. The mRNA produced is translated into proteins that regulate biochemical and physiological processes (Fig. 27.2).

Most peptide hormones and catecholamines bind to membrane receptors linked to heterotrimeric G proteins, initiating a cascade of events involving generation (or inhibition of production) of second messengers such as cAMP, cGMP, and IP_3 that ultimately regulate cellular function. The second messengers can directly act on existing enzymes and other proteins or may activate or induce transcription factors and more indirectly affect cellular function. Growth hormone (GH), on the other hand, binds to a membrane receptor that associates with a monomeric G protein; subsequently, tyrosine kinase activity and various transcription factors are involved in its action.

By definition, hormones (endocrine secretions) are substances that are carried by the blood and act on distal tissues but are part of a larger group of regulatory secretions that also includes **autocrine, paracrine**, and **neuroendocrine secretions** (see Fig. 27.2). The common feature of these substances is that they

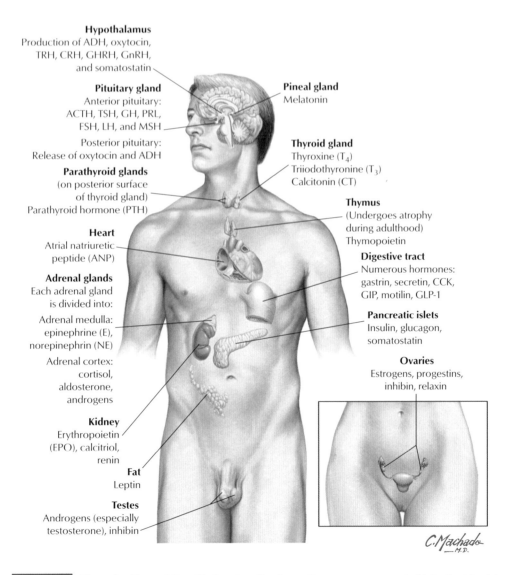

Hypothalamus
Production of ADH, oxytocin, TRH, CRH, GHRH, GnRH, and somatostatin

Pituitary gland
Anterior pituitary: ACTH, TSH, GH, PRL, FSH, LH, and MSH
Posterior pituitary: Release of oxytocin and ADH

Parathyroid glands
(on posterior surface of thyroid gland)
Parathyroid hormone (PTH)

Heart
Atrial natriuretic peptide (ANP)

Adrenal glands
Each adrenal gland is divided into:
Adrenal medulla: epinephrine (E), norepinephrin (NE)
Adrenal cortex: cortisol, aldosterone, androgens

Kidney
Erythropoietin (EPO), calcitriol, renin

Fat
Leptin

Testes
Androgens (especially testosterone), inhibin

Pineal gland
Melatonin

Thyroid gland
Thyroxine (T_4)
Triiodothyronine (T_3)
Calcitonin (CT)

Thymus
(Undergoes atrophy during adulthood)
Thymopoietin

Digestive tract
Numerous hormones: gastrin, secretin, CCK, GIP, motilin, GLP-1

Pancreatic islets
Insulin, glucagon, somatostatin

Ovaries
Estrogens, progestins, inhibin, relaxin

C. Machado
—M.D.

Figure 27.1 **Organization of the Endocrine System** Hormones are secreted into the blood by endocrine organs throughout the body, affecting physiological function at various target sites. *ACTH*, adrenocorticotropic hormone; *ADH*, antidiuretic hormone; *CCK*, cholecystokinin; *CRH*, corticotropin-releasing hormone; *FSH*, follicle-stimulating hormone; *GH*, growth hormone; *GHRH*, growth hormone–releasing hormone; *GIP*, gastric inhibitory peptide; *GLP-1*, glucagon-like peptide-1; *GnRH*, gonadotropin-releasing hormone; *LH*, luteinizing hormone; *MSH*, melanocyte-stimulating hormone; *PRL*, prolactin; *TRH*, thyrotropin-releasing hormone; *TSH*, thyroid-stimulating hormone.

are secreted and have effects on the same cell type (autocrine secretion) or other cell types.

GENERAL ENDOCRINE ROLES OF THE HYPOTHALAMUS AND PITUITARY GLAND

The **hypothalamus** and **pituitary gland** regulate the function of much of the endocrine system (Figs. 27.3 and 27.4). The hypothalamus is located in the ventral diencephalon and is connected by the pituitary stalk to the pituitary gland (**hypophysis**), which resides in the **sella turcica**, a cavity in the sphenoid bone. The pituitary consists of two lobes, known as the **anterior pituitary** (**adenohypophysis**) and the **posterior pituitary** (**neurohypophysis**).

The posterior pituitary is continuous with the hypothalamus and is connected to it by the pituitary stalk, which contains axons originating in hypothalamic nuclei. The posterior pituitary stores and secretes the neurohormones **antidiuretic hormone** (**ADH**; also known as **vasopressin**) and **oxytocin**. These hormones are synthesized mainly in the supraoptic nucleus and paraventricular nucleus of the hypothalamus and are carried by axonal transport to the posterior pituitary.

The anterior pituitary (unlike the posterior pituitary) is not directly connected to the hypothalamus but is connected to it by the vessels of the hypophyseal portal circulation. In addition to synthesizing vasopressin and oxytocin, the hypothalamus synthesizes and releases **hypothalamic releasing and inhibitory hormones** that are carried through the **hypophyseal portal veins** to the anterior pituitary and

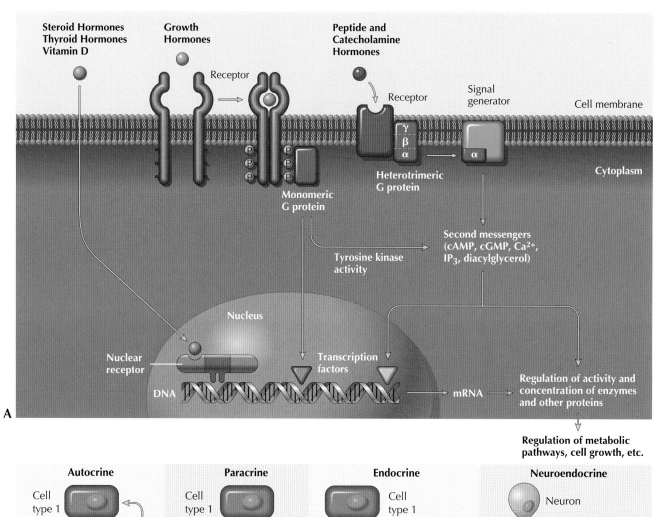

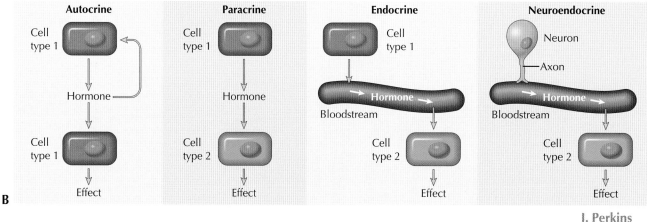

J. Perkins
MS, MFA

Figure 27.2 **Overview of Hormone Action** Hormones act at target cells by binding to specific cell membrane, cytosolic, or nuclear receptors, initiating a cascade of events that produces a physiological change **(A)**. Binding to the receptor may result in generation of second messengers (e.g., cAMP, cGMP, and IP₃) or regulation of gene transcription. True hormones (endocrine secretions) are released by "ductless glands" and are carried by the bloodstream to their sites of action. They are part of a larger group of substances that includes autocrine, paracrine, and neuroendocrine secretions **(B)**.

regulate the secretion of tropic hormones (see Fig. 27.4). This portal circulation allows direct transport of these hypothalamic hormones to the pituitary, without dilution in the general circulation. The major hormones secreted by the anterior pituitary are as follows:

- **Thyroid-stimulating hormone (TSH)**, which stimulates thyroid hormone synthesis and release by the thyroid gland.
- **Adrenocorticotropic hormone**, which stimulates synthesis of adrenocortical steroids.

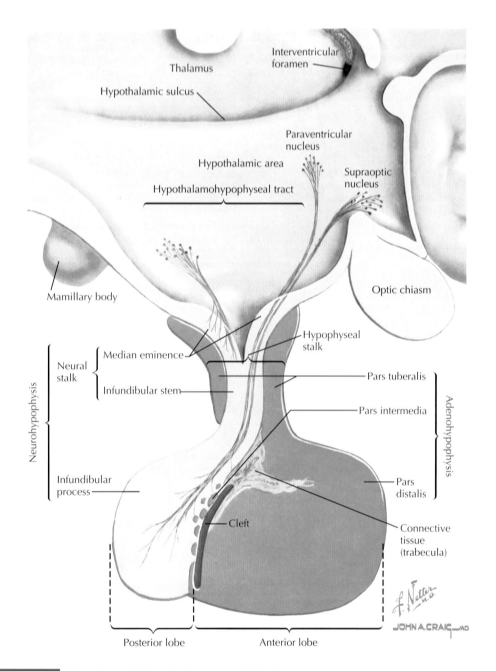

Figure 27.3 **Structure of Hypothalamus and Pituitary** The anterior pituitary (adenohypophysis) and posterior pituitary (neurohypophysis) are derived from different embryonic tissues and function as separate glands. Axons from hypothalamic nuclei extend to the posterior pituitary, where hormones (oxytocin and vasopressin) are stored until released into the systemic bloodstream; other axons from hypothalamic nuclei extend to the median eminence, where they release hormones into the hypophyseal portal circulation, which carries them directly to the anterior pituitary. At the anterior pituitary, these hormones inhibit or stimulate the release of various tropic hormones into the systemic blood.

- **Gonadotropins** (luteinizing hormone (**LH**) and follicle-stimulating hormone (**FSH**), which promote steroidogenesis and gametogenesis by the testes and ovaries.
- **Prolactin**, which stimulates milk production by the breasts.
- **GH**, which promotes the synthesis of **insulin-like growth factors** (**IGFs**) by the liver and other target tissues. IGFs produced by the liver are released into the

circulation, whereas IGFs produced in other tissues act locally, in an autocrine manner. GH has growth-promoting and anabolic effects.

Various anterior pituitary hormones are often referred to as *tropic hormones* because they stimulate secretion of *target gland hormones* (for example, gonadotropins stimulate steroid synthesis and secretion by the gonads).

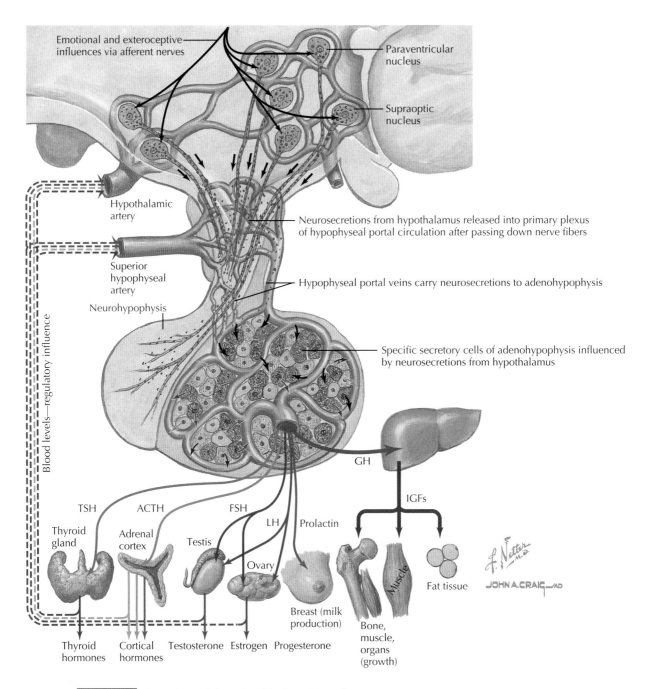

Emotional and exteroceptive influences via afferent nerves

Paraventricular nucleus

Supraoptic nucleus

Hypothalamic artery

Neurosecretions from hypothalamus released into primary plexus of hypophyseal portal circulation after passing down nerve fibers

Superior hypophyseal artery

Neurohypophysis

Hypophyseal portal veins carry neurosecretions to adenohypophysis

Blood levels—regulatory influence

Specific secretory cells of adenohypophysis influenced by neurosecretions from hypothalamus

GH

IGFs

TSH   ACTH   FSH   LH   Prolactin

Thyroid gland   Adrenal cortex   Testis   Ovary   Breast (milk production)   Bone, muscle, organs (growth)   Muscle   Fat tissue

Thyroid hormones   Cortical hormones   Testosterone   Estrogen   Progesterone

Figure 27.4 **Overview of Anterior Pituitary Function**   The anterior pituitary gland is controlled by releasing and inhibitory hormones secreted into the hypophyseal portal circulation; these hormones reach the anterior pituitary directly through this portal circulation without entering the general circulation. Under control of these factors, specific secretory cell types of the anterior pituitary secrete six major tropic hormones (TSH, ACTH, FSH, LH, prolactin, and GH), which act on distal endocrine glands. Tropic hormones and the target gland hormones have feedback effects on these endocrine systems, designed to regulate blood levels of the target gland hormone. *ACTH,* adrenocorticotropic hormone; *FSH,* follicle-stimulating hormone; *GH,* growth hormone; *IGF,* insulin-like growth factor; *LH,* luteinizing hormone; *TSH,* thyroid-stimulating hormone.

FEEDBACK SYSTEMS AND RECEPTOR REGULATION IN THE ENDOCRINE SYSTEM

Feedback systems control blood hormone levels within normal ranges, often with cyclic variations in levels. These feedback systems are an important aspect of homeostasis, providing for regulation of hormone levels and their physiological effects.

The cyclic variations in secretion that occur for many hormones are important in regulating complex processes such as menstrual cycles and in the homeostatic responses to diurnal cycles in activity levels. Furthermore, many hypothalamic and pituitary hormones are released in a pulsatile manner that allows for further fine-tuning of hormone release and action.

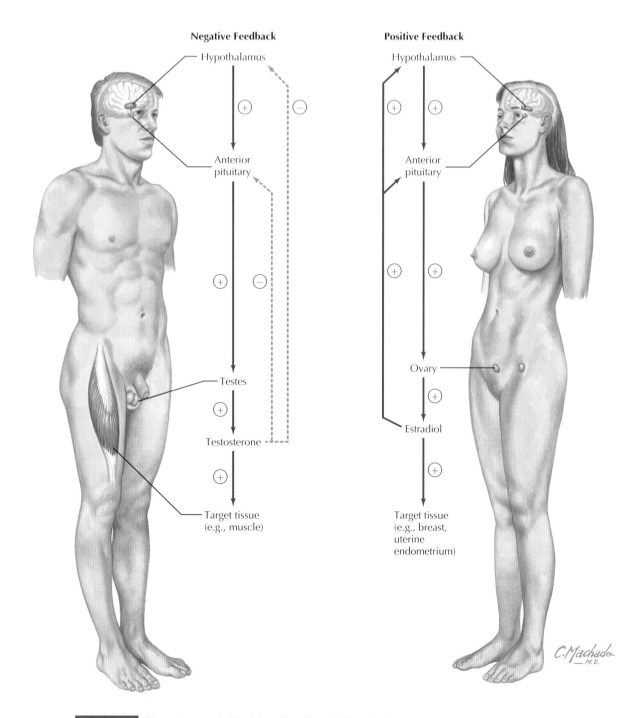

Figure 27.5 **Negative and Positive Feedback Regulation** In most cases, a hypothalamic–pituitary–target gland axis is regulated by negative feedback, whereby the tropic hormone of the anterior pituitary gland has negative feedback effects on the hypothalamus and the target gland hormone has negative feedback effects on both the hypothalamus and the anterior pituitary. Through these mechanisms, illustrated for the hypothalamus-pituitary-testes axis, levels of the target gland hormone are maintained within the normal physiological range. In a few specific cases, positive feedback can also occur. For example, during the late follicular and ovulatory phases of the menstrual cycle, high levels of estradiol actually cause greater secretion of the hypothalamic releasing hormone and tropic hormones in that system, resulting in the surge in pituitary hormone release that is responsible for ovulation at mid cycle.

Negative Feedback

Blood levels of endocrine secretions are controlled within relatively narrow ranges, with normal, often cyclic, variations. Typically, **negative feedback systems** control levels of hormones, whereby increased blood levels of a hormone

inhibit its synthesis, maintaining the normal level (Fig. 27.5). For example, synthesis of the major male sex hormone **testosterone** is stimulated by the pituitary hormone **LH**, which is released in response to a hypothalamic hormone, **gonadotropin-releasing hormone (GnRH)** (see Fig. 27.5). The normal adult male level of testosterone is maintained by negative feedback of testosterone on the production of LH and GnRH, as well as shorter feedback loops (see Chapter 32). In endocrine systems consisting of the hypothalamus, pituitary, and a target endocrine gland, feedback loops are classified as follows:

- **Long-loop feedback**, in which hypothalamic and pituitary hormones in an endocrine axis are inhibited by the target gland hormone, as in the case of testosterone inhibition of GnRH and LH (as well as FSH)
- **Short-loop feedback**, whereby an anterior pituitary hormone inhibits the release of its associated hypothalamic hormone (e.g., inhibition of GnRH by LH)
- **Ultrashort-loop feedback**, in which secretion of a hypothalamic hormone is inhibited by that same hormone

These multiple feedback systems allow for the fine-tuning of hormone levels and participate in the generation of the cyclic variations observed in levels of many hormones.

Positive Feedback

Although negative feedback is the primary homeostatic mechanism in the endocrine system, rare examples of **positive feedback** exist. These positive feedback mechanisms are, by nature, self-limited, as dictated by the need for homeostasis in physiological systems. The prime example of positive feedback occurs during the menstrual cycle (see Fig. 27.5). In the late follicular phase of the cycle, **estradiol** levels rise above a critical point, above which positive feedback occurs. The high estradiol concentration results in a surge in hypothalamic secretion of GnRH and pituitary secretion of LH and FSH, inducing ovulation. Ovulation and transformation of ovarian follicular cells into the **corpus luteum** signals the end of positive feedback.

Receptor Regulation

The cellular response to a hormone is dependent on the presence of specific receptors for the hormone. Although the response to a hormone is dependent on the concentration of the hormone, regulation in the endocrine system can also occur at the level of hormone receptors by altering the number of receptors or their binding affinity for a hormone. In some cases, hormones induce **downregulation** (reduction) of the number of their receptors or of the binding affinity of their receptors as a type of negative feedback. For example, exposure of the ovary to increased levels of LH will result in reduced membrane LH receptors. Hormones may also produce receptor **upregulation**, in which the number or affinity of receptors is increased. As an example, GH increases the number of its receptors in some target tissues. When a hormone affects the number of its own receptors, the process is called **homologous regulation**. In other cases, hormones may affect (upward or downward) the number of receptors for a different hormone. This process is known as **heterologous regulation**.

POSTERIOR PITUITARY HORMONES

ADH and **oxytocin** are both nonapeptides (nine–amino acid peptides) that are derived from preprohormones synthesized by hypothalamic nuclei. Cleavage of the preprohormones produces **neurophysins** along with ADH and oxytocin. The neurophysins function as carrier proteins as ADH and oxytocin are transported in vesicles down the axons of the neurons in which they are synthesized into the posterior pituitary. The vesicles that contain hormone are stored in nerve endings within the posterior lobe of the pituitary. When the hypothalamic neuron is depolarized, the action potential is propagated along the axon to the nerve terminal, where Ca^{2+} influx occurs and results in exocytosis of vesicular contents.

Antidiuretic Hormone

As implied by its name, ADH is secreted during conditions in which water retention ("antidiuresis") is required for homeostasis. The synthesis, release, and physiological actions of ADH are illustrated in Figure 27.6 and described further in Chapters 13 and 19. In addition to its antidiuretic effect, ADH is also a vasopressor, causing contraction of vascular smooth muscle (thus the name "vasopressin"). ADH is released mainly in response to one of two stimuli:

- **Hyperosmolarity of plasma**, detected by osmoreceptors within the hypothalamus
- **Hypovolemia** and **hypotension**, detected by arterial and atrial baroreceptors

Hyperosmolarity (which is caused, for example, by fluid deprivation or dehydration) is normally much more important than hypovolemia or hypotension in terms of stimulation of ADH secretion. Although normal variation of blood pressure is not a significant stimulus for ADH release, significant loss of blood volume (for example, during hemorrhage) causes secretion of the hormone, which participates in blood pressure homeostasis and fluid replenishment under those conditions by stimulating water retention in the kidney and contraction of vascular smooth muscle.

Physiological actions of the hormone are covered in greater detail in Chapter 20. To summarize (see Fig. 27.6), the actions of ADH include effects on renal handling of water—specifically, increased permeability of the late distal tubule and collecting tubule to water—and thus increased water reabsorption by the kidney.

Oxytocin

Oxytocin, the other hormone released by the posterior pituitary, has a role in breastfeeding and in childbirth (Fig. 27.7). Thus oxytocin is released in response to the following actions:

- Breastfeeding
- Cervical and vaginal stimulation

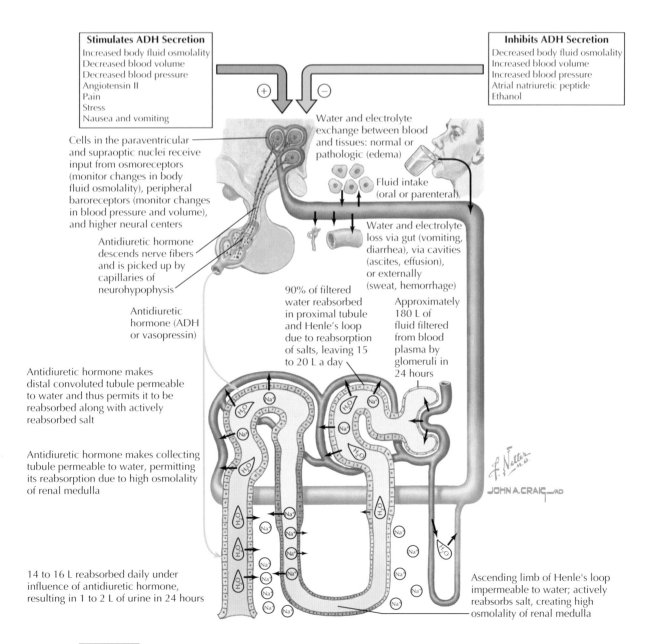

Stimulates ADH Secretion
Increased body fluid osmolality
Decreased blood volume
Decreased blood pressure
Angiotensin II
Pain
Stress
Nausea and vomiting

Inhibits ADH Secretion
Decreased body fluid osmolality
Increased blood volume
Increased blood pressure
Atrial natriuretic peptide
Ethanol

Cells in the paraventricular and supraoptic nuclei receive input from osmoreceptors (monitor changes in body fluid osmolality), peripheral baroreceptors (monitor changes in blood pressure and volume), and higher neural centers

Antidiuretic hormone descends nerve fibers and is picked up by capillaries of neurohypophysis

Antidiuretic hormone (ADH or vasopressin)

Antidiuretic hormone makes distal convoluted tubule permeable to water and thus permits it to be reabsorbed along with actively reabsorbed salt

Antidiuretic hormone makes collecting tubule permeable to water, permitting its reabsorption due to high osmolality of renal medulla

14 to 16 L reabsorbed daily under influence of antidiuretic hormone, resulting in 1 to 2 L of urine in 24 hours

Water and electrolyte exchange between blood and tissues: normal or pathologic (edema)

Fluid intake (oral or parenteral)

Water and electrolyte loss via gut (vomiting, diarrhea), via cavities (ascites, effusion), or externally (sweat, hemorrhage)

90% of filtered water reabsorbed in proximal tubule and Henle's loop due to reabsorption of salts, leaving 15 to 20 L a day

Approximately 180 L of fluid filtered from blood plasma by glomeruli in 24 hours

Ascending limb of Henle's loop impermeable to water; actively reabsorbs salt, creating high osmolality of renal medulla

Figure 27.6 **Posterior Pituitary Function (Antidiuretic Hormone)** Antidiuretic hormone (ADH; also known as vasopressin) is synthesized mainly in the supraoptic nuclei (and also the paraventricular nuclei) of the hypothalamus and is stored and released at the posterior pituitary. Its main function is in water balance; it is released in response to increased osmolarity of extracellular fluid and decreased blood pressure and has the major effect of promoting water reabsorption by the kidney. When ADH levels in plasma are high, a low volume of concentrated urine is produced.

During breastfeeding, oxytocin released by the neurohypophysis causes **"let down"** or **expulsion of milk** from the mammary glands of the breasts. The actual production of milk is stimulated specifically by a different hormone, prolactin. Distension of the cervix and vagina also stimulates the release of oxytocin, and the hormone stimulates **uterine contractions**. However, its actual physiological role in induction or progression of labor is not well established. Oxytocin is also released during orgasm in both sexes and might have a role in sperm transport during ejaculation of the male, as well as transport of sperm within the female reproductive tract. Furthermore, oxytocin appears to stimulate pair-bonding in mammals, including humans, through its central nervous system effects.

ANTERIOR PITUITARY HORMONES

The six major anterior pituitary hormones are peptides, proteins, or glycoproteins, and they are synthesized and secreted by distinct cell types within the adenohypophysis:

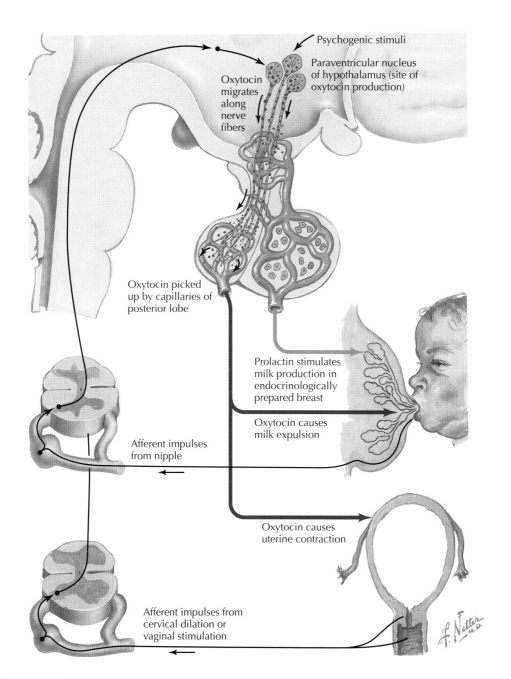

Psychogenic stimuli

Paraventricular nucleus of hypothalamus (site of oxytocin production)

Oxytocin migrates along nerve fibers

Oxytocin picked up by capillaries of posterior lobe

Prolactin stimulates milk production in endocrinologically prepared breast

Oxytocin causes milk expulsion

Afferent impulses from nipple

Oxytocin causes uterine contraction

Afferent impulses from cervical dilation or vaginal stimulation

Figure 27.7 **Posterior Pituitary Function (Oxytocin)** Oxytocin is synthesized mainly in the paraventricular nuclei (and also the supraoptic nuclei) of the hypothalamus and is stored and released at the posterior pituitary. Its main functions are to stimulate milk let-down and uterine contraction.

CLINICAL CORRELATE 27.1
Induction of Labor by Synthetic Oxytocin

Although a physiological role for oxytocin in humans in the induction or progression of labor is not well established, it stimulates strong uterine contractions when used at high doses as a drug. Thus synthetic oxytocin is often used in clinical situations in which artificial induction or stimulation of the progression of labor is medically required. As is generally the case for peptide drugs, it is not effective orally because of cleavage in the gastrointestinal tract and is administered by intravenous infusion.

■ **Gonadotrophs** produce the two gonadotropins (LH and FSH).
■ **Thyrotrophs** synthesize TSH.
■ **Corticotrophs** synthesize adrenocorticotropic hormone.
■ **Lactotrophs** produce prolactin.
■ **Somatotrophs** produce GH.

Synthesis and secretion of these hormones is regulated by hypothalamic hormones that reach the anterior pituitary via the hypophyseal portal circulation, as well as by various feedback loops. In some cases, the hormones are mainly controlled by either a hypothalamic-releasing hormone or an inhibitory hormone; in others, there is dual control by a hypothalamic-releasing hormone and a hypothalamic-inhibitory hormone. The anterior pituitary hormones are discussed in subsequent chapters in the context of their target gland hormones, with the exception of GH and prolactin.

GROWTH HORMONE

GH (also known as **somatotropin**) is a 191–amino acid, single-chain polypeptide with structural similarities to **prolactin** and **human chorionic somatomammotropin** (a placental hormone with GH-like effects, also known as human placental lactogen). As implied by its name, GH is the primary hormone in the human growth process. It is secreted throughout life in a pattern that consists of basal secretion upon which several **daily pulses of secretion** are superimposed. These pulses occur during sleep and are especially pronounced and frequent during the pubertal growth spurt; basal secretion rates are highest in young children. The growth spurt in boys typically begins at about 12 years of age, and growth in height is usually concluded by about 16 or 17 years of age, although significant variability exists between individuals in both parameters. The growth spurt in girls begins about a year earlier and is concluded about 2 years earlier, again with considerable variability.

Regulation of GH Secretion

The secretion of GH by anterior pituitary somatotrophs is regulated by two hypothalamic hormones (Fig. 27.8):

- **GH-releasing hormone (GHRH)**, which stimulates growth hormone release, and
- **somatostatin**, which inhibits GH release.

Most of the effects of GH are not direct but are mediated by the synthesis and release of **somatomedins** by the liver and by stimulation of somatomedin production in specific target tissues. Somatomedins are also known as **IGFs**; the most important of these is **IGF-1**, which is produced by the liver and other tissues in response to GH. IGF-1 binds to receptors on target tissues, producing tyrosine kinase activation, similar to the actions of insulin on its receptor, resulting in intracellular effects.

IGF-1 is produced by the liver and secreted into the bloodstream in response to GH; it is also produced directly at target tissues. Although plasma levels of IGF-1 respond to GH stimulation of the liver, it is believed that local production of IGF-1 at the target tissues, as opposed to circulating IGF-1, is probably more important in mediating the effects of GH on those tissues.

Both GH and IGF-1 exert negative feedback effects on the hypothalamus and anterior pituitary; GHRH inhibits its own release (ultra-short loop feedback). Metabolites also affect GH release directly or indirectly: Glucose and free fatty acids inhibit GHRH release, whereas amino acids stimulate pituitary release of GH.

Effects of GH and IGF-1

The primary roles of GH and IGF-1 are stimulation of growth and development of the body during childhood and adolescence and regulation of metabolism and body composition in adults (see Fig. 27.8). These general effects reflect more specific actions mediated by IGF-1:

- **Stimulation of linear growth** through proliferative effects, amino acid uptake, and protein synthesis in bone and cartilage (in childhood and adolescence)
- **Increase in muscle mass** (anabolic effect) through increased amino acid uptake and protein synthesis in muscle
- **Decreased adiposity** by stimulation of lipolysis
- **Increase in organ size**, which is associated with proliferation and protein synthesis

The actions of IGF-1 on growth of muscle, bone, and other organs involve stimulation of RNA and DNA synthesis, as well its more direct effects on cellular amino acid uptake and synthesis of protein.

Research during the past two decades has led to the discovery of several hormones involved in growth, appetite, and obesity. Ghrelin is a 26–amino acid, acylated peptide hormone produced by the stomach, as well as the arcuate nucleus of the hypothalamus. Its actions include stimulation of hunger and release of GH and may provide a link between food intake and growth stimulation. Its effects on hunger oppose the actions of leptin, a hormone produced by adipose tissue that causes satiety. Current research is focusing on the relationship of these and other hormones to obesity.

GH is often referred to as a **diabetogenic hormone** because of its metabolic effects and opposition of the actions of insulin. GH acts to:

- **elevate blood glucose levels** through inhibition of glucose uptake by muscle and adipose tissue.
- **elevate plasma free fatty acids** through its lipolytic action in adipose tissue.
- **induce insulin resistance** and **elevate plasma insulin levels**.

PROLACTIN

Prolactin is a 198–amino acid, single-chain polypeptide structurally related to GH. It is secreted at low levels except in pregnant and lactating women, in whom the number of lactotrophs in the anterior pituitary is increased and plasma prolactin is elevated. The hormone's main actions are to stimulate breast development and milk production, as suggested by its name.

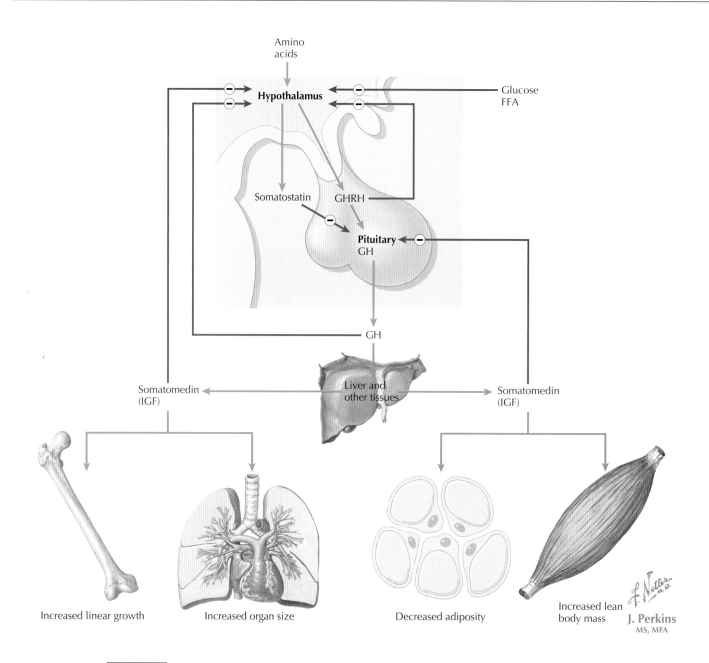

Figure 27.8 **Growth Hormone** Growth hormone (GH) release by the anterior pituitary is controlled by growth hormone–releasing hormone (GHRH) and somatostatin (SS). Note that the illustrated directional (negative or positive) effects of IGF, GH, GHRH, and other factors on the hypothalamus indicate the ultimate positive or negative effect of these factors on pituitary release of GH (through modulation of hypothalamic GHRH and/or SS release). GH has an important role in growth and development of children and regulation of metabolism. Its effects are mediated by somatomedins produced by the liver or by specific target tissues. *FFA,* free fatty acids; *IGF,* insulin-like growth factor.

Regulation of Prolactin Secretion

The hypothalamic hormones involved in the regulation of prolactin secretion by the anterior pituitary are (Fig. 27.9):

- **dopamine** (also referred to as **prolactin inhibitory factor** in this context), which inhibits prolactin release, and
- **thyrotropin-releasing hormone**, which stimulates release of prolactin in addition to TSH.

Prolactin exerts negative feedback effects on its own release by stimulating secretion of dopamine by the hypothalamus.

Effects of Prolactin

In most instances, prolactin synthesis and release is primarily under the inhibitory influence of dopamine. During puberty, prolactin stimulates **breast development,** acting in concert with other hormones. In pregnancy, the final **alveolar and**

CLINICAL CORRELATE 27.2
Growth Hormone Excess and Deficiency

Deficiency of GH in prepubertal children results in short stature and may delay the onset of puberty. GH deficiency may result from a variety of causes, including primary failure to secrete GH, low GHRH production, pituitary or hypothalamic damage, and deficiency in IGF production. Affected children are treated with recombinant human growth hormone therapy. Deficiency in GH can also occur in adults, often as a result of a pituitary tumor. Such deficiency can have multiple manifestations, including loss of muscle mass, weight gain, and psychosocial effects; human GH therapy is beneficial in some cases.

The most common cause of excess GH is a GH-secreting adenoma. These tumors, which grow slowly, are usually diagnosed in midlife; surgery is the most common treatment. GH excess in adulthood produces **acromegaly**, a condition characterized by thickening of the bones of the hands, feet, and jaw, protrusion of the jaw and brow, and enlargement of the tongue (macroglossia), as well as cardiovascular and renal complications, diabetes, and other effects. Prepubertal GH excess is rare and causes **pituitary gigantism**.

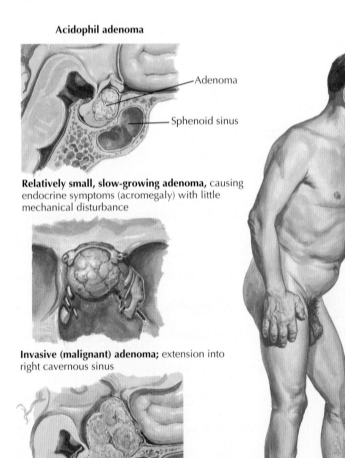

Acidophil adenoma

Adenoma

Sphenoid sinus

Relatively small, slow-growing adenoma, causing endocrine symptoms (acromegaly) with little mechanical disturbance

Invasive (malignant) adenoma; extension into right cavernous sinus

Large acidophil adenoma; extensive destruction of pituitary substance, compression of optic chiasm, invasion of third ventricle and floor of sella

Growth Hormone–Secreting Adenomas The effects of growth hormone–secreting adenomas vary depending on size and growth rate, as well as invasiveness. Large tumors cause destruction of the pituitary and deficiency of other pituitary hormones and may affect the optic chiasm and vision. Growth hormone excess produces acromegaly in adults (at right), with protrusion of the jaw, macroglossia, and other effects.

Short-loop feedback

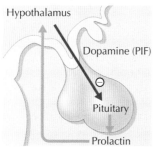

Other modulating factors

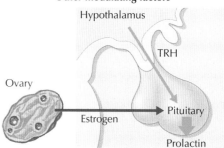

Prolactin-inhibiting factor (PIF), thought to be dopamine, modulates prolactin secretion. Elevated prolactin levels increase PIF secretion and cause feedback inhibition of prolactin secretion (short-loop feedback inhibition). Estrogen and TRH stimulate prolactin secretion.

Breast development

Prolactin

GH
Estrogen

Progesterone

Adrenocorticoids

Prolactin, along with GH, estrogen, progesterone, and adrenocorticoids, is necessary for breast development.

Pregnancy

Prolactin

Estrogen

Progesterone

In pregnancy, elevated prolactin, estrogen and progesterone increase alveolobular development. High estrogen levels inhibit lactation.

Lactation

Prolactin

Oxytocin

Estrogen

Progesterone

Sudden decrease in estrogen and progesterone in presence of prolactin results in milk production. Oxytocin stimulates milk release.

Variations in prolactin levels by age or condition

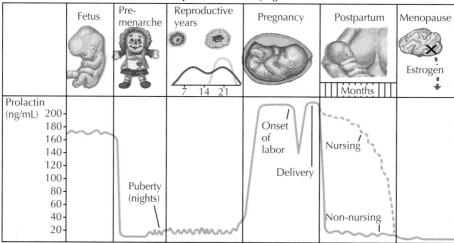

Figure 27.9 **Prolactin** Prolactin synthesis and release by the anterior pituitary is mainly under tonic negative control by dopamine (prolactin inhibitory factor [PIF]). Its major functions are in breast development, pregnancy, and lactation. Its levels are elevated during fetal development, pregnancy, and the postpartum period (if the woman is breastfeeding). *GH,* growth hormone; *TRH,* thyrotropin-releasing hormone.

ductal development of breasts is stimulated by prolactin, estrogen, and progesterone. During pregnancy, prolactin is also produced by the endometrium and other tissues. The high prolactin levels are promoted by estrogen and progesterone, but these hormones also inhibit the lactogenic effects of prolactin.

After parturition, **suckling** promotes prolactin synthesis and release. Estrogen and progesterone levels are reduced and the lactogenic action of prolactin is unopposed. Binding of prolactin to its receptors in the mammary glands results in tyrosine kinase activation, transcriptional regulation, and, ultimately, synthesis of milk proteins. One of the actions of prolactin is inhibition of **gonadotropin-releasing hormone** and therefore suppression of the female reproductive cycle, which accounts for the reduced likelihood of another conception during breastfeeding.

Breastfeeding acts as a partially effective contraceptive measure in the first 6 months after parturition because of its inhibitory effects on the reproductive endocrine axis of the mother. If the baby is fed exclusively by breastfeeding, conception is unlikely during this period as long as the mother has not had a menstrual period. In developing nations, birth rates have sometimes risen early in the process of industrialization because of an increase in formula feeding of infants.

Chapter 28
Thyroid Hormones

Thyroid hormones (TH) have biologic actions in every organ in the body and are critical to proper fetal, postnatal, and pubertal growth and development. Because TH are important in maintaining basal metabolic rate, thyroid excess and deficiency in adults affect a wide swath of physiological processes and can produce various diseases. Treatment for "enlarged neck," or goiter (resulting from hypothyroidism), was described in Chinese herbal medicine almost 5000 years ago.

STRUCTURE OF THE THYROID GLAND

The thyroid is a shield-shaped gland located on the anterior side of the trachea, below the thyroid cricoid cartilages (Fig. 28.1). This gland is composed of right and left lobes and an isthmus and weighs ~20 grams. As with all endocrine glands, it is highly vascularized to supply nutrients for hormone synthesis and blood flow for hormone transport. The functional units of the gland are the **follicles** (Fig. 28.2). The follicular cells secrete thyroglobulin (Tg), the glycoprotein precursor to TH, into the follicular lumen, and the **colloid** is composed of modified Tg. The follicular cells are surrounded by a layer of epithelial cells. Scattered between the follicles are parafollicular cells, which produce **calcitonin**, or "C" cells (see "Main Sites of Calcium Regulation" in Chapter 31). G protein–coupled receptors for thyroid-stimulating hormone (TSH) are found on the follicular cell basal membrane.

SYNTHESIS, RELEASE, AND UPTAKE OF THYROID HORMONES

Synthesis and Release of Thyroid Hormones

The thyroid gland synthesizes two main forms of TH: **triiodothyronine (T_3)** and **thyroxine (T_4)** (Fig. 28.3). Although T_3 is about four times more biologically active than T_4, the amount of T_4 produced and secreted is about 20 times *greater* than the more potent T_3. The synthesis and release of TH is regulated by TSH that is secreted by the anterior pituitary gland (see Chapter 27). TSH binds to its G protein–coupled receptor on the thyroid gland, stimulating production of cAMP, which acts *at each biosynthetic step* (see the following text and Fig. 28.4).

Steps in Thyroid Hormone Synthesis

Within the follicular cells, the elements necessary for TH synthesis are combined in the following steps (Fig. 28.4):

1. **Thyroglobulin** molecules rich in tyrosine are produced in the endoplasmic reticulum, packaged in vesicles by the Golgi, and exocytosed into the lumen of the follicle.
2. **Iodide (I^-)** enters the follicle cell via basolateral Na^+/I^- cotransporters (the I-trap). The iodide exits the cell on the apical side into the follicular lumen via I^-/Cl^- antiporters.
3. In the follicular lumen, I^- is oxidized to iodine by **thyroid peroxidase** and substituted for H^+ on the benzene ring of tyrosine residues of thyroglobulin.
4. Binding of one iodine will form **monoiodotyrosine (MIT),** and binding of two iodine moieties will form **diiodotyrosine (DIT)**. This reaction is termed **organification**. **Thyroid peroxidase** also catalyzes the binding of DIT to another DIT, forming T_4. Some DIT will also bind to MIT, forming T_3. These products remain linked to the Tg.
5. The mature thyroglobulin, containing MIT, DIT, T_4, and T_3 (in order of greater to lesser abundance), is endocytosed back into the follicle cell and can be stored as colloid until it is secreted. TH can be stored *several weeks* while bound to Tg; this phenomenon is different from that seen with other hormones.
6. Proteolysis of the colloid is stimulated by TSH and releases the constituent molecules. MIT and DIT reenter the synthetic pool, and T_3 and T_4 exit via the basolateral membrane into the blood.

Reverse T_3 (rT_3) is also synthesized in the thyroid gland, but in small amounts. It differs from T_3 in the position of the iodine moiety. rT_3 has little biologic activity; only a small amount of rT_3 is present under normal conditions, but high amounts are present in persons with chronic diseases, in the fetus, and during starvation. rT_3 is made in larger amounts when T_4 is processed at end organ tissues.

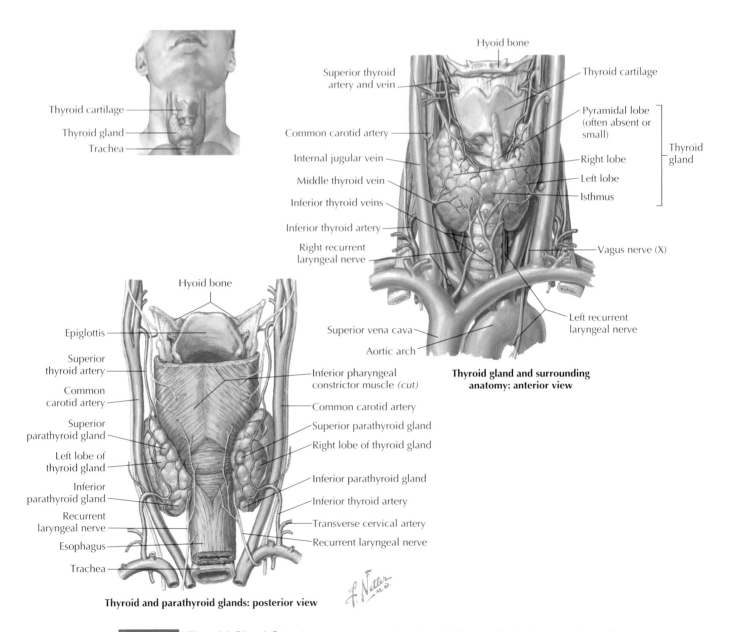

Thyroid cartilage
Thyroid gland
Trachea

Hyoid bone

Superior thyroid
artery and vein

Common carotid artery

Internal jugular vein

Middle thyroid vein

Inferior thyroid veins

Inferior thyroid artery

Right recurrent
laryngeal nerve

Thyroid cartilage

Pyramidal lobe
(often absent or
small)

Right lobe   Thyroid
                  gland
Left lobe

Isthmus

Vagus nerve (X)

Superior vena cava

Aortic arch

Left recurrent
laryngeal nerve

**Thyroid gland and surrounding
anatomy: anterior view**

Hyoid bone

Epiglottis

Superior
thyroid artery

Common
carotid artery

Superior
parathyroid gland

Left lobe of
thyroid gland

Inferior
parathyroid gland

Recurrent
laryngeal nerve

Esophagus

Trachea

Inferior pharyngeal
constrictor muscle *(cut)*

Common carotid artery

Superior parathyroid gland

Right lobe of thyroid gland

Inferior parathyroid gland

Inferior thyroid artery

Transverse cervical artery

Recurrent laryngeal nerve

Thyroid and parathyroid glands: posterior view

Figure 28.1 **Thyroid Gland Structure** The thyroid gland is a highly vascularized structure located anterior to the trachea and inferior to the cricoid cartilage. In about 15% of the population, a small pyramidal lobe is present (as illustrated).

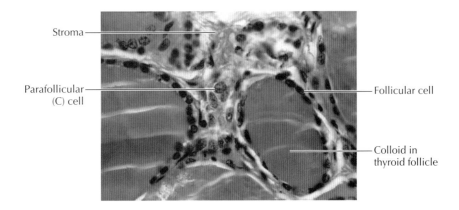

Stroma

Parafollicular
(C) cell

Follicular cell

Colloid in
thyroid follicle

Figure 28.2 **Histologic Features of the Thyroid Gland** Interspersed between follicular cells of the thyroid gland are parafollicular cells, or C cells, which synthesize and secrete calcitonin, a calcium-regulating hormone. (From Ovalle WK, Nahirney PC: *Netter's Essential Histology,* ed 2, Philadelphia, 2013, Elsevier.)

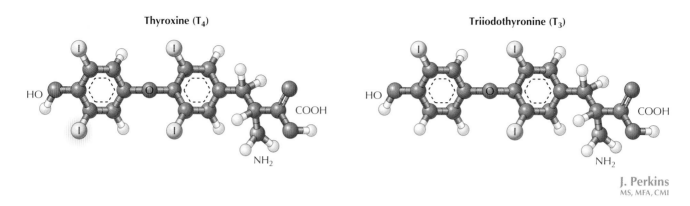

Thyroxine (T$_4$)

HO

COOH

NH$_2$

Triiodothyronine (T$_3$)

HO

COOH

NH$_2$

J. Perkins
MS, MFA, CMI

Figure 28.3 **Structure of T$_3$ and T$_4$**   The two main types of thyroid hormones (TH), thyroxine (T$_4$) and triiodothyronine (T$_3$), differ from each other by the addition of one iodine in T$_4$. The majority of circulating TH is T$_4$, and almost all of the circulating TH is bound to a thyroxine-binding protein.

Propylthiouracil is used to treat hyperthyroidism because it blocks **thyroid peroxidase** and acts at all biosynthetic steps of TH production, from **organification** of iodine to conversion of T$_4$ to T$_3$ in the peripheral tissues.

Although several tissues, including those of the salivary glands, mammary glands, and stomach, can absorb iodine, oxidation can take place only in the thyroid gland because of the presence of thyroid peroxidase.

If TSH is elevated for extended periods, it can also exert a trophic effect on the thyrocytes, causing hypertrophy of the thyroid tissue and **hyperthyroidism.**

- TSH secretion is stimulated by thyroid-releasing hormone that is released by the hypothalamus and low levels of circulating T$_3$ and T$_4$ (the effects of T$_4$ on TSH secretion require its conversion to T$_3$). Estrogens also have a stimulatory effect.
- TSH secretion is inhibited by high circulating T$_3$ or T$_4$, somatostatin, and dopamine. Growth hormone and high levels of cortisol may also inhibit TSH secretion.

TSH ultimately releases T$_3$ and T$_4$ into the blood, where most of the T$_3$ and T$_4$ is bound to proteins, including albumin and **thyroxine-binding globulin**. The thyroxine-binding globulin acts as a plasma reservoir for T$_4$, because T$_4$ will be active only when it is released from the plasma proteins, enters the target cells, and undergoes deiodination to T$_3$. Because of the high affinity for plasma binding proteins to TH, there is relatively little "free" circulating T$_3$ and T$_4$, but the free T$_3$ and T$_4$ is the physiologically and clinically relevant fraction.

Circulating free T$_3$ and T$_4$ provides feedback to both the hypothalamus and pituitary gland to decrease thyroid-releasing hormone and TSH, respectively (see Fig. 28.4). This feedback system is critical to maintaining appropriate levels of TH secretion.

Thyroid Hormone Uptake at Target Tissues

Free T$_4$ and T$_3$ diffuse into target cells, and within the cells the T$_4$ is converted to T$_3$ by **5'-deiodinase,** producing approximately the same concentrations of T$_3$ and rT$_3$. The active T$_3$ then binds to the nuclear thyroid hormone receptor, which forms a complex with the **TH response element** and stimulates gene transcription (Fig. 28.5). TH response elements are found on a variety of genes, including the growth hormone receptor gene, cardiac and sarcoplasmic reticulum Ca^{2+}-ATPase genes, and genes that encode Na$^+$/K$^+$ ATPase subunits. Thus TH can control diverse functions such as growth, heart rate, and general metabolic rate. In general, low to normal levels of TH have anabolic effects and lead to synthesis of other hormones, and high levels of TH have catabolic effects, causing breakdown of proteins and hormones.

Actions of Thyroid Hormones

TH affects virtually all systems and can act at cellular and whole tissue levels, generally to increase metabolism and growth processes (see Fig. 28.5). Within cells, TH promotes

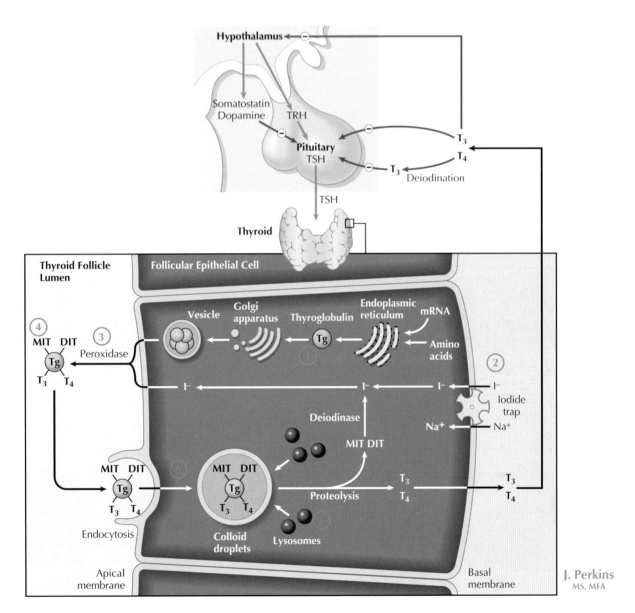

Figure 28.4 **Synthesis and Regulation of Thyroid Hormones** The thyroid gland is composed of follicular epithelial cells that synthesize and store thyroxine (T_4) and triiodothyronine (T_3) and release these hormones into the circulation. The synthesis is controlled by release of thyroid-stimulating hormone (TSH), which is under negative feedback control by the thyroid hormones. Synthesis and storage of the thyroid hormones is outlined here and in the figure: *(1)* In the endoplasmic reticulum, thyroglobulin molecules are produced, packaged in vesicles by the Golgi, and exocytosed into the lumen of the follicle. *(2)* Iodide (I^-) (from the diet) enters the follicle cell via basolateral Na^+/I^- cotransporters (the I-trap). The iodide exits the cell on the apical side into the lumen via I^-/Cl^- antiporters. *(3)* In the follicular lumen, I^- is oxidized to iodine by thyroid peroxidase and substituted for H^+ on the benzene ring of tyrosine residues of thyroglobulin. *(4)* Binding of one iodine will form monoiodotyrosine (MIT), and binding of two iodine moieties will form diiodotyrosine (DIT). This reaction is termed *organification*. Thyroid peroxidase also catalyzes the binding of DIT to another DIT, forming T_4. Some DIT will also bind to an MIT, forming T_3. These products remain linked to the thyroglobulin (Tg). *(5)* The mature Tg, containing MIT, DIT, T_4, and T_3 (in order of greater to lesser abundance), is endocytosed back into the follicle cell and can be stored as colloid until it is secreted. *(6)* Proteolysis of the colloid is stimulated by TSH and releases the constituent molecules. MIT and DIT reenter the synthetic pool, and T_3 and T_4 exit the basolateral membrane into the blood. *TRH,* thyrotropin-releasing hormone.

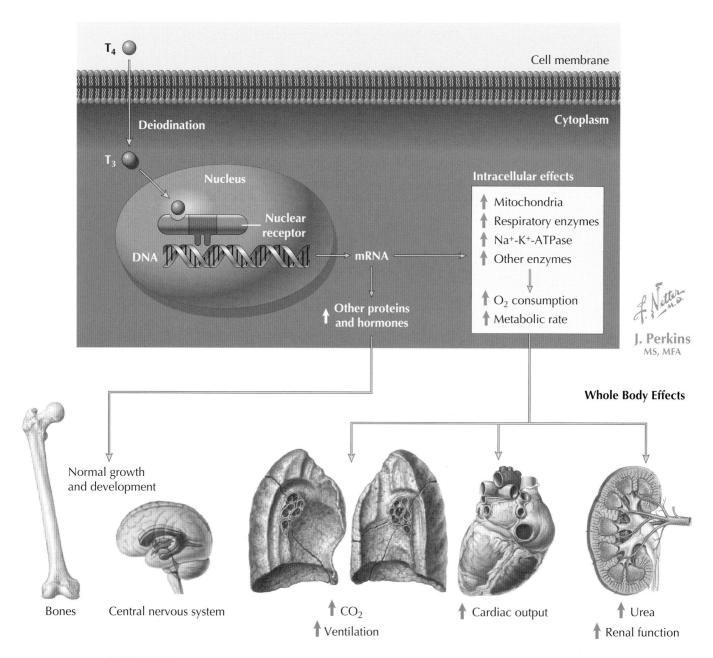

Figure 28.5 **Thyroid Hormone Action** Thyroxine (T_4) is converted to active triiodothyronine (T_3) at target tissue by 5'-deiodinase action. The T_3 binds to nuclear receptors, initiating transcription of a variety of proteins and enzymes. The overall effects of thyroid hormone are to increase metabolic rate and O_2 consumption; the general effects in target organs are illustrated.

production of proteins, as well as other hormones; increases Na^+/K^+ ATPase and other enzymes; and increases the number of mitochondria, which increases O_2 consumption. These actions of TH are seen in the following structures and systems:

- Bone and tissues, contributing to normal growth and development and bone cell proliferation
- Brain and nervous system, contributing to normal growth and development
- Lungs, increasing ventilation

- Heart, increasing cardiac output
- Kidneys, increasing renal function
- Metabolism, stimulating food intake; increasing lipolysis in adipose cells, thus releasing free fatty acids into circulation; decreasing adipose tissue; decreasing muscle mass; and increasing body temperature

Thus TH has major effects on metabolism and growth, and disruptions in its normal rate and process of secretion can have dramatic consequences.

CLINICAL CORRELATE 28.1
Diseases of Thyroid Function: Hypothyroidism

Hypothyroidism, or reduced levels of TH, can result from lack of iodine, as well as autoimmune disease. Although dietary supplementation of iodine occurs in the Western hemisphere, iodine deficiency is a health problem in many countries. **Iodine deficiency** severely reduces organification of Tg and thus TH synthesis. Immune system dysfunction accounts for most cases of thyroid disease in the Western hemisphere. The most common cause of low TH is the production of antibodies to thyroglobulin or thyroid peroxidase (which occurs in persons with **Hashimoto's thyroiditis**). These antibodies act at the thyroid gland to diminish the production and secretion of TH, and they eventually destroy the gland.

- In children, untreated congenital hypothyroidism can result in **cretinism**, which is associated with stunted growth, mental retardation, impaired motor neuron function, and constipation.
- In adults, hypothyroidism results in **myxedema**, which is associated with increased fat deposition, nonpitting edema, cold intolerance, constipation, hypotension, fatigue, and depression, among other symptoms.
- Whether hypothyroidism is caused by iodine deficiency or autoimmune disease, the low circulating TH level results in reduced negative feedback at the pituitary and hypothalamus, resulting in high levels of circulating TSH. Hypothyroidism is treated with synthetic thyroxine.

Appearance of thyroid gland

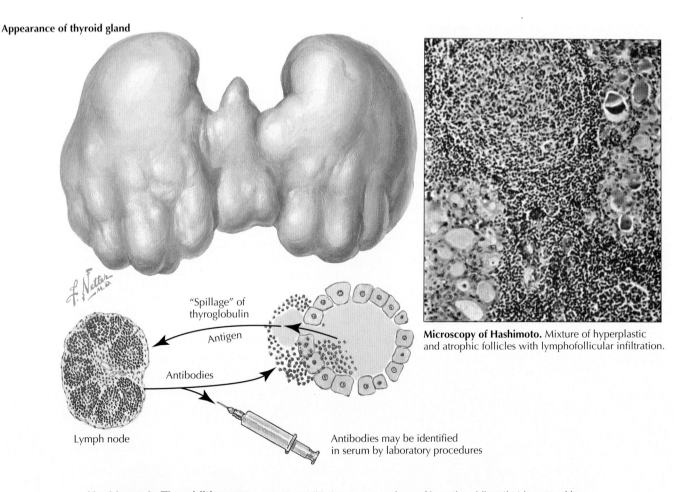

Microscopy of Hashimoto. Mixture of hyperplastic and atrophic follicles with lymphofollicular infiltration.

Hashimoto's Thyroiditis Hashimoto's thyroiditis is a common form of hypothyroidism that is caused by autoimmune antibodies directed against thyroglobulin or thyroid peroxidase. This process results in low TH production (and high circulating TSH) and eventual destruction of the thyroid gland.

CLINICAL CORRELATE 28.2
Diseases of Thyroid Function: Hyperthyroidism

Hyperthyroidism, or overproduction of TH, is associated with a high metabolic rate (30% to 60% above the normal rate), weight loss, increased appetite, tachycardia, hyper-reactive bowels, and muscle weakness. The most common cause of hyperthyroidism is Graves disease, an autoimmune disease in which antibodies produced by immune cells target the thyrocytes.

- **Thyroid-stimulating immunoglobulins** can bind to the TSH receptors on the thyroid gland and produce the same biologic actions as TSH. However, although the elevated circulating TH suppresses pituitary TSH, no feedback regulation of the thyroid-stimulating immunoglobulins occurs, and thus synthesis and release of TH continues. The gland becomes hypertrophied and may form a **goiter** because of the trophic effect of immunoglobulin on the thyrocytes. These immunoglobulins are the cause of **hyperthyroidism** in **Graves' disease**, characterized by elevated levels of circulating TH but a *reduced* level of TSH. Exophthalmus (protruding eyes) can be a symptom of Graves' disease and is caused by deposition of glycoproteins and water behind the eyes.
- Hyperthyroidism may also occur as a result of pituitary or thyroid tumors. In contrast to Graves' disease, **hyperthyroidism** caused by a **TSH-secreting pituitary tumor** is characterized by *elevated levels of both TSH and TH*.
- Although rare, **thyroid storms** can occur when TH levels increase to toxic levels, creating acute awareness of discomfort in the patient. These symptoms can include tachycardia, sleeplessness, dramatic mood swings, and hyperkinesis. If untreated, it can rapidly (within days) progress to congestive heart failure, circulatory collapse, and death.
- Drugs such as propylthiouracil and methimazole can be used to suppress TH synthesis and secretion. Radioiodide ablation of the thyroid or thyroidectomy in cases of cancer may be performed, followed by thyroxine replacement.

Many people with autoimmune thyroid disease have antibodies to both the TSH receptor *and* thyroid peroxidase and have a mix of Graves' disease and Hashimoto's thyroiditis. Initially, the patient will experience hyperthyroidism, followed by hypothyroidism. Approximately 70% of people with Graves' disease also have Hashimoto's thyroiditis. Both diseases occur more frequently in women than in men.

Goiter (an enlarged thyroid gland) can occur in both hypothyroid and hyperthyroid conditions. For example, in both persons with iodine deficiency and persons with a TSH-secreting tumor, elevated TSH levels stimulate thyroid growth. This occurs because, in both cases, elevated TSH has a growth effect on the thyrocytes. In each case, it takes years for a goiter to develop.

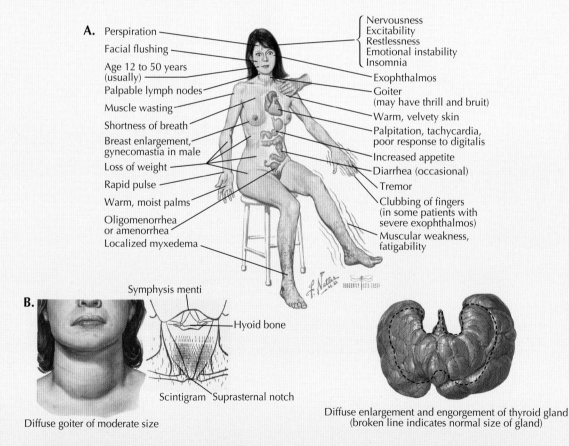

A.
Perspiration
Facial flushing
Age 12 to 50 years (usually)
Palpable lymph nodes
Muscle wasting
Shortness of breath
Breast enlargement, gynecomastia in male
Loss of weight
Rapid pulse
Warm, moist palms
Oligomenorrhea or amenorrhea
Localized myxedema

Nervousness
Excitability
Restlessness
Emotional instability
Insomnia
Exophthalmos
Goiter (may have thrill and bruit)
Warm, velvety skin
Palpitation, tachycardia, poor response to digitalis
Increased appetite
Diarrhea (occasional)
Tremor
Clubbing of fingers (in some patients with severe exophthalmos)
Muscular weakness, fatigability

B.
Symphysis menti
Hyoid bone
Scintigram   Suprasternal notch
Diffuse goiter of moderate size

Diffuse enlargement and engorgement of thyroid gland (broken line indicates normal size of gland)

Clinical Symptoms of Hyperthyroidism A, Hyperthyroidism, such as that in Graves' disease, affects most physiological systems and can increase metabolic rate by 30% to 60%. Elevated TH causes a wide variety of symptoms. **B,** Enlargement of the thyroid gland (goiter). Goiter can be caused by both hypothyroid and hyperthyroid conditions and results from TSH or immunoglobulin-mediated stimulation of hyperplastic growth of the thyrocytes.

Chapter 29

Adrenal Hormones

The paired **adrenal glands** are also known as the **suprarenal glands** because of their location above the kidneys in the retroperitoneal space (Fig. 29.1). These glands consist of inner medullary and outer cortical layers, which produce catecholamines and steroid hormones, respectively. The adrenal medulla and adrenal cortex are derived from embryologically distinct tissues (ectodermal neural crest and endoderm, respectively), and functionally, they can be considered distinct organs. Although the adrenal glands as a whole are sometimes discussed in terms of their role in the stress response, during which both medullary catecholamines and cortical steroids are released, this view of these organs is narrow, because they participate in a wide array of physiological processes.

ADRENAL GLAND STRUCTURE

The human **adrenal cortex** consists of three distinct histological layers (Fig. 29.2):

- The outer **zona glomerulosa**, which synthesizes mineralocorticoid hormones, mainly **aldosterone**
- The middle **zona fasciculata**, which produces the glucocorticoid hormone **cortisol**
- The inner **zona reticularis**, which produces male sex hormones (androgens), mainly **dehydroepiandrosterone** and **androstenedione**

The **adrenal medulla** is found beneath the cortex, at the center of the gland. The adrenal medulla contains **chromaffin cells** that function as postganglionic cells of the sympathetic nervous system, secreting mainly epinephrine (and, in lesser amounts, norepinephrine) into the bloodstream.

SYNTHESIS AND REGULATION OF ADRENAL CORTICAL STEROID HORMONES

The pathways for biosynthesis of adrenal steroids are depicted in Figure 29.3. Cholesterol, either produced de novo in the adrenal cortex or transported to the cortex in the form of low-density lipoprotein cholesterol, is the precursor of these products. Although illustrated in Figure 29.3 as a single, complex pathway, only portions of the pathway exist in each of the layers of the cortex. For example, the cells of the zona glomerulosa contain the enzymes involved in the synthesis of aldosterone, and the cells of the zona fasciculata have the enzymes necessary to produce cortisol.

The hypothalamic-pituitary-adrenal (HPA) axis (see Fig. 29.3) is affected by various physiological states, including stress and anxiety (which activate the axis) and sleep/wake cycles. **Corticotropin-releasing hormone (CRH)**, a 41–amino acid peptide produced by the paraventricular nucleus of the hypothalamus and secreted into the hypothalamic-hypophyseal portal circulation, stimulates the anterior pituitary corticotrophs to release and to synthesize **adrenocorticotrophic hormone (ACTH)**. ACTH stimulates the **conversion of cholesterol to pregnenolone** in the adrenal cortex. This stimulation results in the following effects:

- Increased cortisol synthesis by the zona fasciculata
- Higher androgen production by the zona reticularis (although androgen production is also probably affected by other factors as well)
- *Permissive* effects on aldosterone production by the zona glomerulosa (ACTH is necessary but by itself is not sufficient stimulus for aldosterone synthesis, which is under primary control by other factors)

The 39–amino acid polypeptide hormone ACTH is first synthesized as a 241–amino acid prohormone called **pro-opiomelanocortin (POMC)**. Although ACTH is the main physiologically relevant product of POMC in corticotrophs of the anterior pituitary, POMC is produced in various other sites, including the hypothalamus and brainstem. The multiple products of POMC include ACTH, **β-lipotropin, α–melanocyte-stimulating hormone (α-MSH), β-MSH,** and the endogenous opioid **β-endorphin,** among others. The relevance of endogenous production of some of the products in humans is limited or not well defined in some cases. For example, α-MSH is produced by the pituitary's pars intermedia (see Fig. 27.3) and promotes pigmentation of skin. However, although the pars intermedia produces α-MSH in the fetus, there is little or no intermedia present in adults. Interestingly, because α-MSH consists of the first 13 amino acids of ACTH, when ACTH is produced in excess (for example, by an ACTH-secreting tumor), increased skin pigmentation can occur.

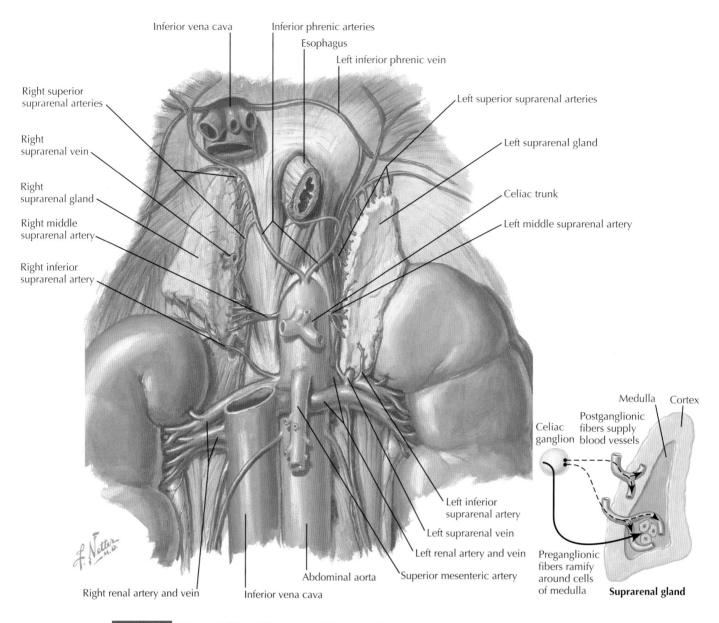

Figure 29.1 **Adrenal Gland Structure** The two adrenal glands are located above the kidneys and below the diaphragm in the retroperitoneal space. The outer cortex of the adrenal gland produces steroid hormones, whereas the inner medulla synthesizes and releases catecholamines.

The predominant adrenal effect of activation of the HPA axis is release of cortisol. ACTH is secreted in a pulsatile manner in response to variations in CRH release, with peak secretion early in the morning, in the hours before awakening, and the lowest secretion just before sleep. Stress (e.g., hypoglycemia, heavy exercise, severe pain, surgery, trauma, and infection) is a major stimulus for activation of the HPA axis and therefore cortisol release.

Negative feedback regulation in the HPA axis involves long-loop feedback by cortisol on the pituitary gland and

hypothalamus and short-loop feedback of ACTH on hypothalamic CRH release (see Fig. 29.3).

ACTIONS OF CORTISOL

Cortisol, which is secreted in response to HPA activation, has a wide array of physiological effects (Fig. 29.4). Like other steroid hormones, it binds to specific cytoplasmic receptors in target cells. The receptor-hormone complex is translocated to the nucleus, where it affects the transcription of specific genes.

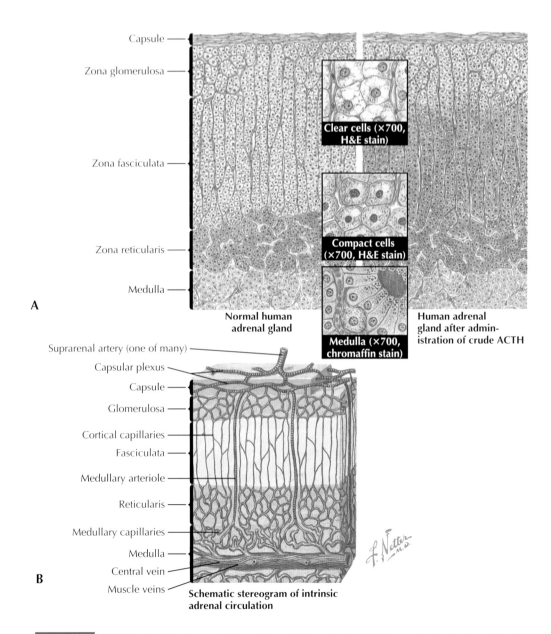

Capsule
Zona glomerulosa
Zona fasciculata
Zona reticularis
Medulla

Clear cells (×700, H&E stain)

Compact cells (×700, H&E stain)

Medulla (×700, chromaffin stain)

A

Normal human adrenal gland

Human adrenal gland after admin- istration of crude ACTH

Suprarenal artery (one of many)
Capsular plexus
Capsule
Glomerulosa
Cortical capillaries
Fasciculata
Medullary arteriole
Reticularis
Medullary capillaries
Medulla
Central vein
Muscle veins

B

Schematic stereogram of intrinsic adrenal circulation

Figure 29.2 **Histologic Features of the Adrenal Gland**   The highly vascularized adrenal glands consist of an outer cortex and inner medulla. The cortex synthesizes the steroid hormones aldosterone, cortisol, and androgens in its zona glomerulosa, zona fasciculata, and zona reticularis, respectively. Adrenocorticotropic hormone (ACTH) administration results in increased cell size and steroid biosynthetic activity mainly in the zona fasciculata but also in the zona reticularis **(A)**. The medullary chromaffin cells synthesize and release catecholamines (mainly epinephrine) into the bloodstream in response to sympathetic nervous system activation. The blood supply to the adrenal gland is provided by the suprarenal arteries **(B)**. *H&E,* hematoxylin and eosin.

The term **glucocorticoid** stems from the fact that cortisol raises blood glucose levels; some of the other products of the steroid synthesis pathways, notably **corticosterone**, also have glucocorticoid activity. Many of the effects of cortisol are **permissive,** meaning that the hormone does not directly cause a particular effect but is necessary for that effect to occur in response to another stimulus. For example, cortisol stimulates the *synthesis* of enzymes involved in gluconeogenesis (i.e., synthesis of glucose from specific gluconeogenic amino acids or from lactate, pyruvate, or glycerol). In general, the effects of glucocorticoids can be categorized as **metabolic, antiinflammatory,** and **immunosuppressive** (see Fig. 29.4).

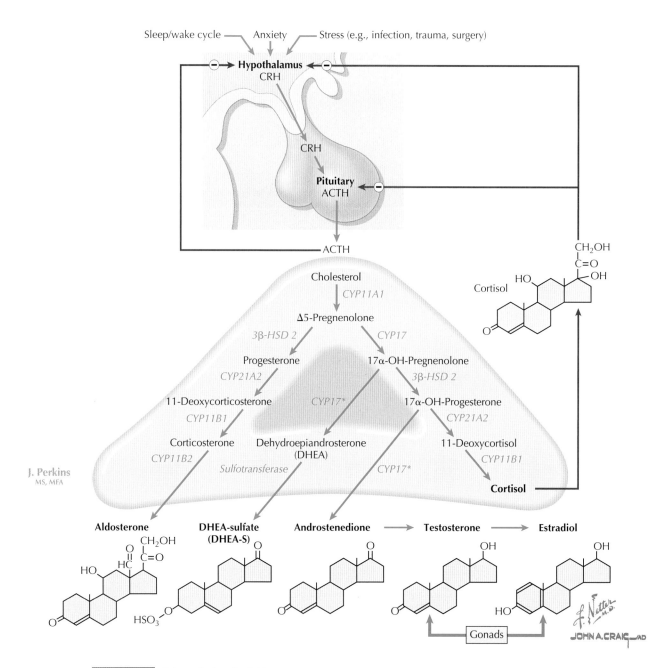

Figure 29.3 **Adrenal Cortical Hormones** Adrenal mineralocorticoid (aldosterone), glucocorticoid (cortisol), and androgens (dehydroepiandrosterone and androstenedione) are synthesized from cholesterol through the biosynthetic pathways illustrated. Adrenocorticotropic hormone (ACTH) stimulates the conversion of cholesterol to Δ5-pregnenolone by the enzyme CYP11A1; the conversion of Δ5-pregnenolone to various products is dependent on additional enzymes in the various zones of the adrenal cortex. The adrenal gland also synthesizes small amounts of other steroids (e.g., testosterone and estradiol). Negative feedback on ACTH and the hypothalamic-releasing hormone corticotropin-releasing hormone (CRH) is accomplished by cortisol.

Secretion of cortisol follows a diurnal pattern, with peak plasma levels occurring in the early morning hours before the onset of physical activity (assuming a normal sleep-wake cycle). It has been widely speculated that this elevation in cortisol may help prepare our bodies for the stresses of our daily activities. This rhythm appears to be controlled by both central and peripheral "pacemakers": the suprachiasmatic nucleus of the hypothalamus acts as the central pacemaker for the HPA axis and the autonomic nervous system (which displays a similar diurnal pattern), while peripherally, the adrenal gland appears to maintain a diurnal pattern of sensitivity to ACTH. Both these central and peripheral factors contribute to the early morning plasma cortisol peak.

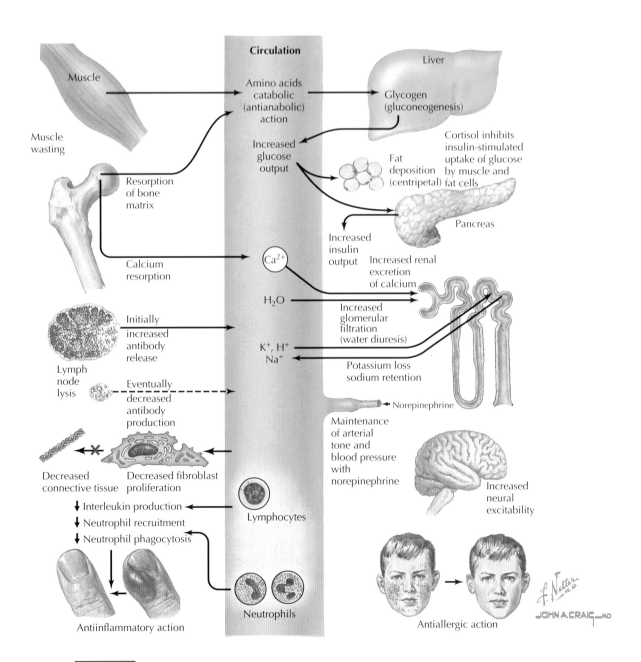

Figure 29.4 **Actions of Cortisol** At normal physiological levels, cortisol has a variety of metabolic effects (see text) and a permissive role for the actions of other hormones. Illustrated are some of the effects of cortisol at pharmacological levels or when secreted in excess (e.g., during chronic stress). It also has mineralocorticoid-like effects on the kidney at high concentrations.

Metabolic Effects of Glucocorticoids

At *normal* levels, cortisol has the following major metabolic effects:

- Stimulation of **gluconeogenesis** (as noted, stimulation of gluconeogenesis is a permissive action of cortisol, which induces synthesis of the enzymes involved in hepatic gluconeogenesis)
- **Protein catabolism** to provide substrate (amino acids) for gluconeogenesis
- **Lipolysis** in adipose tissue
- **Inhibition of insulin-stimulated glucose uptake** by muscle and adipose tissue (because of this action, as well

as stimulation of gluconeogenesis, cortisol is considered a **diabetogenic** hormone)

As a result of these various effects, normal levels of glucocorticoids stimulate gluconeogenesis and, in the well-fed state, storage of glucose in the liver in the form of glycogen. These actions become extremely important in maintaining blood glucose levels during periods of fasting.

As is the case for many hormones, the consequences of excess glucocorticoids are in part but not wholly predictable based on their normal actions. Excess glucocorticoids have the following metabolic effects:

- **Muscle wasting** and **thinning of skin** due to protein catabolism and inhibition of protein synthesis (in excess, cortisol is a **catabolic hormone**)
- **Bone resorption** due to imbalance between osteoclastic and osteoblastic activity
- Deposition of centripetal (truncal) fat, and therefore **centripetal obesity**
- Renal **sodium retention** and **potassium loss** (cortisol binds to mineralocorticoid receptors in addition to glucocorticoid receptors; it is ordinarily converted by the kidney to cortisone, limiting its effects on the renal tubules, but in cortisol excess, mineralocorticoid activity is observed)

Other Effects of Glucocorticoids

Other important actions of glucocorticoids include **anti-inflammatory**, **immunosuppressive**, and **vascular effects**, such as:

- Inhibition of the production of **arachidonic acid metabolites** (e.g., prostaglandin, thromboxane, and leukotrienes) and **platelet-activating factor**
- Reduction in T lymphocytes and their production of **interleukins, interferon**, and **tumor necrosis factor**
- Decreased fibroblast production and their deposition of connective tissue
- An initial increase in **antibody production**, but eventually decreased production with prolonged excess of cortisol
- Increased responsiveness of vascular smooth muscle to norepinephrine (cortisol is important in maintaining normal vascular responses to adrenergic stimulation and, thus, normal blood pressure regulation)

ACTIONS OF ADRENAL ANDROGENS

ACTH stimulates the zona reticularis of the adrenal cortex to synthesize and release androgens, although additional factors are likely involved. The adrenal androgens, dehydroepiandrosterone and androstenedione, have notable effects in females because the adrenal gland is normally the only significant source of androgens in women. Effects of adrenal androgens in females include development of pubic hair, hypertrophy of sebaceous glands (acne), stimulation of libido, and, possibly, inhibition of osteoporosis. These androgens have only a minor role in men, in whom testosterone produced by the testes is the major androgen. However, in both sexes, adrenal androgen production rises in mid childhood, and thus the hormones may contribute to early pubertal changes in both boys and girls (see Chapter 32). **Adrenarche** is the maturation of the adrenal cortex, which occurs between 6 and 10 years of age and results in the three distinct zones of the cortex and, subsequently, production of adrenal androgens. Adrenal androgen levels reach a peak at about 20 years of age and decline through adulthood.

REGULATION AND ACTIONS OF ALDOSTERONE

The mineralocorticoid aldosterone is a regulator of **extracellular fluid volume** and **K^+ homeostasis** (Fig. 29.5). Its actions are accomplished mainly by effects on the collecting ducts and late distal tubules of the kidney, where it stimulates the following effects:

- **Na^+ reabsorption** and, as a result, water retention and **expansion of extracellular fluid volume** (in excess, hypertension results)
- **K^+ excretion** (excess aldosterone will cause hypokalemia)
- **H^+ excretion** (in excess, aldosterone causes metabolic alkalosis)

In addition to the renal tubules, aldosterone also affects Na^+ and K^+ handling by the intestines (specifically, the colon), salivary glands, and sweat glands.

The primary stimuli and inhibitory factors for secretion of aldosterone by cells of the adrenal zona glomerulosa are as follows (see Fig. 29.5):

- **Hyperkalemia** (elevated plasma K^+), which directly acts on the adrenal gland to increase aldosterone secretion; aldosterone reduces plasma K^+ through its renal actions.
- **Angiotensin II**, which stimulates aldosterone secretion. Reduction of blood volume results in renin release by the kidney, which cleaves the plasma protein angiotensinogen to angiotensin I, which is subsequently cleaved by angiotensin-converting enzyme to form angiotensin II (see Section 5, "Renal Physiology"). Aldosterone, by promoting Na^+ and water retention, increases blood volume.
- **Atrial natriuretic peptide**, which inhibits aldosterone secretion. Atrial natriuretic peptide is released by cardiac myocytes when blood volume is elevated (e.g., in congestive heart failure). Inhibition of aldosterone secretion results in reduction of blood volume.

Whereas ACTH is required for synthesis of aldosterone, the aforementioned factors play a more important regulatory role in its synthesis and release. The greatest release of aldosterone, as with cortisol and adrenal androgens, occurs in the early morning hours. The physiological role of aldosterone is discussed in greater detail in Section 5, "Renal Physiology."

THE ADRENAL MEDULLA

In addition to the following discussion, physiological effects of adrenal catecholamines are discussed throughout the various chapters of this book in relation to the physiological features of various target systems and in the specific context of the autonomic nervous system as a whole in Chapter 7.

CLINICAL CORRELATE 29.1
Glucocorticoids as Antiinflammatory and Immunosuppressive Drugs

Based on their antiinflammatory and immunosuppressive properties, cortisol and various synthetic glucocorticoids (for example, dexamethasone) are used therapeutically to suppress serious inflammation and allergic reactions, autoimmune responses, and transplant rejection. However, systemic treatment with steroids is avoided when possible because of a wide array of potentially serious adverse effects, including immunosuppression, muscle wasting, osteoporosis, hyperglycemia, neural excitability and associated psychiatric effects, and, ultimately, adrenal insufficiency (because of negative feedback on the HPA axis). When an antiinflammatory steroid is used systemically, the patient is gradually weaned from the drug to prevent sudden withdrawal before endogenous cortisol production can resume.

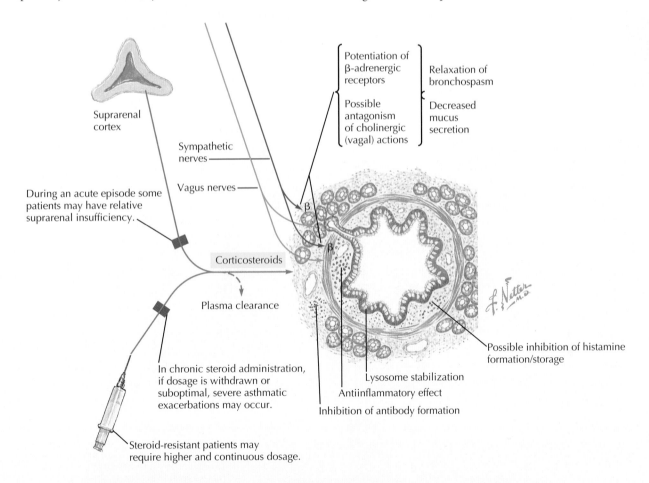

Corticosteroid Actions in Bronchial Asthma Antiinflammatory effects are the basis for use of inhaled corticosteroids as preventive medication in patients who suffer frequent or severe asthma attacks.

The adrenal medulla, which makes up the central portion of the adrenal gland, is functionally distinct from the cortex, and acts as a postganglionic effector of the sympathetic nervous system. The **chromaffin cells** of the adrenal medulla synthesize and release catecholamines into the bloodstream in response to sympathetic nervous system activation. Whereas sympathetic postganglionic nerves mainly release **norepinephrine**, 70% to 80% of the catecholamine release from chromaffin cells is **epinephrine**. As is the case for postganglionic neurons, a number of sympathetic cotransmitters, including **neuropeptide Y**, are also synthesized and released. Release of adrenal catecholamines results in a spectrum of physiological actions throughout the body, with effects differing somewhat in degree compared with effects of norepinephrine released from sympathetic nerves, as a result of the differences in binding affinity of epinephrine and norepinephrine at subtypes of adrenergic receptors (Fig. 29.6).

CLINICAL CORRELATE 29.2
Cushing's Syndrome

Cushing's syndrome is an endocrine disease caused by high levels of blood cortisol. It may be caused by a pituitary or ectopic ACTH-secreting tumor, hyperplasia or tumor of the adrenal gland, or the use of glucocorticoid drugs. The symptoms and signs of Cushing's syndrome, which might be predicted based on the effects of adrenal steroids, include:

- Centripetal obesity, accompanied by muscle wasting in the extremities
- "Moon face" (a rounding of the face)
- Thinning and bruising of the skin
- Skin hyperpigmentation (if ACTH is elevated; see earlier in the chapter)

- Hypertension
- Hyperglycemia and insulin resistance
- Osteoporosis

The dexamethasone suppression test may be useful in diagnosing the cause of hypercortisolism. Although secretion of ACTH by the pituitary (and thus adrenal cortisol production by the adrenal gland) is normally suppressed by administration of dexamethasone (a glucocorticoid drug), pituitary ACTH-secreting tumors have reduced sensitivity to such inhibition. When hypercortisolism is the result of an adrenal tumor, cortisol production is unaffected by exogenous glucocorticoid (dexamethasone) administration. Once the cause of glucocorticoid excess is identified and appropriately addressed (often by surgery), treatment with replacement steroids is usually necessary.

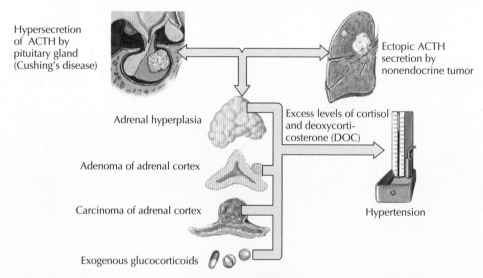

Hypersecretion of ACTH by pituitary gland (Cushing's disease)

Ectopic ACTH secretion by nonendocrine tumor

Adrenal hyperplasia

Excess levels of cortisol and deoxycorticosterone (DOC)

Adenoma of adrenal cortex

Carcinoma of adrenal cortex

Hypertension

Exogenous glucocorticoids

Clinical features

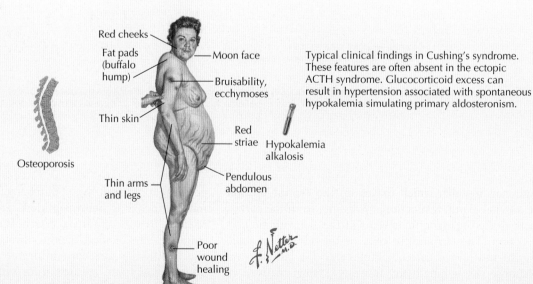

Red cheeks

Fat pads (buffalo hump)

Moon face

Bruisability, ecchymoses

Thin skin

Red striae

Osteoporosis

Hypokalemia alkalosis

Thin arms and legs

Pendulous abdomen

Poor wound healing

Typical clinical findings in Cushing's syndrome. These features are often absent in the ectopic ACTH syndrome. Glucocorticoid excess can result in hypertension associated with spontaneous hypokalemia simulating primary aldosteronism.

Causes of Cushing's Syndrome Cushing's syndrome may result from a variety of causes, all of which result in an elevated plasma glucocorticoid level. *ACTH,* adrenocorticotropic hormone.

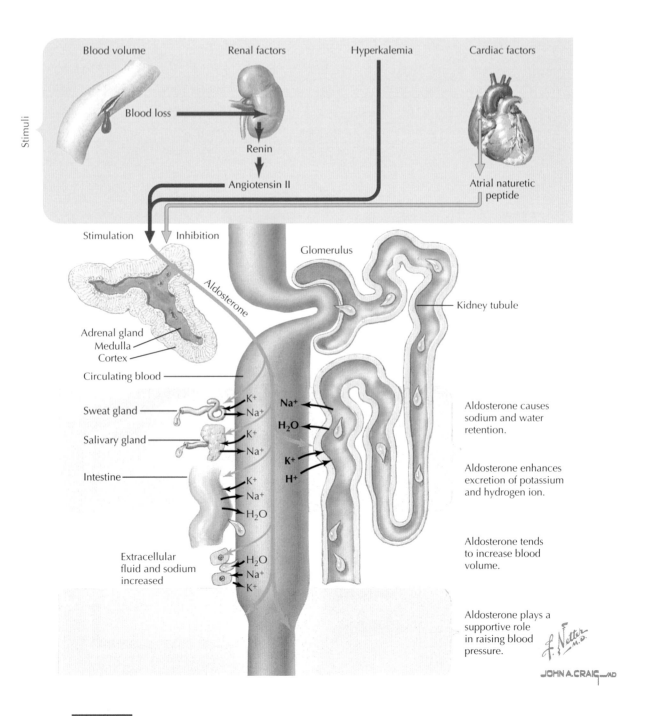

Blood volume   Renal factors   Hyperkalemia   Cardiac factors

Stimuli

Blood loss

Renin

Angiotensin II

Atrial naturetic peptide

Stimulation   Inhibition

Glomerulus

Aldosterone

Adrenal gland
Medulla
Cortex

Kidney tubule

Circulating blood

Sweat gland — K$^+$ / Na$^+$

Salivary gland — K$^+$ / Na$^+$

Intestine — K$^+$ / Na$^+$ / H$_2$O

Na$^+$

H$_2$O

K$^+$

H$^+$

Extracellular fluid and sodium increased — H$_2$O / Na$^+$ / K$^+$

Aldosterone causes sodium and water retention.

Aldosterone enhances excretion of potassium and hydrogen ion.

Aldosterone tends to increase blood volume.

Aldosterone plays a supportive role in raising blood pressure.

f. Netter M.D.

JOHN A. CRAIG—AD

Figure 29.5 **Actions of Aldosterone** The steroid hormone aldosterone has the important functions of regulating extracellular fluid volume and K$^+$ levels. Synthesis and release of aldosterone is promoted by angiotensin II and hyperkalemia and is inhibited by atrial natriuretic peptide. Its action results in water and Na$^+$ retention and K$^+$ and H$^+$ excretion by the kidney; it has similar effects on the intestine, sweat glands, and salivary glands.

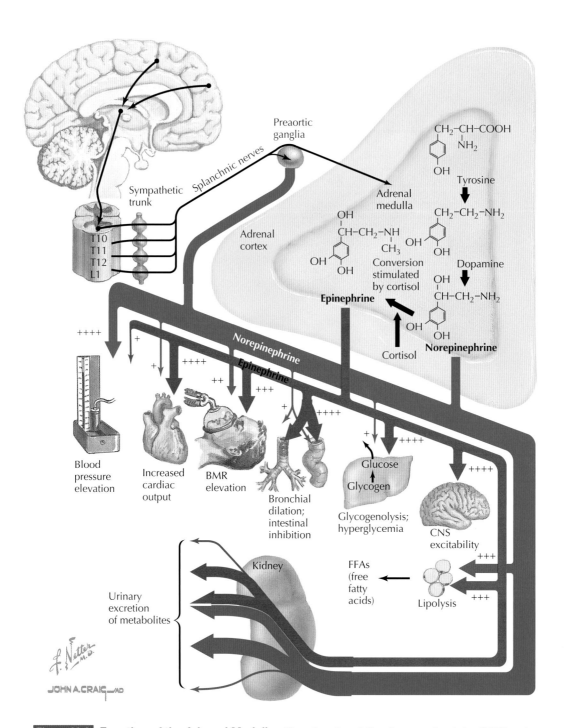

Figure 29.6 **Function of the Adrenal Medulla** The adrenal medulla releases epinephrine (80%) and norepinephrine (20%) into the bloodstream during activation of the sympathetic nervous system. The effects of epinephrine and norepinephrine, including relative magnitude, are illustrated for various sites. *BMR,* basal metabolic rate; *CNS,* central nervous system.

CLINICAL CORRELATE 29.3
Addison's Disease

Addison's disease is an uncommon endocrine disease characterized by failure of steroid synthesis by the adrenal gland as a result of various causes, including autoimmune or tubercular destruction of the gland and rare genetic disorders. As is the case for Cushing's syndrome, effects of deficiency of adrenal steroids might be predicted based on the normal actions of the hormones. Symptoms and signs of adrenal insufficiency include:

- Poor stress tolerance
- Hypoglycemia
- Fatigue and weight loss

- Hyperpigmentation of skin (due to ACTH elevation; see earlier in the chapter)
- Low blood pressure
- Salt appetite

An Addisonian crisis, caused by a severe deficiency of adrenal hormones, is a medical emergency, sometimes marking the onset of the disease. Vomiting and diarrhea, low blood pressure, fainting, loss of consciousness, convulsions, and hypoglycemia are among the signs and symptoms. Patients with Addison's disease require long-term glucocorticoid replacement therapy and sometimes mineralocorticoid treatment as well.

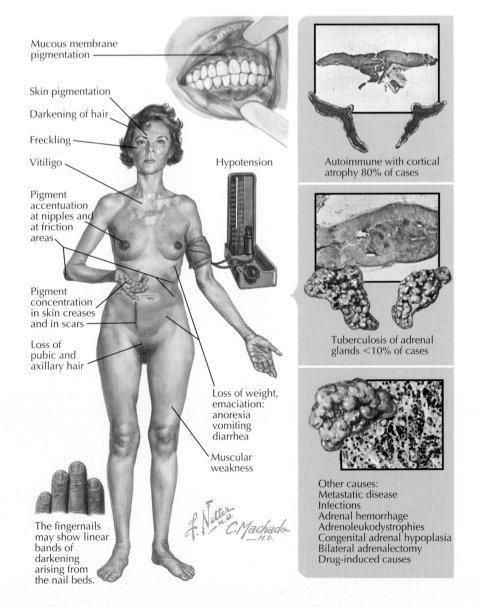

Mucous membrane pigmentation

Skin pigmentation

Darkening of hair

Freckling

Vitiligo

Hypotension

Pigment accentuation at nipples and at friction areas

Pigment concentration in skin creases and in scars

Loss of pubic and axillary hair

The fingernails may show linear bands of darkening arising from the nail beds.

Loss of weight, emaciation: anorexia vomiting diarrhea

Muscular weakness

Autoimmune with cortical atrophy 80% of cases

Tuberculosis of adrenal glands <10% of cases

Other causes:
Metastatic disease
Infections
Adrenal hemorrhage
Adrenoleukodystrophies
Congenital adrenal hypoplasia
Bilateral adrenalectomy
Drug-induced causes

Chronic Primary Adrenocortical Insufficiency (Addison's Disease) In persons with primary adrenocortical insufficiency, adrenocortical steroid production is low as a result of adrenal atrophy, tubercular destruction of the adrenal, or other causes and ACTH level is elevated because of a lack of negative feedback. Symptoms reflect a deficiency of corticosteroids but also reflect an excess of ACTH, which causes pigmentation as a result of its sequence homology with α–melanocyte-stimulating hormone.

30

The Endocrine Pancreas

The pancreas has key roles in both gastrointestinal function and modulation of blood glucose levels. Because the products of the exocrine pancreas (enzymes and buffers) allow digestion and absorption of carbohydrates into the blood and the products of the endocrine pancreas (insulin, glucagon, and somatostatin) regulate the blood glucose levels, the pancreas can be viewed as a total processing unit for controlling glucose entry into the body and the cells.

The primary role of the pancreatic hormones is to maintain an appropriate basal level of glucose in the blood, which in humans is 70 to 90 mg%. To maintain basal levels, the hormones sequester glucose into cells when an influx occurs during feeding (the hypoglycemic effect of insulin) and mobilize glucose out of cells during fasting (the hyperglycemic effect of glucagon). This overall balance is achieved by the integration of glucose metabolic activity primarily within liver, muscle, and adipose tissues, all of which is orchestrated by the pancreatic hormones.

STRUCTURE OF THE PANCREAS

As noted in Section 6, the exocrine pancreas performs crucial digestive and buffering functions. The exocrine pancreas makes up about 99% of the functional cells, with the remainder serving the endocrine function (Fig. 30.1). The primary endocrine portion consists of the **islets of Langerhans**, which are composed of three cell types:

- α Cells produce **glucagon**, which mobilizes glucose stores *into the blood*.
- β Cells produce **insulin**, which stimulates glucose transport *into the cells*.
- δ Cells produce **somatostatin**, which inhibits the secretion of both insulin and glucagon.

In the islets, the cells are organized in such a way that the α cells are interspersed between the β cells. There are many more insulin-producing β cells than α and δ cells.

Pancreatic polypeptide is another endocrine hormone secreted from F cells in the pancreatic islets. Although its exact function is not clear, this peptide is elevated in response to stimuli such as meals high in protein, hypoglycemia, and exercise and acts to inhibit pancreatic exocrine (buffers and enzymes) and endocrine (insulin) secretions. As a person eats, the effects of pancreatic polypeptide modulate the effects of cholecystokinin (see Chapter 24).

Important paracrine (cell-to-cell) interactions also occur within the islets. Because blood forms a capillary network in the islets, insulin (from β cells) is carried to the glucagon-producing α cells in a paracrine manner. Insulin can inhibit its own secretion, as well as glucagon secretion, and somatostatin inhibits both insulin and glucagon secretion. This mechanism allows paracrine modulation of the pancreatic response to blood glucose levels.

Glucose enters cells through two families of glucose transporters, **facilitated glucose transporters (GLUT)** and **sodium-dependent glucose transporters (SGLT)**. Although more than 10 glucose transporters have been identified, the function of several of the newly discovered isoforms is not clear. Of the main transporters:

- **GLUT1** is in all adult cell membranes and is responsible for allowing enough glucose into cells to maintain cellular respiration and viability. It is also found in high concentration in the membranes of the blood-brain barrier, and the transporters are under positive control by circulating glucose (i.e., increased glucose increases the number of GLUT1 transporters).
- **GLUT2** is located on membranes of the brain, liver, and pancreas and facilitates the easy entry of glucose into and out of those tissues. GLUT2 is also located in *basolateral* membranes of the small intestine and renal proximal tubule, allowing glucose to leave the cells. This transporter is independent of insulin.
- **GLUT3** is mainly found in neurons and placentas.

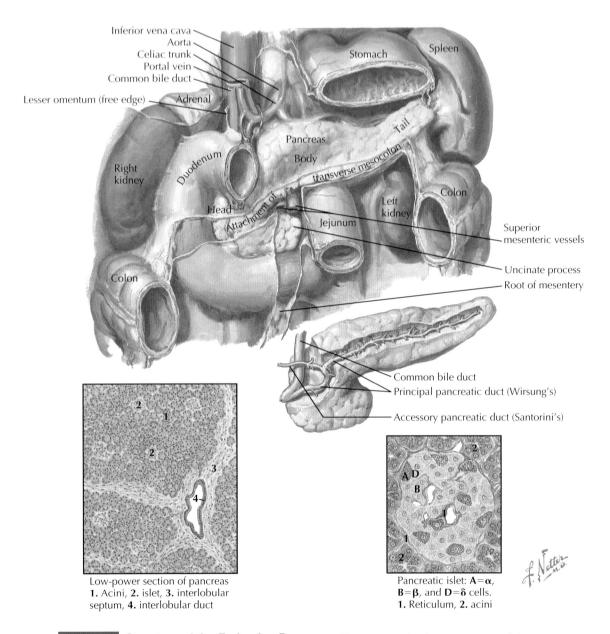

Inferior vena cava
Aorta
Celiac trunk
Portal vein
Common bile duct
Lesser omentum (free edge)
Adrenal
Stomach
Spleen
Pancreas
Tail
Body
transverse mesocolon
Right kidney
Duodenum
Colon
Head
Attachment of
Jejunum
Left kidney
Superior mesenteric vessels
Colon
Uncinate process
Root of mesentery
Common bile duct
Principal pancreatic duct (Wirsung's)
Accessory pancreatic duct (Santorini's)

Low-power section of pancreas
1. Acini, **2.** islet, **3.** interlobular septum, **4.** interlobular duct

Pancreatic islet: **A**=α, **B**=β, and **D**=δ cells.
1. Reticulum, **2.** acini

Figure 30.1 **Structure of the Endocrine Pancreas** The pancreas is a key component of the gastrointestinal tract because of its exocrine function; it also provides the primary control of blood glucose through its production of endocrine hormones. The vast majority of the pancreas (~99%) is composed of acinar cells, which produce and secrete the buffers and enzymes through ducts into the gastrointestinal tract (exocrine function) *(micrograph at bottom left)*. The endocrine pancreas is composed of cells that form the islets of Langerhans. The cells of the islets produce insulin (β cells), glucagon (α cells), and somatostatin (δ cells).

- **GLUT4** is expressed in skeletal and cardiac muscle and adipose tissue and is the **insulin-stimulated glucose transporter.**
- **GLUT5,** the fructose transporter, is independent of insulin.
- **SGLT1** carries glucose with sodium (via secondary active transport) and is found in *apical* membranes of the renal proximal tubule, choroid plexus, and small intestine along with SGLT2. This transporter is independent of insulin.
- **SGLT2** carries glucose with sodium and is found in *apical* membranes of renal proximal tubules, the choroid plexus, and the small intestine along with SGLT1. This transporter is independent of insulin.

SYNTHESIS, SECRETION, AND ACTIONS OF INSULIN

The major control of blood glucose levels occurs through the production, secretion, and action of **insulin**, which is a 51–amino acid peptide hormone formed in the β cells from a prohormone. This prohormone contains within its sequence three segments, representing the A and B chains of the active insulin molecule and a connecting "C" peptide (Fig. 30.2). In the endoplasmic reticulum, disulfide bridges are formed between the A and B chains of the proinsulin, which are still connected by the C-peptide sequence. In the Golgi apparatus, the C peptide is cleaved from the proinsulin, forming active

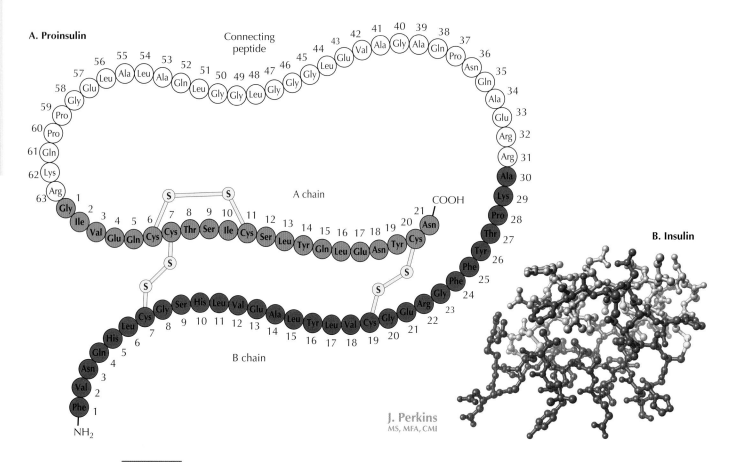

Figure 30.2 **Insulin Structure** **A,** The proinsulin molecule contains within it the peptide sequences for A chain *(blue)* and B chain *(red)* of mature insulin as well as the connecting peptide. Two disulfide bridges are formed between the A chain and B chain sequences within the endoplasmic reticulum. The connecting "C" peptide is cleaved in the Golgi apparatus to yield insulin and C peptide, which are then packaged into secretory granules. The three-dimensional structure of active insulin is illustrated in **B.**

insulin. Both the insulin and C peptide are packaged in granules for secretion into the blood.

> Because both insulin and the **C-peptide fragments** are secreted when contents of the granules that contain insulin are released, the amount of C peptide in the blood is a reflection of insulin production. Clinically, C peptide is used to determine endogenous insulin secretion in diabetic patients receiving insulin injections.

Secretion of insulin is stimulated by blood glucose levels and by gut peptides. The following actions occur when the blood glucose level increases (see Fig. 30.3*A*):

- Glucose enters the pancreatic β cells via **GLUT2** transporters.
- Intracellular glucose metabolism increases the level of ATP, which inhibits K^+ efflux, thus depolarizing the β cells and opening voltage-gated Ca^{2+} channels.
- The Ca^{2+} influx stimulates secretion of the insulin and C peptide into the blood.

In addition to blood glucose, the gut-derived **incretins** (glucose insulinotropic peptide [GIP] and glucagon-like peptide [GLP]-1), increased amino acids and fatty acids in the blood, and locally released acetylcholine also promote insulin secretion (Fig 30.3*B*). A common theme is that all of these elements reflect the fed state as a stimulus for elevating the insulin level. Lastly, glucagon can also stimulate insulin secretion by increasing intracellular Ca^{2+} (via phospholipase C), thus modulating the hyperglycemic effects of glucagon.

Insulin secretion is suppressed when the blood glucose level is low, when the sympathetic nervous system is stimulated (i.e., when norepinephrine and epinephrine levels are elevated), and when the local somatostatin level is elevated.

Overall, insulin promotes entry of glucose into cells, synthesis of glycogen stores, and reduced lipolysis (Fig. 30.4). These anabolic functions ensure that nutrients are stored and thus available to tissues between meals (i.e., when fasting occurs).

The insulin receptor is expressed on most tissues, but the receptor concentration is highest in liver, muscle, and adipose

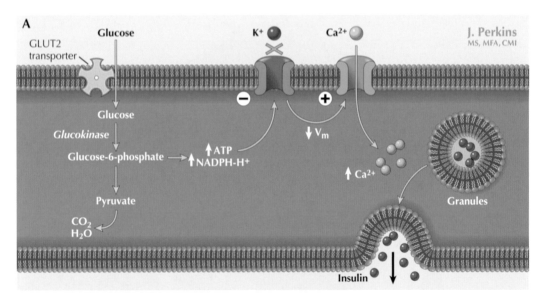

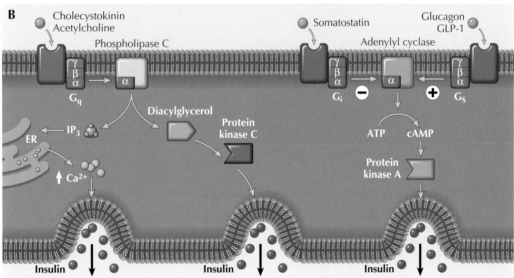

Figure 30.3 **Insulin Synthesis and Release** Glucose is the most important factor in the regulation of insulin synthesis and secretion, although gastrointestinal peptides and local glucagon and somatostatin also contribute to modulation of release. **A**, The action of glucose to increase Ca^{2+} influx, which stimulates insulin secretion. **B**, The receptor-mediated *stimulation* of insulin secretion by local glucagon, gut peptides (cholecystokinin [CCK] and glucagon-like peptide–1 [GLP-1]), and acetylcholine (ACh) and *inhibition* of insulin by local somatostatin. *ER,* endoplasmic reticulum; *GLUT2,* glucose transporter 2; *IP₃,* inositol trisphosphate; *NADPH,* nicotinamide adenine dinucleotide phosphate.

tissue. The insulin receptor consists of two extracellular α subunits and two transmembrane β subunits. Insulin binds to the α subunits and stimulates phosphorylation of the β subunits, as well as phosphorylation of other intracellular proteins. Thus the ability of insulin to stimulate phosphorylation of various intracellular proteins causes its hypoglycemic effect by the following actions:

- Increasing **GLUT4** transporters in membranes, thus allowing efficient glucose entry into cells
- Increasing glycogen production from excess glucose to facilitate storage

- Inhibiting glycogenolysis to maintain glycogen storage
- Inhibiting gluconeogenesis to prevent production and release of glucose back into the blood

All of these factors contribute to the rapid reduction of the blood glucose concentration when insulin is present. Furthermore, insulin also has effects on lipid metabolism through the following actions:

- Inhibiting hormone-sensitive lipase in adipose tissue, which decreases lipolysis and reduces the amount of circulating free fatty acids

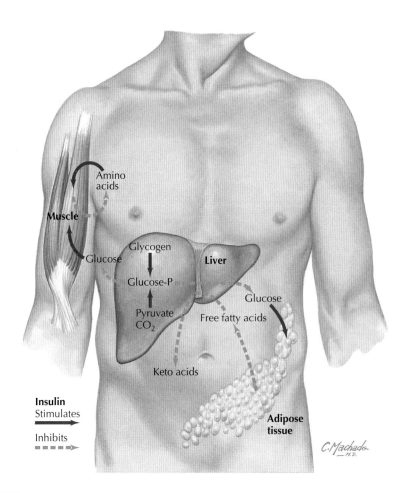

Figure 30.4 **Actions of Insulin** Insulin is considered a "fuel storage" hormone, and therefore insulin promotes the storage of glucose (as glycogen) and fatty acids (as triglycerides in adipose tissue). Insulin stimulates the uptake of glucose into cells via glucose transporter 4 (GLUT4) transporters, and the glucose is used or stored as glycogen. The major glycogen stores are in muscle and liver. Insulin also stimulates fat synthesis and inhibits lipolysis in adipose tissue, which maintains stores of triglycerides and reduces keto acid production. Lastly, insulin stimulates uptake of amino acids into skeletal muscle and storage as protein. The overall result is that insulin decreases plasma glucose, fatty acids, and keto acids.

■ Inhibiting oxidation of fatty acids (especially in the liver), which decreases keto acid formation

Once the receptor phosphorylation occurs, the entire receptor complex is internalized and then is degraded, stored, or reinserted into the membrane. Insulin also has the ability to downregulate its receptors through increased degradation and decreased synthesis.

Insulin is a critical metabolic hormone; without it, most cells cannot take up glucose. The exceptions are the brain, liver tissues, the small intestine, and kidney proximal tubules (which have adequate normal glucose uptake because of the presence of *insulin-independent* glucose transporters) and *exercising* muscle (in which GLUT4 is upregulated independently of insulin activity). Until the discovery of insulin, type 1 diabetes was a fatal disease.

SYNTHESIS, SECRETION, AND ACTIONS OF GLUCAGON

Glucagon is a 29–amino acid peptide hormone produced as a prohormone in the α cells of the islets of Langerhans. As with insulin, intracellular processing results in packaging of active glucagon molecules in dense core granules.

The secretion of glucagon is stimulated primarily by low blood glucose levels, and insulin inhibits glucagon secretion in a paracrine manner within the islets. In addition, modulation of glucagon is observed with ingestion of amino acids, which increases levels of both insulin and glucagon. Glucagon secretion is also inhibited by high levels of glucose and fatty acids in the blood.

In opposition to insulin, glucagon promotes the use of the cellular energy stores through the release of glucose into

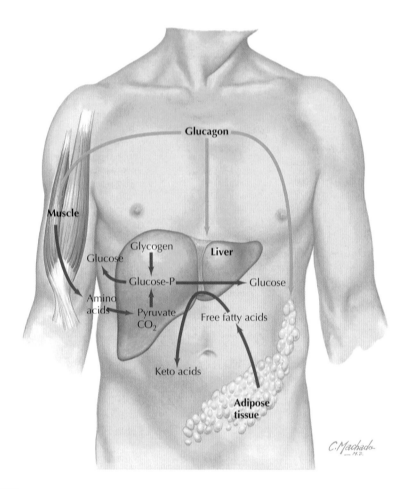

Figure 30.5 **Actions of Glucagon** Glucagon is considered the "fuel mobilization" hormone because it breaks down glycogen, protein, and lipids, thus releasing glucose, amino acids, fatty acids, and keto acids into the blood to serve metabolic demand. Glucagon is stimulated by low blood glucose levels and promotes glycogenolysis and gluconeogenesis in the liver, thus increasing blood glucose levels. Glucagon also stimulates lipolysis and release of fatty acids from adipose tissue, which are oxidized to keto acids in the liver. Lastly, in muscle, glucagon inhibits protein synthesis, thus providing amino acids for conversion to glucose via gluconeogenesis in the liver.

the blood (Fig. 30.5). The G protein–coupled glucagon receptor is a member of the secretin-glucagon family of receptors and shares homology with secretin and GIP receptors. In contrast to the wide distribution of insulin receptors, glucagon receptors are found mainly on the cell membranes in the liver and kidney (although they also are present in adipose, cardiac, pancreatic, and other tissues). The liver has stores of glycogen that can be broken down and released and the ability to oxidize fatty acids at a high rate under the influence of glucagon. Glucagon also stimulates gluconeogenesis in the liver and kidneys.

Binding of glucagon with its receptor stimulates cAMP production, which then activates kinases that phosphorylate various enzymes. These enzymes contribute to the hyperglycemic effect of glucagon by the following actions:

- Inhibiting hepatic glycolysis
- Increasing gluconeogenesis
- Increasing glycogenolysis, breaking down glycogen stores, and releasing the glucose into the blood

These factors produce a rapid increase in blood glucose levels. In addition, glucagon has effects on lipid metabolism by increasing β-oxidation of fatty acids.

In normal periods of fasting between meals, a balance exists between the mobilization of glucose by glucagon and the replenishment of cellular glucose by insulin. However, with prolonged fasting the glucagon effect predominates, and after depletion of glycogen stores, rates of gluconeogenesis and fatty acid oxidation are high. The increased oxidation of fatty acids to acetyl coenzyme A produces **ketone bodies** (acetoacetate and hydroxybutyrate), which can contribute to the acid load of the body (see Chapter 21).

Ω Both insulin and glucagon are stimulated by certain **amino acids.** This phenomenon may seem counterintuitive; however, if a meal is high in protein but low in carbohydrates, insulin will be increased, stimulating both amino acid and glucose uptake into cells, which lowers the plasma glucose concentrations and would produce hypoglycemia *if not for the increase in glucagon.* Thus glucagon modulates the insulin response.

In addition, although both insulin and glucagon reduce plasma amino acid concentrations by increasing uptake into the hepatocytes, their objectives are different. Insulin promotes peptide synthesis, whereas glucagon uses the amino acids for gluconeogenesis, and thus amino acids contribute to the mobilization of glucose in the fasting state.

SYNTHESIS, SECRETION, AND ACTION OF SOMATOSTATIN

The function of **somatostatin** is not well understood, but it is known to play a lesser role in control of blood glucose levels. Somatostatin is a 14 amino acid peptide produced in the δ cells of the islets and acts in a paracrine manner to suppress *both* insulin and glucagon. This action adds one more level of control in the modulation of blood glucose levels. The somatostatin produced in the pancreas is the same hormone that is produced in the brain and gut, and as in these other tissues, it acts locally on adjacent cells.

Somatostatin is secreted in response to all ingested nutrients, as well as glucagon, and is inhibited by insulin.

CLINICAL CORRELATE 30.1
Diabetes Mellitus

Diabetes mellitus is a disease of impaired insulin function that results in hyperglycemia. Diabetes has two classifications that relate to their etiology. **Type 1 diabetes** (T1D) was formerly called juvenile diabetes or insulin-dependent diabetes mellitus, whereas **type 2 diabetes** (T2D) was known as adult-onset diabetes or non–insulin-dependent diabetes mellitus.

T1D is caused by progressive destruction of the pancreatic β cells by autoimmune attack, eventually resulting in minimal release of insulin and hyperglycemia. The autoimmune (T lymphocyte) destruction may be caused by viral infection, but the exact causes are not well understood. The onset of hyperglycemia occurs rapidly and usually becomes evident before the person reaches 20 years of age. The lack of insulin creates two immediate problems: (1) glucose cannot enter the cells efficiently, and (2) the lack of glucose in adipose cells decreases fat synthesis and increases fat breakdown, thus releasing fatty acids into the blood, which are oxidized to ketone bodies by the liver and can result in ketoacidosis (see Chapter 21). The condition is treated by stabilizing the acid-base balance and starting insulin injections. Careful monitoring of blood glucose levels and insulin injections are needed throughout life.

In contrast, **T2D** appears to be a form of **insulin resistance** that results from a reduction in the insulin receptors on target tissues. If insulin is not present (T1D) or the receptors are reduced (T2D), the GLUT4 transporters will not be increased at the target cells and glucose will not efficiently enter the cells, resulting in an elevated level of glucose in the blood. Although insulin resistance has a hereditary component, the downregulation of receptors is exacerbated with obesity. Several ethnic groups have a strong genetic predisposition for T2D, including Mexican Americans, Native Americans in the Southwest, and African Americans. In addition, a higher incidence of T2D occurs in African American and Native American women than in men of the same ethnicity, suggesting a role of sex hormones in the insulin resistance. Regardless of whether a person with T2D was obese or lean prior

to an increase in basal blood glucose levels, a loss of sensitivity to insulin occurs, and as a result the β cells secrete more insulin to maintain euglycemia. Thus, in response to an oral glucose load, a person with prediabetes will display up to twice the insulin secretion and secretion will occur over a prolonged period compared with a nondiabetic person. As the diabetes becomes more pronounced (i.e., with basal hyperglycemia), insulin secretion can decrease, further compounding the hyperglycemia. In cases in which T2D is precipitated by obesity, reduction in weight and increased exercise can help control the hyperglycemia.

Uncontrolled hyperglycemia can result in the following symptoms:

- Polyuria, from the osmotic effect of glucose in the urine
- Polydipsia, to compensate for urinary fluid losses
- Polyphagia, because glucose cannot enter the tissues and thus the tissues are "starved" and hunger is stimulated
- Metabolic ketoacidosis (primarily in persons with T1D)

Long-term uncontrolled or poorly controlled diabetes can lead to or be associated with the following conditions:

- Retinopathy and blindness
- Nephropathy and renal failure
- Hypertension, coronary heart disease, and heart failure
- Cerebrovascular disease
- Peripheral vascular disease

Whereas T1D is treated primarily with insulin injections (along with diet and exercise), T2D can be treated in the following ways:

- Through diet and exercise (exercise increases GLUT4 transporters in cell membranes)
- With insulin injections
- With oral use of sulfonylureas and meglitinides, which increase β-cell insulin secretion
- With use of thiazolidinediones, which increase the sensitivity of muscle and fat tissue to insulin and decrease glucose production in the liver

CLINICAL CORRELATE 30.1
Diabetes Mellitus—cont'd

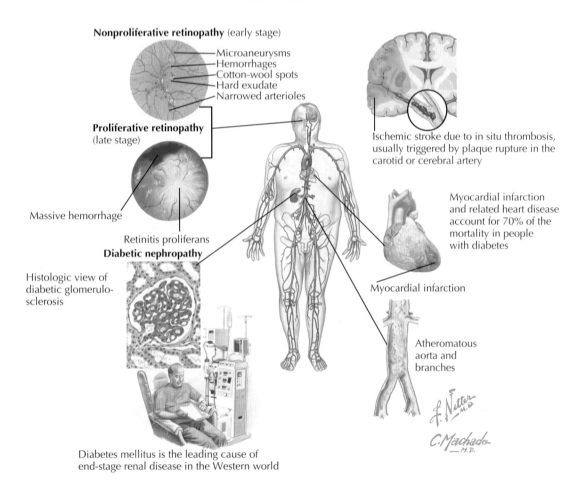

Nonproliferative retinopathy (early stage)
- Microaneurysms
- Hemorrhages
- Cotton-wool spots
- Hard exudate
- Narrowed arterioles

Proliferative retinopathy
(late stage)

Massive hemorrhage

Retinitis proliferans

Diabetic nephropathy

Histologic view of diabetic glomerulosclerosis

Diabetes mellitus is the leading cause of end-stage renal disease in the Western world

Ischemic stroke due to in situ thrombosis, usually triggered by plaque rupture in the carotid or cerebral artery

Myocardial infarction and related heart disease account for 70% of the mortality in people with diabetes

Myocardial infarction

Atheromatous aorta and branches

Microvascular and Macrovascular Complications of Diabetes Vascular complications can occur with either type 1 or type 2 diabetes and include retinopathy (which can lead to blindness), cardiovascular disease, cerebrovascular disease, and diabetic nephropathy. These complications are responsible for the high morbidity in persons with diabetes.

Chapter 31

Calcium-Regulating Hormones

OVERVIEW OF CALCIUM HOMEOSTASIS

As previously discussed in Sections 5 and 6, calcium and phosphate homeostasis are closely linked, primarily because of their roles in bone metabolism. Although several hormones regulate both minerals, this chapter will focus on calcium and its regulation. Plasma calcium concentrations are under tight control, with regulatory systems maintaining normal total plasma calcium concentration at around 9 mg/dL; approximately 40% of the plasma calcium is bound to proteins. Calcium is critical to a myriad of functions, including contraction of cardiac, skeletal, and smooth muscle, bone mineralization, transmission of nerve impulses, and blood clotting. Thus alterations in plasma levels (hypercalcemia or hypocalcemia) can cause severe consequences (see Clinical Correlates 31.1 to 31.4).

Main Sites of Calcium Regulation

The intestines, kidneys, and bone are integral to calcium homeostasis.

- **In the gastrointestinal (GI) tract:** Adults who consume an average diet ingest ~1000 mg of calcium daily. The intestines absorb one third of this amount of calcium, but because additional calcium is lost through salivary and GI secretions, only about 100 to 200 mg of net calcium is absorbed into the blood. **1,25-dihydroxycholecalciferol** (the active form of **vitamin D**) facilitates intestinal calcium absorption.
- **In the bone:** The majority of the body's calcium (~99%) is stored in the skeletal bone, with much of the unmineralized calcium in the **osteoid**. This extracellular pool can re-enter plasma or be used in bone mineralization, and thus calcium readily moves between bone and plasma as part of homeostatic regulation. To a great extent, short-term plasma calcium homeostasis is accomplished through the effects of parathyroid hormone (PTH), which stimulates bone resorption, thus adding calcium to the plasma. Active vitamin D is necessary for the continual process of bone remodeling. It increases entry of calcium into the bone calcium pool, providing substrate for bone deposition; it also increases the osteoclast number, facilitating bone resorption.

- **In the kidney:** The GI system absorbs 100 to 200 mg of calcium daily; thus, to maintain balance, 100 to 200 mg must be excreted by the kidneys. This amount represents about 2% of the filtered load of calcium in the kidneys, meaning that 98% of the filtered calcium is reabsorbed, in part because of stimulation by PTH.

***

Bone remodeling is a constant process involving deposition and resorption of minerals from the exchangeable calcium-phosphate mineral pool. In the bone, **osteoblasts** promote bone deposition, whereas **osteoclasts** are phagocytic cells that promote bone resorption (i.e., absorption of minerals out of bone and into the extracellular fluid). The activity of these two cell types is typically in balance, with bone deposition equaling bone resorption. The ability to continually remodel bone is important and contributes to (1) increasing bone strength when bones and muscles are stressed or exercised and (2) keeping up with bone development in growing children. When this remodeling activity decreases, as seen in elderly persons, bones can become brittle.

Factors That Alter Plasma Calcium Concentration

Forty percent of plasma calcium is bound to calcium-binding proteins—mainly albumin. About 10% is bound to other anions (including bicarbonate, phosphate, and citrate), so only 50% of the calcium is free (ionized) and biologically active. The plasma calcium is under tight control by calcium regulatory hormones—vitamin D, PTH, and to some extent, calcitonin. Hormone-independent changes in plasma calcium concentration can be caused by the following:

- **Acid-base disorders:** In acidosis, albumin is used to buffer excess H^+ in exchange for Ca^{2+}, which increases the amount of free Ca^{2+} in the plasma; conversely, in alkalosis, H^+ is released from albumin in exchange for Ca^{2+}, and the amount of plasma free Ca^{2+} decreases.
- **Changes in plasma protein concentrations:** Because 40% of plasma calcium is bound to proteins, changes in plasma protein concentrations relate directly to the total amount of calcium. Decreases in protein will decrease

the total amount of calcium, and increases in protein will increase the total amount of plasma calcium.

- **Changes in plasma anion concentrations:** Changes in plasma phosphate concentration are especially relevant. If the amount of phosphate increases, the number of calcium-phosphate complexes increases, thus reducing ionized calcium levels. The hormone PTH regulates both calcium and phosphate and helps limit changes in ionized plasma calcium levels.

SYNTHESIS AND ACTIONS OF CALCIUM-REGULATING HORMONES

Parathyroid Hormone

Synthesis of PTH

PTH is the primary hormone regulating plasma calcium concentration. PTH is synthesized by the **chief cells** of the parathyroid glands (see Fig. 28.1) as a 110–amino acid pre-prohormone. It is then cleaved to a prohormone in the endoplasmic reticulum and to the active 84–amino acid peptide hormone in the Golgi apparatus. PTH is then packaged into secretory vesicles in the cytoplasm, and the vesicles are stored until needed (Fig. 31.1).

Actions of PTH

PTH is continually synthesized, with a constant low-level release of PTH into the blood. In response to small decreases in ionized calcium in the plasma, additional PTH is released

into the blood and acts at the bone and kidney (and indirectly at the intestines by increasing renal vitamin D activation) to restore plasma calcium to normal levels (Fig. 31.2). PTH stimulates cAMP synthesis through its G protein–coupled receptors at two major sites to affect plasma calcium and phosphate concentrations.

- **At the kidney:** PTH (1) increases calcium reabsorption at the *distal tubule* and inhibits phosphate reabsorption at the proximal tubule and (2) increases the synthesis of 1,25-dihydroxycholecalciferol, the active form of vitamin D. This conversion is critical, and loss of PTH or loss of kidney function can impair vitamin D metabolism and function. The renal actions of PTH facilitate the rapid restoration of plasma calcium levels.
- **At the bone:** PTH increases bone resorption, releasing calcium into the blood. This action occurs by **osteocytic osteolysis**, through which calcium is rapidly (within 1 to 2 hours) mobilized from the unmineralized calcium phosphate pool, and by **osteoclastic osteolysis**, through which mineralized bone is broken down over about 12 hours. In response to more severe, prolonged decreases in plasma calcium (i.e., days to weeks), PTH can stimulate the proliferation of osteoclasts to further break down mineralized bone. The end result is increased calcium and phosphate in the plasma, at the expense of demineralization of bone.

The chief cells of the parathyroid gland have G protein–coupled calcium-sensing receptors that monitor plasma calcium concentration. This monitoring allows a rapid response when the plasma calcium level changes: when the

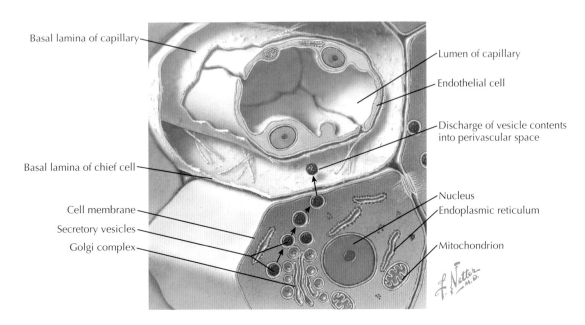

Figure 31.1 **Parathyroid Hormone Secretion** PTH is synthesized in the chief cells of the parathyroid glands. The active hormone is packaged in vesicles and stored in the cytoplasm until it is released. PTH secretion is stimulated by small decreases in plasma calcium levels and produces rapid mobilization of calcium into the extracellular fluid.

CLINICAL CORRELATE 31.1
Effects of Altered Plasma Calcium Concentration

Elevations in plasma calcium (**hypercalcemia**) can result from disorders that increase intestinal calcium absorption, decrease renal calcium excretion, or increase bone resorption. A 30% rise in plasma calcium (to ~12 mg/dL) will depress nervous system activity and cause symptoms that can include increased urination, renal or biliary stone formation, constipation, lethargy, and serious neurologic complications resulting in hyporeflexia. If the amount of plasma calcium continues to rise (above 15 to 17 mg/dL), coma and death can result.

Conversely, **hypocalcemia,** or decreased levels of plasma calcium, will cause hyperexcitability of sensory and motor nerve and muscle cells, resulting in numbness and tingling, as well as muscle twitching and tetany. **Tetany** can occur when the plasma calcium level falls about 30% (to ~6 mg/dL), and death can occur if levels fall to 4 mg/dL. Hypocalcemia can result from vitamin D deficiency, hypoparathyroidism, or dietary calcium deficiency. The effects of inadequate calcium intake are depicted in the following illustration.

As previously noted, changes in plasma calcium can be manifested in disorders involving PTH and vitamin D (see Clinical Correlates 31.2, 31.3, and 31.4).

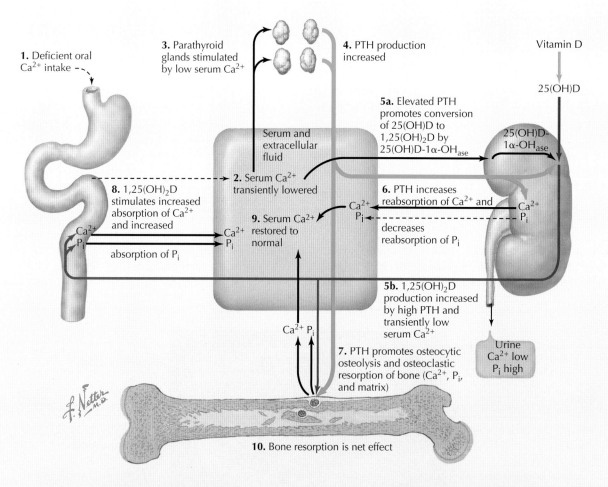

Dietary Calcium Deficiency Inadequate dietary calcium intake *(1)* reduces plasma calcium concentration *(2)*, which stimulates parathyroid hormone (PTH) secretion *(3 and 4)*. PTH increases the renal production of active vitamin D *(5)* (which increases gut absorption of calcium) *(8)*, increases renal calcium reabsorption and decreases renal phosphate reabsorption *(6)*, and increases bone resorption of calcium and phosphate *(7)*. All of these mechanisms serve to increase the plasma calcium concentration so it reaches a normal level *(9)*. When a person has a sustained calcium deficiency (hypocalcemia), the plasma calcium is maintained at a cost of severe bone demineralization *(10)*, and serious symptoms can ensue. P_i, inorganic phosphate.

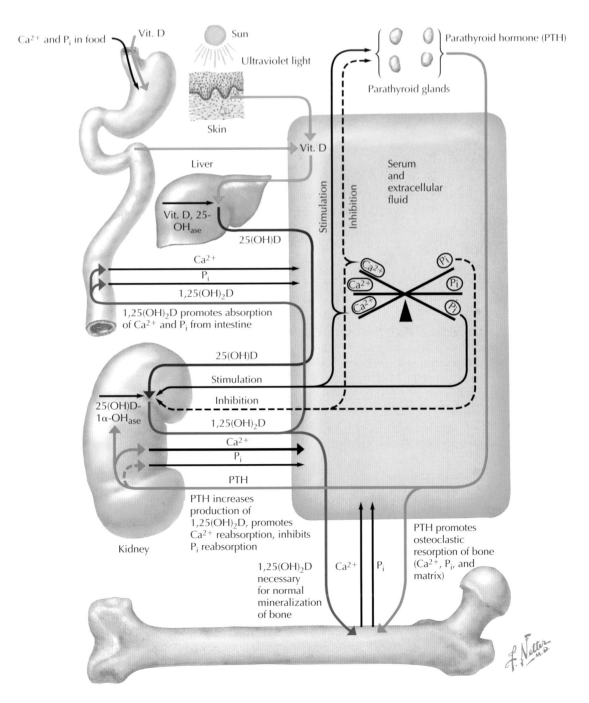

Figure 31.2 **Parathyroid Hormone and Vitamin D Actions on Plasma Calcium Concentrations**
Parathyroid hormone (PTH) is secreted from the parathyroid glands in response to a decrease in plasma ionized Ca^{2+}. PTH acts rapidly (1) on bone to cause resorption and increase plasma Ca^{2+} and (2) on the kidney to increase Ca^{2+} reabsorption and decrease phosphate reabsorption *and* increase production of active vitamin D. The increase in vitamin D increases Ca^{2+} absorption in the gut and promotes bone mineralization. Overall, the rapid effects of PTH increase Ca^{2+} in the blood, restoring homeostasis. P_i, inorganic phosphate.

plasma calcium level decreases, PTH is secreted within seconds, quickly mobilizing calcium from bone and stimulating vitamin D production. Thus PTH has a central role in calcium homeostasis.

> **PTH-related protein** mimics the calcium-mobilizing effects of PTH. It is present in a variety of tissues, including smooth muscle, breast, and the central nervous system, and is believed to facilitate normal bone, dental, and mammary development. PTH-related protein is also secreted from certain tumors, such as breast cancers and renal cell and squamous cell carcinomas, and it accounts for most cases of humoral hypercalcemia of malignancy.

Vitamin D

Synthesis of Vitamin D

Vitamin D, or **cholecalciferol,** is a fat-soluble secosteroid vitamin. Its active form is considered a hormone because it can be produced by the body and acts at nuclear receptors. It enters the blood from dietary sources or is produced in the skin cells by the action of ultraviolet light on 7-dehydroxycholesterol. The initial form of vitamin D is converted to 25-hydroxycholecalciferol in the liver and then undergoes final hydroxylation to the active form, **1,25-dihydroxycholecalciferol,** in the kidneys. Active vitamin D is bound to plasma proteins and acts to increase plasma calcium concentration and promote bone mineralization.

Actions of Vitamin D

When plasma calcium levels decrease, PTH increases and stimulates renal 1-α-hydroxylase to increase formation of active vitamin D. As previously noted, the PTH also increases resorption of calcium phosphate from bone and increases renal calcium reabsorption, with the total effect of increasing plasma calcium levels (Fig. 31.2). The active vitamin D binds to its nuclear steroid receptor and acts at the following sites:

- **Small intestine:** Active vitamin D increases calcium absorption by increasing intracellular **calbindin** (Ca^{2+}-binding proteins) and basolateral Ca^{2+} ATPase. The calbindin promotes calcium entry through the apical calcium channels (TRPV6) into the cells by binding free Ca^{2+} within the cytoplasm, thus enhancing intestinal absorption of calcium. Active vitamin D also increases phosphate absorption; the excess plasma phosphate is then excreted by the kidneys (through PTH action).
- **Bone:** Active vitamin D increases bone remodeling by increasing the number of osteoclasts, and the hormone is necessary for the proper balance between mineralization and resorption.

The synthesis and actions of vitamin D take time and lead to an increase in the availability of calcium from the diet. Thus, whereas in the short term plasma calcium levels are regulated by PTH, the activation of vitamin D by PTH is important in long-term calcium homeostasis. When plasma calcium levels are stabilized, calcium can be restored to the bone by the actions of vitamin D.

Active vitamin D levels in the blood are tightly regulated by feedback control. Under normal conditions, 25-hydroxycholecalciferol concentrations in the plasma remain constant, despite large increases in dietary vitamin D intake. In addition, the 25-hydroxycholecalciferol form can be stored in the liver for several months and released and converted to active vitamin D when needed, allowing another layer of regulation for plasma levels of vitamin D. This overall control of vitamin D levels is important because overproduction of active vitamin D (although rare) can actually cause bone demineralization.

Calcitonin

Calcitonin is a 32–amino acid peptide hormone that is produced in the parafollicular cells of the thyroid gland (see Fig. 28.2, parafollicular [C] cells). It is released in response to elevated plasma calcium levels and acts on G protein–coupled receptors at the following sites:

- **Kidney:** To increase both calcium and phosphate excretion
- **Bone:** To decrease bone resorption by reducing osteoclastic activity

It is not clear whether the physiological effects of calcitonin are important in humans, because loss of the parafollicular cells has no dramatic effect on calcium homeostasis. Calcitonin can be used clinically to treat severe hyperparathyroidism (with hypercalcemia), in conjunction with saline volume expansion and bisphosphonates.

CLINICAL CORRELATE 31.2
Calcium-Related Pathophysiology: Hyperparathyroidism

Hyperparathyroidism is associated with excessive PTH secretion and results in hypercalcemia (high plasma calcium levels) and hypophosphatemia (low plasma phosphate levels).

Primary hyperparathyroidism is usually caused by parathyroid tumors that produce PTH but are not subject to feedback regulation. The elevated PTH increases the renal activation of vitamin D and increases plasma calcium by increasing intestinal absorption, kidney reabsorption, and bone resorption, and it decreases plasma phosphate by increasing urinary phosphate excretion. Although renal calcium reabsorption is elevated, the significantly increased filtered calcium load results in elevated urinary calcium and phosphate excretion, increasing the chance of kidney stone formation. Although many patients are asymptomatic, the elevation in plasma calcium concentration can cause some or all of the following conditions: "stones" (kidney or biliary stones from calcium phosphate and calcium oxalate formation), "bones" (bone pain from excessive resorption), "groans" and "thrones" (from constipation and polyuria), and "moans" or "psychiatric overtones" (from generally not feeling well, depression, and fatigue). Treatment consists of surgical removal of the tumor.

Secondary hyperparathyroidism occurs in response to the primary problem of low plasma calcium levels. In persons with secondary hyperparathyroidism, the low plasma calcium levels are usually caused by vitamin D deficiency and/or chronic renal failure, and the parathyroid glands secrete PTH to correct this problem. Thus PTH levels are elevated but plasma calcium levels are low or normal because of low levels of vitamin D. Treatment is with vitamin D, when indicated.

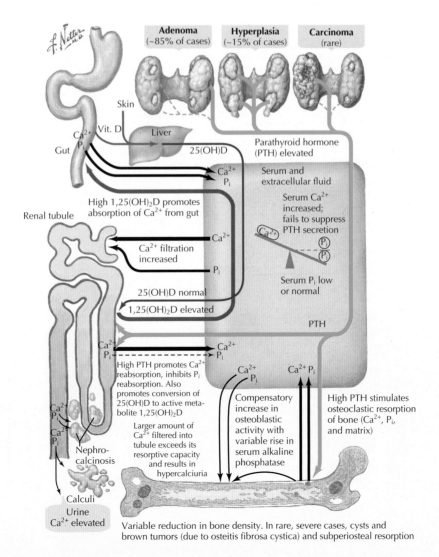

Pathophysiology of Hyperparathyroidism Most cases of hyperparathyroidism arise from parathyroid tumors that produce hypersecretion of PTH. The elevated PTH increases bone resorption, renal calcium reabsorption, and active vitamin D production (which increases intestinal calcium absorption). This process causes hypercalcemia, which can result in multiple symptoms ("stones, bones, groans, thrones, and moans"). P_i, inorganic phosphate.

CLINICAL CORRELATE 31.3
Calcium-Related Pathophysiology: Hypoparathyroidism

Hypoparathyroidism is the deficiency of PTH and results in hypocalcemia (low plasma calcium concentration) and hyperphosphatemia (high plasma phosphate concentration). Hypoparathyroidism is a fairly common occurrence after thyroid or parathyroid surgery, and the decreased level of extracellular fluid calcium increases the excitability of sensory and motor neurons and muscle cells. This excitability can cause sensory effects such as tingling or numbness in the lips and fingers, as well as motor effects, including cramping and twitching. Severe hypocalcemia can lead to tetanic muscle spasms and seizures, so treatment is initiated when the first symptoms are noted. Initial treatment can include intravenous administration of calcium (if plasma levels are dangerously low), followed by oral calcium supplements with active vitamin D. Untreated, chronic hypoparathyroidism can result in cataracts, alopecia (hair loss), and weakened tooth enamel, in addition to the muscle spasms and seizures. Although rare, sustained low plasma calcium concentrations (4 mg/dL) can lead to coma and death. If the condition is life-threatening, the patient is treated with intravenous administration of calcium; otherwise, treatment with vitamin D and oral calcium is effective in maintaining plasma calcium levels.

Pseudohypoparathyroidism has the same overall effect as hypoparathyroidism (i.e., reducing plasma calcium levels and increasing plasma phosphate levels); however, the cause is a primary defect in the G_s protein of the PTH receptors in bone and kidney. Although PTH can bind, the cAMP pathway is not activated by PTH, and thus although PTH levels are *elevated*, renal and bone calcium transport is not activated.

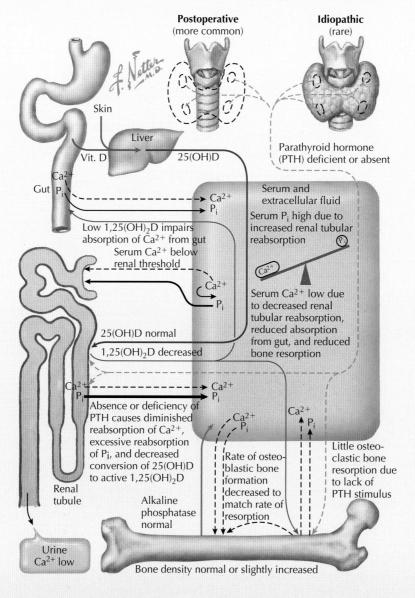

Pathophysiology of Hypoparathyroidism Hypoparathyroidism is usually a result of thyroid or parathyroid surgery, and the lack of parathyroid hormone reduces the formation of active vitamin D, decreases renal calcium reabsorption, and decreases bone resorption. Hypocalcemia develops, and the low plasma calcium levels increase neuronal and muscle cell excitation, causing twitching, cramping, and in extreme cases, tetany. P_i, inorganic phosphate.

CLINICAL CORRELATE 31.4
Calcium-Related Pathophysiology: Vitamin D Deficiency

Vitamin D deficiency is rarely found in regions where vitamin D is supplemented in the diet and people have regular exposure to sunlight. **Rickets** is a disease caused by insufficient vitamin D during childhood. The lack of vitamin D results in inadequate calcium and phosphate for bone mineralization, resulting in stunted growth and malformed bones. In adults, prolonged vitamin D deficiency will decrease new bone mineralization, resulting in bone deformation (**osteomalacia**). This condition is especially evident in weight-bearing bones and can cause a severe bowlegged stance. Although it is rare to see osteomalacia in adults, it can result from chronic steatorrhea, because vitamin D is fat soluble, and most will be lost in the feces. Renal rickets or renal osteodystrophy is a specific form of osteomalacia caused by chronic renal damage. In addition to the direct effects of renal damage on calcium and phosphate handling by the kidney and the endocrine effects of these derangements, the production of active vitamin D is reduced because PTH cannot convert 25-hydroxycholecalciferol to active 1,25-dihydroxycholecalciferol in the damaged kidney.

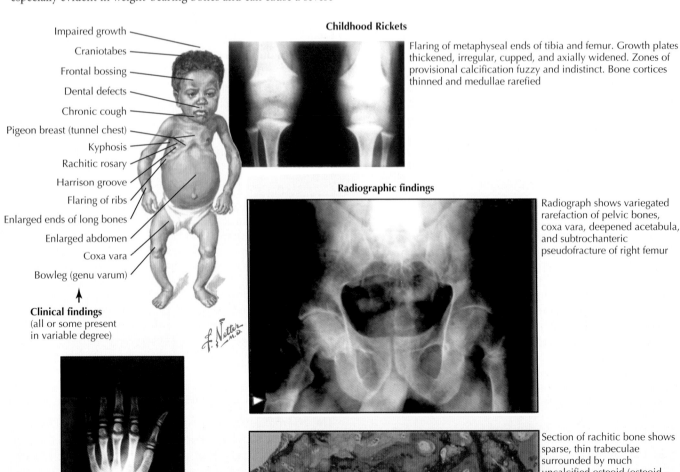

Impaired growth
Craniotabes
Frontal bossing
Dental defects
Chronic cough
Pigeon breast (tunnel chest)
Kyphosis
Rachitic rosary
Harrison groove
Flaring of ribs
Enlarged ends of long bones
Enlarged abdomen
Coxa vara
Bowleg (genu varum)

Clinical findings
(all or some present in variable degree)

Childhood Rickets

Flaring of metaphyseal ends of tibia and femur. Growth plates thickened, irregular, cupped, and axially widened. Zones of provisional calcification fuzzy and indistinct. Bone cortices thinned and medullae rarefied

Radiographic findings

Radiograph shows variegated rarefaction of pelvic bones, coxa vara, deepened acetabula, and subtrochanteric pseudofracture of right femur

Section of rachitic bone shows sparse, thin trabeculae surrounded by much uncalcified osteoid (osteoid seams) and cavities caused by increased resorption

Radiograph of rachitic hand shows decreased bone density, irregular trabeculation, and thin cortices of metacarpal and proximal phalanges. Note increased axial width of epiphyseal line, especially in radius and ulna

Rickets The main manifestation of vitamin D deficiency is in bone mineralization. As depicted in the illustration, in children, vitamin D deficiency results in rickets. In adults, vitamin D deficiency leads to osteomalacia.

Chapter 32

Hormones of the Reproductive System

Sex hormones are involved in fetal development and sexual differentiation, pubertal somatic changes, and reproductive physiology. Although the term **sex hormone** refers to sex steroids (primarily estrogen, progesterone, testosterone, and adrenal androgens), a wider array of hypothalamic, pituitary, gonadal, and placental hormones have important roles in the reproductive system. Endocrine problems involving the reproductive system are relatively common. For example, infertility, which is often a result of an endocrine abnormality, affects approximately 10% of the population.

FETAL DEVELOPMENT OF THE REPRODUCTIVE ORGANS AND DIFFERENTIATION OF GENITALIA

A person's sex can be characterized in terms of genetics (the male XY genotype vs. the female XX genotype), gonads (testes vs. ovaries), and phenotype (external appearance of "maleness" vs. "femaleness"). Although genetic sex is determined at the time of conception, the fetus remains undifferentiated in terms of gonadal sex for the first 5 weeks of development (Fig. 32.1).

Development of Gonadal Sex

Under the control of the *SRY* gene on the Y chromosome, the undifferentiated gonads develop into **testes**; full development of gonadal sex requires several other genes on the Y chromosome and various autosomal chromosomes. By 8 to 9 weeks of development, the **Leydig cells** of the testes begin to secrete **testosterone**. Development of **ovaries**, on the other hand, begins at week 9 of gestation and requires the presence of two X chromosomes. Germ cells within the ovaries give rise to **oogonia**, which proliferate and soon enter meiosis, leading to the **primary oocyte stage**. At this stage meiosis is arrested until it is activated during sexual cycles after puberty. The developing ovary produces **estrogens**.

During the undifferentiated stage of fetal development, two pairs of ducts—the **wolffian ducts** (also known as the **mesonephric ducts**) and the **müllerian ducts**—develop in both sexes (see Fig. 32.1). Under the influence of testosterone, the wolffian ducts develop into male structures, including the

epididymis, vas deferens, and seminal vesicles. Meanwhile, **müllerian inhibitory factor**, a dimeric glycoprotein hormone secreted by the **Sertoli cells** of the fetal testes, causes regression of the müllerian ducts. The combination of these two series of events leads to the development of the male system. In the absence of testes and their secretions, the müllerian ducts persist and develop, forming the fallopian tubes, uterus, and the upper vagina, whereas the wolffian ducts degenerate.

Synthesis and secretion of testosterone by the fetal testes is stimulated initially by a placental hormone known as **chorionic gonadotropin**—the same hormone that is essential in maintaining maternal ovarian progesterone production during pregnancy. After the first trimester, fetal pituitary **luteinizing hormone** (LH) is required to maintain androgen production in a male fetus. Similarly, in a female fetus, estrogens are secreted by the ovaries in response to pituitary LH. After birth, the pituitary gonadotrophs become quiescent until puberty. The role of gonadotropins in pubertal development, reproductive function, and pregnancy is discussed in the following text.

Development of Genital Sex

The external genitalia are also present in an undifferentiated form during early development (Fig. 32.2). Differentiation begins at 9 to 10 weeks of gestation. In the absence of androgens, female genitalia are produced. In the male fetus, testosterone produced by the testes and secreted into the circulation is converted to **dihydrotestosterone (DHT)** in the primitive genital structures; DHT stimulates the formation of the male genitalia. (Conversion of testosterone to DHT is not involved in the actions of testosterone on wolffian duct differentiation.)

Phenotypic Sex

Phenotypic sex is the outward appearance of maleness or femaleness. Whereas genetic sex is based on the presence of the XY or XX genotype, and gonadal sex is determined by specific genes on the X chromosome, as well as several autosomal genes, phenotypic sex develops in response to gonadal hormones. In the presence of testosterone, male genitalia develop; in its absence, the female form of genitalia prevails.

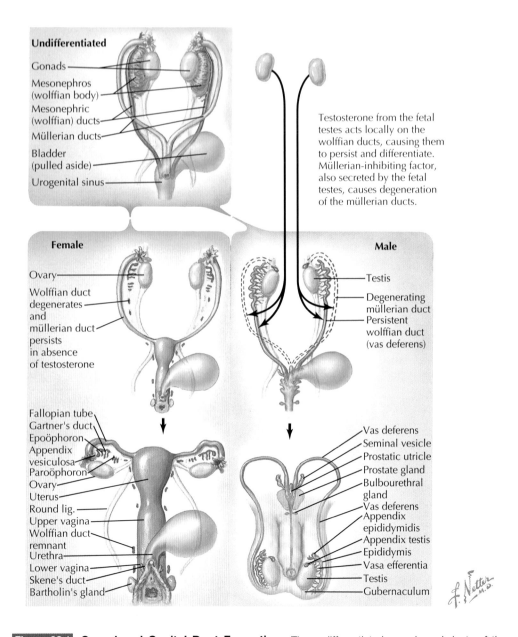

Undifferentiated

Gonads
Mesonephros
(wolffian body)
Mesonephric
(wolffian) ducts
Müllerian ducts
Bladder
(pulled aside)
Urogenital sinus

Testosterone from the fetal
testes acts locally on the
wolffian ducts, causing them
to persist and differentiate.
Müllerian-inhibiting factor,
also secreted by the fetal
testes, causes degeneration
of the müllerian ducts.

Female

Ovary
Wolffian duct
degenerates
and
müllerian duct
persists
in absence
of testosterone

Male

Testis
Degenerating
müllerian duct
Persistent
wolffian duct
(vas deferens)

Fallopian tube
Gartner's duct
Epoöphoron
Appendix
vesiculosa
Paroöphoron
Ovary
Uterus
Round lig.
Upper vagina
Wolffian duct
remnant
Urethra
Lower vagina
Skene's duct
Bartholin's gland

Vas deferens
Seminal vesicle
Prostatic utricle
Prostate gland
Bulbourethral
gland
Vas deferens
Appendix
epididymidis
Appendix testis
Epididymis
Vasa efferentia
Testis
Gubernaculum

Figure 32.1 **Gonad and Genital Duct Formation** The undifferentiated gonads and ducts of the early embryo differentiate into the male or female gonads and duct systems under the influence of various products encoded by the X and Y chromosomes. Notably, a product of the *SRY* gene of the Y chromosome results in differentiation of the gonads into testes. Production of testosterone by the testes results in persistence and differentiation of the wolffian ducts. Müllerian-inhibiting factor secreted by the testes causes müllerian duct degeneration. In the female fetus, in the absence of testosterone, the gonads develop into ovaries and the wolffian ducts degenerate.

Other phenotypic characteristics of femaleness are promoted by the absence of androgens and the presence of estrogen. In persons with androgen insensitivity syndrome, for example, functional androgen receptors are absent, and the female phenotype prevails despite an XY genotype.

PUBERTY

Beginning a year or two before the onset of puberty, production of androgens by the zona reticularis of the adrenal glands commences. The major adrenal androgens are **dehydroepiandrosterone** and **androstenedione**. This event is known as **adrenarche** and is distinct from puberty. Secretion of adrenal androgens results in the following characteristics:

- Appearance of pubic and axillary hair
- Adult body odor
- Acne and oiliness of skin

Hormonal Regulation of Puberty

The actual process of **puberty** is associated with maturation of the hypothalamus and anterior pituitary gland. The

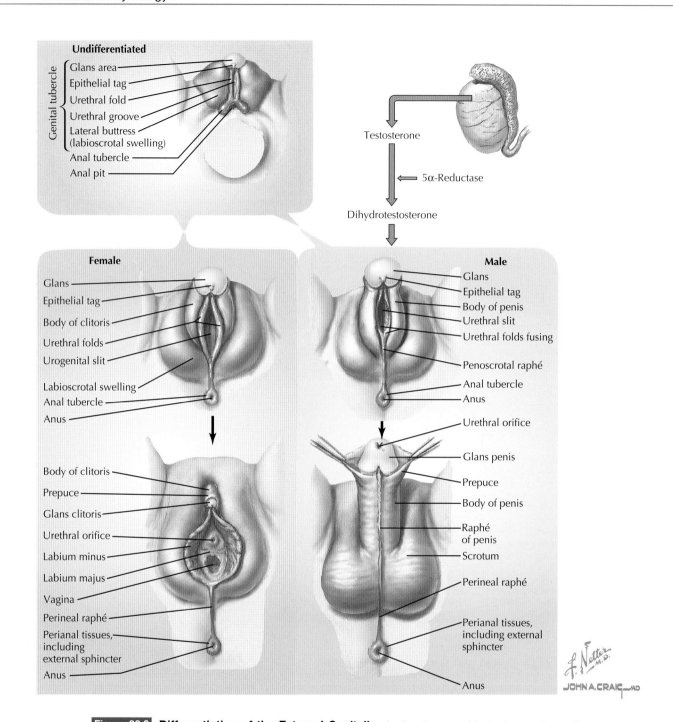

Figure 32.2 **Differentiation of the External Genitalia**   In the absence of testosterone, the undifferentiated external genitalia develop into the female structures. Testosterone, after conversion to dihydrotestosterone, stimulates the formation of male external genitalia from the undifferentiated structures. Homologies between male and female genital structures are color coded.

hypothalamic secretion of the decapeptide **gonadotropin-releasing hormone (GnRH)** increases and becomes pulsatile, as the hypothalamus becomes less sensitive to negative feedback by sex steroids. Thus a pulsatile pituitary secretion of the gonadotropins **follicle-stimulating hormone (FSH)** and **LH** is established, stimulating gonadal maturation and sex hormone synthesis. **Estradiol** (the major estrogen or "estrogenic hormone") and **progesterone** are the predominant sex hormones produced in women, whereas **testosterone** is the main sex steroid of men (the details of the hypothalamic-pituitary-gonadal axis are discussed later in this chapter). Adrenal androgens continue to have a physiological role in puberty and beyond in women, whereas the effects of adrenal androgens are overshadowed by those of testosterone in men. In males, some testosterone is converted to estradiol, and the latter hormone has important functions in male puberty. The following physiological and anatomic changes that occur at puberty are the result of various hormone actions:

- The gonads mature and sex steroids are synthesized and released as a result of increased gonadotropin secretion.
- Other sex organs and genitalia mature, mainly as a result of stimulation by estradiol in females and testosterone in males.
- Secondary sexual characteristics develop (i.e., characteristics of "maleness" or "femaleness" not directly associated with reproduction). These characteristics include gender-specific changes in distribution of body and facial hair, breast development, body fat distribution, muscle development, and pitch of voice. Estrogen is primarily responsible for female characteristics, whereas testosterone is the major stimulus for the development of male secondary sexual characteristics.
- The linear growth spurt occurs. This event is attributed mainly to estradiol in both sexes; estrogenic hormones cause accelerated growth of long bones but also promote the eventual closure of the epiphyseal plates of these bones, and thus the end of growth in height.

The pubertal growth spurt typically begins earlier in girls than in boys. It once was believed that androgens (testosterone and adrenal androgens) were responsible for the rapid increase in height. More recent research, however, has demonstrated that although testosterone increases bone mass and density, estradiol is the hormone primarily responsible for the growth spurt in both sexes. Estradiol stimulates long bone growth but also promotes closure of the growth plates. Females, who enter puberty earlier, attain less adult height than do males.

THE MENSTRUAL CYCLE AND FEMALE REPRODUCTIVE ENDOCRINOLOGY

In late puberty, adolescent girls experience their first menstruation, an event known as **menarche**, and **menstrual cycles** commence, continuing until **menopause** (unless they are interrupted by pregnancy). Menstrual cycles last an average of 28 days and consist of the following three phases (Fig. 32.3 and Video 32.1):

- The **follicular phase** is characterized by proliferation of the endometrium of the uterus and the development of ovarian follicles.
- The **ovulatory phase** is the phase during which one follicle that has fully matured ruptures and releases an ovum.
- The **luteal phase** is characterized by transformation of follicular cells into a **corpus luteum** and further proliferation of the endometrium. Unless implantation of a fertilized ovum takes place, the corpus luteum regresses and menses follows, during which the proliferated

endometrium is sloughed off and bleeding occurs for a period of 3 to 5 days.

Follicular Phase

The development of follicles within the ovary is illustrated in Figure 32.4. During each cycle, after the onset of menses (by convention, "day 1" of the cycle), several primordial ovarian follicles begin to undergo further development under the influence of FSH, and hence the term **follicular phase** is used to describe the first half of the cycle. **Theca interna** cells within the developing follicles are stimulated by LH to secrete **androgens**, which are converted to **estradiol** by the **granulosa cells** of the follicles. This conversion is stimulated by FSH. Estradiol causes endometrial proliferation, as well as development of glands and growth of spiral arteries within the endometrium, in preparation for possible implantation of a fertilized egg. For this reason, the follicular phase is also called the **proliferative phase**. In addition, estradiol promotes secretion of watery cervical mucus, through which sperm can enter the uterus. Ultimately one of the developing follicles predominates and becomes a **mature follicle** (also called a graafian follicle); the other follicles then regress.

During the follicular phase, these hormones are controlled by a negative feedback system (Fig. 32.5A). FSH and LH are secreted in a pulsatile pattern under the influence of pulsatile GnRH secretion by the hypothalamus. The estradiol produced by ovarian granulosa cells exerts negative feedback on the hypothalamic secretion of GnRH and anterior pituitary secretion of gonadotropins. Additionally, granulosa cells of developing follicles secrete a peptide hormone, **inhibin,** that has negative feedback effects, specifically on FSH.

Ovulatory Phase

Toward the end of the follicular phase, estradiol rises to a level at which **positive feedback** is triggered (Fig. 32.5B). A surge in LH and, to a lesser extent, FSH takes place and produces **ovulation** at mid cycle, releasing a mature ovum, which is carried by ciliary action into the fallopian tube (see Fig. 32.3). Interestingly, a mature ovum is produced in alternating ovaries from month to month, but if a woman has only one functional ovary, that one ovary will normally produce a mature ovum monthly.

Luteal Phase

In the ensuing luteal phase of the cycle, the ruptured follicle undergoes involution, forming the **corpus luteum.** Theca interna cells and granulosa cells become theca lutein and granulosa lutein cells, respectively; theca lutein cells continue to produce androgens while granulosa lutein cells produce progesterone, inhibin, and, to a lesser degree, estradiol. The system now reverts to negative feedback, with estrogens, progesterone, and inhibin exerting this feedback (see Fig. 32.5A). Further proliferative and secretory changes take place in the

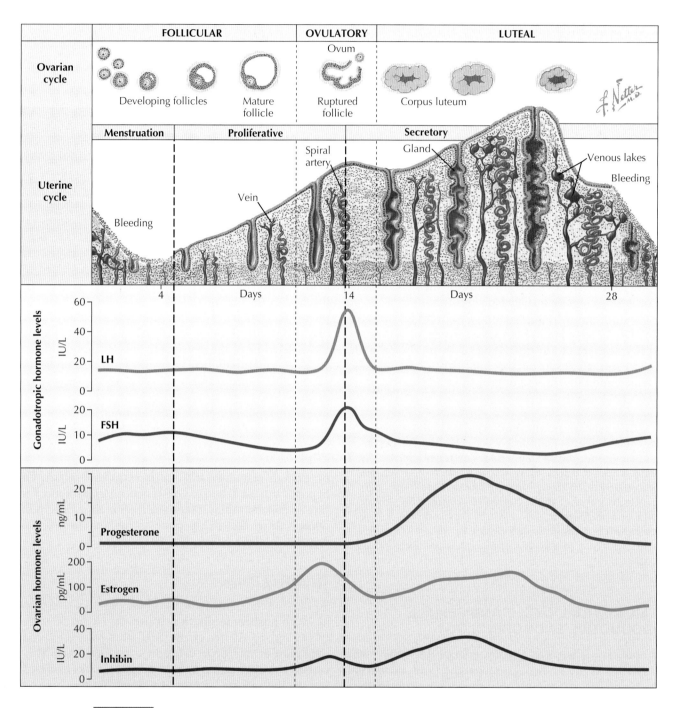

Figure 32.3 **Menstrual Cycle** During the female menstrual cycle, changes that are under the control of the hypothalamus and anterior pituitary gland take place in the ovaries and uterus. During the follicular phase, several primary follicles undergo further development in response to gonadotropins and synthesize androgens, which are converted to estradiol within the developing follicle. Ultimately, one follicle fully matures and the others regress. The uterine endometrium proliferates in response to estradiol. Near mid cycle, estradiol rises to a level that initiates *positive feedback*, and thus there is a surge in luteinizing hormone (LH) and follicle-stimulating hormone (FSH) release by the anterior pituitary, which results in ovulation. During the ensuing luteal phase, the mature follicle becomes the corpus luteum, which secretes progesterone and estradiol. The uterus undergoes further proliferative and secretory changes. Unless pregnancy occurs, endometrial sloughing and menstruation eventually occur, marking the beginning of a new cycle.

endometrium, stimulated by progesterone; the luteal phase is also called the **secretory phase** for this reason. In the cervix, secretions become thicker, making passage of sperm into the uterus more difficult. Conception must take place within a day or two of ovulation, because the ovum is viable for only a short period after release from the graafian follicle (normally, conception takes place while the egg is in transport within a fallopian tube). Toward the end of the luteal phase, unless pregnancy occurs, steroid and inhibin secretion fall and menses results.

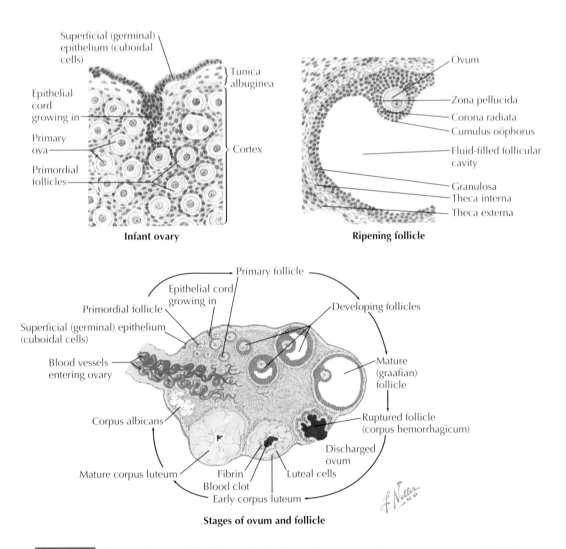

Figure 32.4 **Ovary, Ova, and Follicles** Until puberty, the ovary contains numerous primordial follicles that remain in a dormant state. After puberty, several follicles begin ripening with each menstrual cycle, in stages illustrated in the bottom panel. Only one follicle becomes a mature follicle; the others ultimately regress. After ovulation and release of the ovum, the mature follicle involutes to form the corpus luteum, which persists to the end of the cycle.

Inhibin and activin are related, dimeric proteins with opposite effects on FSH secretion. Inhibin is synthesized by Sertoli cells of the testes and granulosa cells of the ovary and plays a part in the negative feedback on FSH synthesis and secretion by the pituitary gland (see Fig. 32.5). Activin, on the other hand, stimulates FSH synthesis and secretion. Both inhibin and activin are produced by other organs and tissues and have additional functions. For example, activin has roles in wound repair and developmental processes. Inhibin is measured in serum as part of some diagnostic tests; for example, elevated inhibin (specifically inhibin A), along with other changes in maternal serum, may suggest Down syndrome in a fetus.

IMPLANTATION AND PREGNANCY

As discussed, fertilization of the ovum takes place within the fallopian tube. Once the sperm penetrates the ovum, the second meiotic division is completed (all oocytes are arrested in the first meiotic division until a woman reaches puberty; the "mature ovum" released during each cycle has undergone further development and is arrested in the second meiotic division until sperm penetration). Subsequently, mitotic division of the fertilized egg begins and a blastocyst is formed. About 5 days after ovulation, the blastocyst begins to implant in the endometrial lining of the uterus. The sequence of events from follicular maturation to implantation of the blastocyst is illustrated in Figure 32.6.

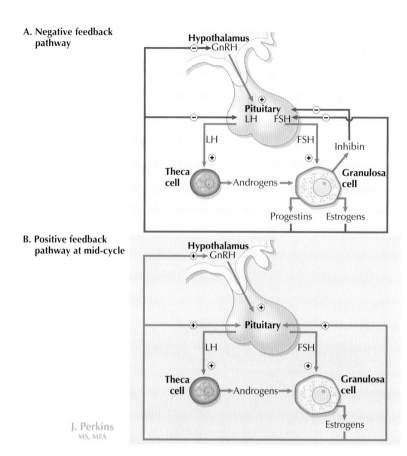

A. Negative feedback pathway

B. Positive feedback pathway at mid-cycle

J. Perkins
MS, MFA

Figure 32.5 **Hormonal Regulation of the Menstrual Cycle** The hypothalamic-pituitary-ovarian axis is characterized by both positive and negative feedback over the course of a menstrual cycle. Initially, gonadotropin-releasing hormone (GnRH) stimulates release of luteinizing hormone (LH) and follicle-stimulating hormone (FSH) by the pituitary; estrogen synthesized by developing ovarian follicles has negative feedback effects on the axis (**A**). However, in the late follicular phase, blood estradiol reaches a high level that initiates positive feedback (**B**) and a surge in LH and FSH release, provoking ovulation. In the luteal phase, the system is again characterized by negative feedback (**A**). Estradiol, progesterone, and inhibin produced by the corpus luteum have negative feedback actions on gonadotropin release.

CLINICAL CORRELATE 32.1
The Rhythm Method and Oral Contraceptives

The average menstrual cycle is 28 days in length (beginning with the first day of menses), but the cycle length may vary both between women and for an individual woman. Ovulation takes place at about mid cycle, but more precisely, it occurs *14 days before the onset of the next menses*. In other words, in a 26-day cycle, ovulation takes place on day 12, but in a 30-day cycle, ovulation takes place on day 16. Sperm are viable in the female reproductive tract for a few days, whereas mature ova are viable for only a short time after release.

For pregnancy to occur, intercourse needs to take place between approximately 5 days before ovulation and at most a day after ovulation. However, because it is not possible to predict exactly when ovulation will take place (because the date of onset of the next menses cannot be predicted with certainty), "rhythm methods" of contraception have low reliability compared with most forms of birth control. In various rhythm methods, intercourse is avoided for several days before and after the predicted date of ovulation. For example, in the Standard Days Method, women whose cycles are between 26 and 32 days in length avoid intercourse between days 8 and 19 of their cycles. In general, the rhythm method has a failure rate of several percent annually when used perfectly, and up to 25% otherwise.

With the use of oral contraceptives, most commonly a combination of an estrogen and a progestin (i.e., a progesterone-like drug), cycles are controlled by the drug. Typically, pills that contain hormones are taken for 21 days, followed by a week of no pills or daily sugar pills. Pregnancy is prevented primarily by inhibition of ovulation as a result of inhibition of gonadotropin release by the oral hormones. Bleeding (menstruation) occurs upon withdrawal from the hormones after 21 days. With perfect use, the annual rate of conception is 0.3%, although the failure rate is actually several percent annually.

CLINICAL CORRELATE 32.2
Abnormal Uterine Bleeding

Normal menstrual bleeding follows the luteal phase of the cycle and is caused by the fall in gonadal hormone levels. Heavy bleeding during menses (menorrhagia) or irregular bleeding during the cycle may be caused by a hormonal imbalance. Abnormal uterine bleeding can also be caused by a variety of disorders of the uterus or adnexa (ovaries and fallopian tubes), as well as systemic disorders in nonpregnant, reproductive-age women. In pregnant women, bleeding may be caused by rupture of a placental blood vessel or may foreshadow an impending miscarriage. Placenta previa, in which the placenta extends over the cervix, is often associated with bleeding after the first trimester of pregnancy. Various causes of abnormal uterine bleeding are illustrated.

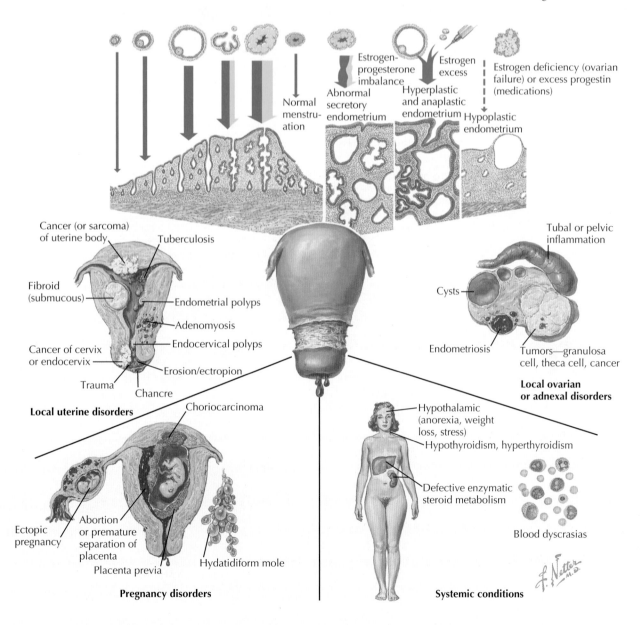

Causes of Abnormal Uterine Bleeding Abnormal uterine bleeding is associated with a variety of disease processes and disorders.

The **placenta** is formed from **trophoblast** cells of the blastocyst and **decidual cells** of the endometrial lining. **Trophoblasts** secrete a hormone known as **human chorionic gonadotropin (HCG)**, which has LH-like actions. HCG "rescues" the corpus luteum from the regression that would otherwise occur at the end of the menstrual cycle, such that synthesis of estradiol and progesterone continues, and thus the proliferated, secretory state of the endometrium is maintained to support the pregnancy. HCG is important in this regard for the first trimester of pregnancy. By the second trimester, the placenta itself secretes large quantities of **progesterone** and **estrogens** (primarily **estriol**) to support the uterus.

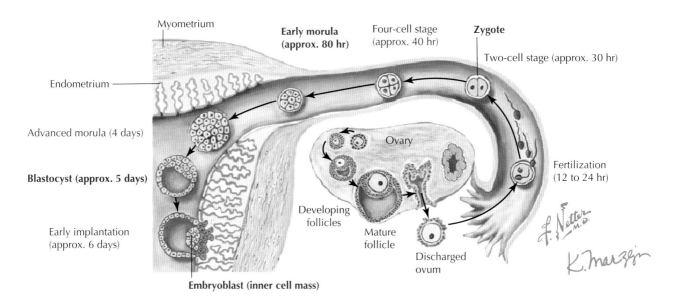

Figure 32.6 **Fertilization and Implantation** Upon rupture of the graafian follicle, the ovum is swept into the fallopian tube. If fertilization occurs, it takes place within the fallopian tube, which transports the ovum or zygote to the uterus. A zygote will have reached the blastocyst stage by day 5 and will implant in the endometrial lining at that time.

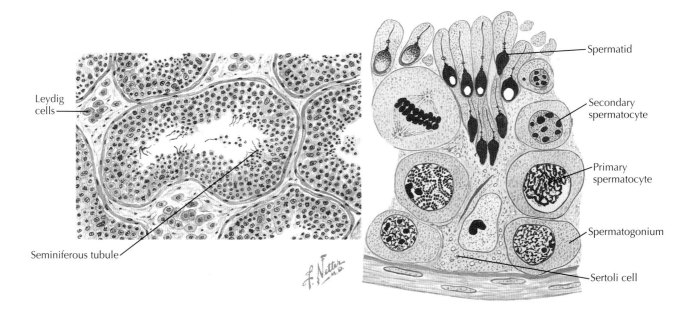

Figure 32.7 **The Testes and Spermatogenesis** The testes contain convoluted seminiferous tubules, where spermatogenesis takes place, and, between the tubules, Leydig cells, which synthesize testosterone in the mature male *(left panel)*. Sertoli cells constitute the epithelium of the tubules *(right panel)*. Differentiation of primary spermatocytes to sperm cells begins between the Sertoli cells and is completed in the epididymis.

MALE REPRODUCTIVE ENDOCRINOLOGY

The testes are the site of gametogenesis and steroidogenesis in males. Spermatogenesis takes place within the convoluted seminiferous tubules of the testes, whereas the **Leydig cells** located between tubules produce testosterone (Fig. 32.7). The hormone **inhibin** is synthesized and released by Sertoli cells in the lining of the seminiferous tubules. Sertoli cells provide the structure upon which male germ cells develop and secrete fluid that supports flow of sperm through the seminiferous tubules to the epididymis, where further sperm maturation takes place, and sperm are stored until ejaculation. Sertoli cell processes form tight junctions that provide a "blood-testis" barrier, sheltering the developing spermatocytes from potential damage by blood-borne substances.

Most pregnancy tests, including the widely available self-administered tests, are based on measurement of HCG. Although blood tests are more sensitive and may detect pregnancy earlier, urine tests are capable of detecting HCG within a few days of implantation of the blastocyst. Serial measurement of HCG also can be useful in monitoring early progress of a pregnancy when risk of ectopic pregnancy or miscarriage is an issue, because HCG normally rises rapidly after successful implantation.

Spermatogenesis is a complex process that begins with mitotic division of spermatogonia to form primary spermatocytes, along with additional spermatogonia. (Whereas women have a fixed number of primary oocytes by the time of birth, the number of germ cells in men is not fixed). Meiosis of primary spermatocytes produces secondary spermatocytes and, subsequently, haploid spermatids, which undergo further differentiation to become spermatozoa. The final product, the mature spermatozoon, includes a head piece, a mitochondria-rich middle section for energy production, and a motile tail section. The head piece has a prominent acrosome containing proteolytic enzymes necessary for penetration of an ovum.

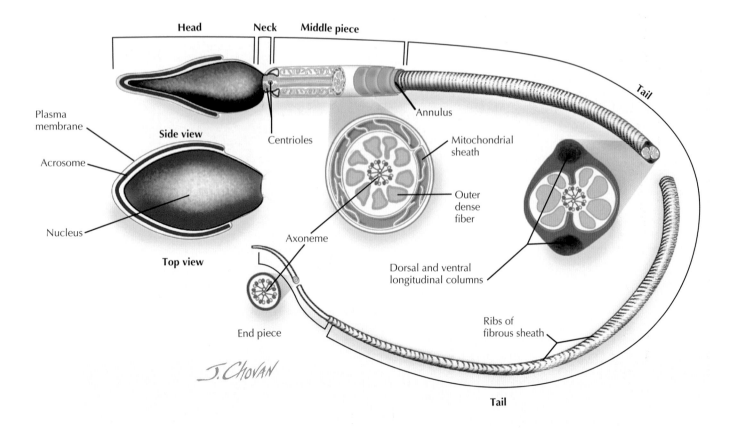

Mature Spermatozoon The morphologic features of mature spermatozoa reflect their motility and ovum-penetrating functions.

Endocrine Regulation of Testicular Function

Testicular function (both gametogenesis and steroidogenesis) is regulated by the hypothalamus and pituitary gland (Fig. 32.8). GnRH is secreted by hypothalamic nuclei into the hypophyseal portal circulation (as in females, GnRH release is pulsatile in males) and subsequently stimulates the release of LH and FSH by anterior pituitary gonadotrophs. LH stimulates the first step in testosterone synthesis (conversion of cholesterol to Δ5-pregnenolone by CYP11A10), whereas FSH acts on Sertoli cells, stimulating the synthesis of **androgen-binding protein**, which subsequently binds testosterone, promoting spermatogenesis in the seminiferous tubules (see Fig. 32.7). Negative feedback in the hypothalamic-pituitary-testis axis occurs by inhibition of GnRH and LH release by testosterone and by inhibition of FSH secretion via inhibin produced by Sertoli cells.

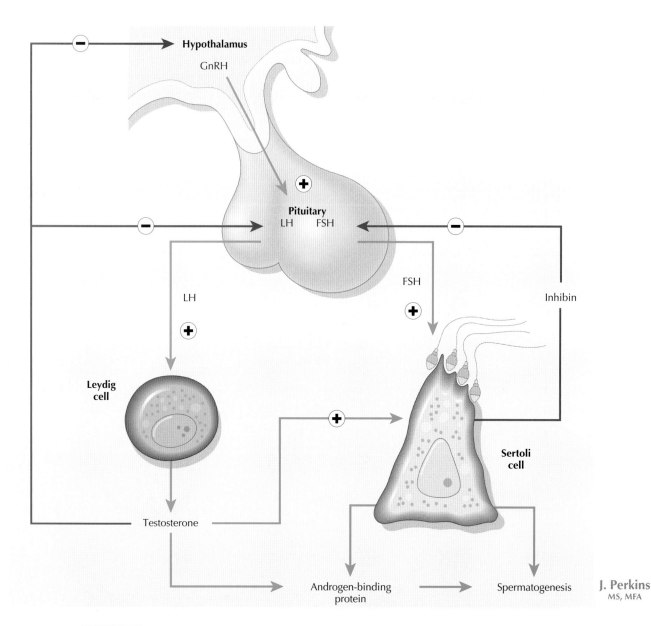

Figure 32.8 **Control of Testicular Function** Gonadotropin-releasing hormone (GnRH) secreted by the hypothalamus stimulates luteinizing hormone (LH) and follicle-stimulating hormone (FSH) secretion by the anterior pituitary. LH stimulates testosterone synthesis by the Leydig cells of the testes, whereas testosterone and FSH are required for spermatogenesis. FSH induces production of androgen-binding protein; androgen is concentrated in the tubules by binding to androgen-binding protein, promoting spermatogenesis. Sertoli cells also produce inhibin, which, along with testosterone, exerts negative feedback effects on the axis (inhibin specifically inhibits FSH secretion).

Nonreproductive Actions of Testosterone

In addition to the previously discussed nonreproductive actions of testosterone at puberty, increased muscle mass, development of the male pattern of hair distribution and baldness, and deepening of the voice are also actions of testosterone. Some of the actions of testosterone are dependent on its conversion to **dihydrotestosterone** (e.g., genital differentiation, prostate development and growth, male hair distribution, and baldness), whereas other actions are direct effects of testosterone.

Sex hormone–binding globulin and **androgen-binding protein** are identical proteins that are produced in the liver and testes, respectively. Sex hormone–binding globulin circulates in the bloodstream and is the major carrier for testosterone and estradiol in the blood (only a small fraction of sex hormones circulates as free hormone). Androgen-binding protein is produced by the testes and serves to concentrate testosterone within the seminiferous tubules, where androgens promote spermatogenesis.

CLINICAL CORRELATE 32.4
Klinefelter's Syndrome

Male infertility has a large number of causes, including blockage of the vas deferens as a result of infection or injury; low sperm count (oligospermia) as a result of infections such as mumps and some sexually transmitted diseases; lack of sperm production as a result of chronic disease, hormonal problems, or injury to the testes; impotence; and exposure to toxins. In some cases, infertility may have a genetic basis. For example, Klinefelter's syndrome is characterized by the genotype 47,XXY. Persons with this syndrome are phenotypic males but have some identifiable physical characteristics, including gynecomastia (breast enlargement), and they are nearly always infertile (see illustration). Persons with this syndrome are hypogonadal, meaning that gonadal hormone levels are low, and as a result, gonadotropin levels are elevated and testicular size is reduced. Klinefelter's syndrome affects 1 to 2 per 1000 men.

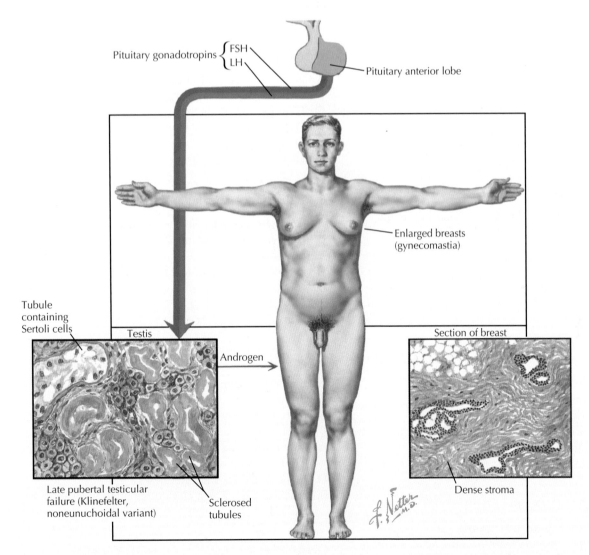

Pituitary gonadotropins { FSH LH

Pituitary anterior lobe

Enlarged breasts (gynecomastia)

Tubule containing Sertoli cells

Testis

Androgen

Section of breast

Late pubertal testicular failure (Klinefelter, noneunuchoidal variant)

Sclerosed tubules

Dense stroma

Nuclear chromatin often positive (female); usually XXY chromosomal pattern but XXXY, XXXXY, XXYY, and mosaic patterns have been described

Testicular Failure in Persons With Klinefelter's Syndrome The presence of two X chromosomes (47,XXY genotype) results in seminiferous tubular dysgenesis and infertility, as well as primary hypogonadism (i.e., low testicular hormone levels, and as a result, high gonadotropin levels). *FSH,* follicle-stimulating hormone; *LH,* luteinizing hormone.

Review Questions

CHAPTER 27: GENERAL PRINCIPLES OF ENDOCRINOLOGY AND PITUITARY AND HYPOTHALAMIC HORMONES

1. Which of the following hormones initiates its actions at target tissues by binding to nuclear receptors?

A. Adrenocorticotropic hormone
B. Vasopressin
C. Epinephrine
D. Insulin-like growth factor–1
E. Thyroid hormone

2. Hormonal actions are regulated by:

A. affinity of receptors for the hormone.
B. the number of receptors for the hormone.
C. the degree of binding of hormone to plasma proteins.
D. the concentration of hormone in the blood.
E. all of the above.

3. Vasopressin (antidiuretic hormone) secretion by the posterior pituitary is stimulated by:

A. decreased body fluid osmolarity.
B. decreased blood volume.
C. atrial natriuretic peptide.
D. ethanol.
E. high blood pressure.

4. The hormone oxytocin:

A. is released directly into the general circulation.
B. is synthesized in the posterior pituitary gland.
C. promotes uterine quiescence.
D. has a well-established physiological role in the induction and progression of labor.
E. is a 191–amino acid peptide.

5. Growth hormone secretion:

A. occurs in pulses, mainly during sleep.
B. is stimulated by a specific hypothalamic neurohormone.
C. is inhibited by somatostatin.
D. results in synthesis of insulin-like growth factor in a variety of target tissues.
E. all of the above.

CHAPTER 28: THYROID HORMONES

6. Thyroid hormone secretion is:

A. inhibited by elevated plasma estrogen levels.
B. stimulated by high plasma triiodothyronine (T_3).
C. inhibited by high thyroxine-binding globulin (TBG).
D. stimulated by high plasma thyroxine (T_4).
E. stimulated by thyroid-stimulating hormone (TSH).

7. In adults, hypothyroidism is associated with which phenotypes?

A. Hypertension, cold intolerance, diarrhea
B. Hypertension, nonpitting edema, cold intolerance
C. Hypotension, higher than normal temperature, diarrhea
D. Hypotension, higher than normal temperature, nonpitting edema
E. Hypotension, nonpitting edema, cold intolerance

8. In the plasma, thyroid hormone:

A. is present primarily in the T_3 form.
B. is present primarily in the T_3 form and is bound to proteins.
C. is completely bound to TBG.
D. is present primarily in the T_4 form and is mostly bound to proteins, including TBG.
E. is present only as free T_3 and T_4.

9. Feedback regulation of thyroid hormone is illustrated by:

A. high circulating levels of free T_3 and T_4, which elevate thyrotropin-releasing hormone (TRH).
B. high circulating levels of free T_3 and T_4, which elevate TSH.
C. high circulating levels of free T_3 and T_4, which reduce TRH.
D. reduced TRH, which elevates TSH.
E. reduced TSH, which elevates circulating levels of free T3 and T4.

CHAPTER 29: ADRENAL HORMONES

10. Stimulation of the adrenal gland by adrenocorticotropic hormone has the most direct effect on release of:

A. epinephrine.
B. norepinephrine.
C. cortisol.
D. androgens.
E. aldosterone.

11. Which portion of the adrenal gland secretes a hormone with antiinflammatory actions?

A. The medulla
B. The zona reticularis
C. The zona fasciculata
D. The zona glomerulosa
E. None of the above

12. Which of the following is a significant action of adrenal androgens in females?

A. Stimulation of libido
B. Hypertrophy of sebaceous glands
C. Development of pubic hair
D. Development of axillary hair
E. All of the above

13. Aldosterone secretion by the adrenal gland is stimulated by:

A. hypokalemia.
B. high plasma Na$^+$ concentration.
C. angiotensin II.
D. atrial natriuretic peptide.
E. all of the above

14. Which of the following is not a typical symptom or sign of Addison's disease?

A. Poor stress tolerance
B. Hyperglycemia
C. Fatigue and weight loss
D. Hyperpigmentation of skin
E. Low blood pressure

CHAPTER 30: THE ENDOCRINE PANCREAS

15. Insulin is necessary for:

A. entry of glucose into tissues via the glucose transporter 1 (GLUT1) transporter.
B. entry of glucose into all tissues.
C. entry of glucose into the brain.
D. entry of glucose into the pancreas.
E. entry of glucose into tissues via GLUT4 transporters.

16. Regulation of insulin release by increased plasma glucose involves which sequence of steps?

A. Entry of glucose into pancreatic β cells through GLUT1 transporters; increased ATP, which decreases K$^+$ efflux; depolarization of the β cell; Ca^{2+} influx; insulin granule release
B. Entry of glucose into pancreatic β cells through GLUT1 transporters; hyperpolarization of the β cell; Ca^{2+} efflux; insulin granule release
C. Entry of glucose into pancreatic β cells through GLUT2 transporters; increased ATP, which decreases K$^+$ efflux; depolarization of the β cell; Ca^{2+} influx; insulin granule release
D. Entry of glucose into pancreatic β cells through GLUT2 transporters; hyperpolarization of the β cell; Ca^{2+} efflux; insulin granule release
E. Entry of glucose into pancreatic β cells through GLUT4 transporters; depolarization of the β cell; Ca^{2+} influx; insulin granule release

17. Factors that stimulate insulin secretion include:

A. glucagon-like peptide-1.
B. amino acids.
C. glucose insulinotropic peptide.
D. fatty acids.
E. all of the above.

18. Insulin is a hypoglycemic hormone because of its ability to:

A. increase gluconeogenesis.
B. reduce glycogen storage in muscle.
C. increase glycogenolysis.
D. increase GLUT1 transporters in brain tissue membranes.
E. increase GLUT4 transporters in adipose tissue and skeletal and cardiac muscle membranes.

CHAPTER 31: CALCIUM-REGULATING HORMONES

19. Decreased plasma calcium levels result in:

A. reduced production of 1,25-dihydroxycholecalciferol.
B. reduced production of 25-hydroxycholecalciferol.
C. increased bone mineralization.
D. increased bone resorption.
E. increased plasma calcitonin.

20. Parathyroid hormone:

A. is released into plasma in response to increases in plasma calcium.
B. acts at the kidneys to increase calcium reabsorption and decrease phosphate reabsorption.
C. acts at the kidneys to decrease calcium reabsorption at the distal tubule.
D. acts at the bone to increase bone mineralization.
E. acts at the bone to decrease osteoclastic osteolysis.

21. Primary hyperparathyroidism is often associated with:

A. development of renal calculi.
B. bone pain.
C. constipation.
D. fatigue.
E. all of the above.

22. Select the **TRUE** statement about plasma calcium.

A. Plasma calcium is always bound to albumin.
B. Ionized calcium in the plasma increases when plasma phosphate is elevated.
C. Plasma calcium is decreased in acidosis.
D. Reductions in plasma protein will increase total plasma calcium.
E. Plasma calcium is elevated by parathyroid hormone.

CHAPTER 32: HORMONES OF THE REPRODUCTIVE SYSTEM

23. During fetal development, the formation of male genitalia is directly stimulated by:

A. testosterone.
B. dihydrotestosterone.
C. dehydroepiandrosterone.
D. androstenedione.
E. estrone.

24. The pubertal growth spurt is mainly stimulated by the hormone:

A. calcitonin.
B. vitamin D.
C. estradiol.
D. testosterone.
E. dihydrotestosterone.

25. Conversion of androgens to estradiol by the follicular granulosa cells is stimulated by which hormone?

A. Luteinizing hormone
B. Follicle-stimulating hormone
C. Oxytocin
D. Growth hormone
E. Prolactin

26. During the prepubertal period, germ cells in women are arrested at which stage of gametogenesis?

A. Oogonia
B. Primary oocyte
C. Secondary oocyte
D. Ovum
E. Zygote

27. The hormone inhibin has direct negative feedback effects on the release of:

A. prolactin.
B. follicle-stimulating hormone.
C. luteinizing hormone.
D. testosterone.
E. gonadotropin-releasing hormone.

Answers

SECTION 1: CELL PHYSIOLOGY, FLUID HOMEOSTASIS, AND MEMBRANE TRANSPORT

1. **B.** Antipyrine diffuses throughout the body water and can be used to measure the volume of total body water; inulin does not enter cells and can be used to measure the ECF volume. Using the indicator-dilution formula, the amount of indicator injected divided by the plasma concentration will result in the volume of the compartment. By this method, the total water (indicated by antipyrine) is 36 L, and the ECF (indicated by inulin) is 12 L; the ICF is 24 L. The ECF (12 L) is one-fourth plasma (3 L) and three-fourths ISF (9 L).

2. **D.** The Starling forces favoring movement out of the capillaries are the capillary hydrostatic pressure (HP_c) and interstitial oncotic pressure (π_i), whereas the forces favoring movement into the capillaries are the interstitial hydrostatic pressure (HP_i) and capillary oncotic pressure (π_c). Thus, $(30 + 8) - (3 + 28) = 7$ mm Hg out of the capillary.

3. **D.** Water will distribute through all compartments; thus both ECF and ICF volumes will increase. With the additional volume, the osmolarity of the compartments will decrease.

4. **A.** The red blood cells will only contract if placed in a hyperosmotic solution; 300 mM NaCl is a 600 mOsm/L solution.

5. **D.** An 80-kg person will have 48 L of TBW. Before expansion, ECF was 16 L and ICF was 32 L. The NaCl solution will remain in the ECF, increasing the ECF compartment to 18 L.

6. **C.** Voltage-gated channels are highly selective, and gating depends on membrane voltage.

7. **D.** The basolateral Na^+/K^+ ATPase, or "sodium pump," uses primary active transport to pump three sodium ions out of cells in exchange for two K^+ ions. The low intracellular Na^+ concentration facilitates apical sodium entry down its concentration gradient through several different secondary active transport symporters (e.g., Na^+-amino acid, Na^+-glucose) and antiporters (e.g., Na^+/Ca^{2+}, Na^+/H^+).

8. **B.** Ouabain and digoxin are cardiac glycosides that inhibit Na^+/K^+ ATPase, allowing equilibration of sodium and potassium across the cell membrane. This stops sodium-dependent transport processes and depolarizes the membrane potential.

9. **C.** The defect in cystic fibrosis is in the chloride channel that transports chloride out of cells (cystic fibrosis transmembrane regulator, *CFTR*).

10. **B.** Smooth muscle contraction occurs following the influx of calcium ions, which bind to calmodulin. The complex then activates myosin kinase, allowing interaction between actin and myosin, and contraction.

SECTION 2: THE NERVOUS SYSTEM AND MUSCLE

1. **A.** Reduced permeability of sodium ion will result in hyperpolarization of the cell. Ordinarily, sodium leakage into the cell contributes to the actual resting membrane potential, making it less negative than would be predicted on the basis of potassium concentration and permeability. Reduction in potassium ion permeability, on the other hand, would cause depolarization of the cell, as would influx of calcium ion or increased extracellular concentration of sodium or potassium ion.

2. **C.** Neuronal action potentials result when the threshold potential is reached and voltage-gated sodium channels open. Although action potentials are evoked in some excitable cells by opening of Ca^{2+} channels, opening of voltage-gated Na^+ channels is responsible for the phase 0 upstroke of the neuronal action potential.

3. **D.** Myelination of an axon results in decreased capacitance and increased membrane resistance. As a result, current travels through the interior of the axon but not through the membrane. Myelination produces a large increase in conduction velocity.

4. **A.** When axonal depolarization reaches the synaptic bouton, voltage-gated Ca^{2+} channels are opened, resulting in influx of Ca^{2+} and vesicular release of neurotransmitter into the synaptic cleft.

5. **D.** A single, large Na^+ and K^+ channel conducts the depolarizing charge during an end plate potential, whereas the neuronal action potential involves multiple ion channels and depolarization is mainly produced by Na^+ influx. Depolarization is rapid in both end plate and action potentials. Depolarization to +40 mV, causation by voltage-gated channels, and repolarization associated with increased K^+ conductance are all characteristics of neuronal action potentials but not end plate potentials.

6. **E.** The neural tissue of the spinal cord is covered by three membranes (meninges). These are the inner pia mater, the middle arachnoid membrane, and the outer dura mater.

7. **B.** Cerebrospinal fluid is produced by the choroid plexus and is reabsorbed at the arachnoid granulations into the venous system as well as into the capillaries of the CNS and pia mater.

8. **E.** Association pathways functionally and anatomically linking the two cerebral hemispheres include the corpus callosum and anterior, posterior, and hippocampal commissures.

9. **C.** Although the pons is also involved in regulation of breathing, the medulla is the major site of autonomic regulation, including regulation and integration of cardiovascular and respiratory functions. The thalamus and hypothalamus are not part of the brainstem.

10. **A.** The medulla oblongata, part of the brainstem, is anatomically continuous with the spinal cord.

11. **A.** Meissner's corpuscles, located in the dermal papillae, especially in the fingertips, palms, soles, lips, face, tongue, and genital (nonhairy) skin, are rapidly adapting receptors with small receptive fields that allow point discrimination and detection of low-frequency stimuli such as flutter. Pacinian corpuscles respond to rapid changes in pressure and vibration; Merkel's disks respond to pressure and touch, especially indentation of the skin; hair follicle receptors respond to movement across the skin surface; and Ruffini's corpuscles respond to stretch of skin and joints.

12. **B.** Photoreceptors respond to light reception through a sequence of events that ultimately results in reduced Na^+ permeability and hyperpolarization of their cell membranes, which inhibits neurotransmitter release by the photoreceptors.

13. **E.** The three ducts of the membranous labyrinth of the cochlea of the inner ear are the scala vestibuli, scala tympani, and scala media. The scala media contains endolymph, which resembles intracellular fluid in its composition. The other two ducts contain perilymph, which resembles extracellular fluid in its composition.

14. **E.** The pattern of pressure changes between the three perpendicularly oriented semicircular canals during angular acceleration of the head depends on the direction of head movement. The bending of cilia as a result of these pressure changes produces hyperpolarization or depolarization and changes in neurotransmitter release by hair cells. As a result, the firing rate of afferent nerve fibers ending on the hair cells is altered. These signals are carried through cranial nerve VIII to the vestibular nuclei of the pons.

15. **C.** Vibration of the tympanic membrane is conducted to the malleus, incus, and stapes (the small bones of the middle ear known collectively as the *ossicles*). The stapes is connected directly to the oval window and transmits sounds to it.

16. **A.** When the patellar tendon of the knee is tapped, muscle spindles in the quadriceps muscle are stretched, producing a reflex in which type Ia afferent nerves are depolarized and conduct the signal to the spinal cord. The nerves synapse with α-motor neurons, which conduct the signal back to the quadriceps, causing contraction. This stretch reflex is a monosynaptic reflex. Golgi tendon reflex is a bisynaptic reflex involved in preventing muscle damage when muscle tension is excessive. The flexor withdrawal reflex is a polysynaptic reflex that occurs in response to painful stimuli.

17. **E.** The corticospinal tract carries fibers that originate in the primary motor cortex, premotor and supplemental motor areas, and somatosensory areas posterior to the motor cortex. It is the most important pathway for fine motor activity controlled by the cortex. The other tracts mentioned are involved in other aspects of motor function, including brainstem control of balance and posture.

18. **C.** The cerebellum coordinates and fine-tunes motion. All the efferent signals from the cerebellar cortex are inhibitory, originating from the Purkinje cells. This inhibitory output is regulated by excitatory input from climbing fibers and mossy fibers by way of the granule cells as well as inhibitory input from interneurons in the cerebellar cortex.

19. **B.** Like the cerebellum, the basal ganglia have an accessory role to the motor cortex. With input from the motor cortex and output to the motor cortex through the thalamus, the basal ganglia function to produce smooth movement and regulate posture.

20. **B.** The superior colliculus of the mesencephalon receives input from the visual nuclei. Through the tectospinal tract, it is involved in control of muscles of the head, neck, and eye.

21. **B.** Preganglionic nerves of the sympathetic nervous system originate in the intermediolateral and intermediomedial cell columns of the thoracolumbar spinal cord (T1 to L3). Their axons extend to the chain ganglia, where they synapse with postganglionic nerves. Preganglionic nerves of the parasympathetic nervous system have cell bodies in the brainstem.

22. **D.** Postganglionic neurons of the parasympathetic nervous system release acetylcholine, which acts at muscarinic receptors at the neuroeffector junction. Postganglionic neurons of the sympathetic nervous system primarily release norepinephrine and epinephrine, which act at a variety of adrenergic receptors on effector organs and tissues. Both sympathetic and parasympathetic preganglionic nerves achieve ganglionic neurotransmission by release of acetylcholine, which acts at nicotinic receptors.

23. **C.** β_2-adrenergic receptors mediate sympathetic bronchial dilation. Constriction of vascular smooth muscle and mydriasis are produced by norepinephrine and epinephrine release by postganglionic sympathetic nerves and subsequent binding to α_1 receptors; increased contractility is mediated by β_1 receptors. Sweat gland secretion is stimulated by atypical sympathetic postganglionic nerve fibers that secrete acetylcholine.

24. **E.** Sweat gland secretion in most areas is a result of acetylcholine release by sympathetic neurons. Acetylcholine acts at muscarinic receptors to cause secretion.

SECTION 3: CARDIOVASCULAR PHYSIOLOGY

1. **E.** Effective hemostatic function requires normally functioning platelets as well as coagulation factors. Vitamin K is required for synthesis of several of the blood coagulation factors.

2. **C.** Hematocrit is the volume percentage of red blood cells in whole blood; 1.6 L is 40% of 4.0 L.

3. **B.** Blood monocytes differentiate to become macrophages when they leave the blood stream and enter the tissues.

4. **A.** Serum is prepared by removing clotting proteins from blood plasma.

5. **B.** The systemic veins normally contain more than 60% of the blood volume and can act as a reservoir for blood. Blood can be mobilized from veins during hemorrhage or dehydration as a compensatory measure to help maintain blood pressure until blood volume can be restored.

6. **C.** Contraction of the left ventricle consists of powerful constriction of the chamber with shortening of the heart along the base-to-apex axis, resulting in the high pressures required to pump blood through the high-resistance systemic circulation.

7. **A.** The coronary circulation receives approximately 4% of the cardiac output at rest; the other parts of the circulation each receive approximately 15% (brain) to 24% (liver and gastrointestinal tract) of the circulation.

8. **C.** The lungs receive all the output of the right ventricle.

9. **E.** Following the plateau of the action potential, gradual inactivation of L-type Ca^{2+} channels leads to activation of K^+ channels and rapid repolarization. Opening of Na^+ channels is responsible for the phase 0 upstroke; this is accompanied by reduced conductance of the inwardly rectified K^+ current. Opening of L-type Ca^{2+} channels is mainly responsible for the plateau (phase 2); inactivation of Na^+ channels and opening of voltage sensitive K^+ channels cause the rapid repolarization to the plateau (phase 1).

10. **E.** A multiple-lead electrocardiogram can yield information on cardiac rhythm and conduction; presence, location, and extent of ischemia; orientation of the heart and chamber size; and abnormalities resulting from altered electrolyte levels or drugs. It does not yield direct information regarding cardiac output, stroke volume, ejection fraction, or other flow-related parameters.

11. **B.** Conduction velocity is a function of the slope of the phase 0 upstroke of the action potential. Velocity is slowed in the atrioventricular node, resulting in a pause between atrial and ventricular depolarization. As a result, there is a pause between atrial and ventricular contraction, allowing atrial contraction to produce the final filling of the ventricles before they contract.

12. **E.** The plateau of the cardiac action potential is sustained by an outward Ca^{2+} current through voltage-sensitive, slow L-type Ca^{2+} channels and inward K+ current through voltage-dependent delayed rectifier K+ channels. The myocytes are refractory to another depolarization during this phase.

13. **B.** Mean arterial pressure is not simply the arithmetic mean of diastolic and systolic arterial pressure because the arterial pressure curve is complex in morphology and diastole is longer in duration than systole. As a rule of thumb, at normal heart rates, mean arterial pressure is approximately equal to the diastolic pressure plus one third the pulse pressure. In this problem, the diastolic pressure is 80 and one third of pulse pressure is 16.7; therefore the mean arterial pressure is predicted to be approximately 97 mm Hg.

14. **A.** The highest pulse pressure is observed in the left ventricle, where normal resting pressure is approximately 120/0 mm Hg. High systolic pressure is required to pump blood through the high resistance systemic circulation, whereas low diastolic pressure allows filling during diastole. Within the pulmonary circuit, the highest pulse pressure occurs in the right ventricle (~25/0 mm Hg at rest).

15. **E.** According to Poiseuille's law, flow is proportional to the fourth power of the radius of a tube. Thus, doubling the radius of the tube will result in an increase in flow by a factor of 16 if other variables are held constant. Doubling the pressure gradient, halving the viscosity of the fluid, or halving the length of the tube will double the flow; doubling the viscosity will reduce flow by 50%.

16. **C.** High viscosity of fluid favors laminar flow. When the hematocrit is reduced, resulting in low blood viscosity, flow may be turbulent, resulting in audible murmurs. Wide tube diameter, rapid velocity of fluid flow, high density of fluid, and pulsatility of flow all favor the development of turbulent flow.

17. **C.** Wall tension (T) is defined by Laplace's law, $T = P_t r$, where P_t is the transmural pressure and r is vessel radius. Reduction of interstitial pressure will result in increased transmural pressure, and thus increased wall tension. Reduced radius, vascular hydrostatic pressure, or transmural pressure will be associated with lower wall tension.

18. **D.** The isovolumetric contraction period begins when ventricular contraction results in ventricular pressure rising to exceed atrial pressure, causing the atrioventricular valve to close. Thus, the beginning of the isovolumetric contraction period for the right ventricle is marked by closure of the tricuspid valve.

19. **A.** During the cardiac cycle, valves on the left side of the heart close before the corresponding valves on the right side (mitral valve closes before the tricuspid, and aortic valve closes before the pulmonic), whereas valves on the right side open before the corresponding valves on the left side (pulmonic valve opens before the aortic, and tricuspid valve opens before the mitral valve).

20. **E.** In respiratory sinus arrhythmia, which occurs primarily in infants and children, heart rate is increased during inspiration and reduced during expiration. This is caused by increased venous return during inspiration, causing stretch of low pressure atrial baroreceptors and reflexive increase in heart rate.

21. **E.** Activation of the sympathetic nervous system will produce all the listed effects. Effects on the heart (increased heart rate and contractility) involve binding of adrenergic neurotransmitter to β_1 receptors; constriction of arteries and veins is produced by binding of the adrenergic transmitter to α_1 receptors.

22. **B.** The force-velocity relationship for cardiac muscle reveals an inverse relationship between velocity of contraction and afterload (force of contraction). With increased contractility, the maximum velocity of

contraction (V_m, which occurs at zero afterload), is increased, as is the maximum force of contraction. Maximum force of contraction occurs during isometric contraction, at the x-intercept of the force-velocity relationship, where velocity is zero.

23. **B.** The vascular function curve represents the relationship between central venous pressure and cardiac output when central venous pressure is the dependent variable; the cardiac function curve represents the relationship when cardiac output is the dependent variable. The intersection of the two curves is at the normal, resting cardiac output and central venous pressure at equilibrium (approximately 5 L/min cardiac output and central venous pressure of approximately 2 mm Hg).

24. **C.** Nitric oxide release by endothelial cells produces vasodilation of the vessel by relaxing underlying smooth muscle. This effect is mediated by the second messenger cyclic guanosine monophosphate, which reduces free intracellular Ca^{2+}, producing the smooth muscle relaxation. Dilation of arterial vessels results in higher capillary hydrostatic pressure downstream. Nitric oxide also inhibits adhesion of platelets to the vascular wall.

25. **A.** In many tissues and organs, if blood flow is increased due to higher perfusion pressure, the expected elevation in flow will be followed by a return in blood flow toward the basal rate. According to the myogenic hypothesis, this autoregulation involves smooth muscle constriction in response to elevated transmural pressure (i.e., in response to stretch).

26. **D.** β_2 receptor binding produces vasodilation, whereas α_1 and α_2 receptor binding is associated with vasoconstriction. β_1 receptors are found in the heart, where the main effects mediated by these receptors are increased heart rate, contractility, and conduction velocity.

27. **A.** Arterial baroreceptors respond to high arterial pressure (and thus stretch) by sending afferent nerve impulses to the central cardiovascular center, resulting in reduced sympathetic efferent activity and increased parasympathetic activity. In addition to high pressures, the baroreceptors also respond to pulse pressure.

28. **C.** Left coronary artery flow is highest during early diastole. Flow is low during systole due to compression of myocardial vessels by the contracting myocardium. As the heart relaxes, this compression is released; this, combined with the effects of vasodilator metabolites that build up in the myocardium during the low flow of systole, results in a large increase in left coronary artery blood flow in early diastole.

SECTION 4: RESPIRATORY PHYSIOLOGY

1. **A.** A rise in pulmonary artery pressure produces passive distension of vessels in the pulmonary microcirculation and opening of some vessels that were previously collapsed (recruitment).

2. **D.** Spirometry measures changes in lung volume (tidal volume, expiratory reserve volume, inspiratory reserve volume, vital capacity, inspiratory capacity), but cannot measure total lung capacity, residual volume, or functional residual capacity. To determine these three values, one of them must be measured indirectly, such as by nitrogen washout, helium dilution, or body plethysmography.

3. **C.** In the standing position, both ventilation and perfusion of the lung are greatest in the bottom portion and poorest in the upper portion of the organ. However, the vertical gradient for perfusion is much greater than the gradient for ventilation. Therefore, the ventilation-to-perfusion ratio is highest toward the top of the lung. The ratio approaches infinity in areas of dead space and zero in areas of shunt.

4. **B.** Diffusion of a gas through a membrane is a passive process that follows Fick's law. It is directly related to the partial pressure gradient, directly related to surface area, directly related to the diffusion constant of the gas, and inversely related to membrane thickness.

5. **C.** In the middle portion of the lung, zone 2, alveolar pressure falls between pulmonary arterial and venous pressures, and the ventilation and perfusion are approximately balanced, resulting in a ratio of approximately 1.

6. **C.** The partial pressure of a gas in the atmosphere is equal to the total atmospheric pressure times the fractional concentration of the gas. In this case, 21% of 700 mm Hg is 147 mm Hg.

7. **B.** Functional residual capacity is lung volume after expiration in normal, quiet breathing. At this point, mechanical forces are in balance, with outward elastic recoil pressure of the chest wall balancing the inward elastic recoil pressure of the lung.

8. **C.** In the respiratory system as a whole, the greatest resistance to flow occurs in the medium-sized airways (fourth to eighth generation). Proceeding down the airways, the diameter of airways decreases whereas the number of tubes increases rapidly. Considering both factors, the resistance is greatest in the medium-sized bronchi (in aggregate).

9. **A.** With progressively greater effort, peak air flow is increased during expiration, but along the downward slope of the expiratory flow-volume curves, airflow is effort independent.

10. **B.** Severe chronic obstructive pulmonary disease is characterized by emphysema, with increased compliance of the lung and decreased elastic recoil of the lung. As a result of the decreased elastic recoil, the equal pressure point forms early during expiration, resulting in trapping of air, and ultimately causing increased total lung capacity, functional residual capacity, and residual volume. In pulmonary function tests, expiratory flow rate and FEV_1 are reduced as a result.

11. **A.** The presence of surfactant at the air-fluid interface of alveoli and small airways results in lower surface tension and therefore increased pulmonary compliance, reducing the work of breathing. Surfactant contains the

phospholipid dipalmitoyl phosphatidyl choline. Surfactant deficiency is responsible for respiratory distress syndrome of the newborn.

12. **A.** An increase in hematocrit will result in a proportional rise in the amount of oxygen bound to hemoglobin in arterial blood. At 100 mm Hg PO_2, hemoglobin is saturated with oxygen, and an increase in PO_2 will only result in a minor rise in oxygen content by raising the small amount of dissolved oxygen. Likewise, because hemoglobin in arterial blood is normally nearly saturated with oxygen, increased alveolar ventilation will have very little effect on oxygen content. An increase in 2,3-DPG or a fall in blood pH will shift the oxyhemoglobin dissociation curve to the right, resulting in a fall in bound oxygen.

13. **C.** The pH is below the normal level of 7.4, indicating acidosis; because PCO_2 is elevated, this is a case of respiratory acidosis (high PCO_2 is the cause of the low pH).

14. **D.** In acute adaptation to high altitude, hypoxemia stimulates respiratory rate. Heart rate is also elevated. 2,3-DPG is elevated in blood, resulting in right shift of the oxyhemoglobin dissociation curve, causing oxygen to more readily dissociate from hemoglobin at the tissue level. Renal compensation will result in elevated plasma bicarbonate level. In the long term, however, increased hematocrit (higher red blood cell count and hemoglobin concentration in blood) is an important compensatory mechanism, resulting in increased oxygen carrying capacity of blood.

15. **B.** Central chemoreceptors respond mainly to changes in arterial PCO_2, which diffuses readily into the cerebrospinal fluid and alters its pH, resulting in stimulation of respiration when arterial PCO_2 is elevated. The blood-brain barrier is largely impermeable to HCO_3^- or H^+. Peripheral chemoreceptors respond to changes in arterial PO_2 and also pH and PCO_2.

16. **D.** The initial, rapid adjustment of respiration during exercise is caused by input from proprioceptive afferents from joint receptors to the respiratory center in the brain, collaterals to the respiratory center from motor pathways for muscle activation, as well as additional undefined factors. The additional elevation of respiration during continuing exercise is caused by feedback systems involving chemoreceptors and changes in body temperature.

17. **E.** In the fetus, hemoglobin F facilitates oxygenation of blood in the placental circulation due to its higher affinity for oxygen.

SECTION 5: RENAL PHYSIOLOGY

1. **C.** The clearance of inulin (C_{in}) is equated with the glomerular filtration rate because inulin is freely filtered, is not reabsorbed or secreted, and all filtered inulin is excreted. Thus, if the clearance of a freely filtered substance is less than C_{in}, it means that overall there was reabsorption (however, it does not indicate whether secretion might also have occurred).

2. **C.** Increased extracellular matrix proteins thicken the glomerular basement membrane, increasing the filtration barrier and thus reducing the permeability to plasma.

3. **A.** Aldosterone is produced in the zona glomerulosa of the adrenal cortex.

4. **B.** Filtration fraction is defined as the glomerular filtration rate (GFR) divided by the renal plasma flow (RPF). The renal plasma flow equals [RBF × (1 − Hematocrit)], or 600 mL/min. Because C_{in} is equated with the GFR, the filtration fraction is 125 mL/min ÷ 600 mL/min, or ~20%.

5. **C.** The renal collecting ducts have principal cells (sensitive to aldosterone) and intercalated cells (which contribute to acid-base homeostasis).

6. **A.** Glucose is 100% reabsorbed in the proximal convoluted tubule via Na^+-glucose cotransporters.

7. **D.** Plasma antidiuretic hormone has no direct effect on renal potassium handling, whereas the other conditions cause either secretion (high dietary or plasma potassium and aldosterone) or enhanced reabsorption (low dietary or plasma potassium and acidosis).

8. **D.** Loop diuretics target the NKCC-2 cotransporters on the thick ascending limb of Henle. When the transporters are blocked, the solutes are carried distally (where there is little sodium reabsorption in the absence of aldosterone), and most of the fluid is excreted as urine. Because this transporter is a key factor in the countercurrent multiplier system, an additional result of the use of loop diuretics is the washing out of the medullary interstitial concentration gradient, which contributes to the sustained diuresis.

9. **B.** Essentially all the filtered sodium is reabsorbed, with the most (~70%) occurring in the proximal tubule (S1-3). Aldosterone acts at the late distal tubule to increase ENaC channels, thereby enhancing sodium reabsorption. Antidiuretic hormone increases water reabsorption in the collecting ducts.

10. **C.** Bicarbonate reabsorption occurs through carbonic anhydrase (CA) activity in the tubular lumen and cell. Bicarbonate and H^+ in the renal tubule exist in rapid equilibrium with carbonic acid; CA catalyzes the transformation of carbonic acid to CO_2 and H_2O, which diffuse into the cells. In the cells, CA action reforms the carbonic acid and thus the H^+ and HCO_3^-. The H^+ is transported back into the lumen, and the HCO_3^- is transported across the basolateral membrane and into blood in exchange for Cl^-.

11. **E.** In the collecting duct cells, binding of antidiuretic hormone to V2 receptors stimulates the insertion of aquaporins into the apical membranes. This causes sodium–free water reabsorption.

12. **A.** Distal sodium reabsorption has no effect on the medullary concentration gradient, whereas the other factors all play significant roles in creating and maintaining the gradient.

13. **D.** Free water clearance (C_{H_2O}) = V − [(U_{osm}/P_{osm}) × V], or +1 mL/min. The positive value implies that water was

cleared in excess of the amount required for iso-osmotic excretion of solutes present in the urine.

14. **D.** Dehydration has no disruptive effect on the interstitial concentration gradient. All other changes (administration of loop diuretics, or increase in glomerular filtration rate, renal blood flow, or vasa recta blood flow) have the potential for reducing the gradient for reabsorption.

15. **C.** Angiotensin II has two direct actions on the kidneys: to increase proximal tubular sodium reabsorption and constrict renal afferent and efferent arterioles. These actions increase sodium and water reabsorption. Renin is secreted from the juxtaglomerular cells in response to *low* sodium concentration and *low* tubular fluid flow rate in the distal tubule. Aldosterone stimulates sodium reabsorption in the late distal tubules and collecting ducts.

16. **A.** The reduction in vascular volume will stimulate sympathetic vasoconstriction and sodium and fluid-retaining systems. Atrial natriuretic peptide is released from cardiac myocytes in response to *increased* atrial stretch during volume expansion. Thus, during dehydration, circulating atrial natriuretic peptide will be low.

17. **E.** Diabetes insipidus (DI) is usually of central origin (nephrogenic DI is rare) and follows trauma, disease, or surgery affecting the hypothalamus or pituitary gland. Central DI involves the loss of antidiuretic hormone, so water channels are not present in the apical membranes of the collecting ducts and urine cannot be concentrated. This leads to massive excretion (3 to 18 L/day) of hypotonic urine.

18. **D.** Reduced sodium concentration in the distal tubule is sensed by the macula densa cells, which stimulate renin secretion from the juxtaglomerular cells (the juxtaglomerular apparatus).

19. **C.** Plasma bicarbonate is low in metabolic acidosis, and the α-intercalated cells of the collecting ducts will increase H^+ secretion.

20. **C.** Net acid excretion is determined by the sum of urinary ammonium and titratable acids minus any excreted bicarbonate. It does not depend on sodium excretion.

21. **B.** The pH is <7.4; therefore the condition is acidosis. Because the low plasma bicarbonate (<24 mEq/L) also reflects acidosis, the disorder is metabolic in origin. The anion gap (AG) can be used to determine whether the acidosis is from acid loading or base loss: AG = Na^+ − (Cl^- + HCO_3^-), or 136 − 114 = 22. The usual anion gap is ~8 to 12, and an increase in the gap reflects the addition of acid. Base loss (as found with diarrhea) would show no change in AG because the loss of HCO_3^- in the stool would be matched by an increase in plasma Cl^-.

22. **D.** In respiratory alkalosis, the high pH will be matched by low P_{CO_2} (indicating a reduction in this plasma "acid" as the primary disturbance). In *uncompensated* respiratory alkalosis, the plasma HCO_3 will be normal.

23. **A.** Phosphoric acid ($H_2PO_4^-$) is the primary titratable acid excreted in the urine. Phosphates make very good buffers in urine because their pK is near the urine pH and there is a large amount (~25% of the filtered phosphate load) of dibasic phosphate (HPO_4^{2-}) available for conversion to titratable acid and excretion.

24. **B.** Glutamine is metabolized in the proximal tubular cells, ultimately yielding two NH_3 (which immediately combine with two H^+ to form two NH_4^+) and two HCO_3^-. This is an important process for acid excretion.

SECTION 6: GASTROINTESTINAL PHYSIOLOGY

1. **D.** The GI tract does not directly regulate systemic blood flow.

2. **E.** Growth hormone has no apparent effect on motility or secretion in the GI tract.

3. **D.** The GI tract has an intrinsic nervous system (enteric nerves), which responds to signals from luminal receptors and hormones as well as extrinsic nerves. When the extrinsic nerve input is severed, motility and secretion continue, but not as efficiently.

4. **B.** Vagal efferent nerves regulate all of the actions EXCEPT salivation, which is controlled by the facial and glossopharyngeal nerves. The vagus does participate in secondary esophageal peristalsis, in conjunction with local enteric nerves.

5. **B.** Gut flora provide a protective function by stimulating the development and maintenance of immune competence in the gut.

6. **B.** Action (or spike) potentials in the GI tract are caused by the influx of calcium into the smooth muscle when the slow waves are depolarized above −40 mV. This mechanism is active throughout the tract, and depolarization can result from local stretch (mechanoreceptors acting on enteric nerves), extrinsic nerves, and peptides.

7. **C.** The undulations are a result of small changes in membrane potential generated by the SIP (*s*mooth muscle, *I*CC, *P*DGFR-α⁺ cell) syncytium. The slow waves occur at different rates from the mid-stomach through the colon.

8. **D.** Peristalsis follows the law of the intestines, where muscle is contracted proximal to the bolus and relaxed distal to the bolus, producing aboral movement of the chyme.

9. **B.** There is no evidence for involvement of the parasympathetic nervous system in the MMC.

10. **B.** While many of the disorders listed can result in her symptoms, the presentation of both thickening of the colon as well as a stricture in the jejunum is indicative of Crohn's disease, which can occur at any point along the small and large intestine.

11. **D.** Gastric inhibitory peptide is an endocrine hormone secreted from cells in the duodenum and jejunum of the small intestine. The other substances are secreted directly into the gastric lumen.

12. **C.** Vagal afferent nerves stimulate gastric acid secretion directly at the parietal cell and indirectly by increasing gastrin-releasing peptide (and, thus, gastrin release) and inhibiting somatostatin release.

13. **E.** Proton pump activity drives acid secretion and is the target for regulatory hormones, peptides, and nerves.

14. **B.** Sodium enters the luminal membrane by all the mechanisms except the Na^+ pump (Na^+/K^+ ATPase), which is located on the *basolateral* membrane and maintains the low intracellular Na^+ concentrations that drive the luminal Na^+ transport mechanisms.

15. **C.** Secretin is a hormone released from S cells in the duodenum and early jejunum in response to acidic chyme. The secretin binds to receptors on the pancreatic acinar cells and stimulates release of electrolyte buffers into the pancreatic duct.

16. **D.** The vagus will stimulate secretions from the parietal cells (HCl, IF) and chief cells (pepsins) into the stomach lumen as well as stimulate gastrin release (from G cells) into the blood. GLP-1 is found mainly in the ileum and is secreted in response to nutrients in the chyme.

17. **B.** The liver performs all the functions noted except vitamin production. It does store a variety of important vitamins and minerals, including vitamin B12 and iron; it also hydroxylates vitamin D.

18. **D.** Obstruction of blood flow through the liver increases the pressure in the portal vein, which brings blood from the intestines to the liver. This eventually results in portal hypertension. The obstruction also reduces bile secretion.

19. **A.** The hepatocytes are the primary site of formation of lipoproteins, which carry fats, protein, and cholesterol to different tissues for processing and storage.

20. **B.** Bile is amphipathic and essential for carrying the hydrophobic lipids through the unstirred water layer adjacent to the enterocytes. The ability to form micelles, with hydrophilic ends of bile oriented outward, and lipophilic ends oriented inward and associated with the lipids, allows the efficient transport of lipids to the intestinal cells.

21. **C.** Villous atrophy will reduce the surface area for absorption of nutrients as well as the brush border proteases and saccharidases that provide the final digestion of proteins and carbohydrates to forms that can be absorbed. Thus, absorption of proteins and carbohydrates would be significantly affected, and production of plasma proteins would be reduced.

22. **D.** Efficient digestion of proteins in the small intestine depends on the pancreatic proteases, which have optimal catalytic activity near physiologic pH (pH between 7 and 8). Low luminal pH will significantly decrease catalytic activity and reduce the amount of proteins digested to constituents that can be absorbed.

23. **D.** Starch is the primary dietary carbohydrate, and digestion begins in the mouth with salivary α-amylase. Only around 25% of digestion occurs preduodenally; pancreatic α-amylase digests the remaining starch to malto-oligosaccharides, and then intestinal brush border saccharidases (maltase, isomaltase, sucrose, and lactase) digest oligosaccharides and disaccharides (sucrose and lactose) to monosaccharides (glucose, galactose, and fructose).

24. **A.** The presence of chyme in the antrum of the stomach and the duodenum stimulates release of gastrin, which is a potent stimulator of gastric acid secretion. It does not promote buffer secretion.

25. **C.** Celiac sprue is an autoimmune reaction initiated by gluten proteins in wheat, resulting in the flattening of the villus lining and expansion of the crypts. The reduction in the upper villi severely decreases the brush border enzyme digestion of carbohydrates and proteins and reduces the surface area for absorption.

26. **B.** Vitamin B12 is an essential nutrient and has multiple mechanisms protecting it from protease activity and facilitating entry into the intestinal cells. There is no limitation to absorption posed by intracellular B12 concentrations.

27. **C.** Because bile recycling and vitamin B12 absorption occur in the last section of the ileum, removing this part of the small intestine will prevent B12 absorption (and bile recycling). The increase in bile excretion will reduce water absorption (because of the osmotic nature of bile).

SECTION 7: ENDOCRINE PHYSIOLOGY

1. **E.** Thyroid hormone, vitamin D, and steroid hormones are lipophilic and readily diffuse into target cells, where they are bound by nuclear receptors, initiating gene transcription. Peptide hormones and catecholamines are bound by membrane receptors, initiating an intracellular signaling cascade that ultimately leads to regulation of cellular function.

2. **E.** The cellular action of hormones depends on the concentration of hormone in the blood as well as the number and affinity of hormone receptors in the cell or on the cell membrane. In addition, some hormones (particularly lipophilic hormones such as steroids) are carried in the blood by binding proteins. It is specifically the free fraction of hormones in the blood that is available for binding by cellular receptors.

3. **B.** Vasopressin promotes increased water absorption by the kidney, and its release is stimulated by decreased blood volume, increased body fluid osmolality, and decreased blood pressure.

4. **A.** Oxytocin, a 9–amino acid peptide, is released directly into the general circulation by the posterior pituitary gland but is synthesized in the paraventricular and supraoptic nuclei of the hypothalamus. It is carried by axonal transport to nerve endings in the posterior pituitary. Its actions include milk expulsion during breastfeeding and uterine contraction, although a physiological role in induction or progression of labor is not well established.

5. **E.** Growth hormone is released by the anterior pituitary somatotrophs, with several pulses occurring during sleep, as well as a basal level of secretion. Its release is stimulated by hypothalamic growth hormone–releasing hormone and inhibited by somatostatin. At the tissues, it acts by stimulating the synthesis of insulin-like growth factor.

6. **E.** Pituitary TSH is a primary stimulus for thyroid hormone release.

7. **E.** Hypothyroidism in adults results in myxedema, which is associated with nonpitting edema, hypotension, fat deposition, cold intolerance, and depression among other symptoms.

8. **D.** Both T_3 and T_4 are released from the thyroid gland, but 20 times the amount of T_4 is released compared with T_3. Most of the T_3 and T_4 is bound to proteins, including albumin and thyroxine-binding globulin.

9. **C.** Feedback control of thyroid hormone production is illustrated by elevated circulating free T_3 and T_4, which reduces both pituitary thyroid-stimulating hormone and hypothalamic thyrotropin-releasing hormone, resulting in reduced thyroid hormone production.

10. **C.** Cortisol synthesis and release by the adrenal cortex is regulated by the anterior pituitary hormone adrenocorticotropic hormone (ACTH). Although ACTH is required for synthesis of all adrenal steroids, androgen and aldosterone synthesis and release are primarily controlled by other mechanisms. Release of catecholamines by the adrenal medulla is controlled by the sympathetic nervous system.

11. **C.** The zona fasciculata secretes cortisol, which has anti-inflammatory and anti-immune effects among its many actions. The zona reticularis secretes androgens, and the zona glomerulosa secretes aldosterone. The adrenal medulla secretes catecholamines.

12. **E.** Dehydroepiandrosterone and androstenedione are the major adrenal androgens. In females, for whom the adrenal gland is the only significant source of circulating androgens, all the listed effects are stimulated by these hormones.

13. **C.** The primary stimuli for aldosterone secretion by the adrenal zona glomerulosa are hyperkalemia and angiotensin II. Aldosterone synthesis and secretion are inhibited by atrial natriuretic peptide.

14. **B.** The symptoms and signs of Addison's disease reflect the lack of steroid hormone production by the adrenals. Thus, poor stress tolerance, hypoglycemia, fatigue and weight loss, hyperpigmentation of skin (due to adrenocorticotrophic hormone elevation), low blood pressure, and salt appetite are observed.

15. **E.** The GLUT4 transporters are the insulin-sensitive transporters present on most cell membranes; insertion of these transporters facilitates glucose entry.

16. **C.** GLUT2 transporters are prominent on the pancreas, small intestine, brain, and liver and facilitate entry of glucose into the cells. In the β-cell of the pancreas, glucose metabolism increases ATP and reduces K^+ efflux. This depolarizes the β cell, causing Ca^{2+} influx and release of the insulin (and C-peptide).

17. **E.** All the factors listed will stimulate insulin secretion.

18. **E.** Insulin has a hypoglycemic effect on the plasma by increasing GLUT4 transporters, which facilitate glucose entry into adipose tissue and cardiac and skeletal muscle, while decreasing gluconeogenesis and glycogenolysis and increasing glycogen storage.

19. **D.** Decreased plasma calcium concentrations will stimulate the release of parathyroid hormone (PTH), which acts at bone to increase resorption, releasing calcium into the blood. PTH also acts at the kidney to increase calcium reabsorption.

20. **B.** PTH is released into plasma in response to reduced plasma calcium levels. At the kidneys, it increases distal calcium reabsorption and reduces phosphate reabsorption in the proximal tubule. It also stimulates the renal conversion of 25-hydroxycholecalciferol to active vitamin D (1,25-dihydroxycholecalciferol), which increases intestinal absorption of calcium.

21. **E.** Primary hyperparathyroidism typically results from parathyroid tumors that have no negative feedback mechanisms and ultimately increases plasma calcium concentrations. The elevated parathyroid hormone (PTH) increases the renal conversion of 25-hydroxyvitamin D to active vitamin D, increasing intestinal calcium absorption. PTH increases bone resorption (can cause bone pain), and the elevated filtered load of calcium to the kidney can cause stone formation (calculi). Symptoms also include constipation and polyuria. Overall, these symptoms can cause fatigue and general malaise. These "stones, bones, moans, and groans" can be characteristic symptoms of hyperparathyroidism.

22. **E.** Parathyroid hormone increases calcium reabsorption in the renal distal tubules, increases bone resorption (releasing calcium into the blood), and activates vitamin D. Activated vitamin D increases intestinal absorption of calcium and stimulates bone mineralization.

23. **B.** Differentiation of the external genitalia produces female genitalia in the absence of androgens. In the male fetus, the testes produce testosterone, which is converted to dihydrotestosterone (DHT) in the primitive genital structures. DHT stimulates formation of male genitalia.

24. **C.** The linear growth spurt that occurs during puberty is mainly attributed to estradiol in both boys and girls. Estrogenic hormones increase linear growth of long bones and also promote the eventual closure of the epiphyseal plates.

25. **B.** In developing follicles, androgen is synthesized and secreted by theca interna cells. The conversion of androgens to estradiol by the follicular granulosa cells is stimulated by follicle stimulating hormone.

26. **B.** During fetal development, germ cells in the ovaries give rise to oogonia, which proliferate and begin meiosis. They become primary oocytes and are arrested at this stage.

27. **B.** Follicle-stimulating hormone (FSH) acts on Sertoli cells of the seminiferous tubules, stimulating androgen-binding protein synthesis. This protein binds testosterone, which promotes spermatogenesis. The Sertoli cells also synthesize inhibin, which inhibits FSH secretion by the anterior pituitary gland.

Index

Page numbers followed by *f* indicate figures; *t*, tables; *b*, boxes.